Wildlife Ecology, Conservation, and Management

To our colleagues Graeme Caughley, Jamie Smith, and Peter Yodzis, who have influenced both our approach to wildlife biology and the writing of this book.

Wildlife Ecology, Conservation, and Management

Third Edition

John M. Fryxell PhD
Department of Integrative Biology, University of Guelph, Guelph, Canada

Anthony R. E. Sinclair PhD
Biodiversity Research Centre, University of British Columbia, Vancouver, Canada

The late Graeme Caughley PhD
CSIRO, Canberra, Australia

WILEY Blackwell

This edition first published 2014 © 2014 by John Wiley & Sons Ltd
First edition © 1994 by Blackwell Science; Second edition © 2006 by Anthony R. E. Sinclair and John M. Fryxell.

Registered office: John Wiley & Sons, Ltd, The Atrium, Southern Gate, Chichester, West Sussex, PO19 8SQ, UK

Editorial offices: 9600 Garsington Road, Oxford, OX4 2DQ, UK
The Atrium, Southern Gate, Chichester, West Sussex, PO19 8SQ, UK
111 River Street, Hoboken, NJ 07030-5774, USA

For details of our global editorial offices, for customer services and for information about how to apply for permission to reuse the copyright material in this book please see our website at www.wiley.com/wiley-blackwell.

Library of Congress Cataloging-in-Publication Data

Fryxell, John M., 1954–
  Wildlife ecology, conservation, and management. – Third edition / John M. Fryxell, Anthony R.E. Sinclair, Graeme Caughley.
    pages cm
  Includes index.
  ISBN 978-1-118-29106-1 (cloth) – ISBN 978-1-118-29107-8 (pbk.) 1. Wildlife management.
2. Wildlife conservation. 3. Animal ecology. I. Sinclair, A. R. E. (Anthony Ronald Entrican)
II. Caughley, Graeme. III. Title.
  SK355.C38 2014
  639.9 – dc23
                                                                              2014013806

A catalogue record for this book is available from the British Library.

Wiley also publishes its books in a variety of electronic formats. Some content that appears in print may not be available in electronic books.

Cover image © Luke Vander Vennen, used by permission.

Set in 9.5/12pt Berkeley by Laserwords Private Limited, Chennai, India
Printed and bound by CPI Group (UK) Ltd, Croydon, CR0 4YY
C9781118291078_120924

Contents

Preface

Modern principles of sustainable management and conservation of wildlife species require a clear understanding of demography, animal behavior, and ecosystem dynamics. Our book weaves together these disparate elements in a single coherent text intended for senior undergraduate and graduate students. The first half provides a solid background in key ecological concepts such as demography, population growth and regulation, competition within and among species, and predator–prey interactions. The second half uses these key ecological concepts to develop a deeper understanding of the principles underlying wildlife management and conservation, including population viability assessment, sustainable harvesting, landscape planning, and ecosystem management.

New quantitative methods, developed over the last 10 years, are now so fundamental to management that we have included them at the most basic levels. Several chapters of the book will be useful to practicing wildlife managers. For example, we have included modern approaches to estimating animal abundance and habitat selectivity, the use of age- and stage-structured data in demography studies, and the use of models as efficient methods for making conservation and management decisions. As a study aid, we have included a wide variety of downloadable computer programs in R and Mathcad on an accompanying website. These are intended to help readers develop a solid understanding of key statistical procedures and population models commonly used in wildlife ecology and management.

In this edition we have arranged the sequence of chapters to reflect the progression from individuals to populations, communities, and ecosystems. Four new chapters have been added to cover rapidly developing topics: effects of climate change on wildlife, the evolutionary response by wildlife populations to rapidly changing conditions, home range use and habitat selection as a consequence of patterns of individual movement, and the importance of corridor use and metapopulation dynamics for wildlife populations living in the highly fragmented landscapes that increasingly characterize the modern world.

Anne Gunn and David Grice were invaluable in bringing together the first edition of this book after Graeme Caughley fell ill. Fleur Sheard prepared the line drawings for that edition. Since then we have continued to benefit from the helpful contributions of a number of people, including Tal Avgar, Andrew McAdam, Cort Griswold, David Grice, Sue Briggs, Andrea Byrom, Steve Cork, Charles Krebs, Graham Nugent, John Parkes, Roger Pech, Laura Prugh, Wendy Ruscoe, Dolph Schluter, Julian Seddon, Grant Singleton, David Spratt, Eric Spurr, Vernon Thomas, and Bruce Warburton. We also thank the Natural Sciences and Engineering Research Council of Canada for continuing support over the years.

Our close friend and colleague, Graeme Caughley, died in 1994. We have retained the substance and spirit of his scholarship, expanding the fields where advances have occurred since the first edition. For this new edition we are indebted to Sue Pennant and Anne Sinclair, who are always willing (if not necessarily eager) to provide a fresh set of eyes for proofreading of the new material.

About the companion website

This book is accompanied by a companion website:

www.wiley.com/go/Fryxell/Wildlife

The website includes:
- Additional resources
- Powerpoints of all figures from the book for downloading
- PDFs of all tables from the book for downloading

1 Introduction: goals and decisions

1.1 How to use this book

This book is structured as two interlocking parts. The first provides an overview of wildlife ecology, as distinct from that portion of applied ecology that is called wildlife management and conservation. The chapters on wildlife ecology (Chapters 2–11) cover such topics as growth and regulation of wildlife populations, spatial patterns of population distribution, and interactions among plants, herbivores, carnivores, and disease pathogens. While these topics are often covered in introductory biology or ecology courses, they rarely focus on the issues of most concern to a wildlife specialist. A solid understanding of ecological concepts is vital in formulating successful wildlife conservation and management policy. In particular, you will need an understanding of the theory of population dynamics and of the relationship between populations, their predators, and their resources if you are to make sensible judgments on the likely consequences of one management action versus another.

The second section deals with wildlife conservation and management (Chapters 12–22). These chapters cover census techniques, how to test hypotheses experimentally, how to evaluate alternative models as tools for conservation and management, and the three major aspects of wildlife management: conservation, sustained yield, and control. In closing, Chapter 22 places the problems of wildlife management into the context of the ecosystem. Species populations cannot be managed in isolation because they are influenced by, and they themselves influence, many other components of the ecosystem. In the long run, wildlife management becomes ecosystem management.

Many of the key issues in wildlife ecology are of a quantitative nature: processes of population growth, spatial distribution, or interactions with the physical environment or other organisms. Coping with these topics demands conceptual understanding of quantitative ecology. Mathematical models are also an essential component of decision-making in both wildlife conservation and management, for the simple reason that we can rarely rely on previous experience to identify the most appropriate choices. Every problem is unique: new species, new sets of challenges and constraints, all taking place in a continually changing physical environment. Mathematical models provide a useful tool for dealing appropriately with these uncertainties. Moreover, mathematical models help to clarify the logic that guides our thinking.

To assist in developing the requisite skills, many of the models and statistical analyses covered in the book can be obtained via a link at Fryxell's departmental Web page (http://www.uoguelph.ca/ib/people/faculty/fryxell.shtml). This provides a set of text files suitable for application using "R," a nonproprietary (i.e. free) software package that has been developed by a hard-working and highly committed group

Wildlife Ecology, Conservation, and Management, Third Edition.
John M. Fryxell, Anthony R. E. Sinclair and Graeme Caughley.
© 2014 John Wiley & Sons, Ltd. Published 2014 by John Wiley & Sons, Ltd.
Companion Website: www.wiley.com/go/Fryxell/Wildlife

of professional scientists and statisticians from all around the world. By learning to perform the examples used to illustrate this book, you will both expand your familiarity with useful mathematical principles and hone the problem-solving skills involved in modern wildlife ecology, conservation, and management. This can prove invaluable in future professional endeavors.

The R package provides a powerful set of integrated tools for numerical computation, statistical analysis, and graphical depiction of data and results. More information about R can be found at the R project homepage, www.r-project.org, while instructions on how to download R can be found at the CRAN repository for R materials, http://cran.r-project.org.

1.2 What is wildlife conservation and management?

The remainder of this chapter explains what wildlife management is, how it relates to conservation, and how it should operate. We discuss the difference between value judgments and technical judgments and how these relate to goals and policies compared to options and actions; we enumerate the various steps involved in deciding what to do and how to do it; and we describe decision analyses and matrices and how they help in evaluating feasible management options.

Wildlife is a word whose meaning expands and contracts according to the viewpoint of the user. Sometimes it is used to include all wild animals and plants. More often it is restricted to terrestrial vertebrates. In the discipline of wildlife management it designates free-ranging birds and mammals, and that is the way it is used here. Until about 25 years ago, "wildlife" was synonymous with *game*: those birds and mammals that were hunted for sport. The management of such species is still an integral part of wildlife management, but increasingly it embraces other aspects too, such as conservation of endangered species.

Wildlife management may be defined for present purposes as the *management of wildlife populations in the context of the ecosystem*. That may be too restrictive for some, who would argue that many of the problems of management deal with people and, therefore, that education, extension, park management, law enforcement, economics, and land evaluation are legitimate aspects of wildlife management and ought to be included within its definition. They have a point, but the expansion of the definition to take in all these aspects diverts attention from the core around which management activities are organized: the manipulation or protection of a population to achieve a goal. Obviously, people must be informed as to what is being done; they must be given an understanding of why it is necessary, their opinions must be canvassed, and their behavior may have to be regulated with respect to that goal. However, the most important task is to choose the right goal and to know enough about the animals and their habitat to ensure its attainment. Hence, wildlife management is restricted here to its literal meaning, thereby emphasizing the core at the expense of the periphery of the field. The broader extension and outreach aspects of wildlife management are dealt with thoroughly in other texts devoted to those subjects (Lyster 1985; Geist and McTaggert-Cowan 1995; Moulton and Sanderson 1999; Vásárhelyi and Thomas 2003).

1.2.1 *Kinds of management*

Wildlife management implies stewardship; that is, the looking after of a population. A population is a group of coexisting individuals of the same species. When stewardship fails, conservation becomes imperative. Under these circumstances, wildlife management shifts to remedial or restoration activities.

Wildlife management may be either *manipulative* or *custodial*. Manipulative management does something to a population, either changing its numbers by direct means or influencing them by the indirect means of altering food supply, habitat, density of predators, or prevalence of disease. Manipulative management is appropriate when a population is to be harvested, when it slides to an unacceptably low density, or when it increases to an unacceptably high level.

Custodial management, on the other hand, is preventative or protective. It is aimed at minimizing external influences on the population and its habitat. It is not aimed necessarily at stabilizing the system but rather at allowing free rein to the ecological processes that determine the dynamics of the system. Such management may be appropriate in a national park where one of the stated goals is to protect ecological processes, and it may be appropriate for the conservation of a threatened species where the threat is of external origin rather than being intrinsic to the system.

Regardless of whether manipulative or custodial management is called for, it is vital that (i) the management problem is identified correctly, (ii) the goals of management explicitly address the solution to the problem, and (iii) criteria for assessing the success of the management are clearly identified.

1.3 Goals of management

A wildlife population may be managed in one of four ways:

1 make it increase;
2 make it decrease;
3 harvest it for a continuing yield;
4 leave it alone but keep an eye on it.

These are the only options available to the manager.

Three decisions are needed: (i) What is the desired goal? (ii) Which management option is therefore appropriate? (iii) By what action is this best achieved? The first decision requires a value judgment, the others technical judgments.

1.3.1 Who makes the decisions?

It is not the function of the wildlife manager to make the necessary value judgments in determining the goal, any more than it is within the competence of a general to declare war. Managers may have strong personal feelings as to what they would like, but so might many others in the community at large. Managers are not necessarily provided with heightened aesthetic judgment just because they work on wildlife. They should have no more influence on the decision than does any other interested person.

However, when it comes to deciding which management options are feasible (once the goal is set) and how goals can best be attained, wildlife managers have the advantage of their professional knowledge. Now they are dealing with testable facts. They should know whether current knowledge is sufficient to allow an immediate technical decision or whether research is needed first. They can advise that a stated goal is unattainable, or that it will cost too much, or that it will cause unintended side effects. They can consider alternative routes and advise on the time, money, and effort each would require. These are all technical judgments, not value judgments. It is the task of the wildlife manager to make them and then to carry them through.

Since value judgments and technical judgments tend to get confused with each other, it is important to distinguish between them. By its essence, a value judgment is neither right nor wrong. Let us take a hypothetical example. The black rat (*Rattus rattus*) is generally unloved. It destroys stored food, it is implicated in the spread of bubonic plague and several other diseases, it contributes to the demise of endangered species, and it has been known to bite babies. Suppose a potent poison specific to this species were discovered, thereby opening up the option of removing this species from the face of the earth. Many would argue for doing just that, and swiftly. Others would argue that there are strong ethical objections to exterminating a species, however repugnant or inconvenient that species might be. Most of us would have a strong opinion one way or the other but there is no way of characterizing either competing opinion as right or wrong. That dichotomy is meaningless. A value judgment can be characterized as hardheaded or sentimental (these are also value judgments), or it may be demonstrated as inconsistent with other values a person holds, but it cannot be declared right or wrong. In contrast, technical judgments can be classified as right or wrong according to whether they succeed in achieving the stated goal.

1.3.2 Decision analysis

In deciding what objective (goal) is appropriate, we consider a range of influences, some dealing with the benefits of getting it right and others with the penalties of getting it wrong. Social, political, biological, and economic considerations are each examined and given due weight. Some people are good at this and others less so. In all cases, however, there is a real advantage, both to those making the final decision and to those tendering advice, to having the steps of reasoning laid out before them as the decision is approached.

At its simplest, this need mean no more than the people helping to make the decision spelling out the reasons underpinning their advice. However, with more complex problems it helps to be more formal and organized, mapping out on paper the path to the decision through the facts, influences, and values that shape it. This process should be explicit and systematic. Different people will assign different values (weights) to various possible outcomes, and, particularly if mediation by a third party is required, an explicit statement of those weights will allow a more informed decision. It helps also to determine which disagreements are arguments about facts and which are arguments about judgments of value.

Table 1.1 is an objective/action matrix in which possible objectives are ranged against feasible actions. The objectives are not mutually exclusive. It comes from the response of the Department of Agriculture of Malaysia to the attack of an insect pest on rice (Norton 1988). It allows the departmental entomologists and administrators to view the full context within which a decision must be made. Each of the listed objectives is of some importance to the department. The next step would be to rank these objectives and score the management actions most appropriate to each. The final outcome is the choice of one or more management actions that best meet the most important objective or objectives. Such very simple aids to organizing our thoughts are often the difference between success and failure.

Another such aid is the feasibility/action matrix. Table 1.2 is Bomford's (1988) analysis of management actions to reduce the damage wrought by ducks on the rice crops of the Riverina region of Australia. The feasibility criteria are here ranked so that if a management action fails according to one criterion there is no point in considering it against further ones. Note how this example effortlessly identifies areas of ignorance that would have to be attended to before a rational decision could be possible.

Our third example of decision aids is the pay-off matrix (Table 1.3). This expresses the state of nature (level of pest damage, in this example) as rows and the options for management action as columns (Norton 1988). The problem is to assess the probable outcome of each combination of the level of damage and the action mounted

Table 1.1 Possible objectives and management actions for public pest management. The initial problem is to assess how each action is likely to meet each objective. (After Norton 1988.)

Actions	Objectives					
	Improve farmers' ability to control pest	Improve farmers' incentives	Strengthen political support	Keep Dept.'s costs low	Reduce damage	Reduce future pest outbreaks
Short Term						
1 Warn and advise farmers						
2 Advise and provide credit						
3 Advise and subsidize pesticides						
4 Advise, subsidize, and supervise spraying						
5 Mass treat and charge farmers						
6 Mass treat at Dept.'s cost						
Medium Term						
7 Intensive pest surveillance						
8 Implement area wide biological control						
9 Training courses for farmers						

Table 1.2 Matrix for examining possible management actions against criteria of feasibility. (After Bomford 1988.)

Control Options	Feasibility Criteria					
	Technically possible	Practically feasible	Economically desirable	Environmentally acceptable	Politically advantageous	Socially acceptable
1 Grow another crop	1	0				
2 Grow decoy crop	1	1	?	1	1	1
3 Predators and diseases	0					
4 Sowing date	1	1	?	1	1	1
5 Sowing technique	1	1	?	1	1	1
6 Field modifications	1	1	?	1	1	1
7 Drain or clear daytime refuges	?	0				
8 Shoot	1	1	?	1	?	1
9 Prevent access, netting	1	1	0	1	1	1
10 Decoy birds or free feeding	?	1	?	1	1	
11 Repellants	1	0				
12 Deterrents	1	1	?	1	1	1
13 Poisons	1	1	?	0		
14 Resowing or transplanting seedlings	1	?	1	?	1	1

1, yes; 0, no; ?, no information.

to alleviate it. Note that the column associated with doing nothing gives the level of damage that will be sustained in the absence of action. It is the control against which the net benefit of management must be assessed. The cells of this matrix are best filled in with net revenue values (benefit minus cost) rather than with benefit/cost ratios, because it is the absolute rather than the relative gain that shapes the decision.

Table 1.3 Pay-off matrix for pest control. (Adapted from Norton 1988.)

| State of Nature | Actions | | | |
| | Do nothing (0) | Pest Control Strategies | | |
		(1)	(2)	(3)
Level of Pest Attack				
Low (L)	Outcome L, 0	Outcome L, 1	Outcome L, 2	Outcome L, 3
Medium (M)	Outcome M, 0	Outcome M, 1	Outcome M, 2	Outcome M, 3
High (H)	Outcome H, 0	Outcome H, 1	Outcome H, 2	Outcome H, 3

1.4 Hierarchies of decision

Before we begin manipulating a wildlife population and its environment, we must ask ourselves why we are doing so and what is it supposed to achieve. In management theory, that decision is usually divided into hierarchical components.

At the bottom, but here addressed first, is the management action. This might be to eliminate feral pigs (*Sus scrofa*) on Lord Howe Island off the coast of Australia. The management action must be legitimized by a technical objective; for example, to halt the decline of the Lord Howe Island woodhen (*Tricholimnas sylvestris*). Above this is the policy goal, a statement of the desired end-point of the exercise, which in this example might be to secure the continued viability of all indigenous species within the nation's National Park system.

In theory, the decisions flow from the general (the policy goal) to the specific (the management action), but in practice this does not work because each is dependent on the others, in both directions. Nothing is achieved by specifying "halt a species decline" as a technical objective unless a set of management actions is available that will secure this Obviously, a management action cannot be specified to cure a problem of unknown cause. All three levels of decision must be considered together, such that the end product is a feasible option.

A feasible option is identified by answering these questions:

1 Where do we want to go?
2 Can we get there?
3 Will we know when we have arrived?
4 How do we get there?
5 What disadvantages or penalties accrue?
6 What benefits are gained?
7 Will the benefits exceed the penalties?

The process is iterative. There is no point in persevering with the policy goal thrown up by the first question if the answer to the second is negative. The first choice of destination should instead be replaced by another, and the process repeated.

Question 3 is particularly important. It requires the formulation of stopping rules. This does not necessarily mean that management action ceases on attainment of the objective, but rather that management action is altered at this point. The initial action is designed to move the system towards the state specified by the technical objective; the subsequent action is designed to hold the system in that state. If we cannot determine when the objective has been attained, either for reasons of logic (ambiguous or abstract statement of the objective) or for technical reasons (inability to measure the state of the system), the option is not feasible.

1.5   Policy goals

Policies are usually couched in broad terms that provide no more than a general guide for the manager. The specific decisions are made when the technical objectives are formulated. However, there are two types of policy goal that the manager must know about in case they clash with the choosing of those objectives.

1.5.1   *The non-policy*

Non-policies stipulate goals that are not clearly defined. They are usually formulated in this way on purpose so that the administering agency is not tied down to a rigidly dictated course of action. Policies are usually formulated by the administering agency whether or not they are given legislative sanction. If the agency has not developed a policy, it may fill the gap with a non-policy that commits it to no specified action. Take for example the goal of "protecting intrinsic natural values." This reads well but is entirely devoid of objective meaning.

1.5.2   *The non-feasible policy*

In contrast to the relatively benign non-policy, the nonfeasible policy can be damaging. Although it may give each interest group at least something of what they desire, sometimes the logical consequence is that two or more technical objectives are mutually incompatible.

An example is provided by the International Convention for the Regulation of Whaling of 1946, which had as its goal "to provide for the proper conservation of whale stocks" and "thus make possible the orderly development of the whaling industry." This pleased both those concerned with conservation of whales and those wishing to harvest whales. Unfortunately, the goal is a nonsense, because, for reasons that are elaborated in Chapter 18, species with a low intrinsic rate of increase are not suitable for sustainable harvesting. The two halves of the policy goal contradict each other. The history of whaling since 1948, in which the blue (*Balaenoptera musculus*), the fin (*B.physalus*), the sei (*B.borealis*), the Brydes (*B.edeni*), the humpback (*Megaptera novaeangliae*), and the sperm (*Physeter macrocephalus*) were reduced to the level of economic extinction, is a direct consequence of the choice of a policy goal that was not feasible.

Another form of the nonfeasible policy is one that is so specific that it actually determines technical objectives and sometimes even management actions. If these are unattainable in practice, the policy goal itself is also unattainable. An example is provided by the now defunct policy to exterminate deer in New Zealand. It was always an impossibility.

1.6   Feasible options

Objectives must be attainable. It is the wildlife manager's task to produce the attainable technical objectives by which the policy goal is defined. In contrast to the goal, which may be described in somewhat abstract terms, a technical objective must be stated in concrete terms and rooted in geographic and ecological fact. It must be attainable in fact, and it should be attainable within a specified time. A technical objective should, therefore, be accompanied by a schedule.

1.6.1   *Criteria of failure*

It follows as a corollary that there must be an easy way of recognizing the failure to attain an objective. The most common method is to measure the outcome against that specified by the technical objective. Another is to compare the outcome with a set of *criteria of failure*, set before the management action is begun. These two methods are not the same. Comparison of outcome with objective can produce assessments such as "not quite" or "not yet." Criteria of failure cannot; they take the form, "the operation will be judged unsuccessful, and will therefore be terminated, if outcome x has not been attained by time t."

1.7 Summary

We view wildlife management as simply the management of wildlife populations. Three important points underlie any management: (i) the management problem is identified correctly; (ii) the goals of management explicitly address the solution to the problem; and (iii) criteria for assessing the success of the management are clearly identified.

Four management options are available: (i) to make the population increase; (ii) to make it decrease; (iii) to take from it a sustained yield; or (iv) to do nothing but keep an eye on it. We have first to decide our goal for the population, which will be largely a value judgment. To help us steer through social, political, and economical influences, we use a decision analysis to reveal these influences and their effects on goals and policies. A series of questions about the selected option must be posed and answered to ensure that it is feasible and that its success or failure can be determined.

Part 1

Wildlife ecology

2 Food and nutrition

<table>
<tr><td>

2.1 **Introduction**

</td><td>

The three main areas of wildlife management (conservation, sustained yield, and control) require knowledge of the food and nutrition of animal populations. Some of the important questions are:

1 Is there enough food to support and conserve a particular rare or endangered species?
2 What is the food supply needed to support a particular sustained yield?
3 Can we alter the food supply so as to provide more effective control of pest populations?

The field of animal nutrition covers subjects such as anatomy, physiology, and ecology, and there are several good reviews of these areas; for example, Hofmann (1973) deals with the anatomy of ruminants, Robbins (1983) addresses the physiology of wildlife nutrition, and Chivers and Langer (1994) review the form, function, and evolution of the digestive system in mammals. From the point of view of wildlife management, however, we are interested in two main types of information if we are to answer the preceding questions: we need to know (i) the availability of the food and (ii) the requirements of the animals. By matching the two sets of information, we can answer these questions. Sections 2.2–2.4 deal with availability, while Sections 2.5–2.9 address animal requirements.

</td></tr>
<tr><td>

2.2 **Constituents of food**

2.2.1 *Energy*

</td><td>

Energy is measured in units of calories or joules (1 cal = 4.184 joules). The energy contents of foods can be found by oxidizing a sample in a bomb calorimeter. Differences in the energy contents of different plant and animal materials result from the differences in their constituents. The energy contents of some common food components are given in Table 2.1. We can see that fats and oils have the highest content (over 9 kcal/g), followed by proteins (around 5 kcal/g) and then sugars and starches (carbohydrates; close to 4 kcal/g). The gross energy of a tissue depends on the combination of these basic constituents, particularly in animals. In plant tissues, energy content remains relatively uniform, in the region of 4.0–4.2 kcal/g. Plant parts with a high oil content, such as seeds (over 5 kcal/g), and evergreen plants with waxes and resins, such as conifers and alpine plants (4.7 kcal/g), are the exceptions (Golley 1961; Robbins 1983).

Energy flow through animals can be measured with isotopes of hydrogen (^{3}H) and oxygen (^{18}O) by the *doubly labelled water method* (Nagy 1983; Bryant 1989). First, water labeled with ^{3}H and ^{18}O is injected and allowed to equilibrate in the animal, this taking 2–8 hours depending on body size. A blood sample is then collected to establish the starting concentrations of the two isotopes. Analysis of ^{3}H is carried out by liquid

</td></tr>
</table>

Wildlife Ecology, Conservation, and Management, Third Edition.
John M. Fryxell, Anthony R. E. Sinclair and Graeme Caughley.
© 2014 John Wiley & Sons, Ltd. Published 2014 by John Wiley & Sons, Ltd.
Companion Website: www.wiley.com/go/Fryxell/Wildlife

Table 2.1 Approximate
energy contents of food
components.
(Source: Robbins, 1983.
Reproduced with
permission of Elsevier.)

Food Component	Energy (kcal/g)
Fat	9.45
Protein	5.65
Starch	4.23
Cellulose	4.18
Sucrose	3.96
Urea	2.53
Leaves	4.23
Stems	4.27
Seeds	5.07

scintillation spectrophotometry and that of ^{18}O by proton activation of ^{18}O to ^{18}F (the isotope of fluoride), with subsequent counting of γ-emitting F in a γ-counter. A second blood sample is collected several days later. The timing of the second collection does not need to be exact, but it should occur when approximately half of the isotope has been flushed from the body. Thus, timing depends on body size and the flow rates of the isotopes. Oxygen leaves the body via CO_2 and water, at a rate measured by dilution of the ^{18}O. Rate of water loss is measured from the dilution of ^3H. Thus, the difference between the total oxygen loss and the oxygen loss in water gives the rate of CO_2 production, which is a measure of energy expenditure. The method and its validation are described by Nagy (1980, 1989).

2.2.2 *Protein*

The term *protein* covers a varied group of high-molecular-weight compounds; these are major components in cell walls, enzymes, hormones, and lipoproteins and are made up of about 25 amino acids linked together through nitrogen–carbon peptide bonds. Most animal species have a relatively similar gross composition of amino acids.

Animals with simple stomachs require 10 *essential amino acids*, these being the forms that cannot be synthesized by the animal and must be obtained in the diet: arginine, histidine, isoleucine, leucine, threonine, lysine, methionine, phenylalanine, tryptophan, and valine. *Nonessential amino acids*, therefore, are ones which can be synthesized in the body. Ruminants, as well as other species that rely on fermentation through the use of microorganisms, synthesize many of the amino acids themselves and so have a shorter list of essential amino acids.

Although there is some variability in the nitrogen content of amino acids (ranging from 8 to 19%), the average is 16%. Thus, in analyzing tissues for *crude protein*, the proportion composed of nitrogen is multiplied by the constant 4.25 (i.e. 100/16). The crude protein content of plant material tends to vary inversely with the proportion of fiber. Since one of the major constituents of fiber is the indigestible compound lignin, fiber content can be used as an index of the nutritive value of the plant food. In many plant tissues, such as leaves and stems, protein and digestible energy content (i.e. the non-fiber component) tend to vary together. However, some plant parts, such as seeds, are high in energy but quite low in protein.

2.2.3 *Water*

The water content of birds and mammals is a function of body weight (W) to the power of 0.98 when comparing across species, but more restricted groups vary in the exponent. Robbins (1983) found the water content of white-tailed deer and several rodents varied as a function of $W^{0.9}$.

Water is obtained from three sources:

1 *Free water*   From external sources such as streams and ponds.
2 *Preformed water*   Found in the food.
3 *Metabolic water*   Produced in the body from the oxidation of organic compounds. Preformed water is high in animal tissues such as muscle (72%) and in succulent plants, roots, and tubers. Because of this, carnivores may not have to drink often; herbivores such as the desert-adapted antelope, oryx, which eat fleshy leaves and dig up roots, can also live without free water (Taylor 1969; Root 1972).

The highest rate of production of metabolic water in animals comes from the oxidation (catabolism) of proteins, due to their initially high water content. Catabolism of fats produces 107% of the original fat weight as water, but the low preformed water content (3–7%) means that the absolute amount produced is less than that from protein (Robbins 1983).

Measures of free water intake from drinking underestimate total water turnover. More accurate methods use the ^{3}H or deuterium oxide isotopes of water: a known sample of isotopic water is injected into an animal, and after a period of 2–8 hours (depending on size of animal) for equilibration, a blood sample is collected; the concentration of isotope in the blood is then measured using a liquid scintillation spectrometer. A second blood sample is collected a few days to a few weeks later, again depending on body size, providing a new value of isotope concentration. Because water is lost through feces, urine, and evaporation, the isotope is diluted by incoming water. Therefore, the rate of dilution is a measure of water turnover. These techniques are described by Nagy and Peterson (1988) and have been used on a wide range of animals, including eutherian mammals, marsupials, birds, reptiles, and fishes.

2.2.4 *Minerals*   Minerals make up only 5% of body composition but are essential to body function. Some minerals (roughly in order of abundance: calcium, phosphorus, potassium, sodium, magnesium, chlorine, sulfur) are present or required in relatively large amounts (mg/g) and are called *macroelements*. Those that are required in small amounts (µg/g) are called trace elements (iron, zinc, manganese, copper, molybdenum, iodine, selenium, cobalt, fluoride, chromium). So far, very little is known about the mineral requirements of wildlife species, but Robbins (1983) has provided a summary of available information. It is assumed that most native species are adapted to their environment and so can tolerate the levels of minerals found there (Fielder 1986). However, some mineral deficiencies have been observed. Selenium deficiency increases the mortality of juvenile, preweaned mammals (Keen and Graham 1989). Flueck (1994) supplemented wild black-tailed deer in California and increased preweaning fawn survival threefold.

Calcium and phosphorus are essential for bones and eggshells. Cervids have a very high demand for these minerals during antler growth. Calcium is also needed during lactation, for blood clotting, and for muscle contraction. Phosphorus is present in most organic compounds. Deficiencies of calcium result in osteoporosis, rickets, hemorrhaging, thin eggshells, and reduced feather growth. Carnivores that normally eat the flesh of large mammals need to chew bone in order to obtain their calcium. Mundy and Ledger (1976) found that the chicks of Cape vultures (*Gyps coprotheres*) in South Africa developed rickets when they were unable to eat small bone fragments. This has an important management consequence: bone fragments from large carcasses are made available to vultures by large carnivores, in this case lions and hyenas; where carnivores are exterminated on ranch land, carcasses are not dismembered and bones are too large for the chicks to swallow. This is a good example of how the interaction of species should be considered in the management and conservation of habitats.

Sodium is required for the regulation of body fluids, muscle contraction, and nerve impulse transmission. Sodium is usually found in low concentrations in plants, so herbivores face a potential sodium deficiency. In areas of low sodium availability, herbivores consume soil or water from mineral licks (Weir 1972; Fraser and Reardon 1980). Carnivores can easily obtain sodium from their food, and so are unlikely to experience sodium deficiency. Isotopic sodium has been used as a measure of the food intake rates of carnivores such as lions (Green *et al.* 1984), seals (Tedman and Green 1987), crocodiles (Grigg *et al.* 1986), and birds (Green and Brothers 1989). This approach is possible because sodium remains at a relatively constant concentration in the food supply. The technique is similar to that for isotopic water described in Section 2.2.3.

Both potassium and magnesium are abundant in plants, and deficiencies in free-living wildlife are therefore unlikely. The same is true for chloride ions and for sulfur. Trace element deficiencies are unusual under normal free-ranging conditions but they occur locally from low concentrations in the soil: there are some reports of iodine and copper deficiencies and of toxicity from too much copper and selenium (Robbins 1983).

2.2.5 *Vitamins*

Vitamins are essential organic compounds that occur in minute amounts in food and cannot normally be synthesized by animals. There are two types of vitamin: fat-soluble (vitamins A, D, E, and K) and water-soluble (vitamin B complex, vitamin C, and several others). Fat-soluble vitamins can be stored in the body. Water-soluble vitamins cannot be stored and hence must be constantly available. Overdose toxicities can arise only from the fat-soluble vitamins.

Vitamin A, a major constituent of visual pigments, can be obtained from β-carotene in plants. Vitamin D is needed for calcium transport and the prevention of rickets. Vitamin E is an antioxidant needed in many metabolic pathways; it is high in green plants and seeds, but decreases as the plants mature. Vitamin K is needed to make proteins for blood clotting. Deficiencies are unlikely to occur because it is common in all foods. The vitamin K antagonist, warfarin, causes hemorrhaging. It is used as a rodenticide.

Little is known about the B-complex vitamins and whether deficiencies occur in free-living wildlife species, although cases of thiamin (B1) deficiency have been reported for captive animals (Robbins 1983). Vitamin C differs from the others in that most species can synthesize it in either the kidneys or liver. Exceptions include primates, bats, guinea pigs, and possibly whales. Vitamin C is not as commonly available as the B vitamins but is found in green plants and fruit. It is absent in seeds, bacteria, and protozoa.

Other physiological constraints that may not be called vitamins nevertheless provide limits to animal nutrition. For example, old-world starlings and flycatchers cannot digest sucrose (Martinez del Rio 1990).

2.3 **Variation in food supply**

2.3.1 *Seasonality*

Food supply varies with season. To some degree, all environments are seasonal, including those of the tropics. Food supply is greatest for herbivores when plants are growing: during the summer at higher latitudes (temperate and polar regions) and during the rainy season in lower ones (tropics and subtropics). Protein in grass and leaves declines from high levels of 15–20% in young growth to as little as 3% in mature flowering grass, or even 2% in dry, senescent grass. Leaves from mature dicots maintain a higher protein content of about 10%. Thus, herbivores such as elk in North America and eland and elephant in Africa will switch from grazing in the growing season to browsing in the non-growing season. Many forest-dwelling Australian marsupials are mycophagous; that is, they prefer to feed on the sporocarps of hypogeous fungi. They feed on dicot fruits and leaves when fungi are about. Growth rates of pouch young in the Tasmanian bettong (*Bettongia gaimardi*) are directly related to periods of sporocarp production (Johnson 1994).

Winter is the main period of stress for animals in higher latitudes. Low temperatures create higher energy demands just when energy is less available. For example, energy intake of moose in Norway declines by 15–30% during winter, resulting in a deficiency of 20–30% relative to their requirements. Energy intake is less ($573\,kJ/kg^{0.75}/day$) in poor habitats than in good ones ($803\,kJ/kg^{0.75}/day$) (Hjeljord *et al.* 1994). Even greater reductions in food intake rates during winter have been recorded for black-tailed deer (Gillingham *et al.* 1997)

Animals adjust their breeding patterns so that their highest physiological demands for energy and protein occur during the growing season. Thus, northern ungulates give birth in spring so that lactation can occur during the growing period of plants, whereas tropical ungulates produce their young during or following the rains, allowing the mother to build up fat supplies to support lactation. Although most birds complete their entire breeding cycle during one season, the timing of breeding is closely associated with food supply (Perrins 1970). Very large birds such as ostrich behave like ungulates and start their reproductive cycle during one wet season so that the precocial chicks hatch at the start of the next (Sinclair 1978).

Carnivores also adapt their breeding to coincide with maximum food supply. Thus, wolves that follow the caribou on the tundra of northern Canada have their young at the time caribou calves are born. Schaller (1972) records that lions on the Serengeti plains of Tanzania have their cubs when the migrant wildebeest are giving birth. In the same area, birds of prey have their young coinciding with the appearance of other juvenile birds and small mammals, which form their prey (Sinclair 1978).

2.3.2 *Year-to-year variation in food supply*

A particular kind of variability in food supply occurs with the production of prolific seed crops by some tree species. This seed is termed *mast*. It occurs when the majority of trees in a region synchronize their seed production. Beech trees (*Fagus*, *Nothofagus*) and many northern-hemisphere conifers (e.g. white spruce, *Picea glauca*) produce their seeds at the same time, these mast years occurring every 5–10 years. Birds that depend on these conifer seeds, such as the crossbill (*Loxia curvirostra*), breed throughout the winter when a mast cone crop occurs. In the following year, when few cones are produced, the crossbills disperse to find regions with a new mast crop, sometimes travelling many hundreds of kilometers (Newton 1972).

Red squirrels (*Tamiasciurus hudsonicus*) also respond to cone masts in white spruce. This species caches unopened cones in food tunnels in the ground and uses them throughout the next winter. Survival of squirrels is high during these mast winters.

An unusual form of variability in food supply occurs in the bamboo species that form the main food of the giant panda (*Ailuropoda melanoleucus*): the bamboo synchronized flowering in much of southern China during the early 1980s (Schaller *et al.* 1985) and then died, and there was little food available for a few years. With the giant panda now confined to a few protected areas, the population suffered from this sudden drop in food supply. Knowledge of such events is important for conservation. It tells us that reserves must be sufficiently diverse in environment, habitat, and food species to avoid the type of restriction in food supply produced by the synchronous flowering of bamboo. Presumably in prehistoric times giant pandas were able to range over a much wider area and so take refuge in regions where bamboo was not flowering. They cannot now move in this way and most of their former range in the lowlands is no longer available.

In the Canadian boreal forest, lynx and great horned owls breed prolifically during the peak of the 10-year snowshoe hare cycle and cease breeding during the low phase (Rohner *et al.* 2001).

2.3.3 *Plant secondary compounds*	Many plants produce chemicals that deter herbivores from feeding upon them. These chemicals are called *secondary compounds*. Their production is associated with growth stage, but this association differs between plant species. Although secondary compounds are found in some grasses (monocots), most are found in dicots. Tannins are low in young oak leaves but abundant in mature leaves (Feeny and Bostock 1968). Conversely, various secondary compounds are abundant in juvenile twigs of willows, birches, and white spruce in Alaska and Canada but sparse in mature twigs of 3 years and older (Bryant and Kuropat 1980). Thus the palatability and availability of food for herbivores differs between seasons and between years because of changes in the concentration of secondary compounds.

There are three major classes of secondary compound: terpenes; soluble phenol compounds; and alkaloids, cardenolides, and other compounds.

Terpenes

These are cyclic compounds of low molecular weight and usually one to three rings. They inhibit the activity of rumen bacteria (Schwartz *et al.* 1980) and are bitter tasting or volatile. Examples are essential oils from citrus fruits, carotene, eucalyptol from eucalyptus, papyriferic acid in paper birch, and camphor from white spruce. Camphor and papyriferic acid act as antifeedants to snowshoe hares (Bryant 1981; Sinclair *et al.* 1988), while α-pinene from ponderosa pine deters tassel-eared squirrels (*Sciurus alberti*) (Farentinos *et al.* 1981).

Soluble phenol compounds

The main groups of these chemicals are the hydrolyzable and condensed tannins (McLeod 1974). They act by binding to proteins and thus make them indigestible. The name "tannin" comes from the action of polyphenols on animal skins, turning them into leather, which is not subject to attack by other organisms, in a process called "tanning."

Tannins are widespread among plant species, occurring in 87% of evergreen woody plants, 79% of deciduous woody species, 17% of annual herbs, and 14% of perennial herbs. Tannins have negative physiological effects on elk (Mould and Robbins 1982) and may determine food selection by browsing ungulates in southern Africa (Owen-Smith and Cooper 1987; Cooper *et al.* 1988) and snowshoe hares in North America (Sinclair and Smith 1984). Domestic goats (*Capra hircus*) learn to avoid young twigs of blackbrush (*Coleogyne ramosissima*) due to condensed tannins (Provenza *et al.* 1990).

Alkaloids, cardenolides, and other compounds

These are cyclic compounds with nitrogen atoms in the ring. They occur in 7% of flowering plants, and some 4000 compounds are known (Robbins 1983). Some alkaloids are nicotine, morphine, and atropine. They have several physiological effects, but act more as toxicants or poisons than as digestion inhibitors. Some alkaloids, such as cardenolides in milkweed (Asclepiadaceae), are sequestered by insects such as the monarch butterfly (*Danaus plexippus*), whose larvae feed on milkweed. These noxious cardenolides act as emetics to birds. Young, inexperienced blue jays (*Cyanocitta cristata*) eat these insects, then regurgitate them; after this, they avoid them (Brower 1984). Cyanogenic glycosides, which release HCN on hydrolysis in the stomach, are sequestered by *Heliconius* butterflies from their passionflower (*Passiflora* species) food plants. These insects are avoided by lizards, tanagers, and flycatchers (Brower 1984).

2.4 Measurement of food supply

2.4.1 Direct measures

The amount of food available to animals may be measured directly. For carnivores, some form of food sampling can be used: insect traps for insectivores; counts of ungulates for large carnivores. For grazing ungulates, McNaughton (1976) clipped grass in exclosure plots to measure the available production for Thomson's gazelle on the Serengeti Plains. Winter food supply for snowshoe hares was estimated from the abundance of twigs with a diameter of 5 mm on its two most common food plants, gray willow (*Salix glauca*) and bog birch (*Betula glandulosa*) (Smith *et al*. 1988; Krebs *et al*. 2001a). Pease *et al*. (1979) used a different approach, feeding a known quantity of large branches to hares in pens and measuring the amount eaten from these. Using this measure as the edible fraction of the large branches, they then estimated the total available biomass of edible twigs from the density of large branches in the hares' habitat.

The most serious problem with direct measures is that they all depend on the assumption that we can measure food in the same way that the animals come across it. It is rare that this assumption is valid: insects that enter pitfall traps or are collected by sweepnets are not the same fraction as is seen by a shrew or bird; ungulate censuses do not indicate which animals are actually available to carnivores, for we can be sure that not all are catchable.

If the food supply is relatively uncomplicated, such as the short green sward that is grazed uniformly by African plains antelopes, then we can measure it in a way resembling the feeding of animals. With woody plants, however, we cannot do so, and in most cases our estimates are simply crude indices of food abundance. Our errors can both over- and underestimate the true availability of food: we may include material that an animal would not eat, so producing an overestimate; or we may overlook food items because animals are better at searching for their own food than we are, so producing an underestimate. We can never be sure on what side of the true value our index lies, unless we calibrate it with another method.

2.4.2 Fecal protein and diet protein

A second method, which has been applied so far only to herbivores, allows the animal to choose its own food and so avoids the problems discussed in the previous section. Diet protein, energy, or other nutrients can be estimated by observing what animals eat and then determining the chemical composition of that diet. These indirect estimates of intake are compared with an estimate of requirements either based upon direct physiological experiments or inferred from the literature. Examples come from reindeer on South Georgia Island (Leader-Williams 1988) and greater kudu (*Tragelaphus strepsiceros*) in South Africa (Owen-Smith and Cooper 1989). Energy intake for the jerboa

Fig. 2.1 Seasonal changes in energy and body weight of the jerboa (*Allactaga elater*) in the cis-Caspian, Russia, during 1985. (a) Percentage energy of forage in the stomach. (b) Daily energy intake, and daily energy requirement. (c) Body weight. (After Abaturov and Magomedov 1988.)

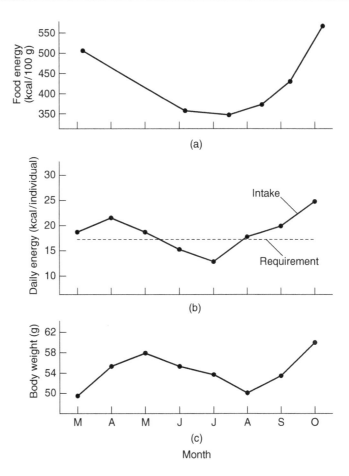

(*Allactaga elater*) in north cis-Caspian, Russia (Fig. 2.1), dropped below requirements in mid-summer, and so body weight declined (Abaturov and Magomedov 1988). Similarly, energy measured from fecal collections showed that the energy intake of moose in Norway during winter dropped below requirements by 25–30% (Hjeljord *et al.* 1994). For greater kudu (Fig. 2.2), energy intake during winter was below requirements, but protein intake was sufficient. In contrast, the protein intake of African buffalo in tropical dry seasons was below requirements (Fig. 2.3).

These indirect measures of food intake can often be inaccurate because they are an amalgam of several different measurements. One way around this is to use a physiological index from the animal to indicate the quality of the food it has eaten. Nitrogen in the feces predicts nitrogen in the diet down to the minimum level of nitrogen balance. If nitrogen intake falls below this level, it is not reflected in the feces, because metabolic nitrogen (from microorganisms and gut cells) continues to be passed out irrespective of intake.

In tropical regions, this relationship has been found for cattle (Bredon *et al.* 1963), buffalo, and wildebeest (Sinclair 1977), and in North America, for cattle, big-horn sheep (*Ovis canadensis*), elk, and deer (Fig. 2.4) (Leslie and Starkey 1985; Howery and Pfister 1990). These relationships apply to ruminants eating natural food. Similar relationships have been found for experimental diets in Australian rabbits (Myers and

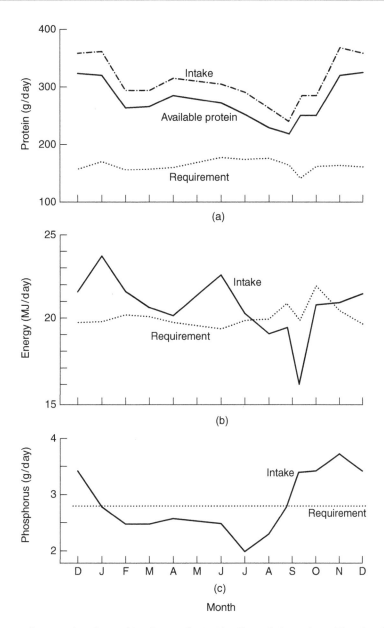

Fig. 2.2 Monthly changes in the estimated daily nutrient intakes of greater kudu relative to estimated maintenance requirements. (a) Crude protein intake (dashed line); available protein (solid line); protein requirement for metabolic turnover, fecal loss, and growth (dotted line). (b) Metabolizable energy intake (solid line); metabolizable energy requirement for resting, activity, and growth (dotted line). (c) Phosphorus intake (solid line); phosphorus requirement (dotted line). (After Owen-Smith and Cooper 1989.)

Bults 1977), snowshoe hares (Sinclair *et al.* 1982), elk, and sheep (Mould and Robbins 1981; Leslie and Starkey 1985), although the slopes of the regression lines differ from the natural diets.

A potential problem with this approach is that plant secondary compounds such as tannin may obscure the relationship by causing higher amounts of metabolic nitrogen to be passed out (Robbins *et al.* 1987; Wehausen 1995). This has been observed in experimental diets with high amounts of these compounds (Mould and Robbins 1981; Sinclair *et al.* 1982). However, these are abnormal situations and when animals are allowed to choose their own diet the relationship holds up. The regression has been determined for only a few species on natural diets, so more work is needed in this area.

Fig. 2.3 The proportion of crude protein in the diet of African buffalo declines below the estimated 5% minimum requirement in the dry season. Estimates from diet selection with 95% confidence limits (solid line); estimates from fecal protein (broken line). (After Sinclair 1977.)

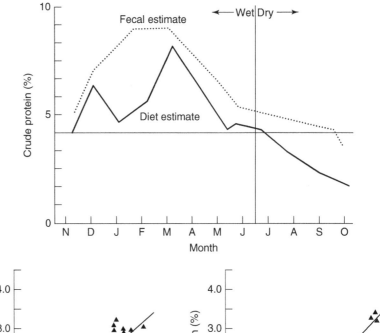

Fig. 2.4 Correlation of dietary nitrogen with fecal nitrogen in (a) elk and (b) black-tailed deer. Nitrogen increases with season. Spring (▲); summer (●); fall (△); winter (○). (After Leslie and Starkey 1985.)

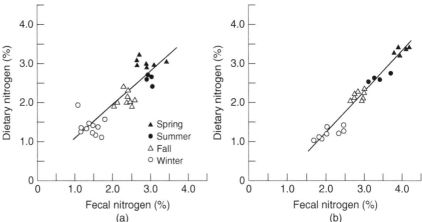

A second potential problem could arise if fecal samples were exposed to the weather and the nitrogen leached out. For white-tailed deer feces in autumn, the bias is minimal if samples are collected less than 24 days after defecation (Jenks *et al.* 1990).

The relationship between fecal nitrogen and dietary nitrogen can be used to estimate whether animals are obtaining enough food for maintenance. In African buffalo, the estimate of dietary nitrogen was compared with estimates of dietary nitrogen from rumen contents (Fig. 2.3). The two are similar.

A comparable approach has related fecal nitrogen directly to weight loss. Thus, Gates and Hudson (1981) found that elk lost weight below about 1.6% fecal nitrogen (Figure 2.5a) during late winter when there was deep snow (Fig. 2.5b).

2.5 **Basal metabolic rate and food requirement**

2.5.1 *Energy flow*

The flow of energy through the body is illustrated in Fig. 2.6. Energy starts as consumption energy or intake energy. Part of this is digested in the gut and passes through the gut wall as digestible energy; the rest is passed out in the feces as fecal energy. Part of the digestible energy is lost in the urine and the remainder, called metabolic or assimilated energy, can then be used for work. The work energy can be divided into two further subtypes: respiration energy, which is used for the basic maintenance of

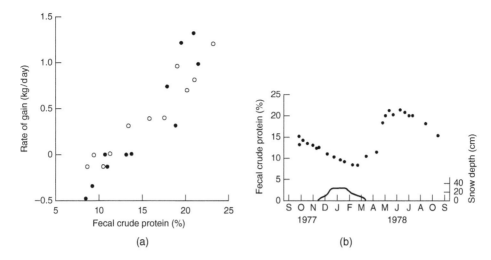

Fig. 2.5 (a) Body weight gain of male elk (●) and calves (○) in Alberta can be predicted from the percentage of fecal crude protein. (b) Seasonal changes in the percentage of fecal crude protein are related to snow depth. (After Gates and Hudson 1981.)

Fig. 2.6 Flow chart of energy through the body.

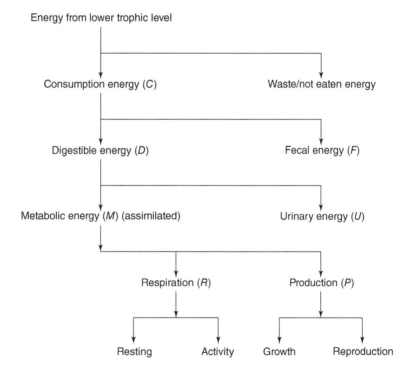

the body (resting energy) and for activity, and production energy, which is used for growth and reproduction.

The flow chart for protein is similar, except that protein is normally used only for production. Protein is not used in respiration except under special conditions of food shortage, when it is broken down (catabolized) to provide energy.

Metabolic energy (M) can be measured in two ways:

1 In the laboratory, by measuring resting energy and activity to obtain the respiration component (R) and from growth and population studies to obtain production (P), so

that:

$$M = R + P$$

2 In the field, by measuring consumption (C) and fecal (F) and urinary (U) outputs, so that:

$$M = C - F - U$$

2.5.2 Basal metabolic rate

Basal metabolism is the energy needed for basic body functions. The energy comes from oxidation of fats, proteins, and carbohydrates to produce water and CO_2. Thus, maintenance energy can be measured from expired air volume and composition because intake air has a stable composition of 20.94% oxygen, 0.03% carbon dioxide, and 79.03% nitrogen. Since 6 moles of CO_2 and water are produced with 673 kcal of heat, the CO_2 in expired air can be used to calculate the rate of energy used for maintenance. Measurements can be obtained either in chambers or from gas masks, and the animal must be in its thermoneutral zone (not shivering, panting, or sweating), resting, and not digesting food. Such conditions give the basal metabolic rate.

When plotted against log of body weight, the basal metabolic rates (BMRs) of different eutherian mammalian groups, such as those in Fig. 2.7, fall on a line whose slope is approximately 0.75. Thus, Kleiber (1947) produced the general equation:

$$BMR = 70\,W^{0.75}$$

where BMR is in kcals per day and W is body weight in kilograms. This is an average over all mammals. Specific groups may differ: desert-adapted mammals have lower rates, marine mammals higher ones. Large nonpasserine birds are similar to eutherians, but the smaller passerines are 30–70% higher. The constant 70 also differs; in marsupials it is 48.6 and in the echidna (a monotreme) 19.3 (Robbins 1983). McNab (1988) predicted that animals feeding on lower-energy foods should lower their BMR. Experiments in Chile with the burrowing rodent (*Octodon degus*) fed on low- or high-fiber diets have confirmed this prediction (Veloso and Bozinovic 1993).

Fig. 2.7 Relationship of basal metabolic rate and body weight in different groups of small mammals. (After Clutton-Brock and Harvey 1983, which is after Mace 1979.)

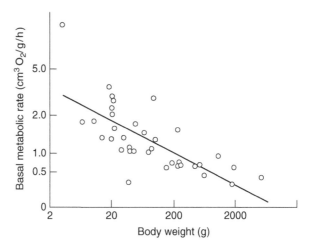

Hibernating mammals, such as ground squirrels, can lower their body temperatures to a few degrees above ambient temperature, but no lower than about 0°C. Hummingbirds can lower their body temperature to about 15°C, in a process called torpor. Both hibernation and torpor save energy (Kenagy 1989; Kenagy *et al.* 1989).

So far, we have discussed resting or maintenance requirements. Activity adds a further energy cost to maintenance. Standing is on average 9% more costly than lying for mammals and 13.6% for birds (Robbins 1983). The cost of locomotion is similar for bipedal and quadrupedal animals (Fedak and Seeherman 1979). Cost of locomotion (LC), expressed as kcal/kg/km, declines linearly with increasing log body size. Thus:

$$LC = 31.10 W^{-0.34}$$

where W is in grams.

Hence, the cost of moving is higher per unit body mass for smaller species and juveniles.

Average daily metabolic rate (ADMR, the sum of resting and activity rates) is approximately $2 \times$ BMR in captive mammals, but it is difficult to measure for free-living animals. For captive passerines, ADMR is $1.31 \times$ BMR, and for captive nonpasserines, $1.26 \times$ BMR. As a rough approximation, free-living birds and small mammals have a metabolic rate two to four times BMR.

2.5.3 *Variation in food requirements*

The ADMR and other average measures of metabolic rate hide seasonal fluctuations in food and energy demands. The costs of reproduction add considerably to those for normal daily activity. In the red deer and the wildebeest, the rut imposes a considerable energetic cost upon males, which spend several weeks fighting, defending territories, and herding females while eating very little (Sinclair 1977; Clutton-Brock *et al.* 1982). Males put on large amounts of body fat before the rut and use it to cover the extra energy requirements it imposes. Mule deer males (Fig. 2.8a) deposit kidney fat in autumn and use it during mating in November (Anderson *et al.* 1972).

Female mammals use additional energy for lactation and to grow a fetus. Like males, they accumulate body fat, especially in the mesentery and around the kidneys, before birth and lactation. During the last third of gestation, metabolic costs are twice ADMR, and during lactation they are three times ADMR. In female mule deer (Fig. 2.8b), fat is built up in autumn and early winter and used during gestation, birth, and lactation between late winter and summer. Thus, the timing of reproduction in ungulates is influenced in part by the need to obtain good food supplies and to build up fat reserves.

2.6 **Morphology of herbivore digestion**

2.6.1 *Strategies of digestion*

Carnivores and omnivores digest their food in the stomach and small intestine. The small intestine is relatively short in these species. Herbivores, which make up most (about 90%) of the mammals (Björnhag 1987), need to digest large amounts of fairly indigestible cellulose and hemicellulose, and have thus adapted the gut to increase retention time. One strategy is to evolve a much longer small intestine. An exception is the giant panda, which evolved from bears and has retained the short intestine. In this species, organic matter digestibility is only 18%, one of the lowest recorded (Schaller *et al.* 1985). Another adaptation is to use microorganisms (bacteria, fungi, protozoa), which digest cellulose through fermentation. Plant material must be retained in a fermentation chamber long enough for the microorganisms to cause fermentation. Squirrels eat high-energy foods such as seeds, fruits, and insects and so do not need such

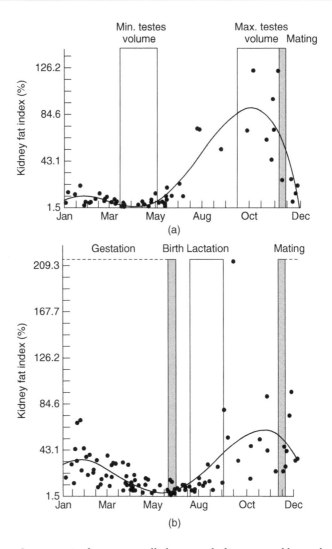

Fig. 2.8 Seasonal changes in the kidney fat index of mule deer are closely associated with reproduction and season. (a) Males; (b) females. (After Anderson *et al.* 1972.)

mechanisms. Some species have unusually low metabolic rates and hence longer retention times. Most are arboreal folivores: koalas (*Phascolarctos cinereus*) (Dawson and Hulbert 1970), sloths, and hyraxes (Rubsamen *et al.* 1979; Björnhag 1987). Reviews of digestive adaptations can be found in Hume and Warner (1980), Hornicke and Björnhag (1980), Robbins (1983), Björnhag (1987), and Chivers and Langer (1994).

2.6.2 *Ruminants*

True ruminants, which include the bovids (cattle, sheep, antelopes), cervids (deer), tylopodids (camels), and giraffes, have an extension of the stomach divided into three chambers. One of these is the rumen, which acts as the fermentation chamber. Plant food is gathered without chewing and stored in the chamber during a feeding period. This is followed by a rumination period, during which portions of compacted food (bolus) are returned to the mouth for intensive chewing. In this way, coarse plant material is broken down mechanically and made available to the microorganisms for fermentation. The amount of fiber in the food determines how coarse it is; the coarser the food, the longer the process of grinding and fermentation. There is a limit to how coarse the food can be before fermentation takes so long that the animal uses more energy than it gains. On average, a ruminant retains food in the gut for about 100 hours.

Microorganisms break down cellulose into short-chain fatty acids and proteins into amino acids and ammonia, which they use to produce more microorganisms. The host animal obtains its nutrients by digesting the dead microorganisms in the stomach and short intestine. The system is efficient, and digestibilities of organic matter and protein are around 65–75% are achieved for medium- to good-quality food (i.e. relatively low in fiber). Another advantage is that nitrogen can be recycled as urea. A disadvantage is that microorganisms digest nutrients that could be used directly by the host, which leads to a loss of energy through production of methane. Another is that ruminants cannot digest very high-fiber diets.

2.6.3 *Hindgut fermenters*

In contrast to the foregut fermenters, or ruminants, a number of animal groups have developed an enlarged colon or caecum, or both, for fermentation. Large animals (over 50 kg) are in general colon fermenters, while small ones (under 5 kg) that feed on fibrous food are caecum fermenters.

Colon fermenters

In most cases, both the colon and the caecum are enlarged to hold fiber for microbial digestion. There is little separation of material into small particles and microbes on the one hand and fiber on the other, and there is little evidence that microbial proteins are digested and absorbed, although fatty acids can be absorbed.

Animals in this group are perissodactyls (horses, rhinos, tapirs), macropods (kangaroos), and perhaps elephants, wombats (*Vombatus ursinus*), and dugongs (*Dugong dugon*). These are all large animals and so do not need to ingest large amounts of energy and protein per unit of body weight (see Section 2.5.2). Since food material can be retained in the gut for longer periods in large animals, the rate of passage may be slow enough to allow fermentation and absorption of fatty acids to take place. None of these animals eat their feces, a practice called *coprophagy*.

Caecum fermenters

Small animals (less than 5 kg) have a relatively high metabolic rate. Those species which consume high-fiber diets such as grasses and leaves need to use the microbial protein produced by hindgut fermentation. They do so by coprophagy. In conjunction with this process, there is a sorting mechanism in the colon that separates fluids, small particles of food, and microbes from the fiber. The fluids and microbes are returned by antiperistaltic movements to an enlarged caecum for further fermentation and digestion. This mechanism therefore retains the nutrients long enough for fermentation. It is necessary because small animals cannot hold food material long enough to allow fermentation under normal passage rates.

Dead microbial material is passed out in the form of special soft pellets, *caecotrophs*, and these high-nutrient feces are eaten directly from the anus, a behavior called *caecotrophy*. The sorted high fiber is passed out as hard pellets, which are not re-ingested.

Animals that both ferment food in the caecum and practice caecotrophy include myomorph rodents (voles, lemmings, brown rat), lagomorphs (hares, rabbits), some South American rodents (coypu, guinea pig, chinchilla), and some Australian marsupials, such as the ringtail possum (*Pseudocheirus peregrinus*) (Chilcott and Hume 1985).

Two marsupials, the koala and the greater glider (*Petauroides volans*), feed on arboreal leaves and have caecal fermentation and a colonic sorting mechanism (Cork and Warner 1983; Foley and Hume 1987). Neither practice caecotrophy. At least in the koala, both the metabolic rate and the passage time are slow enough that caecotrophy is not necessary.

Björnhag (1987) identifies four strategies employed by small mammals that feed on plants:

1 Eat only highly nutritious plant parts, such as seeds, berries, birds, and young leaves. Squirrels fall into this group.
2 Have a low metabolic rate for your size so that fermentation is prolonged. Koala and tree sloths are examples of this group.
3 Separate digesta in the colon and retain easily digestible food particles plus microorganisms so as to allow fermentation of fibrous material being sorted and passed out.
4 Separate and retain only the microorganisms so as to allow rapid fermentation.
Both 3 and 4 recirculate protein-rich fecal material by re-ingestion through caecotrophy. Examples are voles, lemmings, and lagomorphs. Foley and Cork (1992) review these strategies of digestion and their limits.

2.7 Food passage rate and food requirement

The passage rate of food through an animal depends on the *retention time*, which is the mean time an indigestible marker takes to pass through. Various markers can be used, including dyes, glass beads, radioisotopes, and polyethylene glycol. Certain rare earth elements (samarium, cerium, and lanthanum) bind to plant fiber and provide useful markers by which to measure fiber passage times (Robbins 1983).

The rate of food intake by herbivores depends on the nutritive quality of the food. For example, in domestic sheep (Fig. 2.9) and white-tailed and mule deer, intake rate first increases and then decreases as the energy quality of food declines (Sibly 1981; Robbins 1983). This relationship occurs because both energy and protein are inversely related to fiber content.

Estimation of fecal protein can be used as a means of determining whether a population is obtaining enough food, because protein intake is related to the amount of protein in the feces (see Fig. 2.4). This method has been used to predict the change in body weight of elk (see Fig. 2.5b) (Gates and Hudson 1981) and to monitor food requirements in snowshoe hares (Sinclair *et al.* 1988).

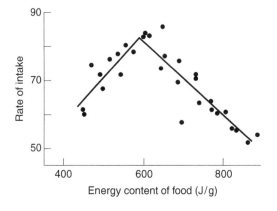

Fig. 2.9 Relationship of intake rate to energy content of food in domestic sheep. Below an energy content of 590 J/g, intake rate falls because of a finite gut capacity and declining fermentation rates. Rate of intake is dry matter/day/body weight$^{0.75}$. (After Sibly 1981, which is after Dinius and Baumgardt 1970.)

2.8 Body size and diet selection

The gut (i.e. the rumen, large intestine, and caecum, as well as the crops of humming-birds and the cheek pouches of heteromyid rodents) has a capacity that is a linear function of body weight ($W^{1.0}$) (Clutton-Brock and Harvey 1983; Robbins 1983). Energy requirements, however, are a function of metabolic body weight ($W^{0.75}$). Thus, the difference between the exponents ($W^{1.0}/W^{0.75} = W^{0.25}$) means that a larger animal can eat more food relative to requirement than a smaller animal. This can be expressed in two ways: (i) on the same quality of diet, a larger animal needs to eat less food per unit of body weight than does a smaller one; and (ii) a larger animal on a lower-nutrient diet can extract the same amount of nutrient per unit of body weight as a smaller animal on a higher-nutrient diet. Thus, larger animals can eat higher-fiber diets, a feature that allows resource partitioning in African ungulates (Bell 1971; Jarman and Sinclair 1979).

Jarman (1974) extended the relationship between body size and diet in African ungulates to explain interspecific patterns of social and anti-predator behavior. We can identify five categories, from selective browsers to unselective grazers:

1 Small species (3–20 kg), solitary or in pairs, which are highly selective feeders on flowers, birds, fruits, seed pods, and young shoots. Their habitats are thickets and forest, which provide cover from predators. There is little sexual dimorphism and both sexes help in defending a territory. This group includes duikers (*Cephalophus* species), suni antelope (*Nesotragus moschatus*), steinbuck (*Raphicerus campestris*), dikdik (*Madoqua* species), and klipspringer (*Oreotragus oreotragus*).

2 Small to medium species (20–100 kg), which can be both grazers and browsers, but are very selective of plant parts, as in 1. Their habitat is riverine forest, thicket, or dense woodland. Group size is larger, from two to six, comprising one male and several females. There is some sexual dimorphism. Predators are avoided by hiding and freezing. They are usually territorial and include lesser kudu (*Tragelaphus imberbis*), bushbuck (*T.scriptus*), gerenuk (*Litocranius walleri*), reedbuck (*Redunca* species), and oribi (*Ourebia ourebia*).

3 Medium-size species (50–150 kg), which are mixed feeders, changing from grazing in the rains to browsing in the dry season. Habitats are varied and range from dense woodland and savanna to open flood plains. There is one male per territory. Female group size is variable (6–200). Females do not defend a territory, but wander through the male territories. Nonterritorial male groups are excluded from territories and behave like female groups. Females have a large home range, which is smaller in the dry season than in the wet season. This group is sexually dimorphic in the extreme. Predators are avoided by group vigilance and by running. Species typical of this group include impala, greater kudu, sable, kob (*Kobus* species), lechwe, and gazelles (*Gazella* species).

4 Medium to large species (100–250 kg), which are grazers, selecting high-quality grass leaves. Males are single and territorial or form large bachelor groups. Female groups range from six to many hundred. They have a large home range, often divided into wet- and dry-season ranges separated by a considerable distance. Habitats are generally open savanna and treeless plains. Predators are avoided by group vigilance and running. Sexual dimorphism is present but less extreme than in 3. This group includes wildebeest, hartebeest, topi (*Damaliscus korrigum*), and Grevy's zebra.

5 Large species (> 200 kg), which are unselective grazers and browsers of low-quality food. Habitats are closed woodland and open savanna. Movements are seasonal.

Males are nonterritorial and form a dominance hierarchy. Females form groups of ten to several hundred and have a large home range. Active group defense against predators is shown by African buffalo and African elephant, while other species use group vigilance and running to avoid predation. This group includes Burchell's zebra (*Equus burchelli*), giraffe (*Giraffa camelopardalis*), eland (*Taurotragus oryx*), gemsbok (*Oryx gazella*), and Roan antelope (*Hippotragus niger*).

Jarman's (1974) categories relate body size inversely to food supply, because low-quality food is more abundant. This allows species to form larger groups in order to avoid predators, and the size of group then determines how a male obtains his mate. In small species, males keep females in their territories year-round, and this may be the only way of finding females in estrus. When female groups are larger (group 3), females cannot remain within one territory. Males thus compete for territories within the females' home range, providing an opportunity for mating as females move through the territory. These territories are for mating and not to provide year-round food.

Finally, interspecific competition for male mating territories may have led to larger males with elaborate weapons. Since these selection pressures have not operated on the females, which have remained at a smaller size, sexual dimorphism develops. Thus, we see a connection between body size, food quality, group size, predator defense, and mating system.

In other groups of animals, gut size can be phenotypically plastic, varying with food availability and season, particularly in birds (Piersma and Lindstrom 1997). Thus, garden warblers (*Sylvia borin*) migrating over the Sahara reduce their gut size and hence food intake (McWilliams *et al.* 1997).

2.9 Indices of body condition

2.9.1 *Body weight and total body fat*

Body weight and fat reserves affect survival and reproduction in mammals (Hanks 1981; Dark *et al.* 1986; Gerhart *et al.* 1996) and birds (Johnson *et al.* 1985). Body weight can be measured directly for small mammals such as the jerboa (Fig. 2.1) and for birds. Weight changes seasonally in response to changes in food supply and hence intake.

Body weight is a function of genetic determinants, age, and the amount of fat and protein stored in the body. In order to monitor fat and protein changes, it is better to remove the effects of body size (the genetic and age components). This can be done by using a ratio of weight to some body measure that is a function of size. Thus, for cottontail rabbits (*Sylvilagus floridanus*), Bailey (1968) found a relationship between predicted body weight (PBW, in g) and total length (L, in cm) such that:

$$\text{PBW} = 16 + 5.48\ (\text{L}^3).$$

The condition index for the rabbits would then be the ratio of observed weight to predicted weight. A similar relationship has been found between foot length and weight for snowshoe hares (O'Donoghue 1991). Murray (2002) found that the bone-marrow fat of snowshoe hares was predicted by the ratio of body weight to foot length. Kruuk *et al.* (1987) used a general equation for European otters (*Lutra lutra*) in which the index of body condition (K) was related to mean body weight (W, in kg) and body length (L, in m) by:

$$\text{K} = \text{W}/(\text{a L}^n)$$

where $a = 5.02$ and $n = 2.33$.

At the other extreme of size, blubber volumes of fin whales (*Balaenoptera physalus*) and sei whales (*B.borealis*) have been calculated from body length, girth, and blubber thickness measured at six points along the carcass (Lockyer 1987).

Although body weight alone is a satisfactory measure of condition for such birds as sandhill cranes (*Grus canadensis*) and white-fronted geese (*Anser albifrons*) (Johnson *et al.* 1985), it is usually better to account for body size using some measure such as wing length, tarsus length, bill length, or keel length.

In female mallards (*Anas platyrhynchos*), fat weight (F), an index of condition, is related to body weight (BW) and wing length (WL) by:

$$F = (0.571 \, BW) - (1.695 \, WL) + 59.0$$

A similar relationship holds for males (Ringelman and Szymczak 1985).

In maned ducks (*Chenonetta jubata*) in Australia, female body weight and total fat increase before laying. Some 70% of the fat is used during laying and incubation. Protein levels, however, do not change (Briggs 1991). Among northern-hemisphere ducks, there are four general strategies for storing nutrients prior to laying:

1 Fat is deposited before migration and is supplemented by local foods on the breeding grounds, as demonstrated by mallard.
2 Reserves are formed entirely before migrating to the breeding area, as in lesser snow geese (*Chen caerulescens*).
3 Reserves are built up entirely on the breeding grounds with no further supplementation, as illustrated by the common eider (*Somateria mollissima*).
4 Body reserves are both formed in the breeding area and supplemented by local food, as seen in the wood duck (*Aix sponsa*) (Owen and Reinecke 1979; Thomas 1988).
Both ducks and game birds can alter the lengths of their digestive systems in response to changes in food supply. Under conditions of more fibrous diets during winter, gut lengths increased in three species of ptarmigan (Moss 1974), gadwall (*Anas strepera*), and mallard (Paulus 1982; Whyte and Bolen 1985).

In passerine birds, energy is stored in various subcutaneous and mesenteric fat deposits, and protein is stored in flight muscles. The latter varies with total body fat, as in the yellow-vented bulbul (*Pycnonotus goiavier*) in Singapore (Ward 1969) and the house sparrow (*Passer domesticus*) in England (Jones 1980). In the gray-backed camaroptera (*Camaroptera brevicaudata*), a tropical African warbler, both total body fat and flight-muscle protein vary in relation to laying date (Fig. 2.10) (Fogden and Fogden 1979).

As in body-weight measures, flight-muscle weights are corrected for body size by dividing by a standard muscle volume (SMV). Davidson and Evans (1988) used the formula:

$$SMV = H(L \times W + 0.433 \, C^2)$$

for shorebirds of the genus *Calidris*, where H is the height of the keel of the sternum, L is the length of the keel of the sternum, W is the width of the raft of the sternum (one side only), and C is the distance from the keel to the end of the coracoid.

Direct measures of body weight are feasible with birds and small mammals but impractical for large mammals, for which some other index of body condition and food reserves must be used. These have been reviewed by Hanks (1981) and Torbit *et al.* (1988). Large mammals store fat subcutaneously, in the gut mesentery, around

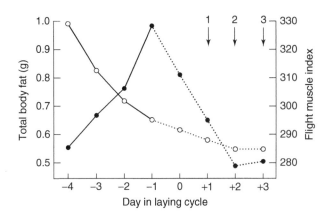

Fig. 2.10 Total body fat
(●) and flight muscle
index (○) are related to
laying day in the female
gray-backed camaroptera
(a tropical African
warbler). Eggs are laid
on days 1–3. Flight
muscle index is the ratio
of (lean dry flight muscle
weight)/(flight muscle
cord^3). The broken line
indicates estimates.
(After Fogden and
Fogden 1979.)

the kidneys and heart, and in the marrow of long bones. The fat stores are used up in that order (Mech and DelGiudice 1985). Because of this sequential use, no single fat deposit is a perfect indicator of total body fat. In caribou, for example, a combination of body mass and a visual index of condition provided the best predictor of fatness (Gerhart *et al.* 1996). Particular fat stores are of interest for specific purposes, such as reproduction (kidney fat) or starvation (bone marrow fat). For these purposes, they provide a reasonable guide for managers, and total body fat is less useful.

2.9.2 *Kidney fat index*

Ungulates accumulate fat around the kidney and in other places in the body cavity in anticipation of the demands of reproduction. We saw in Section 2.5.3 how the fat reserves of mule deer change according to the stage of the reproductive cycle (Fig. 2.8), the timing of these changes differing between the sexes.

Although there is little relationship between kidney fat and total body fat in some species (Robbins 1983), others, such as most African ruminants, show a close relationship (Smith 1970; Hanks 1981). For white-tailed deer, the percentage of fat in the body is related to the kidney fat index (KFI) by (Finger *et al.* 1981):

$$\% \text{ fat} = 6.24 + 0.30 \, \text{KFI}$$

In mule deer, both the weight of kidney fat and the KFI are correlated with total body fat (Anderson *et al.* 1969, 1972, 1990; Torbit *et al.* 1988).

Although a similar relationship was found for the brush-tailed possum (*Trichosurus vulpecula*) in the South Island of New Zealand (Bamford 1970), a better correlation was found between total body fat and mesenteric fat.

KFI has been measured in two ways:

1 The kidney is pulled away from the body wall by hand and the surrounding connective tissue and embedded fat tears away along a natural line posterior to the kidney. A cut along the mid-line of the kidney allows the connective tissue to be peeled away cleanly. The KFI is the ratio of connective tissue plus fat weight to kidney weight summed for both kidneys.

2 The connective tissue is cut immediately anterior and posterior to the kidney, so that only the fat immediately surrounding the kidney is used. This has the small advantage of being more objective than 1, but the much greater disadvantage of discarding most of the tissue where fat is deposited, so that much of the relevant variability in fat deposition is lost. Hence, the first method may be the more useful index.

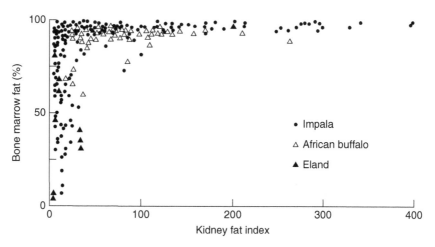

Fig. 2.11 Kidney fat index (the ratio of kidney-plus-fat weight to kidney weight) is depleted almost entirely before bone marrow fat declines in ruminants. (After Hanks 1981.)

2.9.3 *Bone marrow fat*

Bone marrow fat does not decline until after most of the kidney fat has been used (Fig. 2.11) in temperate and tropical ungulates and in some marsupials (Ransom 1965; Bamford 1970; Hanks 1981). In mule deer, marrow fat changed most rapidly at very low levels of total body fat (Torbit *et al.* 1988). Consequently, a decline in bone marrow fat reflects a relatively severe depletion of energy reserves and therefore provides an index of severe nutritional stress, as was found for starving proghorn antelope (Depperschmidt *et al.* 1987).

Mobilized marrow fat is replaced by water. Hence, the ratio of dry weight to wet weight of marrow is a good measure of fat content. A number of studies on both temperate and tropical ruminants (Hanks 1981) indicate as a close approximation that:

% marrow fat = % dry weight − 7.

Dry weight of marrow is measured from the middle length of the marrow in one of the long bones, avoiding the hemopoitic ends. The method has been used on wildebeest (Fig. 2.12) and deer (Klein and Olson 1960) to establish whether they had died from lack of food.

Broad categories of marrow fat content in ruminants are provided by the color and texture of the marrow (Cheatum 1949). This method is quick (it avoids collection

Fig. 2.12 Marrow fat of wildebeest (broken line) dying from natural causes in Serengeti is related to season and the percentage of crude protein in their diet (solid line, with 95% confidence limits). (After Sinclair 1977.)

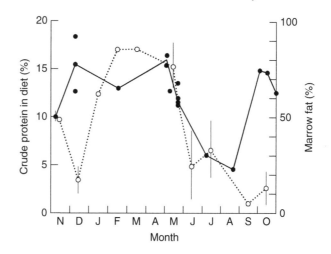

of marrow) and is sufficient to determine whether an animal has been suffering from undernutrition at death (Verme and Holland 1973; Kirkpatrick 1980; Sinclair and Arcese 1995). The categories with approximate fat values are:

1 *Solid, white, and waxy*   The marrow can stand on its own and contains 85–98% fat. Such animals are not suffering from undernutrition.
2 *White or pink, opaque, gelatinous*   The marrow cannot stand on its own and covers a broad range of fat values (15–85%). Such animals have depleted fat reserves.
3 *Yellow, translucent, gelatinous*   The clear, gelatinous appearance is distinctive, and indicates there is less than 15% fat, and often only 1%. Such animals are starving.

2.9.4 *Bioelectrical impedance*

Bioelectrical impedance analysis uses an electrical current passed through anaesthetized animals in a bioelectrical impedance plethysmograph. Resistance (R) and reactance (X_c) of the current are recorded and are related to impedance (Z) by:

$$Z = \sqrt{(R^2 + X_c^2)}$$

In wombats (*Lasiorhinus latifrons*), resistance is a good predictor of total body fat (Woolnough *et al.* 1997). The technique has also been applied to seals (Gales *et al.* 1994) and bears (Farley and Robbins 1994).

2.9.5 *Blood and urine indices of condition*

Blood parameters as indices of condition and food intake are potentially useful for living animals that are too large to be easily weighed. However, blood characteristics are not well known for most species. More work is needed. Different parameters have been examined in different studies. Plasma non-esterified fatty acid, protein-bound iodine, and serum total protein were all related to nutrition in Australian tropical cattle (O'Kelly 1973). Body condition of moose has been related to various sets of blood parameters, ranked according to their sensitivity (Franzmann and LeResche 1978). Starting with the single best parameter, sensitivity increased with the addition of other measures: (i) packed cell volume (PCV), (ii) PCV plus hemoglobin content (Hb), (iii) PCV, Hb, Ca, P, and total blood protein (TP), and (iv) PCV, Hb, Ca, P, TP, glucose, albumin, and β-globulin.

Protein loss from the body was strongly correlated with body weight loss in white-tailed deer on experimentally restricted diets. Serum urea nitrogen and the ratio of urinary urea nitrogen to creatinine were the best blood and urine indicators of undernutrition and protein loss (DelGiudice and Seal 1988; DelGiudice *et al.* 1990). For example, serum urea nitrogen is a good indicator of recent protein intake in white-tailed deer (Brown *et al.* 1995). Similarly, the ratio of urinary urea nitrogen to creatinine provided a reasonable predictor of physiological responses to nutrition in this species (DelGiudice *et al.* 1996)

2.9.6 *Problems with condition indices*

We have already noted that it is generally impractical to obtain measures of total body fat in large mammals. Various single indices such as kidney fat and bone marrow fat have been used, but these are useful for specific purposes and cannot be used over the whole range of total body fat values. Kidney fat is more appropriate for estimating the upper range of body fat values, while bone marrow fat represents the lower values. A combination of six indices of body fat deposits in carcasses has been proposed (Kistner *et al.* 1980). This method is useful for complete carcasses but cannot be used for animals dying naturally because the soft parts are usually eaten by predators and scavengers, or else they decompose. Under these conditions, the only index that remains uniformly useful is bone marrow fat.

Bone marrow fat as an index is biased towards the low-body-fat values. It cannot reflect changes in the higher levels of body fat, so very fatty bone marrow does not necessarily mean an animal is in good condition (Mech and DelGiudice 1985).

Many studies use some form of visual index of condition. However, studies in which total body fat has been measured directly find poor correlations with body condition indices (Woolnough *et al.* 1997).

Although blood indices may be useful as a means of assessing condition and nutrition in living animals, they require careful calibration. Many of the blood characteristics are influenced by season, reproductive state, age, sex, and hormone levels. More importantly, they can be altered rapidly by the stress of capture and handling. All of these factors could act to obscure and confound changes in nutrition.

All estimates of body condition taken from a sample of the live population are poor indicators of the nutritional state of the population for two reasons. First, such samples are biased towards healthy animals, because those in very poor condition are either dead or dying and not available for sampling. Second, the age groups that are most sensitive to density-dependent restriction in food supply – the very young and very old – form a small proportion of the live population. Thus, even a strictly random sample of the population will include a majority of healthy animals and consequently the mean value of condition will be very insensitive to changes in food supply. Therefore, it is unlikely that one can assess whether a population is regulated by food supply or by predators based solely on body condition samples of the live population. To make this assessment, one should look at the conditions of the animals that have died.

2.10 **Summary**

Nutrition and feeding behavior underlie many critical issues in wildlife ecology and management, such as the adequacy of food supplies for endangered species and the potential yield in response to harvesting. For carnivores, the nutrient composition of prey is usually well balanced to their specific needs, whereas in herbivores the foods eaten may be deficient in key nutrients, such as protein or sodium. Many plant tissues defend themselves against herbivory using poisons, protective structures such as spines, or chemicals that bind to ingested proteins, making them unavailable for digestion. In herbivores, it can also prove difficult to assess food availability in a meaningful way, because the plant tissues eaten represent only a small fraction of the plant biomass present. Various animal-based measures, such as fecal nutrient composition, have been developed to assess food availability and body condition from the herbivore's point of view. Nutritional constraints often vary disproportionately with body size. Many aspects of the behavior and ecology of wildlife species are closely tied to seasonal and spatial variation in food availability, including social organization, spacing patterns, breeding synchrony, and mating system.

3 Home range and habitat use

3.1 Introduction

In this chapter we introduce the pivotal concept of the *home range* and consider how it relates to patterns of *habitat use*. The home range for an individual refers to the geographic area utilized for all normal activities, linked together through animal movement (Burt 1943). This is likely of course to depend on the time frame under consideration, so home ranges can be specified on a daily, seasonal, annual, or even lifetime basis. In order to meet an individual's full range of physiological and ecological requirements over an extended period, a suitable home range typically must have an adequate supply of food and water, shelter, breeding sites, and often locations that are secure against predators and parasites. The spatial locations associated with common sets of these attributes are often termed *habitats*, so we can refer to "nesting habitat," "shelter habitat," or "refuge habitat from predation."

Obviously, such a complex set of requirements is unlikely to be met routinely at every place in the environment, so home range locations often yield important insight into the requirements for life, particularly those resources (see Chapter 2) that have the greatest influence on survival and reproductive potential. As such, home range analysis provides wildlife biologists with one of the most important tools by which to assess ecosystem quality. Habitat attributes that reliably indicate home range use can offer insight into key ecological features of importance from the individual's point of view.

Comparison of home range size across different species or among populations of the same species can help us to understand better what ecological factors most strongly limit population abundance. These factors can be used to identify the *ecological niche*, central in any discussion about community structure (see Chapter 6) but also vitally important in defining suitable candidates for habitat restoration, improvement projects, and conservation. In species whose home ranges of individuals rarely overlap, the reciprocal of home range size essentially defines a social carrying capacity that naturally regulates population density (see Chapter 5). Even in species with extensive individual home range overlap, the locations of suitable habitats across the landscape can offer useful insights into potential population size, a characteristic that can be of enormous practical use with respect to species reintroduction or recovery programs (see Chapter 4). In other words, patterns of habitat use within the home range link to many different aspects of wildlife ecology and play a useful role in many conservation and management initiatives.

Wildlife Ecology, Conservation, and Management, Third Edition.
John M. Fryxell, Anthony R. E. Sinclair and Graeme Caughley.
© 2014 John Wiley & Sons, Ltd. Published 2014 by John Wiley & Sons, Ltd.
Companion Website: www.wiley.com/go/Fryxell/Wildlife

3.2 Estimating home range size and utilization frequency

There are a wide variety of ways in which home ranges can be estimated, but all start with a requirement for spatial field data. In rare circumstances, such data might be estimated from direct observation of individually recognizable individuals over continuous time, such as in many primate studies. More commonly, wildlife biologists rely on remote sensing apparatuses, such as radio collars, to provide the spatial coordinates of animal locations sampled at regular time intervals. Until recently, such data were laborious and enormously time-consuming to gather for field biologists, with substantial effort required to get close enough to radio-collared animals to allow reliable positioning of directional antennae. Multiple fixes of this sort were then used to estimate coordinates using triangulation methods derived from basic trigonometry. An inevitable consequence of the uncertainty in determining the direction of the telemetry signal relative to the receiver was that even the most careful field research yielded positional data ("fixes") of dubious reliability.

In more recent years, however, there has been an explosion of new technologies for the gathering of animal location data using *global positioning systems* (GPS) identical to those used in modern cell phones. GPS radio collars determine their own spatial positions at pre-assigned intervals through linkage with a set of satellites circling the globe. Such data are typically much more precise than hand-gathered radio-telemetry data, often yielding fixes within 10 m of the true location and sometimes within 1 m. This remarkable precision, coupled with major improvements in the number of fixes possible even for multiple organisms, has opened exciting new avenues of research and allowed for much more reliable information on which to base home range assessment.

As an example, consider the fix data shown in Fig. 3.1 for a gray wolf in northern Ontario, Canada (Kittle 2014). Sequential positions of the study animal at 24-hour

Fig. 3.1 Sequential radio-telemetry fixes from GPS radio-telemetry fixes for a gray wolf from Canada taken every 5 h over the course of a year.

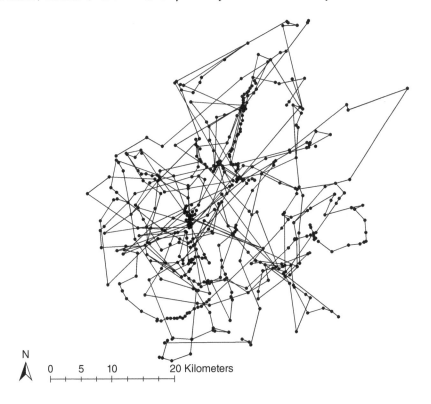

N

0   5   10   20 Kilometers

intervals are connected by lines. This example demonstrates several common features of positional data: first, the researcher knows only that at some point over the course of a single day the study animal transferred between sequential locations, not typically how long it remained or if it remained at all at any single site; second, while on occasion the individual moved a large distance between sequential fixes, short "moves" outnumber long "moves"; third, much of this individual's time was spent in a small fraction of the available area − that is, it seemed to prefer some locations to others. The typical goal of home range analysis is to characterize the overall pattern suggested by the collection of fixes, measure the areal coverage, and identify correlations between habitat elements in areas of heavy versus light use to assess selectivity.

Once one has gathered spatial fixes for a given individual, there are a variety of ways to estimate home range size, differing largely in the underlying assumptions about the individual's position during the intervals between fixes. The oldest method is simply to connect the outermost fixes, forming what is termed a *convex polygon* (Fig. 3.2a). This has several virtues, simplicity being paramount. It does not come heavily loaded with preconceptions about the normal patterns of home range use. Moreover, it is inclusive, in that all fixes are part of the home range. On the negative side, convex home range polygons are often thought to be too inclusive, with the border including some regions that might not be used at all. Most importantly, the convex polygon method offers little insight into identification of areas of intense use.

The most common alternative to convex polygons is what is known as a *kernel estimator* for the putative home range (Kie *et al.* 2010). This can take essentially two forms: a bivariate kernel tries to fit as many fixes as possible within an ellipse centered over the observed fixes, whereas an adaptive kernel is a complex surface draped over the observed fixes (Fig. 3.2b). Contoured isopleths drawn on the surface demarcate areas of similar frequency of visitation. One can immediately see the advantages of the kernel method. The complex pattern of folding allows the researcher to discriminate between

Fig. 3.2 Alternate home ranges estimated from daily GPS radio-telemetry fixes for a gray wolf in Canada based on (a) convex polygon, (b) adaptive kernel, (c) Brownian bridge kernel, and (d) local weighted polygon methods (Kittle 2014). For subplots (b) and (c), areas of intense use are lighter in tone, whereas areas of minor use are shaded darkly.

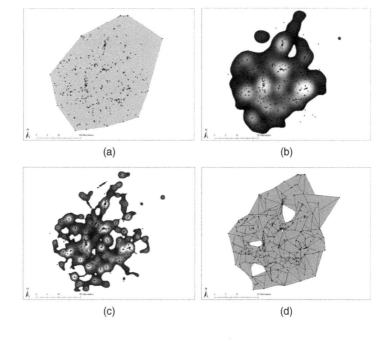

(a)

(b)

(c)

(d)

those areas that were used versus those not used at all. More importantly, locations of intense use are nicely captured by peaks in the fitted surface. The height of the kernel home range at any spatial location is proportionate to the utilization probability, so the kernel itself represents a *utilization distribution* in space. This is an improvement on simply assuming that all parts of the home range are equally important, because we know that animals almost always spend disproportionate amounts of time in different parts of the home range (Fig. 3.1). Improved understanding of the most heavily utilized parts of the home range can often lead to insights about the most important factors, in a sense giving us an animal's-eye-view of the things that matter ecologically.

One slightly disturbing feature of kernel home ranges is that they have no absolute limit, because, for mathematical reasons associated with fitting a complex surface to point data, the fitted surface extends without limit in all directions, albeit with low levels of use. To circumvent this logical impossibility, scientists routinely truncate the kernel home range at the point where the surface includes 95 or 99% of the observed fixes (a so-called isopleth), eliminating the infinity conundrum. The advanced software required to fit such complex home ranges to data are included with some commonly used GIS packages (e.g. Spatial Analyst, available as a toolbox in ARCmap) or in R statistical packages freely available on the Internet (e.g. adehabitat, available at http://cran.r-project.org). The quality of kernel-estimated home ranges obviously depends on the accuracy and the number of fixes, as well as on the degree to which fixes are clumped across the landscape.

A recent modification of the kernel approach is termed *Brownian bridge kernel* estimation (Horne *et al.* 2007). Like other kernel methods, it fits a complex surface to the collection of fixes, accommodating local variation in home range use. Unlike other kernel estimators, however, the Brownian bridge approach estimates the probability of space use by an animal with typical movement characteristics during the intervals between fixes, assuming so-called Brownian movement during each interval (Fig. 3.2c) (Brownian movement refers to a random pattern of movement, such as that of small pollen grains suspended in a water droplet). This helps to identify corridors of likely importance linking sites used repeatedly by the study animal. On the negative side, presumption of purely Brownian movement adds to an already ponderous set of assumptions.

A fourth approach that has recently come into its own is termed *local convex hulls* (Getz and Wilmers 2004). This is a methodological approach that conceptually lies somewhere between the simplicity of convex polygons and the extreme complexity of kernel home range estimators. Simple geometric shapes (convex hulls) are constructed around subsets of the fix data (Fig. 3.2d), thus permitting researchers to construct a sequence of home ranges for different seasons. This can be helpful in understanding more complex patterns of movement or for comparing patterns of home range use for multiple animals over short time intervals. R code for the calculation of local convex hulls is available over the Internet.

3.3 Estimating habitat availability and use

The pattern of animal fixes that generates a home range estimate and indicates the degree of utilization of parts of the home range can also be used to define habitat usage. It is rarely possible to know all the relevant attributes of every spatial position of an animal's home range. Nonetheless, it is often possible to make educated guesses about the most important features using *geographic information systems* (GIS). This is a means of linking complex geographical information on physical structure, topographic relief, biological features, and human-made landscape elements into a computerized database. One important feature of GIS is rapid and simple construction of tailor-made "maps" that are readily accessible from a computer screen. This allows users to sift rapidly through complex spatial information in a visual context. Just as importantly, GIS allows the user to identify and measure spatial interrelationships among variables that would be exceedingly difficult to perform in the field. For example, one can rapidly calculate the sizes of forest stands of similar species composition, measure the distance of each of these stands from the nearest road, and calculate what fraction of the stands falls within the home range of a wildlife species of interest. From the point of view of assessing habitat selection, GIS also offers a convenient means of random sampling of geographic features across complex landscapes. GIS is clearly a technological break-through in the analysis of wildlife habitat needs and is transforming the way that we think about conservation and management issues.

The remarkable capacity afforded by GIS data does not come without cost. First, someone has to gather spatial data and map them in the first place. The quality of that initial data collection (sometimes termed "ground-truthing") and how recently it has occurred have a strong bearing on the utility of the GIS database. For example, it is quite common to rely on GIS databases to assess the degree of wildlife prefer-ence for specific vegetation communities, such as forest stand types. Such habitats are often defined in rather crude terms, representing a few predominant species that can be identified from aerial photos or limited samples taken at a limited number of acces-sible sites, rather than field data sampled extensively across the landscape. The aerial coverage of each habitat type is then based on extrapolation from a limited number of ground-truthed sites to a much larger landscape. Different forest stands have different spectral reflectance characteristics, allowing remote-sensing specialists working with a satellite image to fill in a GIS database, much as a child would fill in a line drawing in a coloring book. Statistical extrapolation of such derived variables is always tenuous and the reliability of GIS databases is affected accordingly.

Sometimes GIS data cannot provide information on the actual resources used by animals. For example, the rich mix of forbs used by browsing deer may be poorly pre-dicted by forest stand type. One way around this is to use satellite-generated estimates of plant abundance, such as NDVI (Normalized Difference Vegetation Index), rather than plant community types. Alternatively, changes in NDVI from one satellite pass to the next can be used to ascertain areas of rapid vegetation growth, often indicative of the most nutritious forage (Pettorelli *et al.* 2005; Bischof *et al.* 2012). Such variables lump all vegetation classes together, so they inevitably yield a rather coarse view of the world. The information in GIS databases may also be out of date and therefore irrele-vant to current conditions. Human population growth, industrial development, habitat disturbance due to natural or anthropomorphic causes, and ecological processes such as community succession are rapidly changing habitat availability in many if not all landscapes.

Finally, many relevant variables are dynamic in nature, which is difficult to accom-modate in static GIS databases. Plant biomass grows over the summer or wet season and then declines over the winter or dry season in most ecosystems. Prey distribution can shift markedly from month to month and even in cases where prey are resident, local snow conditions influencing access to those prey may vary considerably over time and space across large landscapes. From the prey point of view, the spatial distribution of predation risk is itself highly dynamic, depending on the manner in which predators use their home range. All of these factors introduce uncertainty into wildlife habitat assessment. Statistical modeling can never eliminate the uncertainties introduced by inadequate habitat data.

3.4 Selective habitat use

By now it should be apparent that there are good reasons for wildlife species to choose habitats carefully, in order to enhance the opportunities for feeding while reducing the risk of being eaten. Moreover, most species have a suite of other needs to meet, including obtaining shelter from inclement weather, gaining access to water, and locating suitable breeding sites, such as cavities in dead trees or burrows. Quantification of specific habitat needs is known as *habitat assessment* and is an important area of wildlife ecology. Much of this interest derives from practical benefits: knowing precisely which wildlife habitats are essential allows appropriate management decisions regarding alternate forms of land use. Moreover, good understanding of habitat requirements can improve the odds of success when wildlife species are reintroduced to areas from which they were extirpated.

There are many ways to quantify wildlife habitat selectivity (Aarts *et al.* 2008). All such methods rest on a common assumption: that selectivity can be determined by comparing use to availability of alternative habitat types. This is often assessed by taking S = use/availability. If the proportion of occurrences in a particular habitat type exceeds the proportion of the landscape composed of that habitat then S > 1 and positive selectivity (preference) is implied. If use is less than proportionate availability, on the other hand, then S < 1 and negative selectivity (avoidance) is implied.

The simplest way to evaluate such logic is through contingency tables. The number of actual occurrences in a given habitat is compared with the number expected under the null hypothesis of nonpreference. If 30% of the environment is composed of grassland and 70% of woodland then the null hypothesis will be that of 500 radio-telemetry fixes of elk, we should expect to get 150 (= 0.3 × 500) fixes in grassland and 350 in woodland. One can test statistically whether the actual observed values significantly deviate from this expectation using a χ-squared test. A similar analysis can be used to determine whether animals preferentially choose diets or other resources.

A commonly used variant on this simple contingency test is known as *composition analysis* (Aebischer *et al.* 1993). Using this approach, one converts the proportions of use for each habitat i to a log-ratio using the formula $y_i = \ln(u_i/u_1)$, where u_1 is the proportion of use of one arbitrarily chosen habitat type. Log-ratios of availability are similarly calculated as $z_i = \ln(a_i/a_1)$. Preference is then indexed by the difference between the log-ratios ($p_i = y_i - z_i$) for all habitat types. If these log-ratios are available for a number of individuals then it is possible to rank the habitats in terms of mean preference and to use a MANOVA to test whether patterns of preference differ from what one might have expected at random. The strength of composition analysis is that the individual becomes the logical sampling unit, rather than fixes, which are rarely truly independent of one another and thus bias statistical tests. The primary weakness, however, is that selectivity can only be assessed for those habitats that are actually used to some degree, hence complete avoidance presents a serious problem.

While simple tests based on use versus availability can help us identify important habitat needs, they are not very useful for predicting patterns of occurrence across the landscape. For such purposes, wildlife biologists use a different approach, termed a *resource selection function* (Manly *et al.* 1993; Boyce and McDonald 1999). Resource selection functions offer a flexible means of quantifying the degree of habitat preference. Complex combinations of categorical and continuous variables can be readily accommodated using this method. Moreover, the method can use a GIS to locate, manipulate, and analyze habitat data of interest.

Perhaps the easiest way to understand the most typical resource selection procedure is to walk through an example. The rufous bristlebird is a threatened passerine species living in coastal areas of Australia. Gibson *et al.* (2004) used GIS to evaluate critical habitat needs for bristlebirds in a site with competing land use interests (biodiversity values versus mining). Along a series of trails bisecting the study area, Gibson *et al.* recorded the presence (scored with a 1) or absence (0) of bristlebirds, noting the exact geographic coordinates of each positive identification made. They later transferred these sightings to a GIS, overlaying digitized topographic data on aspect, slope, and elevation, as well as spatially explicit data on hydrology and vegetation complexity derived from multispectral remote-sensing imagery. The probability that a habitat is used ($w(x)$) is given by the following logistic regression model:

$$w(x) = \frac{\exp(\beta_0 + \beta_1 X_1 + \ldots \beta_k X_k)}{1 + \exp(\beta_0 + \beta_1 X_1 + \ldots \beta_k X_k)}$$

where the logistic regression coefficients $\beta_1 - \beta_k$ measure the strength of selection for the k different habitat variables (symbolized by X_i) over the full set of sample units. The S-shaped logistic function $w(x)$ is bounded between 0 and 1 and represents a probability of usage, given the set of habitat characteristics within a spatial unit. Given the descriptive nature of both the data on bristlebird presence or absence and habitat variables derived from the GIS, Gibson *et al.* elected to use model evaluation (Chapter 15; Burnham and Anderson 1998) rather than classical hypothesis testing (Chapter 14). They found that there was a positive association between bristlebird presence and vegetation vertical complexity, but negative associations between bristlebird presence and "elevation," "distance to creek," "distance to the coast," and "sun incidence." This suggests that bristlebirds require densely vegetated stands in close proximity to coastal fringes and drainage lines. Such habitats composed approximately 16% of the study area, demonstrating how resource selection can help in the assessment of land use priorities for wildlife conservation in a planning context.

There are many variations on this basic statistical design, discussed in detail in the comprehensive treatise by Manly *et al.* (1993). If one can visit all spatial units and know for sure whether each has been used or not, the logistic function $w(x)$ can be treated as a true probability of occurrence. An example of this might be the study of squirrel nests in a small wooded area at a time of year in which nests are readily observable from the forest floor. While it might require a good deal of work, one could probably say with reasonable certainty that spatial units without a nest were not used as nesting habitats.

In most cases in the contemporary literature, however, resource selection is based on sampled radio-telemetry data, used to identify used habitats and sample available habitat randomly from a larger set of sites. While this kind of study design unambiguously identifies sites that were actually used by the marked population, one cannot be sure that sites without fixes were not used at least some times by at least some individuals (Keating and Cherry 2004). This means that recorded "0"s would actually be "1"s if the biologist had completely accurate data. Under these circumstances, the proper procedure is to use logistic regression to estimate the coefficients of $w(x)$, discard the denominator and intercept β_0, and use the resulting modified formula $z(x) = \beta_1 X_1 + \ldots + \beta_k X_k$ to estimate the relative magnitude of use ($z(x)$) for each habitat. In other words, the model cannot be treated as a true probability function, but the resource selection function is still appropriately scaled to preferences (Keating and Cherry 2004) and therefore useful in assessing conservation or management objectives.

3.5 Using resource selection functions to predict population response

Resource selection can in principle be used to evaluate the potential success of reintroduction programs (Boyce and McDonald 1999). This approach has been used, for example, to predict the potential for successful reintroduction of gray wolves to different parts of the United States (Mladenoff *et al.* 1995; Mladenoff and Sickley 1998). Data for existing wolf populations were first used to determine the suite of critical habitat variables for wolves and to relate local wolf densities to habitat features. GIS data were then fed into the resource selection models to predict the potential of different areas to support gray wolves. The model has been validated against data for an expanding wolf population in Wisconsin, demonstrating that this approach can be a useful planning tool.

Resource selection functions are also a powerful means of linking habitat characteristics with spatially realistic models of population viability. For example, Akçakaya and Atwood (1997) used logistic regression to develop a habitat suitability model for the threatened California gnatcatcher (*Polioptila californica*) in the highly urbanized environment of Orange County, California. Gnatcatcher distribution data were mapped on to a GIS map. Numerous geographical habitat features were then evaluated and a resource selection probability function was developed on the basis of the strongest suite of variables. Suitable habitat fragments were mapped on to the Orange County landscape and this spatial configuration was modeled as a metapopulation in order to evaluate the long-term viability of gnatcatchers (see Chapters 16 and 22). This is a valuable way of evaluating the conservation needs of threatened populations. It is particularly appropriate for species utilizing fragmented landscapes, because it gives useful insights into the ecological implications of alternative land use policies and planning scenarios.

3.6 Sources of variation in habitat use

A number of factors can have a strong influence on habitat selectivity and use. First and foremost, estimates of habitat selectivity are strongly influenced by the spatial scale at which the assessment is made. A useful set of habitat use categories is in common use by wildlife biologists (Johnson 1980). Type 1 habitat selection is based on the entire range occupied by a given species relative to the unoccupied area. Type 2 habitat selection is based on the full home range chosen by an individual relative to the area encompassed by the species range. Type 3 habitat selection is based on the use of specific habitats within the home range of a single individual. Finally, Type 4 habitat selection is based on the selective use of particular food types by a single individual within a single resource "patch."

Although not commonly done, analyses conducted at multiple spatial scales often find quite different patterns of selection at each scale (DeCesare *et al.* 2012). It is not entirely clear what this means, but it has been suggested that the most limiting demographic factor will elicit selection at the most coarse scale (Rettie and Messier 2000). So, giant pandas might be expected to coarsely select for bamboo forest, because that is a crucial, nonreplaceable food resource, but to select for ancillary features such as slope, aspect, or stem density at finer spatial scales. At the very least, repetition of resource selection studies at multiple scales is more likely to identify robust habitat predictors that happen to have their greatest effect at different spatial scales.

Even if one maintains a similar spatial scale, patterns of selectivity will often be intrinsically sensitive to variation in habitat availability in the local environment, potentially clouding the ability of use-availability sampling designs to identify preference (Beyer *et al.* 2010). This is an inevitable outcome of the mathematics used to estimate preference coefficients combined with spatial heterogeneity in habitat distribution that occurs routinely in nature. It is also a likely consequence of animals valuing choices differently when resources are common as opposed to rare (Mysterud and Ims 1998). One remedy is to repeat resource selection studies at a variety of spatial scales or to repeat studies at sites with different levels of habitat availability, if only to appreciate the robustness of statistical inferences. It is also advisable to use some form of cross-validation, a statistical method of resampling one's dataset to evaluate the robustness of conclusions. Caution in accepting resource selection functions at face value is often well warranted.

Given that there are differences in the intrinsic suitability of habitats, due to variation in resources, cover, and risk from predators, one might expect animals to concentrate in the most favorable habitats. It has been long appreciated, however, that habitat selectivity tends to vary with densities of both consumers and their resources (Rosenzweig 1991). Birth rates tend to fall and mortality rates to climb as forager density increases (see Chapter 5). As a consequence, habitat suitability is often negatively associated with density. Density-dependent decline in habitat suitability can arise from a variety of causes, including resource depletion, direct interference among individuals, disease transmission, and elevated risk of predation.

At low consumer densities, individuals tend to be highly selective of habitats, but the range of used habitats expands as consumer density increases (Rosenzweig 1991). This is often interpreted as evidence of adaptive changes in foraging decisions that should have positive fitness consequences (Rosenzweig 1981, 1991; Brown 1988; Brown *et al.* 1994). Animals that do not expand their range of acceptable options when preferred items (habitats) are scarce face an opportunity cost that can be deadly (Brown 1988).

Density-dependent decline in habitat suitability can be extended to multiple habitats. Individuals should concentrate in the best habitat until the density there reduces in suitability to that of the next best alternative (Fig. 3.3). Thereafter, both habitats should receive equal use. The resulting pattern of distribution among alternative habitats is known as the *ideal free distribution* (Fretwell and Lucas 1970). It is free in the sense that every individual is presumed equal and capable of choosing the best option available. It is ideal in the sense that all individuals are presumed to have perfect knowledge about the relative suitability of each habitat on offer. Hence, it would not pay for any individual to deviate from the ideal pattern of distribution, because their fitness would be compromised. This is a prime example of an *evolutionarily stable strategy* (ESS; Maynard Smith 1982). Once adopted by all individuals in a population, no mutant or deviant strategy could do better. Hence, the ESS will be favored by natural selection.

The ideal free hypothesis predicts that most individuals should be found in preferred habitats when forager population density is low, spilling over into less preferred habitats when forager density is high. This pattern has been demonstrated several times in

Fig. 3.3 Schematic diagram of the ideal free distribution. As density in the preferred habitat 1 increases, suitability declines to a point indicated by the light broken line where it equals that in the poorer habitat 2 (60 units). At this point it pays some individuals to use habitat 2.

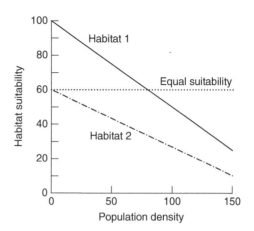

Fig. 3.4 Use of preferred habitats by three different bird species declines as population size increases. (After Sutherland 1996.)

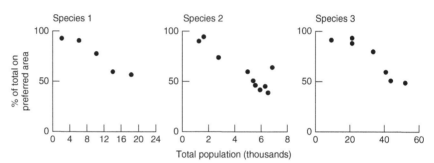

different bird species (Fig. 3.4). One of the earliest examples was Brown's (1969) pioneering studies of great tits (*Parus major*) in the woodlands near Oxford, UK. Brown showed that adult birds nested predominantly in woodland habitats in years with low bird abundance, expanding outwards into less attractive hedgerows only when densities were high. Krebs (1971) tested the assumption that this distribution stemmed from differences in fitness by experimentally removing birds from woodland habitats, resulting in vacancies that were filled rapidly by former hedgerow "tenants."

A powerful way to test the ideal free hypothesis is to compare the feeding rates of individuals in different patches with different rates of food delivery. Milinski (1979) delivered food at differing rates to the two ends of an aquarium and measured the consequent pattern of distribution of sticklebacks. The ideal free hypothesis predicts that once they have determined the rate of food delivery at each end of the tank, the density of fish at each end should be proportional to the same. In other words, delivering twice as much food to one end of the tank should lead to two-thirds of the fish congregating in the food-rich patch. The sticklebacks redistributed themselves in precisely this manner (Milinski 1979). Similar results have been recorded in continuous-food-input experiments with numerous other species, including mallard ducks (Harper 1982), cichlid fish (Godin and Keenleyside 1984), and starlings (Inman 1990). Measurements in the field have been less supportive. However, animals in preferred habitats generally obtain higher rates of food intake than those relegated to poorer habitats (Sutherland 1996). Researchers frequently find that individuals vary in the quantity of food that they acquire, with more dominant or larger individuals securing more of the food delivered than lower-ranking individuals. This hierarchy suggests that although animals are capable of adjusting their behavior in predictable ways to accommodate the presence of other competitors for scarce resources, differences in dominance status tend to maintain differences in fitness (Sutherland 1996).

One way to accommodate these effects is through a modified model known as the *ideal despotic distribution* (Fretwell 1972). This model assumes that individuals choose the best habitat possible on the basis of their dominance status. In extreme cases, this process can lead to *territoriality*, in which an individual, pair, or social group defends an area of exclusive use against all intruders. The most dominant individuals choose first, followed by others in rank order of their dominance status. Under these conditions, individuals of lower status might well choose to split their time between two habitats offering similar levels of suitability, whereas high-ranking individuals invariably choose the best habitat. More importantly, the ideal despotic distribution predicts that there will be disparities in food intake, mortality risk, and reproductive success among individuals. These differences dissolve when we focus on individuals of similar rank.

3.7 Movement within the home range

One nagging flaw in simple studies of habitat selectivity is that not all parts of an animal's home range are physically accessible from every other part of the home range at every moment. For example, imagine one has a dataset composed of daily telemetry fixes for a caribou occupying a circular home range with a radius of 24 km. If an animal never travels more than an 8 km straight-line distance in a day, only a small fraction of the home range can be exploited from any particular position in any given day. As a result, not all habitat types will be equally accessible, even if we lump data over the entire annual cycle.

Wildlife biologists have recently taken enormous strides in overcoming this intrinsic bias. The key is in thinking about the movement process itself. One simple way to characterize movement is simply to measure the distance between subsequent fixes over constant time intervals (termed steps) and the degree to which an animal deviates in direction from one step to the next (termed the turn angle). Such a distribution of steps and turn angles for elk in southern Ontario, Canada is shown in Fig. 3.5. Like most organisms, elk tend to make many short "steps" punctuated by a small number of long ones. While searching for food patches, elk tend to move in fairly straight trajectories, whereas they tend to turn tightly and take shorter steps once they have located a patch with abundant food (Hazell 2006; Fryxell *et al.* 2008).

By randomly sampling from such distributions of step length and turn angle, one can hazard a reasonable guess about the locations that an animal could have plausibly visited as it moved from one point in time to the next. This procedure is very similar to that described for resource selection functions, except that the set of available habitat types is defined by the points that are plausible in light of the observed movement process, hence the method is called a *step selection function*. By taking a random sample of 200 points realistically arrayed around every elk fix, Fortin *et al.* (2005) were able to estimate which local habitat variables were most highly preferred by elk in Yellowstone National Park. Using this procedure, they demonstrated that elk tend not to move away from roads and preferentially move towards stands of aspen trees or open grasslands when wolves are rare in the vicinity, but move preferentially towards coniferous forest when wolves are locally common. This suggests that elk are able to make a subtle assessment of the relative value of food versus predation risk. When risk is high, elk tend to choose sites with heavy forest cover, presumably for concealment, whereas when risk is low they choose sites that offer better opportunities to feed on grasses or young aspen saplings, which are their preferred food resources.

This general approach using distributions of movement steps and turn angles has come into wide use in a variety of other contexts, allowing behavioral ecologists to predict better patterns of dispersal from newly-established versus long-standing populations (Ovaskainen *et al.* 2008), alternation in movement behavior between exploratory and encamped "phases" (Morales *et al.* 2004), and identification of seasonal and daily cycles of movement behavior for animals of different social rank (Wittemyer *et al.* 2008). A useful new framework for developing such integrated models of animal and plant movement is nicely outlined in Nathan *et al.* (2008).

Mechanistic movement models can also be used to develop a deeper understanding of the critical factors shaping animal home ranges. For example, behavioral studies

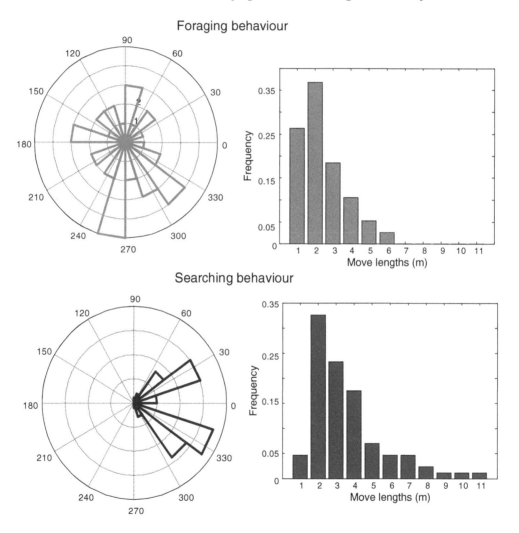

Fig. 3.5 Distribution of step lengths and turn angles for Ontario elk while foraging (top) and while searching for new foraging sites (bottom) and Yellowstone elk (next page). (Source: Hazell, 2006. Reproduced with permission of Megan Hazell, AMEC E & I, Toronto, Ontario, Canada.)

Fig. 3.5 *(Continued)*

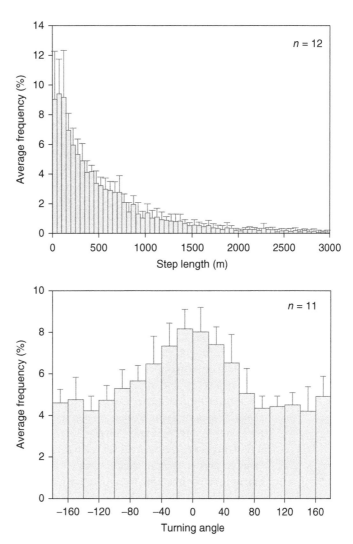

of coyote home range behavior suggest that individuals use scent-marking both to establish their own presence and as a means of assessing the identities of intruders on to their home range. The typical response to the presence of a foreign scent-mark is to mark on top of the intruder's scent, but then move away towards the center of the home range. By the same token, the size of movement "steps" tends to be small when prey abundance is high but large when prey abundance is low. As a result, coyotes tend to linger for long periods of time in parts of the home range with lots of prey (rodents and hares) but quickly move on to better opportunities when prey are scarce, a process termed *area-restricted search* (Smith 1974; Benhamou 1992).

Combining the process of habitat selection with that of movement, Moorcroft and coworkers (Moorcroft and Lewis 2006; Moorcroft *et al.* 2006) developed a predictive model of the home range that would eventually be formed by a population of coyotes that obeyed these mechanistic movement rules. Their model did a good job of explaining observed patterns of coyote home ranges in Yellowstone National Park.

Fig. 3.6 Home range use by Yellowstone coyotes in relation to prey (shading) and neighboring individuals. (Source: Moorcroft *et al.*, 2006. Reproduced with permission of The Royal Society.)

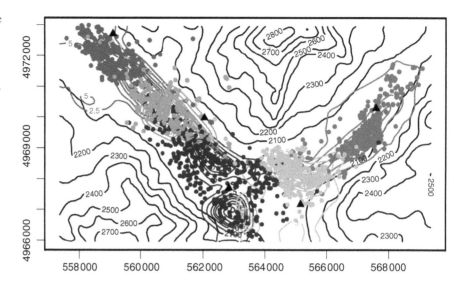

The coyote home ranges are clustered in the valley bottoms, because that is where prey are most abundant in this montane landscape (Fig. 3.6). No single coyote can monopolize the valley, however, because of the movement response to neighboring individuals. As a result, coyote home ranges become evenly strung out, like beads on a necklace, due to the coyotes' aversion to home range overlap but common interest in exploiting spatially clumped prey.

While mechanistic in nature, Moorcroft and Lewis's (2006) movement model does not necessarily maximize fitness of the individual. Other researchers have however taken this tack in trying to develop predictive home range models from first principles. For example, Mitchell and Powell (2004) developed an explicit model of trade-offs in time versus energy gained for a single individual black bear gathering a variety of resources across a complex landscape. Their model predicts that there are diminishing returns for an animal that tries to utilize too large a home range, as a result of both time wasted in travel from the home range center and resource depletion by neighboring animals. Hence, low home range overlap could be a natural consequence of animals striving to maximize their net energy gain, rather than avoidance of competitors.

In a follow-up to the underlying logic of these spatial movement models, Moorcroft and Barnett (2008) showed that the equilibrium spatial distribution of any animal following mechanistic movement rules should be scaled to the resource selection function. This suggests that animal home ranges, movement patterns, and habitat preferences are all intertwined as a bundle – opening exciting new opportunities for synthetic work in the future.

3.8 Movement among home ranges

The traditional view of home ranges is predicated on the notion that animals have well-defined homes in the first place. This is true, of course, for many species. On the other hand, some wildlife species have no fixed address at all, but rather wander over a large area each year. With this kind of movement variation, it is often convenient to consider whether patterns of movement are repetitive (i.e. is there a tendency to return to places visited before or not?) and whether movements of multiple individuals are coordinated. A cyclical pattern of movement by individuals among two or more non-overlapping home ranges at different seasons of the year is termed *migration*, whereas *nomadism* refers to unpredictable patterns of movement among multiple non-overlapping home ranges (Mueller and Fagan 2008). Animals whose movements fall within a single boundary all year long are termed *residents*. Within each of these categories there are examples of species that move as a group and share the same home ranges at different points in time, as well as of those whose movements are independent of each other (Fig. 3.7).

A well-known example of a migratory species that moves en masse from one location to another in a periodic manner (Fig. 3.7b,c) is the Serengeti wildebeest, whose annual migration takes it from the arid Serengeti Plains to the moist woodlands of the Tanzania–Kenya border and back again. Offspring production takes place in a 3–4 week period while the wildebeest are on the plains; the precocious young then travel with their mothers, even during lactation. Recent models suggest that movements of Serengeti wildebeest track the wave of new grass growth triggered by monsoonal rainfall patterns, which sweep across the Serengeti landscape in an annual cycle

Fig. 3.7 Conceptual examples of (a) resident behavior, (b,c) migration, and (d,e) nomadism. (Source: Mueller and Fagan, 2008. Reproduced with permission of John Wiley & Sons Ltd.)

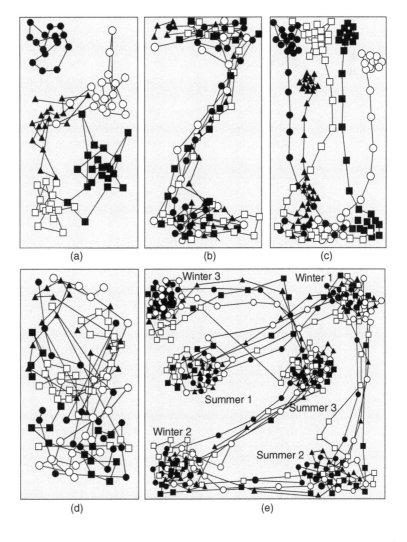

(Holdo *et al.* 2009). By relocating periodically to areas with new grass growth, wildebeest obtain more energy over the annual cycle than would be possible for an animal remaining in any single part of the migratory pathway. A secondary benefit is that wildebeest also make themselves less available to predators, such as lions and hyenas, which are far less mobile and are unable to track herds across the entire annual pathway (Fryxell *et al.* 1988).

There are lots of other species that migrate en masse, including red crabs on Christmas Island, hawks and vultures migrating from North to South America each winter, and jellyfish in the Pacific Ocean. Other migrations are conducted by individual animals moving independently of each other, such as Monarch butterflies; although the majority of the population has an overwintering site high in the forested mountains outside Mexico City, multiple generations of offspring fan across North America during the spring and summer, before a single generation of butterflies returns home to Mexico in the autumn (Flockhart *et al.* 2013). This pattern of movement allows different generations of Monarch butterfly to find appropriately aged stands of their single food source: milkweed plants. Return to Mexico allows individuals to overwinter as part of a truly spectacular aggregation.

A similar pattern of migration is often seen in many ungulates in montane environments. For example, elk, red deer, mule deer, and pronghorn antelope move from low elevations occupied during the winter to high-elevation pastures in the summer (Bischof *et al.* 2012; Monteith 2011; Middleton *et al.* 2013). By taking advantage of variation in the timing of snow melt and subsequent green-up at different elevations, it is possible for migrants to improve access to the most nutritious forage available across the montane landscape (Hebblewhite *et al.* 2008; Bischof *et al.* 2013). Although different individuals fan out to independent summer home ranges, migrants must often pass through narrow valley corridors as they move upwards in elevation. Such narrow corridors can become critical habitat if competing human concerns, such as oil rigs or other industrial developments, block the passageway. Such blockages are increasingly common on a crowded planet, where migrants must share limited real estate with humans. This is a pressing concern in montane environments, where suitable terrain for human and wildlife use is in particularly short supply (Berger *et al.* 2006; Sawyer *et al.* 2009). There is also growing evidence that while elevational migration improves access to food resources, it can also expose migrants to heightened risk of mortality en route, perhaps due to increased exposure to multiple predator territories or to a lack of familiarity with refugia along the way (Nicholson *et al.* 1999; Hebblewhite *et al.* 2008, 2012; Middleton *et al.* 2013).

Many of the most impressive movements are conducted by nomadic species, whose movements take them unpredictably across a vast landscape (Fig. 3.7d,e). A spectacular example is the Mongolian gazelle, which roams unpredictably across the steppes of eastern Mongolia. This grazing species has a well-defined preference for pastures that are at an intermediate level of abundance, rather than the tallest grass swards (Mueller *et al.* 2007), probably because of trade-offs between cropping rate and nutritional quality. In semi-arid environments, the location of suitable pastures is uncertain, because rainfall is highly unpredictable in both space and time. Monitoring of radio-marked animals (Mueller *et al.* 2011), supported by herd distributions conducted from lengthy ground transects (Mueller *et al.* 2008), suggests that gazelles continually shift nomadically across this unpredictable landscape, lingering at sites with suitable pasture much more reliably than at poorer sites (Fig. 3.8), so that a dynamic landscape feature

Fig. 3.8 Observed (triangles) vs predicted (shading) distribution of Mongolian gazelles. (Source: Mueller *et al.*, 2007. Reproduced with permission of John Wiley & Sons Ltd.)

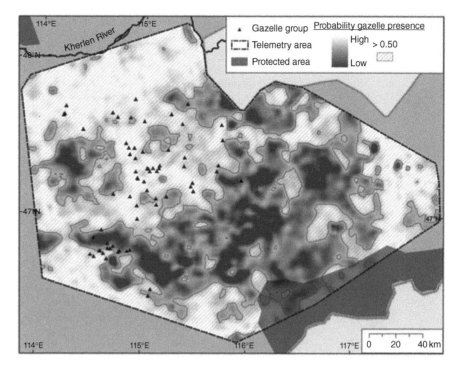

(grass biomass) represents a continuous habitat variable that is well correlated with home range use, at least at short timescales. Human pastoralists in the same region adopt a similar land use strategy, moving nomadically with their herds of yaks, sheep, and goats in search of short-lived patches of suitable pasture.

Fitting home range models to migratory or nomadic species obviously presents challenges. For a seasonal migrant, one might demarcate seasonal ranges, restricting habitat assessment to particular periods of the year. For truly nomadic species, home range analysis has to be conducted at multiple times per year in order to provide a useful picture of the area meaningfully occupied and for habitat assessment at any point in time. It is still possible to assess patterns of space use and habitat utilization, however, provided that the spatial frame of reference is well specified and fits the animal's lifestyle (Mueller *et al.* 2007).

3.9 **Summary**

Animals tend to move within a restricted home range that defines those areas used in their effort to acquire resources, seek suitable mating sites, and avoid predators. There are various ways in which home ranges can be identified from sequential location data obtained from marked animals, including simple convex polygons, adaptive or fixed kernel distributions, and local convex hulls. Comparisons across animal taxa suggest that home range size has an allometric relationship with body size, with different relationships in herbivores and carnivores, as well as in animals living in aquatic versus terrestrial environments.

The nonrandom positioning of home ranges in the larger landscape and the preferential use of parts of the home range suggest preferential use of specific habitats. We discuss various ways that habitat data are typically gathered for use in preference studies, as well as the telemetry data that are now routinely applied. Habitat selection can be measured in a variety of ways, including comparisons of use versus availability and more complex nonlinear resource selection functions. The methodology, strengths, and limitations of each of these statistical approaches are briefly examined.

Habitat selection is shaped by a variety of behavioral and ecological variables, including availability, consumer density, social processes, movement mechanics, optimal foraging, and anti-predatory behavior. Fitness-based movement rules, heterogeneous resource distribution (as well as risk), and social interactions can be linked to predict patterns of habitat use and home range shape.

4 Dispersal, dispersion, and distribution

4.1 Introduction

This chapter explores some of the reasons why populations are found where they are. We describe the finer-scaled pattern as the dispersion and the broader scale as the distribution. We offer examples of how different factors such as temperature and seasonality limit the distribution of wildlife. We then discuss the causes for dispersal, and finally methods for modeling rates of dispersal of populations.

Dispersal is the movement an individual animal makes from its place of birth to the place where it reproduces. Dispersal is not to be confused with *migration* (movement backward and forward between summer and winter home ranges) or *local movement* (movement within a home range). The terms "immigration" and "emigration" are used in mark–recapture studies to mean movement into and movement out of a study area of arbitrary size and location. "Migration" is used by population geneticists to mean "the movement of alleles between semi-isolated subpopulations, a process that by definition involves gene flow between subpopulations" (Chepko-Sade *et al.* 1987). Although these uses differ from their ecological ones, the difference is usually obvious from context and causes little confusion.

Dispersion is the pattern of spatial distribution taken up by the animals of an area. Dispersions may be fixed if the animals are sessile but more commonly they change with time under the influence of a changing dispersion of resources. A dispersion at a given time may be changed by dispersal or local movement, or both.

The *distribution* of a population or species is the area occupied by that population or species. It is depicted as the line drawn around the dispersion. The distribution can be subdivided into gross range and breeding range, and it can be mapped at different scales.

4.2 Dispersal

Dispersal is an action performed by an individual (Johnson and Gaines 1990). An animal either disperses or remains within its maternal home range. If it disperses, it may move only a sufficient distance to bring it to the nearest unoccupied and suitable area within which to establish its own home range, or it might move a considerable distance, crossing many areas that look suitable enough, before settling down.

The mechanism of dispersal can also vary. The individual may be pushed out of the maternal home range by a parent or it may move without any prompting save for that supplied by its genes. The young of some species never meet their parents

Wildlife Ecology, Conservation, and Management, Third Edition.
John M. Fryxell, Anthony R. E. Sinclair and Graeme Caughley.
© 2014 John Wiley & Sons, Ltd. Published 2014 by John Wiley & Sons, Ltd.
Companion Website: www.wiley.com/go/Fryxell/Wildlife

(e.g. frogs, reptiles, the mound-building birds of the family Megapodidae) and so must provide their own motivation. In mammals at least, there are two forms of dispersal that have been recognized (Stenseth and Lidicker 1992). *Presaturation* dispersal is seen in some species of small mammals and involves juveniles leaving their natal range even when the density of the population is low. The mechanism is either that the juveniles leave voluntarily, their behavior being innately determined by their genes (e.g. in Belding's ground squirrels, *Spermophilus beldingi*; Holekamp 1986), or that adults forcibly exclude the juveniles. *Saturation* dispersal is seen in many large mammals (Sinclair 1992). In this case, dispersal occurs when a population reaches a threshold density determined by food limitation. It is density-dependent (see Chapter 5 for an explanation of this mechanism), such that population density remains the same in the initial area. Examples of this have been described for Himalayan tahr (*Hemitragus jemlahicus*) as they spread through the Southern Alps of New Zealand (Parkes and Tustin 1985) and for wood bison (*Bison bison*) as their population increases through their former range in the boreal forest of Canada (Larter *et al.* 2000). (See Section 4.6 for models of range expansion.)

The likelihood of dispersal differs markedly between individuals in a population. Fig. 4.1 shows a sample of distances dispersed by juvenile kangaroo rats (*Dipodomys spectabilis*) (Jones 1987), a solitary, nocturnal, grain-eating desert rodent. Females averaged 29 m and males 66 m, but the majority of individuals did not disperse at all. Jones (1987) reports that adults of this species do not disperse much: 70% of adult males and 61% of adult females remained in one mound for the rest of their lives. Juvenile females of red deer (*Cervus elaphus*) seldom disperse, instead adopting home ranges that overlap those of their mothers. In contrast, males leave the natal home range between the ages of two and three, mostly joining stag groups in the vicinity (Clutton-Brock *et al.* 1982).

Patterns of dispersal are related to the type of mating system (Greenwood 1980, 1983; Dobson 1982; Greenwood and Harvey 1982). Thus, in mammals, females are concerned with obtaining resources, while males compete for mates. In general, in promiscuous and polygynous species, males disperse, because they are more likely to find new mates by doing so, while females are philopatric (i.e. remain at their birth site), because they are more likely to find food in areas they know well. Both sexes disperse in monogamous species. Among higher vertebrates, one sex is more prone to dispersal than the other. Thus, in mammals, males are the dispersers, whereas in birds it is the females that disperse, although there are exceptions for both groups: females

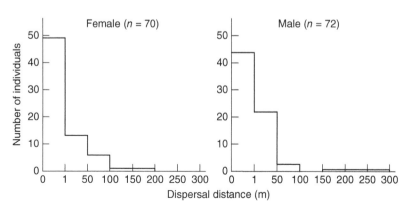

Fig. 4.1 The frequency distributions of distances dispersed by juvenile kangaroo rats. (Data from Jones 1987.)

are the dispersers in wild dogs (*Lycaon pictus*) and zebra (*Equus burchelli*), while in fishers (*Martes pennanti*) and wolves both sexes disperse equally (Arthur *et al.*1993; Boyd and Pletscher 1999).

The causes of dispersal fall into three broad categories: competition for mates, avoidance of inbreeding, and competition for resources (Johnson and Gaines 1990). In polygynous species, females invest more in each offspring than do males, and so their reproductive success is determined by resource competition. Male reproductive success is limited by the number of mates they can find, so competition for mates is important.

Inbreeding avoidance is often cited as a cause of dispersal on theoretical grounds (reviewed in Thornhill 1993; see Chapter 20 of this book for an explanation of the genetics of inbreeding depression). Inbreeding depression was observed in a captive wolf (*Canis lupus*) population (Laikre and Ryman 1991), but there was no evidence of inbreeding depression or avoidance in a social carnivore, the dwarf mongoose (*Helogale parvula*) (Keane *et al.* 1996). In general, the occurrence of inbreeding depression depends on the species (Waser 1996). There are some instances where inbreeding avoidance has been found, as in some species of birds (Pusey 1987; Keller *et al.* 1994), primates (Pusey 1992), rodents (Hoogland 1982), and marsupials (Cockburn *et al.* 1985). However, there are many instances where populations occur in small numbers and inbreeding is not avoided, but it has no deleterious effects (Keane *et al.* 1996). In other cases, there are multiple causes of dispersal (Dobson and Jones 1985).

Dispersers tend to have lower survival than those that remain in their natal area. In arctic ground squirrels (*Spermophilous parryii*), survival of philopatric juveniles was 73%, whereas survival of dispersing squirrels was in the range of 25–40%. Also, survival declines with the distance of dispersal, due to the increasing probability of being caught by predators (Byrom and Krebs 1999). The survival of dispersing ferrets (*Mustela furo*) in New Zealand was 100% where predators had been removed experimentally, compared with only 19–71% in areas where predators were present (Byrom 2002). However, survival of dispersing male San Joachin kit foxes (*Vulpes macrotis mutica*) was higher than that for philopatric males (Koopman *et al.* 2000), indicating exceptions to the rule.

4.3 Dispersion

Dispersions may be random, clumped, or spaced. The most common is a *clumped dispersion* (sometimes called a *contagious dispersion*). If an area is divided into quadrats and the frequency distribution of animals per quadrat is recorded, the variance of that distribution will equal its mean if the animals are randomly distributed (a Poisson distribution), will be greater than the mean if the animals are clumped, and will be less than the mean if the animals space themselves.

Scale is important when dispersions are considered, because two or more orders of dispersion may be imposed upon each other: randomly distributed clumps of animals, for example. In these circumstances, a quadrat in a grid of small quadrats will either include part of a group or will miss a group: its count will be of many animals or of no animals. When the grid comprises large quadrats, an average quadrat will contain several groups of animals and the variation in counts between quadrats will be less marked. The dispersion is the same whether the quadrats used to sample it are large or small, but in this case the clumping as measured by the variance/mean ratio will appear to be more intense when quadrats are small.

An alternative to characterizing dispersion in terms of the frequency distribution of quadrats containing 0, 1, 2, etc. animals each is to run up the frequency distribution of nearest-neighbor distances or of the distances between randomly chosen points and the nearest animal to each. The problem of quadrat size does not then arise, because no quadrats are involved, but no simple measure is presently available for distributions of distances that clearly differentiates between classes of dispersion, as a result of the wealth of possible dispersions. However, J.M. Cullen and M. Bulmer (in Patterson 1965) provide a formula for calculating the random distribution of interindividual (or intergroup) distances in a known area. Given the same number of individuals N, distributed randomly with respect to each other in the same area A, the proportion (p) of individuals having their nearest neighbor at a distance x is given by the expression:

$$p_x = \exp[(-\pi N/A)((x - 0.5a)^2] - \exp[(-\pi N/A)((x + 0.5a)^2]$$

where a is the unit of measurement used. The number at distance x is Np_x. Thus, if one observes 200 birds in an area of 2 km radius ($A = 12.57 \times 10^6\,\text{m}^2$), and observations are in units of 50 m ($= a$), then the expected frequency of distances at the nearest interval ($x_1 = 25\,\text{m}$) is 23.5, that at the next interval ($x_2 = 75\,\text{m}$) is 55.2, and so on until the sum of Np_x equals 200. We see that the increments of x must start with the first one equal to $0.5a$ (mid-point of the first interval) and then increase in increments of a (25, 75, 125, 175, etc.). By comparing this frequency of distances with the observed frequency, one can identify clumped or overdispersed distributions.

Dispersion is affected by the *home range* of an individual; that is, the area used during its normal daily activities. Traditionally, home ranges are estimated from radiotelemetry locations (usually > 30 locations are required) using computer software packages. Habitat type affects range area (Relyea *et al.* 2000), as does the individual's gender (McCullough *et al.* 2000). Some species have tight habitat preferences, their dispersion reflecting where that habitat is to be found. Others are more catholic in their requirements and will therefore be distributed more evenly across the landscape. The ecology of the dispersion is important. Dispersion can be measured more directly, however, by the average distance between locations (Conner and Leopold 2001). We considered the concept of home range in Chapter 3, in which we outlined methods for determining the key determinants of home range use.

When we design surveys to count wildlife (see Chapter 12), we must pay attention to a species' dispersion and allocate our sampling units accordingly. We explore this practical aspect of dispersion more fully in Section 12.4.

4.4   Distribution

Krebs (2001) considered that "the simplest ecological question one can ask is simply: Why are organisms of a particular species present in some places and absent in others?" There are several interesting ways that this question can be answered. We start with a consideration of the ultimate limits of a species' range, before going on to consider the distribution of introduced or invading species and finally patterns of occupancy in spatially subdivided populations (metapopulations).

Fig. 4.2 shows three hypothetical distributions, not as a map but as a plot within a range of mean annual temperature and mean annual rainfall. For species A, temperature and rainfall act independently of each other in setting limits to distribution. A single mean temperature and a single mean annual rainfall are all one needs to predict whether or not the species will be in a given area.

Fig. 4.2 Three hypothetical envelopes of adaptability of a species to temperature and moisture: A, the two factors act independently; B, the level of one factor influences the effect of the other; C, the effect of each factor varies according to the level of the other. (After Caughley *et al.* 1988.)

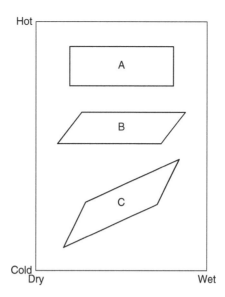

The distribution of species B is also determined by temperature and rainfall, but this time in an asymmetric interactive manner. Distribution is determined absolutely by an upper and a lower limiting temperature, but it is demarcated within those bounds by rainfall, whose effect varies with temperature. High rainfall is tolerated only in hot areas and low rainfall only in colder areas, where evaporation is reduced.

The distribution of species C is controlled by a symmetric interaction of rainfall and temperature. The species' tolerance of high temperatures increases with increasing annual rainfall and its tolerance of rainfall with temperature. This is a two-way interaction.

A known range of tolerance to one or more factors, such as temperature and rainfall, does not translate directly into a map of distribution because the factors may interact as in examples B and C of Fig. 4.2, where the level of one factor determines the effect of another. Whether distribution is determined by one or several factors depends critically on the geographic dispersion of the levels of each factor.

4.4.1 Range limited by temperature

Temperature can limit the distribution of animals directly by affecting their physiology and indirectly by affecting resources. Some distributions can be described empirically by temperature contours (isotherms). Thus, the southern limit to northern-hemisphere seals is set by sea surface temperatures of < 20 °C (Lavigne *et al.* 1989). The reason is unclear, but most seals breed in regions of high marine productivity, and these are largely restricted to high latitudes. Similarly, the penguins of the southern hemisphere inhabit seas with temperatures < 23 °C. Most penguin species inhabit latitudes between 45 and 58° S, where marine productivity is high (Stonehouse 1967). They reach the equator at the Galapagos Islands off the Pacific coast of South America, but only because those shores are bathed by the cold Humboldt Current.

The northern limit for rabbits (*Oryctolagus cuniculus*) in Australia is marked by the 27 °C isotherm. This temperature coincides with high humidity, and the combination of the two causes resorption of embryos, so that the animals cannot breed.

Cold is clearly an important factor limiting species in the Arctic and subarctic. Although the Arctic is an important breeding ground for birds, most leave during winter. Only four North American species can withstand the cold to reside year-round in the Arctic: the raven (*Corvus corax*), the rock ptarmigan (*Lagopus mutus*), the snowy owl (*Nyctea scandiaca*), and the hoary redpoll (*Acanthis hornimanni*) (Lavigne *et al*. 1989). Amphibians and reptiles are particularly affected by temperature. The American alligator (*Alligator mississipiensis*) cannot tolerate temperatures below 5 °C. Although several species of amphibians and reptiles tolerate freezing temperatures, in general there is a negative relationship between the number of species and the latitude. The direct effect of cold in limiting the distribution of these groups is probably less important than the availability of hibernation sites remaining above lethal temperatures (Lavigne *et al*. 1989).

Movements of large mammals can be affected by temperature. In the Rocky Mountains, several ungulates, such as moose (*Alces alces*), elk, and deer, move downhill for the winter. Sometimes a temperature inversion in winter positions a warmer air layer above a colder one, and in these conditions Dall sheep (*Ovis dalli*) in the Yukon climb higher rather than lower.

The limiting effects of temperature are demonstrated by historic changes in the ranges of several species. Temperatures increased in the northern hemisphere from 1880 to 1950. The breeding ranges of herring and black-headed gulls (*Larus argentatus*, *L.ridibundus*) moved north into Iceland, and that of green woodpeckers (*Picus viridis*) extended into Scotland. Temperatures have declined since 1950 and the breeding ranges of snowy owls and ospreys (*Pandion haliaetus*) have moved south (Davis 1986). On the American prairies, the warming period was associated with severe droughts in the 1930s. As a result, the cotton rat (*Sigmodon hispidus*) has spread north (Davis 1986). Further changes in distribution of these and many other wildlife species are anticipated in the future, as a result of global warming (see Chapter 11).

Cold temperatures themselves may be less important than the consequent changes in snow pack. Caribou must expend greater amounts of energy in exposing ground lichens when snow develops a crust (Fancy and White 1985). Further north, on Canada's High Arctic Islands, the warming temperatures of spring melt the surface snow. The water trickles through the snow pack and freezes when it hits the frozen ground, forming an impenetrable layer. The caribou abandon feeding in these areas, and may migrate across the sea ice to areas where the wind has blown the shallow snow away (Miller *et al*. 1982)

Deep snow limits other species too. North American mountain sheep (*Ovis canadensis*, *O.dalli*) are usually found on cold windswept ledges in winter, where there is little snow. Deer (*Odocoileus* spp.) are limited by snow cover of moderate depths (< 60 cm), whereas moose can walk through meter-deep snow (Kelsall and Prescott 1971). Both move to coniferous forest in late winter, because the snow is less deep there (Telfer 1970; Rolley and Keith 1980).

The stress of cold temperature has resulted in various adaptations to conserve energy, the most notable being the hibernation of ground squirrels during winter and the dormancy and lowering of body temperature of bears. Hummingbirds also lower their body temperature overnight, to about 15 °C, or when resting in cold conditions, a state called *torpor*. The limiting effect of temperature on ground squirrels operates indirectly through soil type, slope, and aspect. Squirrels need to dig burrows deep enough to avoid the cold, and this requires sandy, friable soil. They also need to avoid being swamped by meltwater in spring, so their burrows are situated on slopes, where water can drain away. Similarly, in Australia the distribution of rabbits within the 27 °C isotherm is influenced by soil type, soil fertility, vegetation cover, and distribution of water (Parer 1987).

4.4.2 *Range limited by water loss and heat stress*

High temperatures are often combined with high solar radiation and restricted water supplies. In high-rainfall areas, the last factor is important to restricting distribution, while in arid regions all three have interrelated effects on animals. These effects are expressed as heat loads built up in the body, and various adaptations have developed to overcome them.

Adaptations to high temperatures include behavioral responses, such as using shade in the middle of the day and restricting feeding to the hours of darkness. Both eland (*Taurotragus oryx*) and impala (*Aepyceros melampus*) reduce heat stress by feeding at night in East Africa (Taylor 1968a). At the driest times of year, both species boost water intake by switching from grazing grasses and forbs to browsing on succulent shrubs (Taylor 1969; Jarman 1973).

Solar radiation restricts the movements of animals that are large and have dark coats. Elephant and buffalo are examples that seek shade in the heat of day to cool off (Sinclair 1977). Coat color and structure can reduce heat loads. The lighter, tan-colored coat of hartebeest (*Alcelaphus buselaphus*) reflects 42% of short-wave solar radiation, as against only 22% for the darker coat of eland. In both species, re-radiation of long-wave thermal radiation is greater than absorption, representing 75% of total heat loss (Finch 1972).

High heat loads can be avoided by sweating when water is abundant. African buffalo, eland, and waterbuck use sweating for evaporative cooling (Taylor 1968a; Taylor *et al.* 1969b). Buffalo keep body temperature in the range 37.4–39.3°C and allow them to rise to 40°C only when water is restricted. They cannot reduce water loss from sweating when water is restricted (Taylor 1970a, 1970b). Waterbuck show similar physiological adaptations. When water is restricted for 12 hours at 40°C ambient (environmental) temperature, they lose 12% of their body weight, compared with the 2% for beisa oryx (*Oryx beisa*), which is a desert-adapted species (Taylor *et al.* 1969b). As a consequence, both buffalo and waterbuck must remain within a day's walk of surface water.

Large animals can afford to lose water by sweating but smaller animals such as the gazelles cannot. They employ panting instead, as do species in arid areas (e.g. the beisa oryx) and species on open plains with high solar radiation (e.g. wildebeest) (Robertshaw and Taylor 1969; Taylor *et al.* 1969a; Maloiy 1973).

Some species can adapt to extreme arid conditions by allowing their body temperature to rise before they start panting: up to 43°C for Thomson's gazelle (*Gazella thomsonii*) and 46°C for Grant's gazelle (*G.granti*) (Taylor 1972). Other adaptations for water conservation include restriction of urine output, concentration of the urine, and reabsorption of water from the feces. Dikdik, a very small antelope that lives in semi-arid scrub away from water, had the lowest fecal water content and the highest urine concentration of all antelopes, followed by hartebeest, impala, and eland (Maloiy 1973).

Grazing ungulates in Africa are restricted to areas within reach of surface water and all show behavioral adaptations such as night feeding or migration (Sinclair 1983). Those that can do without water are all browsers (Western 1975). Beisa oryx and Grant's gazelle select hygroscopic shrubs (*Disperma* species). They eat them at night because they contain only 1% free water in the day but absorb moisture from the air at night to boost the water content of the leaves to 43% (Taylor 1968b).

Perhaps not apparent at first sight is the restricted availability of water for wildlife in cold regions. Not only are many of these regions deserts, as their rainfall is low, but during winter moisture is only available as snow, and valuable energy is needed to melt it. Arctic mammals go to some lengths to conserve water. Caribou recycle nitrogen to reduce the formation of urine, thereby conserving water.

4.4.3 *Range limited by day length and seasonality*

The distribution of many North American birds is limited at northern latitudes by season length: the number of days available for breeding above a certain temperature. This is another aspect of temperature limitation. However, the southern boundary is limited by day length: the number of hours available for feeding themselves and their young (Emlen *et al.* 1986; Root 1988).

Seasons are highly predictable in the northern temperate latitudes of North America and Eurasia, and many birds and mammals have evolved a response to *proximate factors* (i.e. the immediate factors affecting an animal), particularly day length (photoperiod), which trigger conception and result in the production of young during optimum conditions. Such conditions are the *ultimate factors* (i.e. the underlying selection pressure) to which an animal is adapted by breeding seasonally (Baker 1938).

Increasing photoperiod determines the start of the breeding season in many bird species (Perrins 1970), while declining photoperiod triggers the rut in caribou; the rut is so synchronized that most conceptions occur in a mere 10-day period starting around the first day of November (Leader-Williams 1988). Moose and elk also have highly synchronized birth seasons (Houston 1982), which suggests photoperiodic control of reproduction.

Among tropical ungulates, only the wildebeest is known to respond to photoperiod. In southern Africa, it uses solar photoperiod to synchronize conceptions, but near the equator, where solar photoperiod varies by only 20 minutes in the year, it is cued by a combination of lunar and solar photoperiod (Spinage 1973; Sinclair 1977).

In variable environments with less-predictable seasons, as in the tropics and arid regions, animals tend not to use photoperiod to anticipate conditions but rather adjust their reproductive behavior to the current conditions. Thus, tropical birds begin breeding when the rainy season starts, responding to the increase in insect food supply and the growth spurt of the vegetation (Sinclair 1978). In some arid areas, such as Western Australia, the seasonality of rain is relatively predictable but its location is not. Emus (*Dromaius novaehollandiae*) there travel long distances searching for areas that have received rain (Davies 1976), as do male red kangaroos (*Macropus rufus*) (Norbury *et al.* 1994).

Most ungulates produce their young during the wet season in Africa and South America but put on fat prior to giving birth. This fat is then used during lactation, the period when the energy demands on the female are highest (Ojasti 1983; Sinclair 1983). Therefore, nutrition in the seasonal tropics becomes both the proximate and ultimate factor determining the timing of births. An example is provided by the lechwe (*Kobus lechee*), an African antelope that lives on seasonally flooded riverine grasslands (Fig. 4.3). During the peak of the floods, animals are confined to the less preferred surrounding woodlands. The greatest area of flood plain is exposed at the low point in the flood cycle and it is at this time, corresponding with the greatest availability of food, that births take place. In Zambia, the peak of births occurs in the dry season,

Fig. 4.3 The numbers of lechwe, a flood plains antelope of southern Africa (●, left axis), increase on the flood plains as water recedes in the Chobe River exposing the greatest area of high-quality food. The recruitment of newborn per 100 females (○, right axis) shows that births occur at this time. (After Sinclair 1983, which is after Child and von Richter 1969.)

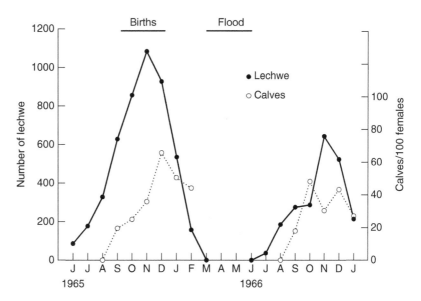

3 months after the rains; in the Okavango Swamp of Botswana it occurs in the middle of the wet season, 9 months after the previous rains; but both occur when the swamp grasslands are most available.

4.4.4 *Range limited by biotic factors*

So far we have discussed range limitation by abiotic environmental factors. However, these abiotic factors can interact with biotic processes such as predation and competition to further limit a species' range. For example, the geographic distribution of arctic fox (*Alopex lagopus*) occurs largely in the tundra of the Holarctic and is separate from that of the more southerly red fox (*Vulpes vulpes*). However, their ranges overlap in some areas of North America and Eurasia. The northern limit of the red fox's range is determined directly by resource availability, which is determined by climate. The southern limit of the arctic fox's range is determined by interspecific competition with the more dominant red fox (Hersteinsson and Macdonald 1992).

4.5 **Distribution, abundance, and range collapse**

A major pattern in ecology is the positive relationship between the range of a species and its abundance. In general, locally abundant species have wide ranges and rare species have narrow ranges (Brown 1995; Gaston and Blackburn 2000). This observation has led to Rapoport's rule, namely that *the latitudinal extent of a species' range increases towards the poles* (Rapoport 1982). This general pattern is modified by species richness, rainfall, vegetation, and land surface, as seen in studies of birds (Gentilli 1992) and mammals (Pagel *et al.* 1991; Letcher and Harvey 1994; Smith *et al.* 1994).

Of great importance in conservation management is what happens to a species range when the population declines. One expects that population densities tend to be higher at the center of a population's range than at the periphery. Geographic ranges should collapse from the outside, with the center being the last to go (Brown 1995). Analyses of range contractions in a wide variety of animals and plants suggest that populations often collapse first in the center, however, leaving isolated fragments on the periphery (Lomolino and Channell 1995; Channell and Lomolino 2000), with these collapses being due to the variety of causes outlined by Caughley (1994). Thus, peripheral populations not only provide a refuge for endangered species but also represent genetical and morphological varieties that differ from central populations (Lesica and Allendorf 1995).

4.6 Species reintroductions or invasions

Many species of wildlife have been eliminated from their traditional range, for one reason or another. This can even happen to common species, such as the plains bison (*Bison bison*). Europeans arriving in North America encountered millions of bison on the Great Plains. In remarkably short order, this massive population was nearly extirpated, through a combination of commercial hunting by Europeans and subsistence hunting by aboriginal groups, competition with livestock, and fencing off of migration routes (Isenberg 2000). Since the turn of the century, the plains bison has been reestablished by wildlife authorities to parts of its former range, though in nothing like its former abundance. Such reintroductions are becoming ever more common.

In other cases, species have naturally recovered from catastrophic decline, expanding into their former range. A well-documented example is the California sea otter (*Enhydra lutris*; Lubina and Levin 1988). This species was nearly exterminated throughout its Pacific Coast range through overharvesting by fur traders in the late 19th century, before a moratorium on harvesting was signed in 1911. A small relict population of otters survived in an inaccessible part of the California coast south of Monterey Bay and provided the nucleus for gradual spread of the population back northward and southward along the coast.

Whether intentional or accidental, such reintroductions have some fascinating characteristics that have important bearing on their successful conservation. Key among these is the interplay between demography and patterns of movement.

4.6.1 Diffusive spread of reintroduced species

Although there are many elegant ways to model patterns of movement by invasive or reintroduced species (Turchin 1998), simple random walk models can often predict the pattern of spread surprisingly well. We first consider what is meant by a random walk, then use this algorithm to develop a simple model of population distribution.

What pattern would emerge over time for a single individual that moved randomly every day of its life? We will assume that this hypothetical animal can only move forwards, backwards, or sideways, one step at a time. We further assume that each of these events is as probable as its remaining where it is. To model this, we need to sample randomly from a uniform probability distribution (see Box 4.1).

Box 4.1 Modeling a random walk in space.

We first need to randomly sample a large number of values uniformly distributed between 0 and 1, assigning each random number on this interval to either the variable *p* or the variable *q*. We then use these probabilities to mimic forward versus backward movement using the variable *x* and side-to-side movement using the variable *y*, according to the logic shown below. As a result, an animal will move left one-third of the time and right one-third of the time, and will stay on track one-third of the time. Similar probabilities correspond to forward, backward, and stationary outcomes.

$$X_{t+1} = \begin{vmatrix} X_t - 1 & \text{if } p_t < 0.333 \\ X_t + 1 & \text{if } p_t > 0.333 + p_t > 0.667 \\ X_t & \text{otherwise} \end{vmatrix}$$

$$y_{t+1} = \begin{vmatrix} y_t - 1 & \text{if } q_t < 0.333 \\ y_t + 1 & \text{if } q_t > 0.333 + q_t < 0.667 \\ y_t & \text{otherwise} \end{vmatrix}$$

Fig. 4.4 Hypothetical trajectory over 100 time steps for a single individual following a random walk, starting from the origin (0,0).

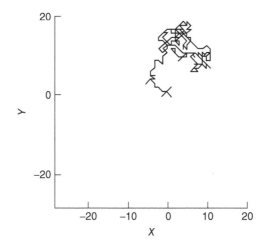

For this kind of random walk model, most trajectories tend to find their way back to a position not far from the initial starting point (Fig. 4.4). In other words, walking randomly is not a very effective means of getting anywhere new. This is a useful null model, however, setting an extreme standard against which we might evaluate the movements of real organisms. The random walk model is perhaps most plausible at large spatial scales, such as for dispersing juveniles, on which animals have no past experience with local conditions.

We can readily expand this kind of model to a group of individuals (Case 2000). To keep it simple, we will concentrate on only one spatial dimension, such as for sea otters dispersing up and down the coast of California. Let's say that there are 100 individuals released at a central position "0" and that each individual has a 20% probability of moving left and a 20% probability of moving right, with position along this axis indicated by the variable x. This probability we will term d for dispersal. Local changes in the densities of individuals can be modeled in the following manner:

$$N_{x,t+1} = N_{x,t} - 2dN_{x,t} + dN_{x-1,t} + dN_{x+1,t}$$

The local population loses $2 \times d \times N$ individuals due to movement in either direction, but gains $d \times N$ individuals from each adjacent site. We need to repeat this exercise over the full range of distance intervals.

The output of this model demonstrates two important features (Fig. 4.5). First, the spatial distribution of individuals in the population begins to take on a bell-shaped or normal distribution over time. Second, the rate of spread is initially fast, but slows over time. This is because movement away from the release point is balanced to a considerable degree by movement backwards. This slower movement away becomes more pronounced over time because the distribution is getting flatter. When dynamics are driven purely by random motion, the population range spreads at a rate proportional to $\sqrt{time}$. If we repeat this simulation with a larger fraction of dispersers (say $d = 0.3$), the rate of spread will increase accordingly. The rate of spread is proportional to $\sqrt{d}$.

We can use our random walk model to derive the differential equation that defines diffusive changes in local density over a continuous gradient of space and time:

$$\Delta N_x = N_{x,t+1} - N_{x,t} = -2dN_{x,t} + dN_{x-1,t} + dN_{x+1,t}$$

Fig. 4.5 Variation in the population density of individuals over time, when those individuals redistribute themselves every time step (t) according to an unbiased random walk.

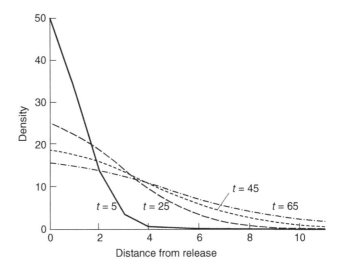

We then rearrange terms on the right-hand side of the equation:

$$\Delta N_x = d[(N_{x-1,t} - N_{x,t}) - (N_{x,t} - N_{x+1,t})]$$

The rate that individuals accumulate at site x depends on the degree of difference between the density gradient below the site and the density gradient above it. In other words, it is not the gradient itself but the rate of change of the density gradient over space that dictates the rate of diffusive movement. Mathematicians refer to the rate of change of the density gradient as the "second derivative." If this occurs over short enough intervals of time and space then the result is the following differential equation (called the "diffusion equation in one dimension"):

$$\frac{dN(x,t)}{dt} = D\frac{\partial^2}{\partial x^2}N(x,t)$$

The solution to this equation is the normal distribution:

$$N(x,t) = \frac{N_0}{\sqrt{4\pi Dt}} \exp\frac{-(x-\mu)^2}{4Dt}$$

where t is the time since the animals were released, μ is the initial position (usually 0), and D is the diffusion coefficient. This reflects how fast individuals tend to diffuse away from an initial point of release. We discuss how to calculate it presently. This equation may look familiar – it is closely related to the normal (sometimes called "Gaussian") probability distribution. The variance in spatial locations is given by $\sigma^2 = 2Dt$.

   The easiest way to estimate D is to estimate the mean-squared displacement of the individuals in the population over time. We simply measure the distance of a given individual from its original release point, square this displacement to get rid of positive versus negative values, sum the squared displacements for all individuals, and divide this sum by the total sample size to estimate mean squared displacement. D is then calculated by dividing the mean squared displacement by $2t$.

In the more typical case of diffusion in two dimensions (x and y, centered at the release point), these equations are slightly altered:

$$\frac{dN(x,y,t)}{dt} = D\left[\frac{\partial^2}{\partial x^2}N\left(x,y,t\right) + \frac{\partial^2}{\partial y^2}N(x,y,t)\right]$$

$$N(r,t) = \frac{N_0}{\sqrt{4\pi Dt}}\exp\frac{-r^2}{4Dt}$$

where r is the distance (i.e. radius) from the release point. Despite the slight change in formula, this equation also predicts that the range occupied by the population is proportional to $\sqrt{time}$. This is a very useful prediction that differs from other models of population spread, as we shall shortly see.

4.6.2 Spread of reintroduced species: diffusion plus exponential growth

As we discuss in Chapter 5, a newly reintroduced population is likely to have plenty of resources with which to grow and multiply. This logically leads to geometric or exponential growth, at least in the initial period following release. Unrestricted population growth can be readily incorporated in our random walk model. We simply multiply the local population by the finite rate of growth (in this case, let us say that $\lambda = 1.05$):

$$N_{x,t+1} = \lambda N_{x,t} - 2dN_{x,t} + dN_{x-1,t} + dN_{x+1,t}$$

The rate of spread now seems to be much more consistent over time (Fig. 4.6) than was the case for diffusive movement alone (Fig. 4.5). In fact, the population range now spreads at a rate proportional to t and d, because the population grows fastest where density is highest. This relationship tends to create a rapid rate of change in the density gradient, which we have already suggested fuels a high rate of diffusion. The net result is a population that explodes over both time and space.

Another interesting feature of the diffusion + exponential growth model is that a standing wave of animals spreads over time across the landscape (Fig. 4.6), rather than the gradually eroding "mountain" seen in the pure diffusion model (Fig. 4.5). This

Fig. 4.6 Variation in the population density of individuals over time, when those individuals redistribute themselves every time step (t) according to an unbiased random walk. Unlike Fig. 4.5, the population is also growing at an annual rate of $\lambda = 1.05$.

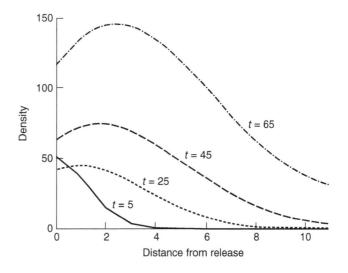

Fig. 4.7 Spatial spread
over time of a small
population of muskrats
introduced into the
countryside near Prague.
(After Elton 1958.)

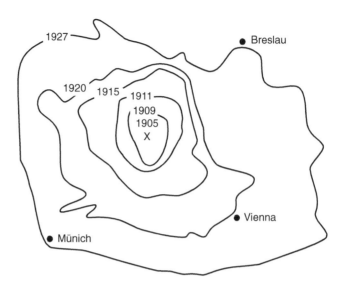

wavelike form of spread is echoed in most models that incorporate population growth
as well as diffusive movement, such as those with logistic growth and predator–prey
models. The velocity with which this wave rolls across the landscape is identical in
virtually all such models: $v = 2\sqrt{\lambda D}$.

4.6.3 *Empirical tests of diffusion theory*

We should be able to discriminate between alternative models of population spread
by looking at population range versus time. If the increasing radius of population distribution were to become slower over time then this deceleration would be consistent
with a pure diffusion process, in which population growth is not involved. On the other
hand, a constant increase in radial spread of the population would be most consistent
with the diffusion + exponential growth model.

Skellam (1951) made this comparison using data on a population starting with five
muskrats (*Ondatra zebithica*) translocated into the countryside near Prague in 1905
(Fig. 4.7). Skellam's analysis, supported by more rigorous analysis by Andow *et al.*
(1990), clearly demonstrated that the radial spread of muskrats increased linearly over
time, at a rate of 11 km/year (Fig. 4.8), thus supporting the exponential diffusion
model.

Fig. 4.8 Radial spread
(measured as the square
root of area) of the
population of muskrats
introduced into the
countryside near Prague.
(After Skellam 1951.)

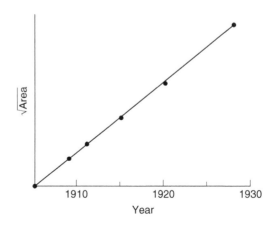

Similar analysis of the naturally recovering population of California sea otters also supports the diffusion + exponential growth model (Lubina and Levin 1988), although the pattern is more complex. Radial spread to the north is slower than that to the southern California coast. Moreover, there seems to be a dramatic jump in the distance dispersed per year as the otters move into sandy coastal areas with less of their preferred rocky habitat.

Since these early days, there has been considerable development of alternate models of population spread. These recognize directional bias on the part of the disperser, changes in disperser motivation, and heterogeneous environmental effects on dispersal tendency (Turchin 1998). Nonetheless, the simple model of diffusive spread combined with exponential growth often does a tolerably good job of predicting patterns of population spread over time. These successful predictions suggest that both rapid population growth at the wavefront and some degree of randomness in the pattern of movement contribute heavily to observed patterns of spread in many wildlife species. Both the theory and the empirical mechanisms underlying animal movement across complex ecological landscapes are now developing rapidly, because both have important conservation implications.

4.7 Summary

The distribution is the area occupied by a population or species, the dispersion is the pattern of spacing of the animals within it, and dispersal, migration, and local movement are the actions that modify dispersion and distribution. Dispersion and distribution are states; dispersal, migration, and local movement are processes. The edge of the distribution is that point at which, on average, an individual just fails to replace itself in the next generation. Its position may be set by climate, substrate, food supply, habitat, predators, or pathogens. The limiting factor can often be identified by the trend in density from the range boundary inward.

Dispersal plays a key role in dictating the rate of spread of a species reintroduced into a new area or of one recovering from catastrophic decline. Diffusion models are often an effective means of modeling the spread of reintroduced species, particularly if they incorporate both demographic and random walk processes. We demonstrate the logical basis for the simplest random walk and diffusion models.

5 Population growth and regulation

5.1 **Introduction**

In this chapter we first describe the fundamental characteristics of population growth and how these vary with body size. We consider the range of processes that can constrain population growth through *population limitation* and ultimately serve to stabilize populations through the density-dependent process of *natural regulation*. We then analyze the processes that can cause fluctuations and population cycles, using models to develop our understanding of them. Finally, we examine one of the major causes of regulation: competition between individuals for resources, or *intraspecific competition*. Other causes of regulation, such as predation, parasites, and disease, will be dealt with in Chapters 7–9.

5.2 **Rate of increase**

If a population comprising 100 animals on (say) January 1 contains 200 animals on the following January 1 then obviously it has doubled over one year. What will be its size on the next January 1 if it continues to grow at the same rate? The answer is not 300, as it would be if the growth increment (net number of animals added over the year) remained constant each year, but 400, because it is the growth rate (net number of animals added, divided by numbers present at the beginning of the interval) that remains constant. Thus the growth of a population is analogous to the growth of a sum of money deposited at interest with a bank. In both cases, the growth increment each year is determined by the rate of growth and by the amount of money or the number of animals that are there to start with. Both grow according to the rules of compound interest and all calculations must therefore be governed by that branch of arithmetic.

Populations decrease as well as increase. The population of 100 animals on January 1 might have declined to 50 by the following January 1, in which case we would say that the population has halved. If its decline continues at the same rate it will be down to 25 on the next January 1. Halving and doubling are the same process operating with equal force, the only difference being that it is running in opposite directions. The terms by which we measure the magnitude of the process should reflect that equivalence. This is poorly achieved by simply giving the multiplier of the growth, 2 for a doubling and 0.5 for a halving, and it becomes even more confusing when percentages are used. We need a metric that gives exactly the same figure for a halving as for a doubling, but with the sign reversed. This will make it obvious that a decrease is simply a negative rate of increase.

Wildlife Ecology, Conservation, and Management, Third Edition.
John M. Fryxell, Anthony R. E. Sinclair and Graeme Caughley.
© 2014 John Wiley & Sons, Ltd. Published 2014 by John Wiley & Sons, Ltd.
Companion Website: www.wiley.com/go/Fryxell/Wildlife

This is achieved by expressing the rate of increase, positive or negative, as a geometric rate according to the following equation:

$$N_{t+1} = N_t \lambda = N_t e^r$$

where N_t is population size at time t, N_{t+1} is the population size a unit of time later, e is the base of natural logs taking the value 2.7182817, and r is the exponential rate of increase. The *finite rate of increase* (λ) is the ratio of the two censuses:

$$\lambda = N_{t+1}/N_t$$

and therefore the *exponential rate of increase* is:

$$r = \log_e(N_{t+1}/N_t) = \log_e \lambda$$

Let us test this on a doubling and a halving.
  With a doubling:

$$\lambda = 200/100 = 2$$

and so:

$$r = \log_e \lambda = 0.693.$$

With a halving:

$$\lambda = 50/100 = 0.5$$

and so:

$$r = \log_e \lambda = -0.693.$$

Thus, a halving and a doubling both provide the same exponential rate of increase, 0.693, but in the case of a halving with the sign reversed (i.e. −0.693). This makes the point again that a rate of decrease is simply a negative rate of increase.
  The finite rate of increase (i.e. the growth multiplier λ) and the exponential rate of increase r must each have a unit attached to them. In our example, the unit is a year, and so we can say that the population is multiplied by λ per year. The exponential rate r is actually the growth multiplier of $\log_e$ numbers per year. That is something of a mouthful, so we say that the population increased at an exponential rate r on a yearly basis. Note that λ and r are simply different ways of presenting the same rate of change; they do not contain independent information.
  Unlike the finite rate of increase, the exponential rate of increase can be changed from one unit of time to another by simple multiplication and division. If $r = -0.693$ on a yearly basis then $r = -0.693/365 = -0.0019$ on a daily basis. This simplicity is not available for λ.
  These equations are simplified to embrace only one unit of time. They can be generalized to:

$$N_t = N_0 e^{rt}$$

where N_0 is the population size at the beginning of the period of interest and N_t is the population size t units of time later. The average exponential rate of increase over the period is:

$$r = (\log_e(N_t/N_0))/t$$

which can also be written as:

$$r = (\log_e N_t - \log_e N_0)/t$$

It would be of a waste of data to use only the population estimates at the beginning and end of the period to estimate the average rate of increase between these two dates. If intermediate estimates are available, they can and should be included in the calculation to increase its precision. The appropriate technique is to take natural logarithms of the population estimates and then fit a linear regression to the data points, each comprising $\log_e N$ and t. A linear regression takes the form $y = a + bx$, where y is the dependent variable (in this case logged population size) and x is the independent variable (in this case time measured in units of choice). Our equation thus becomes:

$$\log_e N = a + bt$$

where a is the fitted value of $\log_e N$ when $t = 0$ and b is the increase in $\log_e N$ over one interval of time. Note that this is the definition of r, so $r = b$. The equation for the linear regression may thus be rewritten in the following manner:

$$\log_e N = a + rt.$$

This can be converted back to the notation used in the example where rate of increase was measured between only two points by designating the start of the period as time 0:

$$\log_e N_t = \log_e N_0 + rt$$

which with a little rearranging converts to:

$$r = (\log_e N_t - \log_e N_0)/t$$

as before. Fig. 5.1 shows such use of linear regression to estimate the rate of increase of the George River caribou herd in eastern Canada, yielding $r = 0.11$ (Messier *et al.* 1988).

5.2.1 *Intrinsic rate of increase*

The rate of increase of a population of vertebrates usually fluctuates gently most of the time, around a mean of zero. If conditions suddenly become more favorable, the population increases, the environmental improvement being reflected in a rise in fecundity and a decline in mortality.

The environmental change might be an increase in food supply, perhaps a flush of plant growth occasioned by a mild winter and a wet spring. The rate at which the population increases is then determined by two things: the amount of food available and the intrinsic ability of the species to convert this extra energy into enhanced fecundity and diminished mortality. It thus depends on an environmental effect and

Fig. 5.1 Exponential population growth of the George River caribou herd, as discussed in the text. (After Messier *et al.* 1988.)

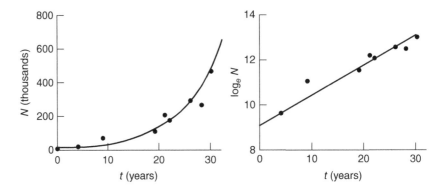

an intrinsic effect, but neither is without limit. From the viewpoint of the animal, both are constrained. There comes a point at which the animal has all the food it can eat, any further food having no additional effect on its reproductive rate and probability of survival. Similarly, an animal's reproductive rate is constrained at the upper limit by its physiology. Litters can be only so big and the interval between successive litters cannot be reduced below the gestation period. The potential rate of increase can never be very high, irrespective of how favorable the environmental conditions are, if the period of gestation is long (e.g. 22 months for the African elephant (*Loxodonta africana*)). All species, therefore, have a maximum rate of increase, which is called their *intrinsic rate of increase* (Fisher 1930) and symbolized r_m. This is a particularly important parameter in estimating sustainable yield (see Chapter 18). Populations do not attain this maximum very often. It requires a very high availability of food and a low density of animals, such that there is negligible competition for the food. These conditions are most closely approached when a population is in the early stage of active growth subsequent to the release of a nucleus of individuals into an area from which they were formerly absent. Fig. 5.2 gives the intrinsic rates of increase of several mammals.

Fig. 5.2 Intrinsic rate of increase of mammals plotted against body weight. (After Caughley and Krebs 1983.)

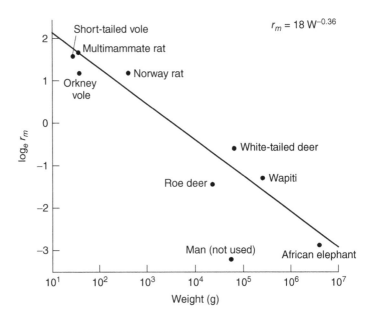

Table 5.1 Expected intrinsic rates of increase r_m on a yearly basis for herbivorous mammals, as estimated from mean adult live weight.

Weight (kg)	r_m
1	1.50
10	0.65
100	0.29
1000	0.08

Alternatively, the rate can be estimated from the initial stages of growth of a population recovering from overhunting. This would work for blue whales (*Balaenoptera musculus*), for example, which are presently recovering from intense overharvesting between about 1925 and 1955 (Cherfas 1988).

Intrinsic rate of increase r_m tends to vary with body size. The following relationship has been calculated for herbivorous mammals (Caughley and Krebs 1983; Sinclair 1996):

$$r_m = 1.5W^{-0.36}$$

where W is mean adult live weight in kilograms. Table 5.1 gives r_m calculated according to this equation for a range of body weights. In the absence of other data, it provides an approximation that can be used to make a first estimate of sustained yield (see Chapter 18).

5.3 Geometric or exponential population growth

In 1798, Thomas Malthus recognized that populations have an intrinsic tendency towards exponential or geometric growth, just as a bank account at fixed interest grows geometrically with the amount of money it contains. The growth of such populations can be calculated as either a continuous or a discrete process. For simplicity, we will concentrate on discrete time representations of population growth. Strictly speaking, such models are most applicable to organisms whose patterns of deaths and births follow a seasonal or annual cycle of events, which includes most wildlife species. Consider, for example, a population whose finite growth rate $\lambda = 0.61$ and whose initial density N_0 is 1.5. The geometric growth model predicts subsequent changes in density over time according to the following equation:

$$N_t = N_0 \lambda^t$$

The outcome depends on whether λ is larger or smaller than 1: when $\lambda < 1$ (Fig. 5.3), there is a decelerating pattern, while when $\lambda > 1$ (Fig. 5.4), there is an accelerating pattern. The geometric model can be readily translated into the exponential model:

$$N_t = N_0 e^{r_{max}t}$$

Hence, it is straightforward to shift between representations of population dynamics in continuous versus discrete time. Such simple models are most appropriate for small populations introduced into a new environment or for a short period following a perturbation. For example, the George River caribou herd in eastern Canada grew exponentially at a rate of $r = 0.11$ during a 30 year period following recovery from overharvesting (Messier *et al.* 1988).

5.4 Stability of populations

If we look at long-term records of animal populations, we see that some populations remain quite constant in size for long periods of time. Records of mute swans (*Cygnus olor*) in England from 1823 to 1872 (Fig. 5.5) show that although the population

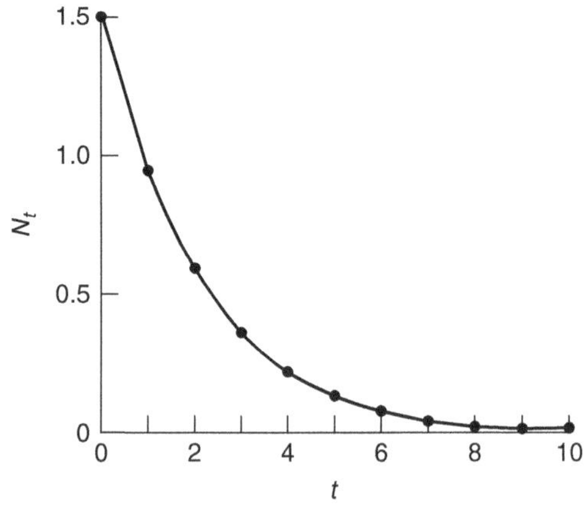

Fig. 5.3 Population changes according to the geometric model with $\lambda = 0.61$.

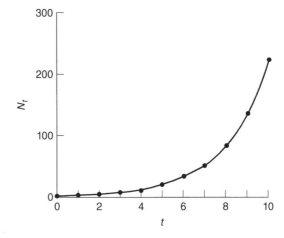

Fig. 5.4 Population changes according to the geometric model with $\lambda = 1.65$.

Fig. 5.5 Some populations remain within relatively close bounds over long time periods. The mute swan population of part of the River Thames in England (estimated by total counts) shows a steady level or gentle increase despite some perturbations due to severe winters, for example in 1946–1947 and 1963–1964. (Data from Cramp 1972.)

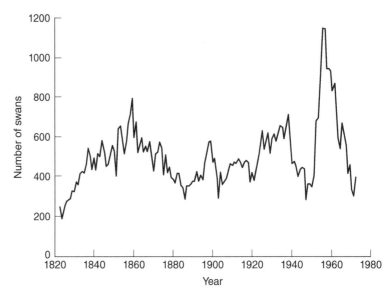

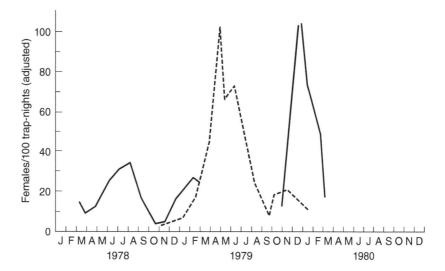

Fig. 5.6 Density indices for old female house mice on contour banks and in stubble fields of rice crops in southeastern Australia. Broken lines distinguish the crop cycle cohort of 1978–1979 from those of 1977–1978 and 1979–1980. The extent of the peak in January 1980 is unknown due to a poisoning campaign. (After Redhead 1982.)

fluctuates, it remains within certain limits (190–1150). Other populations, such as those of insects and house mice (*Mus domesticus*) in Australia (Fig. 5.6), fluctuate to a much greater extent and furnish no suggestion of an equilibrium population size. Nevertheless, such populations do not always go extinct and remain in the community for long periods. Occasionally one finds unusual situations where populations show regular cycles. The snowshoe hare (*Lepus americanus*) in northern Canada has the clearest (Fig. 5.7), as indicated by the furs collected by trappers for the Hudson Bay Company over the past two centuries (MacLulich 1937).

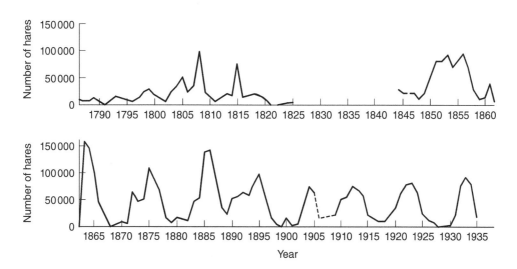

Fig. 5.7 Snowshoe hares in the boreal forest of Canada show regular fluctuations in numbers with a 10-year periodicity. Data are from the Hudson Bay Company fur records up to 1903 and questionnaires thereafter. (After MacLulich 1937.)

This relative constancy of population size, or at least fluctuation within limits, is in contrast to the intrinsic ability of populations to increase rapidly. The fact that population increase is limited suggests that there is a mechanism that slows down the rate of increase and so regulates the population.

5.5 The theory of population limitation and regulation

5.5.1 Density dependence

Populations have inputs of births and immigrants and outputs of deaths and emigrants. For simplicity, we will confine discussion to a self-contained population having only births B and deaths D per unit time.

If either the proportion of the population dying increases or the proportion being born decreases as population density increases then we define these changes as being *density-dependent*. The underlying causes of the changes in these rates are called *density-dependent factors*.

Births and deaths as a proportion of the population $(B/N_t, D/N_t)$ can be related to the instantaneous birth b and death d rates in the following way. The change in population per unit time is:

$$N_{t+1} - N_t = B - D$$

The instantaneous rate of increase r is given by:

$$r = b - d$$

The finite rate of increase R is given by:

$$\lambda = N_{t+1}/N_t = e^r$$

Therefore:

$$e^{b-d} = (N_{t+1}/N_t) = (B - D + N_t)/N_t$$

If $d = 0$, $D = 0$ then:

$$e^b = (B + N_t)/N_t = (1 + (B/N_t))$$

and:

$$b = \ln(1 + (B/N_t))$$

Similarl,y if $b = 0$, $B = 0$, and D/N_t is much less than 1 then:

$$d = \ln(1 + (D/N_t))$$

If B and D fall in the range of 0–20% of the population then b and d are nearly linear on N. They remain approximately linear even if B and D are 20–40% of N. This range covers most of the examples we see in nature, so for our purposes we can say that D/N_t and B/N_t change with density in the same way as do b and d, and both go through the origin.

Fig. 5.8 Model of density-dependent and density-independent processes. (a) Birth rate, b, is held constant over all densities while mortality, d, is density-dependent. The population returns to the equilibrium point, K, if disturbed. The instantaneous rate of increase, r, is the difference between b and d. (b) As in (a) but b is density-dependent and d is density-independent. (c) Both b and d are density-dependent. (d) d is curvilinear so that the density dependence is stronger at higher population densities.

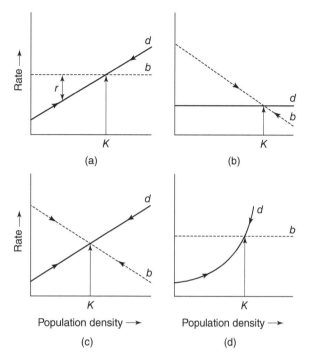

In Fig. 5.8a we plot b against density (or population size) N as a constant so that it is a horizontal line. If we now plot d as an increasing function of density, we see that where the two lines cross, $b = d$ and the population is stationary at the equilibrium point K. The difference between the b and d lines represents r, which declines linearly as density increases, in the same way as it does for the logistic curve (see Section 5.6). In Fig. 5.8a, the decline in r is due solely to d being density-dependent. Since b (or B/N_t) is constant in this case, we describe it as *density-independent* (i.e. it is unrelated to density). In real populations, density-independent factors such as weather can affect birth and death rates randomly. Rainfall acted in this way on greater kudu (*Tragelaphus strepsiceros*) in Kruger National Park in South Africa, causing mortality of juveniles (Owen-Smith 1990).

We can apply the same arguments if we assume that b is density-dependent and d is density-independent (Fig. 5.8b) or if both are density-dependent (Fig. 5.8c). So far we have assumed that the density-dependent factor has a linear effect on rate of increase, as in the logistic curve. However, density-dependent mortality is more likely to be curvilinear, as in Fig. 5.8d.

5.5.2 *Limitation and limiting factors*

In Fig. 5.9 we take the argument a little further. Let us assume a constant (density-independent) birth rate b. Shortly after birth, a density-independent mortality d_1 (depicted here as a constant) kills some of the babies so that inputs are reduced to b_1. There follows a density-dependent mortality d_2, and the population reaches an equilibrium at K_3. If mortality d_1 had not occurred (or was smaller), the equilibrium population would be at K_1. Therefore, the presence or absence of the density-independent factor causing d_1 alters the size of the equilibrium population.

The strength or severity of the density-dependent factor is indicated by the slope of d_2. If the density-dependent factor becomes stronger, for example to produce d_3 instead

Fig. 5.9 Model showing that the equilibrium point, K, can vary with both density-dependent and density-independent processes. Birth rate, b, is held constant over all densities. In sequence, a density-independent mortality d_1 reduces the input to the population to b_1. There follows a density-dependent mortality d_2 or d_3. The intercept of b or b_1 with d_2 or d_3 determines the equilibrium (K_1-K_4).

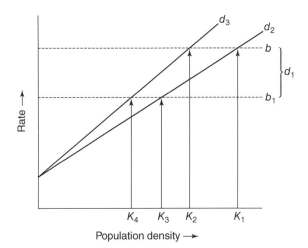

of d_2, the slope becomes steeper and the equilibrium population drops from K_3 to K_4 (or from K_1 to K_2 if d_1 is absent). Thus, altering the strengths of density-dependent factors also alters the size of the equilibrium population.

We define the process determining the size of the equilibrium population as *limitation* and the factors producing this as *limiting factors*. We can see therefore that both density-dependent and density-independent factors affect the equilibrium population size; they are thus both limiting factors. Any factor that causes mortality or affects birth rates is a limiting factor.

5.5.3 *Regulation*

Populations are often disturbed from their equilibrium K by temporary changes in limiting factors (a severe winter or drought or an influx of predators might reduce the population; a mild winter or good rains might increase it). The subsequent tendency to return to K is largely due to the effect of density-dependent factors, and this process is called *regulation*. Therefore, *regulation is the process whereby a density-dependent factor tends to return a population to its equilibrium*. We say "tends to return" because the population may be continually disturbed, so that it rarely reaches equilibrium. Nevertheless, this tendency to return to equilibrium results in the population remaining within a certain range of sizes. Superficially, it appears as if the population has a boundary to its size and fluctuates randomly within this boundary. However, it is more constructive to picture random fluctuations in both the density-independent (d_1) and density-dependent (d_2) mortalities as the shaded range in Fig. 5.10a. This results in a fluctuation of the equilibrium population indicated by the range of K. Fig. 5.10a shows that this range of K is relatively small when the density-dependent mortality is strong (steep part of the curve); Fig. 5.10b shows the much greater range of K (which we see in nature as fluctuations in numbers) when the density-dependent mortality is weak. Note in Fig. 5.10a and b that differences in the amplitudes of fluctuations are due to changes in the strength of the density-dependent mortality, as we have held density-independent (random) mortality constant in this case.

5.5.4 *Delayed and inverse density dependence*

Some mortality factors do not respond immediately to a change in density but act after a delay. Such *delayed density-dependent factors* might be predators whose populations lag behind those of their prey or a drop in food supply causing the delayed action of starvation. Both causes can have a density-dependent effect on the population, but

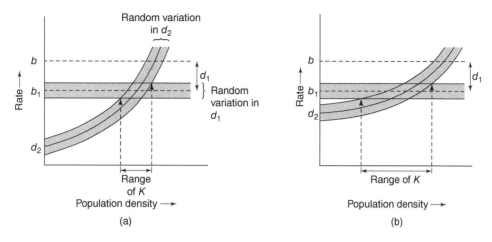

Fig. 5.10 Random variation in the mortalities d_1 and d_2 (indicated by the shaded area) are the same in (a) and (b). In (a) there is stronger density dependence at the intercept of b_1 and d_2 than in (b), and this difference results in a smaller range of equilibria, K, in (a) than in (b).

this effect is related to density at some previous time period rather than the current one. For example, a 34-year study of white-tailed deer in Canada indicated that both the population rate of change and the rate of growth of juvenile animals were dependent on population size several years previously, rather than to current population size (Fryxell *et al.* 1991). A similar relationship was found with winter mortality of red grouse (*Lagopus lagopus*) in Scotland. Delayed density dependence is indicated when mortality is plotted against current density, and the points show an anticlockwise spiral if they are joined in temporal sequence. These delayed mortalities usually cause fluctuations in population size.

Predators can also have the opposite effect to density dependence, termed an *inverse density-dependent* or *depensatory* effect. They take a decreasing proportion of the prey population as it increases, allowing the population to increase faster as it becomes larger. Conversely, if a prey population is declining for some reason, predators will take an increasing proportion and so drive it down even faster towards extinction. In either case we do not see a predator–prey equilibrium. We explore this further in Chapter 9.

5.5.5 Carrying capacity

Carrying capacity is one of the most common phrases in wildlife management. It conveys a variety of meanings, however, and unless we are careful and define it precisely, it can be a source of confusion (Caughley 1976, 1981). Some of its more common uses are discussed in this section.

Ecological carrying capacity
This can be thought of abstractly as the K of the logistic equation, which we derive later in this chapter (Section 5.8). In reality it is the natural limit of a population set by resources in a particular environment. It is one of the equilibrium points that a population tends towards through density-dependent effects from lack of food, space (e.g. territoriality), cover, or other resources. As we discussed earlier, if the environment changes briefly, it deflects the population from achieving its equilibrium and so produces random fluctuations about that equilibrium. A long-term environmental change

can affect resources, which in turn alters K. Again, the population changes by following or tracking the environmental trend.

There are other possible equilibria that a population might experience through regulation by predators, parasites, or disease. Superficially, they appear similar to the equilibrium produced by a lack of resources, because if the population is disturbed through culling or weather events it may return to the same population size. To distinguish the equilibria produced by predation, by resource limitation, and by a combination of the two, we need to know whether predators or resources or both are affecting b and d.

Economic carrying capacity

This is the population level that produces the maximum offtake (or maximum sustained yield) for culling or cropping purposes. It is this meaning that is implied when animal production scientists and range managers refer to "livestock carrying capacity." We should note that this population level is well below the ecological carrying capacity. For a population growing logistically, its level is half of K (Caughley 1976).

Other senses

We can define carrying capacity according to our particular land use requirements. At one extreme, we can rate the carrying capacity for lions on a Kenya farm or wolves on a Wyoming ranch as zero (i.e. farmers cannot tolerate large predators killing their livestock). A less extreme example is seen where the aesthetic requirements of tourism entail reducing the impact of animals on the vegetation. Large umbrella-shaped *Acacia tortilis* trees make a picturesque backdrop to the tourist hotels in the Serengeti National Park, Tanzania. In the early 1970s, elephants began to knock these trees over. Whereas elephants could be tolerated at ecological carrying capacity in the rest of the park, in the immediate vicinity of the hotels the "carrying capacity" for elephants was much lower and was determined by human requirements for scenery.

5.5.6 *Measurements of birth and death rates*

Birth rates are inputs to the population. Ideally we would like to measure conception rates (*fecundity*), pregnancy rates in mammals (*fertility*), and births or egg production. In some cases it is possible to take these measurements, as in the Soay sheep of Hirta (Clutton-Brock *et al.* 1991). Pregnancies can be monitored in a variety of ways, including ultrasound, x-rays, blood protein levels, urine hormone levels, and rectal palpation of the uterus (in large ungulates). In many cases, however, these are not practical for large samples from wild populations.

Births can be measured reasonably accurately for seal species, in which the babies remain on the breeding grounds throughout the birth season. Egg production, egg hatching success, and fledgling success can also be measured accurately in many bird populations. However, in the majority of mammal species birth rates cannot be measured accurately, either because newborn animals are rarely seen (as in many rodents, rabbits, and carnivores) or because many die shortly after birth and are not recorded in censuses (as in most ungulates). In these cases we are obliged to use an approximation to the real birth rate, such as the proportion of the population consisting of juveniles first entering live traps for rodents and rabbits, or juveniles entering their first winter for carnivores and ungulates. These are valid measures of recruitment.

Death rates are losses to the population. Ideally they should be measured at different stages of the life cycle to produce a life table (see Chapter 13). Once sexual maturity is reached, age classes often cannot be identified and all subsequent mortality is therefore lumped together as "adult" mortality. Mortality can be measured directly by using mortality radios, which indicate when an animal has died, as Boutin *et al.* (1986) and Trostel *et al.* (1987) did for snowshoe hares in northern Canada. Survivorship can be calculated over varying time periods by the method of Pollock *et al.* (1989).

Mortality caused by predators can also be measured directly if the number of predators (numerical response) and the amount eaten per predator (functional response) are known (see also Chapters 7 and 9). Such measurements are possible for those birds of prey that regurgitate a single pellet containing the bones of their prey each day. With appropriate sampling, the number of pellets indicates the number of predators, while prey per pellet shows the amount they eat. This method was used for raptors (in particular the black-shouldered kite, *Elanus notatus*) eating house mice during mouse outbreaks in Australia (Sinclair *et al.* 1990).

5.5.7 *Implications*

We should be aware of a number of problems associated with the subject of population limitation and regulation:

1 Much of the literature uses the terms "limitation" and "regulation" in different ways. In many cases they are used synonymously, but the meanings differ between authors. Since any factor, whether density-dependent or density-independent, can determine the equilibrium point for a population, any factor affecting b or d is a limiting factor. It is therefore a trivial question to ask whether a certain cause of mortality limits a population: it has to. The more profound question is how mortality and fecundity factors affect the equilibrium.

2 Regulation requires, by our definition, the action of density-dependent factors. Density dependence is necessary for regulation, but it may not be sufficient. First, the particular density-dependent factor that we have measured, such as predation, may be too weak, and other regulating factors may be operating. Second, some density-dependent factors have too strong an effect and thus cause fluctuations, rather than a tendency towards equilibrium (see Section 5.9).

3 The demonstration of density dependence at some stage in the life cycle does not indicate the cause of the regulation. For example, if we find that a deer population is regulated through density-dependent juvenile mortality, this does not give us any indication as to the cause of the mortality. Correlation with population size is merely a convenient abbreviation that hides underlying causes. Density itself is not causing the regulation; the possible underlying factors are competition for resources, competition for space through territoriality, and an effect of predators, parasites, and diseases (see Section 5.10).

5.6 **Evidence for regulation**

There are three ways of detecting whether populations are regulated. First, as we have seen in Section 5.5.3, regulation causes a population to return to its equilibrium after a perturbation. Perturbation experiments should therefore detect this return. Similarly, natural variation in population density, provided it is of sufficient magnitude, can be used to test whether per capita growth rates decline with density (see Chapters 15, 16, and 18). Second, if we plot separate and independent populations at their natural carrying capacity against some index of resource (often a weather factor), there should be a relationship. Third, we can try to detect density dependence in the life cycle.

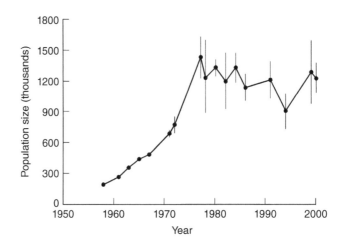

Fig. 5.11 The wildebeest population in the Serengeti increased to a new level determined by intraspecific competition for food after the disease rinderpest was removed in 1963. (After Mduma *et al.* 1999 and unpublished data.)

5.6.1 *Perturbation experiments*

If a population is moved experimentally to either below or above its original density and then returns to this same level, we can conclude that regulation is occurring. An example of downward perturbation is provided by the northern elk herd of Yellowstone National Park (Houston 1982). Prior to 1930, population estimates ranged between 15 000 and 25 000. Between 1933 and 1968, culling reduced the population to 4000 animals. Culling then ceased and the population rebounded to around 20 000 (Coughenour and Singer 1996). This result is consistent with regulation through intraspecific competition for winter food (Houston 1982), since there were no natural predators of elk in Yellowstone until the return of wolves in the early 1990s.

Density is usually recorded as numbers per unit area. If space is the limiting resource (as it might be in territorial animals), or if it is a good indicator of some other resource such as food supply, numbers per unit area will suffice in an investigation of regulation. However, space may not be a suitable measure if density-independent environment effects (e.g. temperature, rainfall) cause fluctuations in food supply. It may be better to record density as animals per unit of available food or per unit of some other resource.

The Serengeti migratory wildebeest experienced a perturbation (Fig. 5.11) when an exotic virus, rinderpest, was removed. The population increased fivefold from 250 000 in 1963 to 1.3 million in 1977 and then leveled out (Mduma *et al.* 1999). This example is less persuasive than that of the Yellowstone elk because the pre-rinderpest density (before 1890) was unknown, but evidence on reproduction and body condition suggests that rinderpest held the population below the level allowed by food supply, a necessary condition for a perturbation experiment implicating a disease.

A case of a population perturbed above equilibrium is provided by elephants in Tsavo National Park, Kenya (Laws 1969; Corfield 1973). From 1949 until 1970, the population was increasing, due in part to immigration from surrounding areas in which human cultivation had displaced the animals. A consequence of this artificial increase in density was depletion of the food supply within reach of water. In 1971, the food supply ran out and there was starvation of females and young around the water holes. After this readjustment of density, the vegetation regenerated and starvation mortality ceased.

5.6.2 *Mean density and environmental factors*

A population that is uninfluenced by dispersal and unregulated (i.e. has no density-dependent factors affecting it) will fluctuate randomly under the influence of weather and will eventually drift to extinction (DeAngelis and Waterhouse 1987).

Just by chance, there may for a time be a correlation between density and environmental factors. However, if we take many separate populations, the probability that all of them are simultaneously correlated with an environmental factor by chance alone is very small. Therefore, if we find a correlation between mean densities from independent populations and environmental factors, a strong inference can be made that weather is influencing some resource for which animals are competing, resulting in regulation about some equilibrium point.

An example of this approach is shown in Schluter's (1988) study of seed-eating finches in Kenya (Fig. 5.12): finch abundance from various populations is correlated with seed abundance. Other examples of density correlated with weather factors are given in Sinclair (1989).

5.6.3 *Examples of density dependence*

As discussed in Section 5.5.7, density dependence is a necessary but not sufficient requirement for demonstrating regulation. There are an increasing number of studies in the bird and mammal literature demonstrating density-dependent stages in the life cycle. For birds (Fig. 5.13a), the long-term study of great tits (*Parus major*) in Oxford, UK has shown that winter mortality of juveniles is related to the number of juveniles entering the winter (McCleery and Perrins 1985). In contrast, (Fig. 5.13b) it is early chick mortality in summer that is density-dependent for the English partridge (*Perdix perdix*) (Blank *et al.* 1967).

For mammals, density-dependent juvenile mortality has been recorded for red deer on the Isle of Rum, UK (Clutton-Brock *et al.* 1985) (Fig. 5.14a), for reindeer in Norway (Skogland 1985) (Fig. 5.14b), for feral donkeys (*Equus asinus*) in Australia (Choquenot 1991), and for greater kudu in South Africa (Owen-Smith 1990). Adult mortality was density-dependent for African buffalo in the Serengeti (Sinclair 1977). In each case, the cause was lack of food at critical times of year. Reproduction is known to be

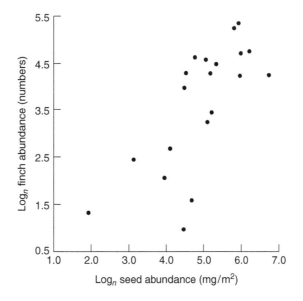

Fig. 5.12 The total abundance of seedeating finches in savanna habitats of Kenya is related to the abundance of the food supply. Such a positive relationship in unconnected populations may demonstrate regulation. (After Schluter 1988.)

Fig. 5.13 Examples of density-dependent mortality in birds. (a) Great tit (*Parus major*) overwinter mortality (log of [juveniles in winter/first year breeding population]) plotted against log juvenile density in winter. (After McCleery and Perrins 1985.) (b) Chick mortality of European partridge (*Perdix perdix*) (measured as log hatching population/log population at 6 weeks) plotted against log hatching population, in Hampshire, UK. (After Blank *et al.* 1967.)

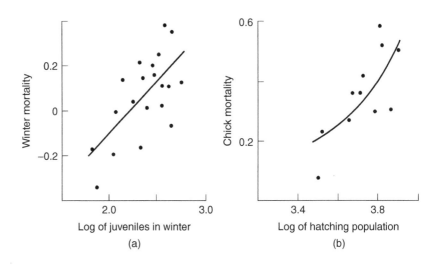

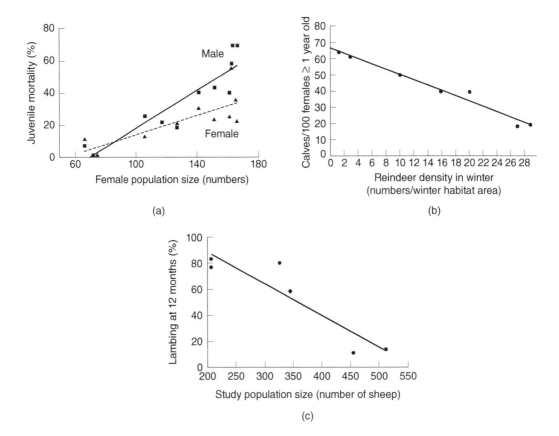

Fig. 5.14 Density dependence in large mammals. (a) Juvenile mortality of male and female red deer on the Isle of Rum, UK. (After Clutton-Brock *et al.* 1985.) (b) Juvenile recruitment per 100 female reindeer older than 1 year in Norway. (After Skogland 1985.) (c) Fertility rate of 1-year-old Soay sheep on St. Kilda island. (After Clutton-Brock *et al.* 1991.)

Fig. 5.15 The proportion of a red grouse population in Scotland which disappears over winter (August–April) is related to population density in the previous August in a complex way. Mortality varied according to whether the population was increasing or decreasing. By joining the points sequentially an anticlockwise cycle is produced, indicating a delayed density-dependent effect in the cause of the mortality. By plotting the percentage disappearance against density 1 year earlier, a closer fit can be obtained for a regression line. Thus the delay is 1 year. Numbers at the points are years. (After Watson and Moss 1971.)

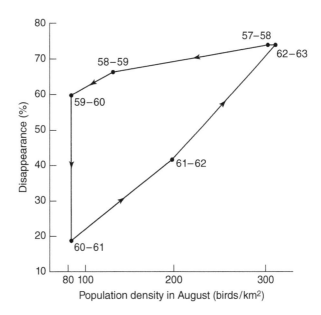

density-dependent in both birds (Arcese *et al.* 1992) and mammals (Clutton-Brock *et al.* 1991). Fig. 5.14c shows that the proportion of Soay sheep that give birth at 12 months of age declines with density. Fowler (1987) reports over 100 studies of terrestrial and marine mammal populations in which density dependence was detected.

Delayed density dependence has been recorded in winter mortality of snowshoe hares in the Yukon and in overwinter mortality of red grouse in Scotland (Watson and Moss 1971) (Fig. 5.15). For the hares, the delay appears to have been due to a lag of 1–2 years in the response of predator populations to changing hare numbers (Trostel *et al.* 1987), while for the grouse the delay came from the density response to food conditions in the previous year (see Section 5.10.3).

5.7 Applications of regulation

Causes of population change can be divided into (i) those that disrupt the population and often result in "outbreaks," which can be either density-dependent or density-independent, and (ii) those that regulate and therefore return the population to original density after a disturbance. These are always density-dependent.

Knowledge of regulation may be useful for management of house mice (*Mus domesticus*) plagues in Australia. In one experimental study (Barker *et al.* 1991), mice in open-air enclosures were contained by special mouse-proof fences. The objective was to create high densities, mimicking plague populations, in order to test the regulatory effect of a nematode parasite (*Capillaria hepatica*). It turned out that the effect of the parasite could not be tested because other factors regulated the population and thus obscured it. The replicated populations declined simultaneously. Why did this happen? By dividing up the life cycle into stages, the researchers found that late juvenile and adult mortality were strongly density-dependent but that other stages, including fertility and newborn mortality, were not. This allowed them to discount causes that would affect reproduction and focus more closely on what was happening among adults, particularly in terms of social interactions.

Other studies suggest that mouse populations in Australia may be regulated by predators, disease, and juvenile dispersal (Redhead 1982; Sinclair *et al.* 1990). Under conditions of superabundant food following good rains, the reproductive rate of females increased faster than the predation rate and an outbreak of mice occurred. The implication of these results for management is that if reproduction can be reduced, for example through infections of the *Capillaria* parasite, then predation may be able to prevent outbreaks even in the presence of abundant food for the mice.

5.8 Logistic model of population regulation

At the beginning of this chapter, we derived the geometric and exponential models of population growth. In 1838, Pierre-Francois Verhulst published a paper that challenged the assumption of unlimited growth implicit in these models. Verhulst argued that the per capita rate of change (dN/Ndt) should decline proportionately with population density, simply due to a finite supply of resources being shared equally among individuals. If each individual in the population gets a smaller slice of the energy "pie" as N increases, this will prevent them from devoting as much energy to growth, reproduction, and survival as would be possible under ideal conditions. Changes in demographic parameters lead to corresponding changes in the finite rate of population growth λ_t or its equivalent exponential rate r_t, where t denotes a specific point in time. Other factors, such as risk of disease, shortage of denning sites, and aggressive interactions among population members may also cause the rate of population growth to decline with population size. The simplest mathematical depictions of such phenomena are commonly termed "logistic" models.

There are numerous ways to represent logistic growth. For simplicity, we will focus on population growth modeled in discrete time, which is often a reasonable approximation for species that live in a seasonal environment. One of the most commonly used forms is called the Ricker equation, in honor of the Canadian fisheries biologist, Bill Ricker, who first suggested its application to salmon stocks (Ricker 1954):

$$N_{t+1} = N_t e^{r_{max}\left(1 - \frac{N_t}{K}\right)}$$

The Ricker logistic equation represents the exponential rate of increase under ideal conditions as r_{max}, with a proportionately slower rate of increase with each additional individual added to the population. When the rate of increase has slowed to the point that births equal deaths, the population has reached its carrying capacity K. These two population parameters (r_{max} and K) dictate how fast the population recovers from any perturbation to abundance.

A population growing according to the logistic equation will have slow growth when N is small, will grow most rapidly when N is of intermediate abundance, and will grow slowly again as N approaches carrying capacity K (Fig. 5.16). This kind of sigmoid or S-shaped pattern is often termed *logistic* growth.

At first, it may seem somewhat counterintuitive that a proportional decline in per capita demographic rates should produce the nonlinear growth pattern seen in Fig. 5.16. The reason lies in the fact that population changes are dependent on both population size and per capita growth rate, in much the same way that the growth of a bank account depends on both the money already in the account and the interest rate. When a population is small, the per capita rate of change will tend to be large, in fact close to r_{max}, either because birth rates are high or because mortality rates are low. Nonetheless, the population will still display a slight change from one year to the

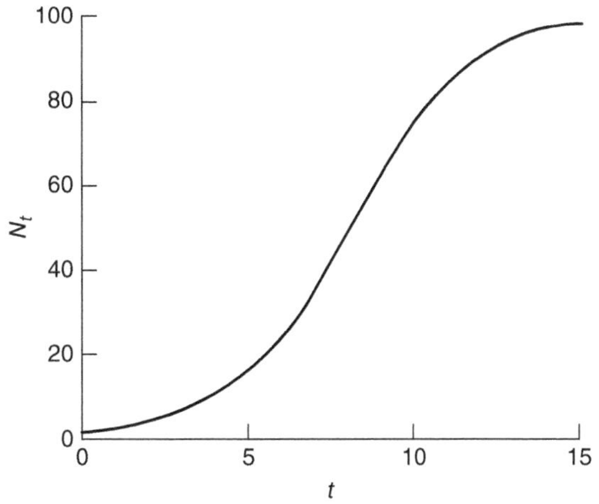

Fig. 5.16 Population growth according to the logistic equation, with $r_{max} = 0.5$, initial population density $N_0 = 1.5$, and carrying capacity $K = 100$.

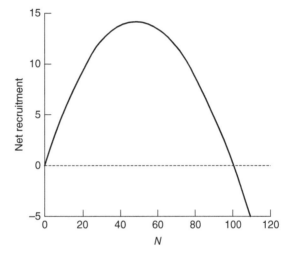

Fig. 5.17 Net recruitment $(N_{t+1} - N_t)$ as a function of population density N_t, according to the Ricker logistic growth model, with $r_{max} = 0.5$ and $K = 100$.

next. At the other end of the spectrum, even when N is enormous the population will similarly display only modest change from year to year. This is because the per capita rate of growth is small, due either to low birth rates or high mortality rates. It is only when the population is of intermediate size and growing at intermediate per capita rate that growth is maximized (Fig. 5.17).

Population data displaying the classic sigmoid pattern of change are rare. They will only be seen when a population is reduced to very low initial density and then monitored closely over an extended period. So, logistic growth will not be obvious in most populations that we might see around us in nature, which are presumably close to their carrying capacity. In some cases, however, populations have been perturbed (reduced) to low densities, giving us a rare glimpse of logistic growth in the field. For example, as we discussed earlier, the Yellowstone elk herd has been aggressively culled at various times in the past, particularly in the late 1960s. Cessation of culling operations, stimulated by a new policy of natural regulation in US National Parks, led to a subsequent pattern of elk recovery reminiscent of the sigmoid pattern predicted by the logistic model (Fig. 5.18). Similarly, release of the Serengeti wildebeest population from the

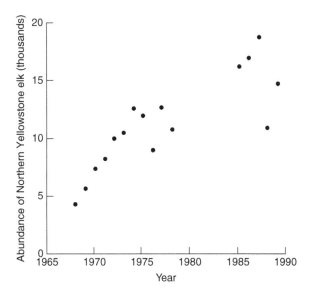

Fig. 5.18 Population dynamics of northern Yellowstone elk between 1968 and 1989. (Data from Coughenour and Singer 1996.)

exotic disease rinderpest led to a subsequent sigmoid pattern of change (Fig. 5.11) reminiscent of the logistic model. Indeed, perturbation is an important ingredient in detecting natural regulation and logistic growth because it gives us evidence to work with, unlike populations kept close to their ecological carrying capacity. We demonstrate how to estimate the parameters for the Ricker logistic model, and compare it to other possible population growth models, in Chapter 15.

5.9 Stability, cycles, and chaos

Paradoxically, the same density-dependent processes that are responsible for natural regulation can also induce population fluctuations, at least under special circumstances. One way that this can happen is when the maximum rate of growth is particularly high. For example, consider the dynamics of a hypothetical population whose maximum rate of increase $r_{max} = 3.3$ and carrying capacity $K = 100$ (Fig. 5.19). In this case the population does not increase smoothly over time and level off at the carrying capacity but rather fluctuates erratically over time, with no apparent repeated pattern. Such a pattern of population change is known as *deterministic chaos* (May 1976). It arises because the population grows so fast that it tends to overshoot the carrying capacity, a process known as *overcompensation* (May and Oster 1976). Once above the carrying capacity, the net recruitment is negative (Fig. 5.19), so the population declines rapidly. Repetition of this boom–bust pattern of overshooting and subsequent decline results in the erratic fluctuations of deterministic chaos seen in Fig. 5.19. For lower rates of increase ($2.0 < r_{max} < 2.7$), the pattern of fluctuation will be regular cycles rather than deterministic chaos, but the underlying cause is still overcompensation.

The underlying cause of instability due to overcompensatory density dependence can be better appreciated by plotting the population dynamics over time on a graph with N_t on the horizontal axis and N_{t+1} on the vertical (Fig. 5.20). The diagonal identifies potential points of equilibria, at which $N_{t+1} = N_t$. We will also plot the recruitment curve. Dynamics are plotted by starting at a particular value of N_0, projecting upwards to the recruitment curve (which identifies the next year's population density), and

Fig. 5.19 Simulated dynamics over time of two different populations growing according to the Ricker logistic equation, with $r_{max} = 3.3$ and $K = 100$. The first population was initiated at a density of 2.0 individuals per unit area, whereas the second population was initiated at a slightly higher density of 2.1 individuals per unit area. The rapid divergence in population dynamics due to slight changes in starting conditions is typical of deterministic chaos.

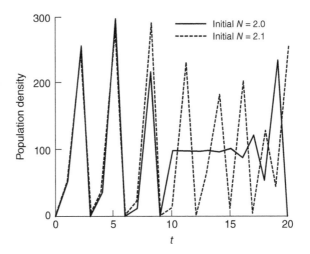

Fig. 5.20 Plot of predicted recruitment (N_t+1) relative to N_t (the heavy curve), equilibrium line at which $N_t+1 = N_t$ (thin broken line), and trajectory of population dynamics over time for a simulated population following the Ricker logistic model, with $r_{max} = 1.3$ and $K = 100$ (thin solid line).

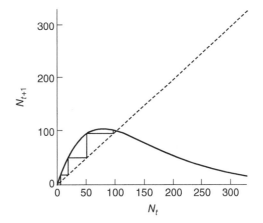

then projecting horizontally to the dotted equilibrium line, before repeating the process. At modest values of r_{max}, the recruitment curve is low and has a shallow angle of incidence as it intersects the equilibrium line. The result is that the population trajectory becomes pinched between the recruitment curve and the equilibrium line as it converges on K. This leads to stability.

Now, let us consider the pattern arising when $r_{max} = 3.3$ (Fig. 5.21). The recruitment curve has a pronounced hump and intersects the equilibrium line at a sharp angle (> 90°). The recruitment curve is so sharply peaked that recruitment events tend to overshoot the carrying capacity. This leads to the population collapsing to well below the carrying capacity, where the boom–bust cycle begins anew. In this way, the population never reaches an equilibrium, despite the fact that there is strong density dependence. This example demonstrates overcompensation, and it occurs when the angle of incidence of the recruitment curve exceeds 90° as it approaches the equilibrium line (May 1976; May and Oster 1976).

A diagnostic feature of deterministic chaos is that slight changes in starting conditions lead to quite different population dynamics over time. In Fig. 5.21, the simulated dynamics of the two hypothetical populations, begun at slightly different densities,

Fig. 5.21 Plot of predicted recruitment (N_t+1) relative to N_t (the heavy curve), equilibrium line at which $N_t+1 = N_t$ (thin broken line), and trajectory of population dynamics over time for a simulated population following the Ricker logistic model, with rmax = 3.3 and $K = 100$ (thin solid line).

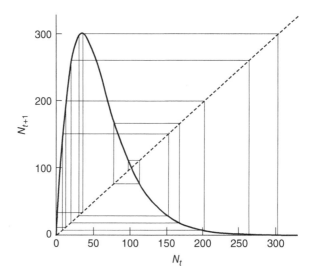

become quite different later on, illustrating their sensitivity to initial conditions. The two populations go through similar changes in the first few years but rapidly diverge thereafter, displaying different patterns of fluctuation.

We have thus far limited our discussion to the simplest pattern of density dependence: linear changes in per capita rates of reproduction or survival. We saw earlier (Figs 5.8 and 5.10) that there is no reason to expect natural regulation to be linearly density-dependent. Some wildlife biologists have even argued that this may be the exception rather than the rule (Fowler 1981); adult mortality in Serengeti wildebeest is a good example (Mduma *et al.* 1999).

5.10 Intraspecific competition

Regulation can occur through a number of mechanisms, such as predation or parasitism, but the most common cause is competition between individuals for resources. Such resources might be food, shelter from weather or from predators, nesting sites, or space to set up territories. We have seen some examples already in Figs 5.13 and 5.14.

5.10.1 *Definition*

Intraspecific competition occurs when individuals of the same species utilize common resources that are in short supply; or, if the resources are not in short supply, when the organisms seeking the resource nevertheless harm one another in the process (Birch 1957).

5.10.2 *Types of competition*

When individuals use a resource so that less of it is available to others, we call this type of competition *exploitation*. This includes both removal of a resource (consumptive use), as when food is consumed, and occupation of a resource (pre-emptive use), as when resources such as nesting sites are used. Individuals competing for food need not be present at the same time: one ungulate can reduce the food supply of another that arrives later.

Another type of competition involves the direct interaction of individuals through various types of behavior. This is called *interference* competition. One example of behavioral interference is the exclusion of some individuals from territories; another is the displacement of subordinate individuals by dominants in a behavioral hierarchy.

5.10.3 *Intraspecific competition for food*

Experimental alteration of food supply

Food addition experiments provide the best evidence for intraspecific competition. Krebs *et al.* (1986) supplied extra food to snowshoe hares in winter from 1977 to 1985. This raised the mean winter density fourfold at the peak of the 10-year population cycle. Similarly, Taitt and Krebs (1981) increased the density of vole populations (Fig. 5.22) by giving them extra food. The elk population at Jackson Hole, Wyoming, is kept at a higher level than would otherwise be the case by supplementary feeding in winter (Boyce 1989). These examples show that food is one of the factors limiting density.

The dense shrubland (chaparral) of northern California contains two shrubs, chamise (*Adenostema taxiculatum*) and oak (*Quercus wislizenii*), that are preferred food for black-tailed deer (*Odocoileus hemionus*). These shrubs resprout from root stocks after burning to provide the new shoots that the deer consume. Taber (1956) showed that on plots thinned by experimental burning, herbaceous food supply increased to 78 kg/ha from the 4.5 kg/ha found on control plots, while the shrub component increased from 165 to 460 kg/ha. Deer densities consequently increased from 9.5 km^{-2} on the experimental controls to 22.9 km^{-2} on the treatment plots, while fertility increased from 0.77 to 1.65 young per adult female.

Red grouse (*Lagopus lagopus*) live year round on heather (*Calluna vulgaris*) moors in Scotland. Their diet consists almost entirely of heather shoots. Watson and Moss (1971) describe experiments in which some areas were cleared of grouse, fertilized with nitrogen in early summer, and then left to be recolonized. Fertilizing increased the growth and nutrient content of the heather. The size of their territories did not differ between fertilized and control areas when the grouse set them up in autumn. However, territorial grouse that had been present all winter reared larger broods on the fertilized than on the control areas, indicating that reproduction was affected by overwinter nutrition. Territory sizes did decline in the following autumn, and densities increased, showing the 1-year lag of density response to nutrition. On other areas, old heather was burned every 3 years, creating a higher food supply of young regenerating heather. Territory size on these plots decreased (as density increased) in the same year as the treatment, so there was a more immediate response than on the fertilized plots.

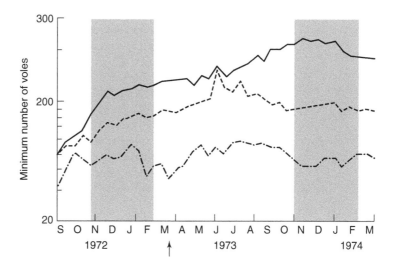

Fig. 5.22 The numbers of Townsend's voles on trapping grids increase in proportion to the amount of food that is provided, indicating that intraspecific competition regulates the population. Dashed-dotted line: control; dashed line: low food addition; solid line: high food addition; shaded area: winter. (After Taitt and Krebs 1981.)

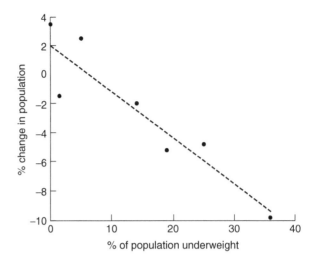

Fig. 5.23 The percentage change in a wood pigeon population in England is related to the proportion of the population that is underweight. (Data from Murton *et al.* 1966.)

Direct measures of food

Snowshoe hare populations in the boreal forests of Canada and Alaska reach high numbers every 10 years or so. Measurement of known food plants and feeding experiments suggest that the animals run short of food at peak numbers (Pease *et al.* 1979). Other measures, such as the amount of body fat (Keith *et al.* 1984) and fecal protein levels (Sinclair *et al.* 1988), also identify food shortage at this time (see Section 2.9).

African buffalo graze the tropical montane meadows of Mount Meru in northern Tanzania, keeping the grass short. Grass growth rates and grazing offtake were measured by use of temporary exclosure plots. Growth in the rainy season was more than sufficient for the animals, but in the dry season available food fell below maintenance requirements (Sinclair 1977).

Murton *et al.* (1966) measured the impact of wood pigeons (*Columba palumbus*) on their clover (*Trifolium repens*) food supply. Food supply was measured directly by counting clover leaves in plots. Pigeons consumed over 50% of the food supply during winter. They fed in flocks, those at the front obtaining more food than those in the middle or at the back. The proportion of underweight birds (under 450 g) was related directly to the overwinter change in numbers (Fig. 5.23) and inversely related to the mid-winter food. Thus, competition within flocks resulted in some animals starving, and the change in numbers was related to the proportion that starved.

Indirect measures of food shortage

Indirect evidence for competition for food comes from indices of body condition (see Section 2.9). The last stores of body fat that are used by ungulates during food shortages are in the marrow of long bones such as the femur. Bone marrow fat can be measured directly by extraction with solvents. However, since there is an almost linear relationship between fat content and dry weight (Hanks 1981) (see Section 2.9.3), it is easier to collect a sample of marrow from carcasses found in the field and oven dry it. A cruder but still effective method, introduced by Cheatum (1949), is to describe the color and consistency of the marrow.

Other fat stores, such as those around the heart, mesentery, and kidney, are used up before the bone marrow fat starts to decline (see Section 2.9). The relationship between kidney and marrow fat holds for many ungulate species (see Fig. 2.11). If both kidney and marrow fat can be collected, a range of body conditions can be recorded. However, often the marrow fat is all that is found in carcasses, because scavengers have eaten the internal organs.

Klein and Olson (1960) used bone marrow condition indices to conclude that deer in Alaska died from winter food shortage, as did Dasmann (1956) for deer in California. Similarly, migratory wildebeest in the Serengeti that died in the dry season were almost always in poor condition, as judged by the bone marrow, and this was correlated with the protein level in their food (see Fig. 2.12). This dry-season mortality was density-dependent and was sufficiently strong to allow the population to level out (Sinclair *et al.* 1985; Mduma *et al.* 1999).

Problems with measurement of food supply

To determine whether competition for resources such as food is the cause of regulation, we need to know what type of food is eaten, how much is needed, and how much is available. What is needed must exceed what is available in order for competition to occur. The types of food eaten form the basis for many studies on diet selection, sometimes called *food habit studies*. These in themselves do not tell us what is needed in terms of digestible dry matter, protein, and energy. We should note that such requirements are unknown for most wild species and we have to use approximations from other, often domestic, species. The amount of food available to animals is particularly difficult to assess because we are unlikely to measure potential food in the same way as does the animal. Animals are likely to be far more selective than our crude sampling, and so we are likely to record more "food" than the animal sees. Our measures of food supply are often seriously flawed, which is one of the reasons why direct evidence for intraspecific competition for food is rare. There is far more indirect evidence for competition, provided by indicators such as body condition.

5.11 Interactions of food, predators, and disease

The effect of limited food on population demography can go beyond the direct effects of undernutrition: there can also be synergistic interactions with predation and disease. Animals may alter their behavior when food becomes difficult to find in safe areas, searching increasingly in areas where they are at risk of predation in order to avoid eventual starvation (Lima and Dill 1990; McNamara and Houston 1987). This is called *predator-sensitive foraging* and has been observed in snowshoe hare feeding (Hik 1995; Hodges and Sinclair 2003). Such behavior can result in increased predation well before starvation takes effect, as seen in wildebeest (Sinclair and Arcese 1995).

Disease can also interact synergistically with food, pathological effects suddenly becoming apparent at a certain, sometimes early, stage of undernutrition (see Chapter 11). Sometimes food, disease, and predators all interact. Wood bison numbers in the Wood Buffalo National Park, Canada, switch suddenly from a high-density food-regulated state to a low-density predator-regulated one when diseases such as tuberculosis and brucellosis affect the population (Joly and Messier 2004).

5.12 Summary

Regulation is a biotic process that counteracts abiotic disturbances affecting an animal population. Two common biotic feedback processes are predation and intraspecific competition for food. These are called density-dependent factors if they act as negative feedbacks. Negative feedback imparts stability to the population. Disturbances are provided by fluctuating weather or other environmental conditions or by chance effects on reproduction and survival. These are called density-independent factors and will cause populations to drift to extinction if there are no counteracting density-dependent processes operating. For wildlife management, it is necessary to know (i) the causes of the density-dependent processes that stabilize the population and of fluctuations and instability, and (ii) which age and sex groups are most influenced by these stabilizing or destabilizing processes.

One way to understand such effects is to model density-dependent changes in population growth rate using logistic models. Application of such models shows that whereas density dependence is often stabilizing, overcompensatory density dependence can itself encourage population fluctuation, beyond the degree we would expect from demographic or environmental stochasticity. A common cause of regulation is intraspecific competition for food.

Competition occurs if the needs of the population exceed availability. To measure such competition, we need to know how much food is available and how much is needed, and whether it is density-dependent. Food can also interact with predation and disease to regulate populations.

6 Competition and facilitation between species

6.1 **Introduction**

Species do not exist alone. They live in a community of other species, some of which will interact. Interaction between two species can take various forms: competition, commensalism (facilitation), mutualism (symbiosis), predation, and parasitism are the main ones. These are defined by the way each species affects the other, as shown in Table 6.1. In *competition*, each species suffers from the presence of the other, although the interaction need not be balanced. With *commensalism*, one species benefits without affecting the other, while in *mutualism* both benefit. These can be thought of as the converse of interspecific competition. With *predation* and *parasitism*, one species benefits to the disadvantage of the other. We shall discuss predation in Chapter 7 and parasitism in Chapter 8; we confine ourselves here to interspecific competition and mutualism.

6.1.1 *Definition*

Interspecific competition is similar to *intraspecific competition*. It occurs when individuals of different species utilize common resources that are in short supply; or, if the resources are not in short supply, when the organisms seeking them nevertheless harm one another in the process (Birch 1957).

6.1.2 *Implications*

Interspecific competition deals with cases in which there are two or more species present, and we should be aware of a number of implications arising from this:

1 Competition must have some effect on the fitness of the individuals. In other words, resource shortage must affect reproduction, growth, or survival, and hence the ability of individuals to get copies of their genes into the next generation.

2 Although it is necessary for species to require common resources (i.e. overlap in their requirements), we cannot conclude there is competition unless it is also known that the resources are in short supply or that the species affect each other.

3 The amount of a resource such as food that is available to each individual must be affected by what is consumed by other individuals. Thus, two species cannot compete if they are unable to influence the amount of resource available each to the other, or to interfere with the species obtaining the resource.

4 Both exploitation and interference competition (see Section 5.10.2) can occur between species, although interference between species is relatively uncommon.

Wildlife Ecology, Conservation, and Management, Third Edition.
John M. Fryxell, Anthony R. E. Sinclair and Graeme Caughley.
© 2014 John Wiley & Sons, Ltd. Published 2014 by John Wiley & Sons, Ltd.
Companion Website: www.wiley.com/go/Fryxell/Wildlife

	Species 1		
Species 2	+	0	−
+	Mutualism	Commensalism	Predation/parasitism
0	Commensalism		(Amensalism) competition
−	Predation/parasitism	Competition	Competition

Table 6.1 Types of interaction.

6.2 Theoretical aspects of interspecific competition

6.2.1 Graphical models

To obtain an understanding of what might be the expected outcome of a simple and idealized interspecific competition, we return to the logistic equation:

$$dN_1/dt = r_{m1}N_1(1 - N_1/K_1)$$

The term in parentheses, $(1 - N_1/K_1)$, describes the impact of individuals upon others of the same species and on the population growth rate dN_1/dt. We must now add a term representing the impact of the second species N_2 on species 1. The equation for the effect of species 2 on the population growth of species 1 is:

$$dN_1/dt = r_{m1}N_1(1 - (N_1 + a_{12}N_2)/K_1)$$

where r_{m1} is the intrinsic rate of increase of species 1.

The ratio N_2/K_1 represents the abundance of species 2 relative to the carrying capacity (K_1) of species 1. It is a measure of how much of the resource used by species 2 would have been used by species 1. The *coefficient of competition* a_{12} measures the competitive effect of species 2 on species 1. If we define the competitive effect of one individual of species 1 upon the resource use of an individual of its own population as unity then the coefficient for the effect of the other species is expected to be less than unity. We expect this because individuals will compete more strongly with those similar to themselves than with the dissimilar individuals of other species. This does not always occur: when two species differ greatly in size, an individual of the larger species (l) may consume far more of a resource than one of the smaller species (s), and in this case the a_{sl} could be greater than unity. The converse effect of species 1 on species 2 is denoted by the coefficient $a_{2\,1}$ in the equation for the other species:

$$dN_2/dt = r_{m2}N_2(1 - (N_2 + a_{21}N_1)/K_2)$$

These two formulae are called the Lotka–Volterra equations, named after the two authors who produced them (Lotka 1925; Volterra 1926a). We can examine the implications of the equations graphically by plotting the numbers of species 2 against those of species 1, as in Fig. 6.1a. First we plot the conditions for species 1 when dN_1/dt is zero. There are two extreme points: one at which N_1 is at K_2 so that N_2 is zero and one at which N_1 is zero because species 2 has taken all of the resource. This latter point can be found from our second equation by setting dN_1/dt to zero and rearranging so that it simplifies to:

$$N_1 = K_1 - a_{12}N_2$$

If the resource is taken entirely by species 2 then:

$$N_1 = 0 \text{ and } N_2 = K_1/a_{12}$$

Fig. 6.1 Isoclines for the Lotka–Volterra equations. (a) At any point on the isocline $dN_1/dt = 0$. This indicates where the number of species 1 is held constant for different population sizes of species 2. Species 1 increases to the left of the isocline, but decreases right of it. (b) The isocline where $dN_2/dt = 0$. This shows where the population of species 2 is held constant at different values of species 1. Species 2 increases below the line, but decreases above it.

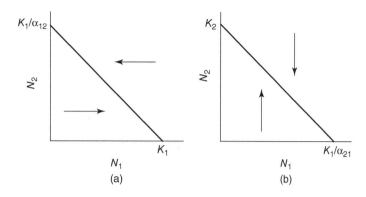

Of course, there can be any combination of N_1 and N_2 that gives $dN_1/dt = 0$, as can be seen from the diagonal line joining these two extreme points. To the left of this line, dN_1/dt is positive, so that N_1 increases, and to the right it is negative and N_1 decreases, as indicated by the arrows. At all points on the line (called an *isocline*), the population is stationary. Exactly the same reasoning produces the equivalent diagram for species 2 (Fig. 6.1b). Below the line (isocline), N_2 increases, and above it N_2 decreases.

With these two diagrams describing the competitive abilities of the two species independently, we can now predict the outcome of competition between them. If we put the two diagrams in Fig. 6.1 together, as in Fig. 6.2a, we see that K_1 is larger than K_2/a_{21}. The latter term is the number of species 1 required to drive species 2 to extinction, and since it is possible for species 1 to exist at higher numbers than this level (i.e. at

Fig. 6.2 The relationship of the two species' isoclines determines the outcome of competition. (a) Species 1 increases at all values of species 2 so that species 1 wins. (b) The converse of (a) such that species 2 wins. (c) In the region where the isocline of species 1 is outside that of species 2, species 1 wins and vice versa, so that either can win. (d) A stable equilibrium occurs because all combinations tend towards the intersection of the isoclines.

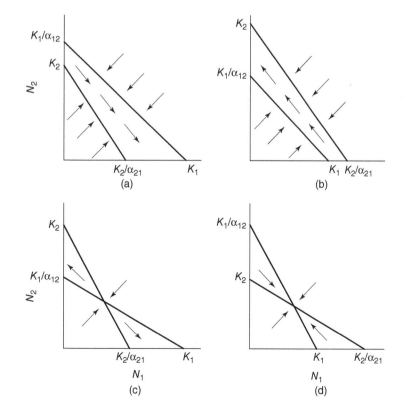

K_1), species 1 will drive species 2 down. On the other axis, we see that K_2, which is the maximum number of species 2 that the environment can hold, is less than is necessary to drive down species 1. Therefore, species 2 always loses when the two species occur together, as can be seen by the resultant arrows, and by the fact that the species 1 isocline is always outside that of species 2.

This outcome is not the only possible solution, as it depends on the relative positions of the two isoclines shown in Figs 6.2b,c,d. Fig. 6.2b is the converse of Fig. 6.2a, so that species 2 always wins. In Fig. 6.2c, we see that $K_2 > K_1/a_{12}$ and $K_1 > K_2/a_{21}$, so that, depending on the exact combination of the two population sizes, either can win. Where the two isoclines cross, there is an equilibrium point, but this is unstable in the sense that any slight change in the populations will cause the system to move to either K_1 or K_2 and the extinction of one of the species. In nature, we would never see such an equilibrium.

Fig. 6.2d also shows the two isoclines crossing, but in this case $K_2 < K_1/a_{12}$ and $K_1 < K_2/a_{21}$ (i.e. individuals of the same species affect each other more than do individuals of the other species, and neither is capable of excluding the other). This also means that intraspecific competition is always greater than interspecific competition. Hence, the arrows show that whatever the combination of the two populations, the system moves to the equilibrium point, which is therefore stable. This situation can only occur if there is some form of separation in the resources that the two species use, which we call *niche partitioning* (see Section 6.6).

6.2.2 *Implications and assumptions*

1 We can see from the figures that the outcome of competition depends upon the carrying capacities (K_1 and K_2) and the competition coefficients (a_{12} and a_{21}), according to the Lotka–Volterra model. The intrinsic rate of increase has no influence on which species will be the eventual winner.

2 Coexistence occurs when intraspecific competition within both species is greater than interspecific competition between them.

3 These equations can be expanded to include the effects of several species on species 1 by summing the $a \times N$ terms. This assumes that each species acts independently on species 1.

4 There are several other assumptions underpinning the logistic equation, such as constant environmental conditions leading to constant r and K, and there being no lags in competing species' responses to each other. Furthermore, the competition coefficients are constant: the intensity of competition does not change with size, age, or density of the competing species.

These assumptions mean that the Lotka–Volterra equations, like the logistic one, are simplistic and idealized. It is unlikely that the assumptions hold, although they may be approximated in some cases. The real value of these models is that they show how it is possible for coexistence to occur in the presence of competition, and that exclusion is not necessarily predetermined but may depend on the relative densities of the competing species.

6.3 **Experimental demonstrations of competition**

6.3.1 *Perturbation experiments*

Much of the work in ecology has assumed that competition has occurred and is necessary for the coexistence of species, and competition is one of the major assumptions in Darwin's Theory of Natural Selection. Nevertheless, it is necessary to demonstrate that interspecific competition does actually take place. One of the most direct approaches is to carry out a *removal experiment* whereby one of the species is removed, or reduced in number, and the responses of the other species are recorded. If competition has been

Fig. 6.3 Population densities of deermice (*Peromyscus*) on two areas from which voles (*Microtus*) were removed in the years 1973 and 1974. On area *F* only deermice were monitored after voles were allowed to recolonize. Deermice were absent or in low numbers before and after the vole removal, but high during the removal. (Data from Redfield *et al.* 1977.)

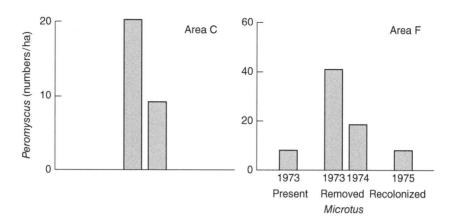

operating, we would expect that either the population, the reproductive rate, or the growth rate of the other species would increase.

Forsyth and Hickling (1998) showed from an incidental removal experiment through hunting that Himalayan tahr (*Hemitragus jemalahicus*) are associated with declining populations of chamois (*Rupicapra rupicapra*). Competition appears to occur through behavioral interference, with the larger tahr excluding chamois. Another experiment, illustrated in Fig. 6.3, examined the competitive effect of voles (*Microtus townsendii*) on deermice (*Peromyscus maniculatus*). Deermice normally live in forests, but one race on the west coast of Canada can also live in grassland, the normal habitat of voles. Redfield *et al.* (1977) removed voles from three plots and compared the population response of the deermice there with those on two control areas. On one control there were no deermice, on the other 4.7/ha. All the removal areas showed increases in deermouse numbers, one going from 7.8/ha before removal to 62.5/ha 2 years later. At the end of the study, when they stopped removing voles, these animals recolonized, reaching densities of 109/ha, while deermice numbers dropped to 9.4/ha. In another experiment, instead of removing voles, Redfield disrupted their social organization by altering the sex ratio so that there was a shortage of females, but the density remained similar to those of the controls. In this area, deermice numbers increased from nearly zero to 34/ha. This result suggests that it was *interference competition* due to aggression from female voles that excluded the deermice, because the density and food supply remained the same.

A similar type of experiment was conducted on desert rodents in Arizona by Munger and Brown (1981). They excluded larger species from experimental plots while allowing smaller species to enter. Plots were surrounded by a fence and access was controlled by holes cut to allow only the smaller species to enter. There were two types of small rodent: those that ate seeds (granivores) and those that ate a variety of other foods as well (omnivores). Munger and Brown predicted that if there was exploitation competition for seeds between the large and small granivores then the latter should increase in number in the experimental plots, while the omnivore populations should stay the same; if, however, the increased density of granivores was an artifact of the experiment (e.g. by excluding predators) then the number of small omnivores should also increase. Fig. 6.4 shows that after a 1-year delay, small granivores reached and maintained densities that averaged 3.5 times higher on the removal plots than on the controls, but the small omnivores did not show any

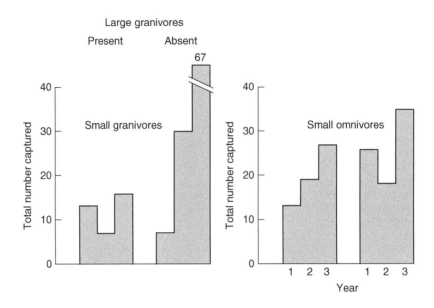

Fig. 6.4 Exclusion of large granivorous rodents resulted in an increase in the small granivorous rodent population relative to control areas, indicating competition. Small omnivorous rodent numbers did not increase significantly, indicating lack of competition. (Data from Munger and Brown 1981.)

significant increase. These results are consistent with the interpretation that there was competition between large and small granivorous rodents.

Although these examples produced results consistent with the predictions of inter-specific competition, there was no attempt to measure the competition coefficients. However, Abramsky *et al.* (1979) carried out a similar removal experiment on the shortgrass prairie in Colorado, in which a competition coefficient was measured. In this case, voles (*M. ochrogaster*) were removed and the response of deermice (*P. maniculatus*) was recorded. Fig. 6.5 shows the negative relationship between the number of deermice present in the removal plot and the number of voles present in the previous sampling period 2 weeks earlier, as expected if competition were acting. To measure the competition effect *a* of voles on deermice, the Lotka–Volterra equation was used. At equilibrium, $dN_1/dt = 0$, and so:

$$K_1 = N_1 + a \times N_2 \times (2^{0.75})$$

Fig. 6.5 Number of deermice (*Peromyscus maniculatus*) known to be alive at time *t* plotted against number of voles (*Microtus ochrogaster*) known to be alive at time $t - 1$ ($t - 1$ is the sample period 2 weeks earlier than period *t*). (After Abramsky *et al.* 1979.)

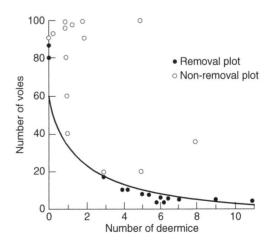

Table 6.2 Mean dry weight of subaquatic invertebrates available to mallard ducklings and the rate of duckling food intake in Calnes with and without fish in Sweden. (Source: Pehrsson, 1984. Reproduced with permission of John Wiley & Sons Ltd.)

	Year	Lakes without fish	Lakes with fish
Mean dry weight (May–June)	1977	119.8 (21.0)	45.3 (13.7)*
	1978	159.0 (9.9)	26.5 (4.8)*
Duckling feeding (food items/min)	1977	12.4 (0.6)	9.5 (0.5)**
	1978	20.4 (5.1)	7.9 (0.7)*

*$P < 0.01$;
**$P < 0.001$.

where K_1 is the carrying capacity of the environment for individuals of deermice when alone, N_1 is the number of deermice, N_2 is the number of voles, and $2^{0.75}$ is the conversion factor, which standardizes species in terms of their metabolic rates.

The body weight W of voles is about two times that of deermice and the basal metabolic rate is taken as $M = W^{0.75}$ (see Section 2.5.2). Using various combinations of N_1 and N_2, an average estimate of $a = 0.06$ is obtained.

Properly designed removal experiments are difficult to carry out for practical reasons, so it is not surprising that they have not yet been performed with large mammals.

6.3.2 Natural experiments

An easier approach uses natural absences or combinations of species to observe responses that would be predicted from interspecific competition. For example, mallard ducks (*Anas platyrhynchos*) breed on oligotrophic (low-nutrient) lakes in Sweden (Pehrsson 1984). Some of these lakes contain fish, while others do not. In lakes with fish, the density of mallards is lower, mean invertebrate food size is lower, and emerging insects are significantly smaller. In an experiment in which ducklings were released, their intake rate was higher on lakes without fish (Table 6.2). This result implies competition between ducks and fish.

Another type of natural experiment is illustrated by the distributions of two gerbilline rodent species (Abramsky and Sellah 1982). One, *Gerbillus allenbyi*, lives in coastal sand dunes and is bounded in the north by Mount Carmel. In the same region, the other species, *Meriones tristrami*, is restricted to nonsandy habitats. In the coastal area north of Mount Carmel, *M. tristrami* occurs alone and inhabits several soil types, including the sand dunes. Abramsky and Sellah suggested that *M. tristrami* colonized from the north and was able to bypass Mount Carmel, whereas *G. allenbyi* colonized from the south and could not pass the Mount Carmel barrier. In the region of overlap, south of the barrier, interspecific competition had excluded *M. tristrami* from the sand dunes. They tested this hypothesis by removing *G. allenbyi* from habitats where the two species overlapped and found that there was no significant increase in *M. tristrami*. They concluded that there was no present-day competition occurring. Instead, they suggested that competition in the past had resulted in a shift in habitat choice so that there was no longer any detectable competition.

Islands are sometimes used to look at the distributions of overlapping species, because on some islands a species can occur alone while on others it overlaps with related species. The theory of interspecific competition would predict that when alone, a species would expand the range of habitats it uses (a process we call *competitive release*), while on islands where there are several species the range of habitats would contract (*competitive exclusion*). A good example of this is seen in ground doves on New Guinea and surrounding islands (Diamond 1975). On the larger island of New Guinea there are three species, each with its own habitat (Fig. 6.6): *Chalcophaps indica*

Fig. 6.6 Habitats of ground doves (*Chalcophaps* and *Gallicolumba*) on islands off New Guinea demonstrate "competitive release." (After Diamond 1975.)

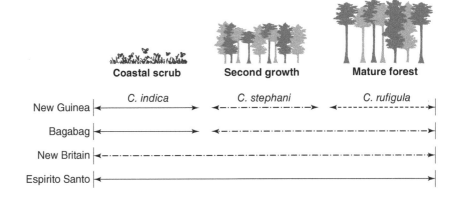

in coastal scrub, *C. stephani* in second-growth forest further inland, and *Gallicolumba rufigula* in the interior rainforest. On the island of Bagabag, *G. rufigula* is absent and *C. stephani* expands into the mature forest. On some islands (New Britain, Karkar, Tolokiwa), only *C. stephani* occurs, and it uses all habitats, while on the island of Espiritu Santo, only *C. indica* occurs, and this species also expands into all habitats. It is assumed that this habitat expansion has resulted from competitive release through the absence of the other potential competitors.

6.3.3 *Interpreting perturbation experiments*

Perturbation experiments are designed to measure responses of populations that would be predicted from interspecific competition theory. We should be aware, however, that there are two types of perturbation (Bender *et al.* 1984). One, called a *pulse* experiment, involves a one-time removal of a species. The rate of return by the various species to the original equilibrium is then measured. This requires accurate measurements of rates of population increase, which in practice is not easy, and in fluctuating environments is very difficult. As a consequence, few of these experiments are carried out.

The other type of perturbation is the continuous removal or *press* experiment. Let us assume that species 1 is reduced to a new level and kept there. Other species are allowed to reach a new equilibrium and it is this level that is observed. This type of perturbation avoids the necessity of measuring rates, but there are other problems. If there are more than two species in a community, which in most cases there are, an increase in another species' population is neither a necessary nor a sufficient demonstration of competition. First, species 1 and species 2 may not overlap, and so not compete, but still affect each other through interactions with other competing species: this is *indirect competition*. Second, the two species may be alternative prey for a food-limited predator. Changes in the population of species 1 could affect that of species 2 by influencing the predator population: this has been termed *apparent competition* (Holt 1977) and we will discuss it again in Section 6.8.

All of the examples we have discussed in this section are *press* experiments. Strictly speaking, in order to demonstrate competition unequivocally we would need to know that: (i) resources were limiting, (ii) there was overlap in the use of the resources, (iii) other potential competitors were having a negligible effect, and (iv) predator populations were not responding to the experiment. In few cases have all these conditions been met. Because of these difficulties, an entirely different approach to the study of interspecific competition has measured the pattern of overlap in the use of resources. We now consider this approach.

6.4 The concept of the niche

Different species that adopt the same role in the community often evolve similar morphological and behavioral adaptations. This place in the community is called the *niche*, defined by Elton (1927) as the functional role and position of the organism in its community. (We provide the modern definition later.)

For practical reasons, the niche has come to be associated with use of resources. Thus, we can plot the range and frequency of seed sizes eaten by different bird species as a hypothetical example in Fig. 6.7a. Species that exploit the outer parts of the resource spectrum use a broader range of resources, because they are less abundant. Some species, such as 2, 3, and 4 in Fig. 6.7a, overlap, while others, such as 2 and 5, do

Fig. 6.7 (a) Hypothetical frequency distribution of seeds of different sizes indicating the range and overlap of potential niches for granivores. (b) Range of seed sizes eaten by British finches feeding on herbaceous plants. Seeds are in five size categories A to E. The finches are redpoll (*Carduelis flammea*), linnet (*C. cannabina*), greenfinch (*Chloris chloris*), and hawfinch (*Coccothraustes coccothraustes*). (After Newton 1972.)

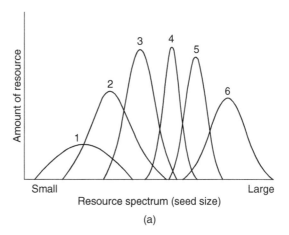

(a)

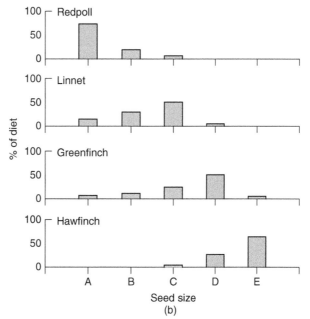

(b)

Fig. 6.8 Hypothetical frequency distribution of species 1 and species 2 along two parameter gradients: (a) moisture; (b) temperature. Outline of the species distributions when considering the two parameters simultaneously shows niches that can be either distinct (c), or overlapping (d).

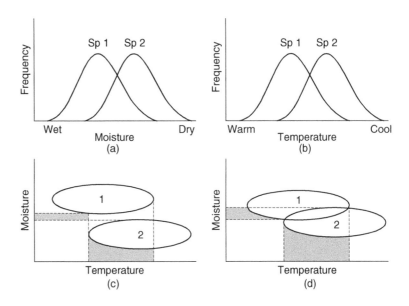

not. Overlap is necessary (but not sufficient) to demonstrate competition. An example (Fig. 6.7b) is provided by the range of seed sizes eaten by finches in Britain (Newton 1972). In this case, we see that, contrary to the theoretical distribution proposed in Fig. 6.7a, there is a broader range of seed sizes eaten by these finches in the middle range than by birds eating seeds at the extremes.

So far we have considered only one resource axis; that is, one variable, such as seed size. When we consider two or more axes the picture becomes less clear-cut in terms of overlap. Take two species, 1 and 2, which overlap along two axes, such as moisture and temperature, as in Fig. 6.8a,b. If we plot the outline of the two species distributions by considering the two axes simultaneously, we see that it is possible for the two distributions to be distinct (Fig. 6.8c) or to overlap (Fig. 6.8d). Which one occurs depends on whether individuals show *complementarity* (i.e. overlap on one axis but not on the other; Fig. 6.8c) or overlap simultaneously on both axes (Fig. 6.8d).

An example of complementarity is shown in Fig. 6.9. DuBowy (1988) examined the resource overlap patterns in a community of seven North American dabbling ducks, all of the genus *Anas*, by plotting habitat overlap against food overlap for pairs of species. In winter, when it is assumed that resources are limiting, points for pairs were below the diagonal line (Fig. 6.9a), indicating complementarity: pairs with high overlap in one dimension had low overlap in the other. In contrast, during summer, species pairs showed high resource overlap in both dimensions (several points are outside the line), indicating that species fed on the same food at the same place. In summary, the change in niche of these duck species from summer to winter resulted in lower overlap and, by implication, lower competition at a time when we would expect that resources would be limiting. Note, however, that neither the lack of resources nor interspecific competition was demonstrated, merely that the results conform to what we would predict if competition had been acting.

Green (1998) found complementarity in ducks along habitat and feeding behaviour axes. In dabbling and diving ducks in Turkey, he found that pairs with similar habitats had dissimilar feeding mechanisms.

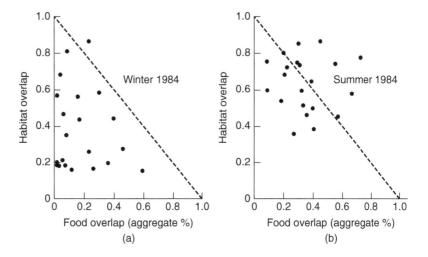

Fig. 6.9 Resource overlaps in seven species of dabbling ducks. Below the broken line there is complementarity in overlaps. (a) In winter, high habitat overlap between pairs of species tends to be associated with low food overlap, demonstrating complementarity. (b) In summer there is simultaneous overlap in both dimensions. (After DuBowy 1988.)

We have considered only two dimensions of a niche so far, but clearly the niche must include every aspect of the environment that would limit the distribution of the species. We cannot draw all these dimensions on a graph, but we could perhaps imagine a sort of sphere or volume with many dimensions, which could theoretically describe the complete niche. Hutchinson (1957) described this as the *n-dimensional hypervolume*. This is the *fundamental niche* of the species and is defined by the *set of resources and environmental conditions that allow a single species to persist in a particular region* (Schoener 1989; Leibold 1995). This suggests that the niche is in some way discrete. However, resource measures are usually continuous, so the discreteness does not come from these. Rather, it comes from the constraints of the species in terms of their morphology, physiology, and behavior: a species is more efficient at using some combinations of a resource than others, while different species have different combinations in which they are most efficient. These peaks of efficiency, then, are the adaptive peaks exhibited by a species (Schluter 2000).

The fundamental niche is rarely if ever seen in nature, because the presence of competing species restricts a given species to a narrower range of conditions. This range is the observed or *realized niche* of the species in the community. It emphasizes that interspecific competition excludes a species from certain areas of its fundamental niche. In terms of the Lotka–Volterra diagrams (see Fig. 6.2), the weaker competitor has no realized niche in Figs 6.2a and 6.8b, and for Fig. 6.8d parts of the fundamental niche are not used.

The difference between the two types of niche can be seen in a study by Orians and Willson (1964) of red-winged blackbirds (*Agelaius phoeniceus*) and yellow-headed blackbirds (*Xanthocephalus xanthocephalus*). Both species make their nests among reeds in freshwater marshes in North America, and, if alone, both will use the deep-water parts of the marsh (there is greater protection from mammalian predators there). However, when the two species occur together, the yellow-heads exclude the red-winged blackbirds, which are then restricted to nesting in the shallow parts. Thus, the fundamental niche for nesting red-winged blackbirds is the whole marsh, but the realized niche is the shallow-water reed bed. Coexistence occurs from the partitioning of the resource (nesting habitat) and the divergence of realized niches.

6.5 The competitive exclusion principle

In 1934, Gause stated that "as a result of competition two species hardly ever occupy similar niches, but displace each other in such a manner that each takes possession of certain kinds of food and modes of life in which it has an advantage over its competitor" (Gause 1934). In short, two species cannot live in the same niche, and if they try to then one will be excluded; further, coexisting species live in different niches. This is known as the *principle of competitive exclusion*, or Gause's principle (Hardin 1960), and has become one of the fundamental tenets of ecology. It proposes that species can coexist if adaptations arise to effectively partition resources. Examples of such adaptations include the use of different micro-habitats, different components of prey, different ways of feeding, different life stages of the same prey, different time periods in the same habitat, the taking advantage of disturbance, and interference competition (Richards *et al.* 2000). Therefore, Gause's principle has become the basis for studies of resource partitioning and overlap as a way to measure interspecific competition.

There are, however, two serious problems with Gause's statement. The first is that it is a trivial truism, because we have already identified the two coexisting populations as being different by calling them different species, and therefore, if we look hard enough, we are likely to find differences in their ecology as well. This is called a tautology: having defined the species as being different, it should be no surprise to find they are different.

The second problem is that the principle is untestable. It cannot be disproved because either result (exclusion or coexistence) can be attributed to the principle. To disprove the principle, it is necessary to demonstrate that the niches of two species are identical. Yet, as we can see from Fig. 6.8, what appears to be overlap, even complete overlap, may not be so when an additional axis is taken into account. Since we can never be sure that we have measured all relevant axes in describing the niches of two species, we can never be sure that the two niches are the same. Hence, we cannot disprove the principle.

6.6 Resource partitioning and habitat selection

6.6.1 *Habitat partitioning*

Despite these problems with the competitive exclusion principle, it underlies the numerous studies of habitat partitioning among groups of coexisting species. Lamprey (1963) described the partitioning of habitats by species of savanna antelopes in eastern Africa. A similar study by Ferrar and Walker (1974) showed how various antelopes in Zimbabwe used the three habitat types of grassland, savanna, and woodland (Fig. 6.10). In both cases there was partitioning as well as overlap.

Similar studies by Wydeven and Dahlgren (1985) show partitioning of both habitat and food in North American ungulates (Fig. 6.11). In Wind Cave National Park, elk and mule deer have similar winter habitat choices, as do pronghorn and bison, but these pairs have very different diets. For example, the diet of bison contains 96% grass, as against 4% for pronghorn.

Interspecific overlap in the diet niches of two sibling bat species (*Myotis myotis, M. blythii*) of Switzerland shows niche partitioning: *M. myotis* feeds largely on ground insects (carbid beetles) and *M. blythii* on grass-dwelling insects (bush crickets). This allows coexistence within the same habitats (Arlettaz *et al.* 1997). MacArthur (1958), in a now classic paper, described the different feeding positions of five species of warblers within conifer trees in the northeastern United States. They varied in both height in the tree and use of inner or outer branches. Nudds *et al.* (1994) found habitat partitioning in dabbling ducks in both Europe and North America. Species with a high density of filtering lamellae in their bills (fine filter feeders) tended to live in deep water

Fig. 6.10 Habitat partitioning and overlap by ungulates in Kyle National Park, Zimbabwe. The width of the boxes reflects the degree of preference. (After Ferrar and Walker 1974.)

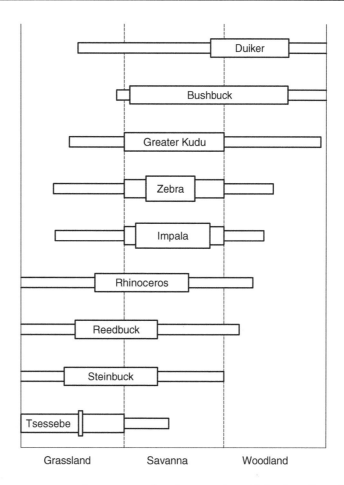

Grassland   Savanna   Woodland

with short, sparse vegetation, compared to those species with few lamellae, which lived in shallow water with tall, dense vegetation.

6.6.2 Limiting similarity

As we have mentioned, it should not be surprising that species divide up the resources available to them. However, Gause's principle implies that there should be a limit to the similarity of niches allowing coexistence of two species. Earlier studies predicted values of limiting similarity based on theoretical arguments (MacArthur and Levins 1967). If the distance between the mid-points of species distributions along the resource axis is d and the standard deviation of the curves (such as those in Fig. 6.7a) is w (the relative width) then limiting similarity can be predicted from the ratio d/w. However, various assumptions make this approach unrealistic: such as that the curves must be similar and normally distributed and must exist along only one resource axis.

Pianka et al. (1979) asked, how much would niches overlap if resources were allocated randomly among species in a community? A frequency distribution of niche overlaps generated from randomly constructed communities is shown in Fig. 6.12. This can be compared with distributions of observed overlap in the diets of desert lizards from 28 sites on three continents (Fig. 6.13). Those in the Kalahari Desert of southern Africa showed the greatest degree of overlap (because one food type, termites, made up a large amount of the diet), and those in Australia the least. In no case

Fig. 6.11 (a) Diet and (b) winter habitat use of elk, mule deer, pronghorn, and bison in Wind Cave National Park, South Dakota. Where habitat choice is similar, there are major differences in diet. (Data from Wydeven and Dahlgren 1985.)

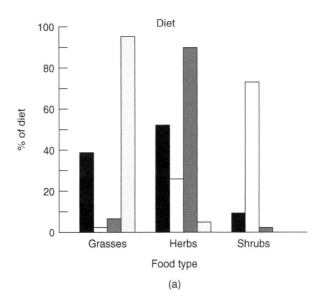

(a)

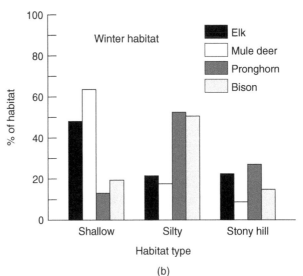

(b)

were observed distributions similar to the random distributions: there were far more species pairs with small overlaps than would be predicted by chance, implying that interspecific competition was causing niche segregation.

6.6.3 *Habitat selection from field data*

As we have seen, species usually differ from each other in choosing different resources such as food types, habitats, and so on. We call this choice *habitat selection*. One approach to measuring the competition coefficients has been to look at the variation of a species density in different habitats. The variation in density due to habitats and other resources is estimated by statistical procedures such as multiple regression (Crowell and Pimm 1976) and then the remaining variance can be attributed to interspecific competition with another identifiable species. An example of this approach

Fig. 6.12 Frequency distribution of niche overlaps in 100 randomly constructed communities with 15 species and five equally abundant resource states. (After Pianka *et al.* 1979.)

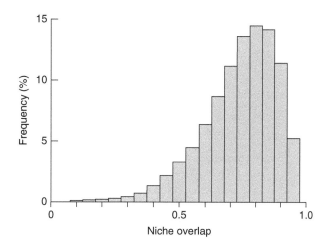

Fig. 6.13 Distributions of observed overlap in the diets of desert lizards. Dietary overlap is higher in Kalahari lizards, where termites make up 41% of the diet. (After Pianka *et al.* 1979.)

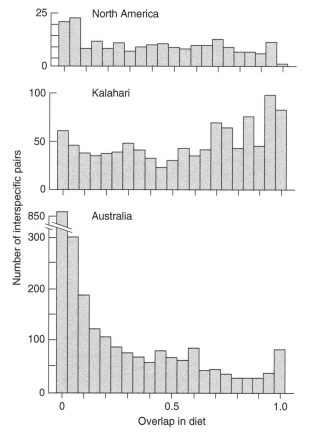

is given by Hallett (1982) in a study of 10 desert rodent species in New Mexico. He measured habitat variables related to common plants, such as number of individuals, plant height, distance to nearest plant from trap, and percentage cover. Regression analysis was used to partition the variance in capture frequency at trap stations due to habitat variables and competitors. Competition was observed within one group of

Table 6.3 Matrix of competition coefficients for the *Perognathus–Peromyscus* guild for each year. Entries are the partial regression coefficient after removal of the effects of the habitat variables. The coefficients are the effects of the column species (independent variable) on the row species (dependent variable).

	1971		1972			1973		
	PP	PI	PP	PI	PE	PP	PI	PE
PP	...	−0.43 *	...	−0.17	−0.42 *	...	−0.12	−0.82 *
PI	−0.17	...	−0.09	...	0.05	−0.05	...	−0.39 *
PE	NI	NI	−0.09	0.19	...	−0.12 *	−0.10	...

*Significant (P < 0.05).
NI, not included in the analysis; PP, *Perognathus penicillatus*; PI, *Perognathus intermedius*; PE, *Peromyscus eremicus*.

three species, *Perognathus intermedius*, *Perognathus penicillatus*, and *Peromyscus eremicus* (Table 6.3). Although the competitive effects differed from year to year, they were not random, and the inhibitory effects were not symmetrical; thus, *P. eremicus* always had a greater effect on the other two species than the reverse, and similarly *P. intermedius* had a greater effect on *P. penicillatus* than vice versa.

Abramsky (1981) used a similar regression method to look at interspecific competition and habitat selection in two sympatric rodents, *Apodemus mystacinus* and *A. sylvaticus*, in Israel. Plotting the densities of the two species in different habitats against each other (Fig. 6.14) indicated a negative relationship and suggested that there might be interspecific competition operating. However, he found that species abundances could have been the result of habitat differences alone; the effect of the presence of the other species was negligible in this case, implying no competition.

There are problems with the regression method, some of which are outlined by Abramsky *et al.* (1986). One is that if sympatric populations of different species differ greatly in average abundance then estimates of their variance and regression coefficients will be distorted. In turn, estimates of competition will be unreliable. A second problem lies in the assumption of constant competition coefficients; if competition is weak when populations are close to equilibrium (which we assume is when regressions from field populations are estimated) but strong when disturbed from equilibrium (the situation in perturbation experiments) then regression analysis is likely to miss competitive effects, while experiments will indicate their presence. A third problem is that we can never be sure that we have accounted for all the variability in density

Fig. 6.14 The negative correlation in numbers of two woodmice species (*Apodemus mystacinus* and *A. sylvaticus*) in Israel. (After Abramsky 1981.)

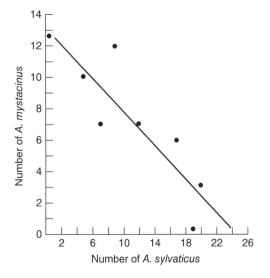

from environmental factors at various sites; there may be some factor that has been overlooked that can account for the remaining variability, instead of attributing it to interspecific competition.

6.6.4 The theory of habitat selection

Since species prefer to use some habitats over others, we ask how this choice changes when resources become limiting. There are two hypotheses that we should consider. We start with the *theory of optimal foraging*, which predicts that when resources are not limiting, species should concentrate their feeding on the best types of food or the best types of habitat and ignore the others, no matter how abundant they are (see Chapter 10). When resources are limiting, a species should expand its niche to include other types of food, habitat, and so on. This is the expected response under intraspecific competition.

When two species are present, one might expect both to respond to declining resources by *expanding their niches* and so increasing the overlap. However, Rosenzweig's (1981) *theory of habitat selection* introduces a second hypothesis. This predicts that when resources are limiting, species should *contract their niches* as a result of interspecific competition. He considered two different situations: we start by assuming that there are two species, 1 and 2, and two habitats, A and B. In the first case, called the *distinct preference* case, both species use both habitats, but each prefers to use a different one (i.e. species 1 treats A as the better and B as the poorer habitat, while species 2 does the reverse). In the second case, the *shared preference* case, again both species use both habitats, but now both treat the same habitat A as the better one and B as the poorer.

In either case, we should first consider the habitat choice of a species when no competitors are present. Under conditions of abundant food, for example, a species should confine itself to its preferred habitat. As density increases and food become limiting through the feeding of other individuals, the species will continue to remain in the better habitat (A) so long as the fintake rate is greater than it would be in the poorer habitat (B). At some point, density in habitat A increases so that intake rate falls until it equals that in habitat B; at this stage the species should not confine itself to A but should expand its habitat use in such a way that densities keep intake rates similar in the two habitats. The intake rate at which a species changes from one to two habitats is called the *marginal value*.

We now consider what happens when there is a competitor present and resources are limiting. The outcome depends on which of the two preference cases occurs. In the distinct preference case, each species will confine itself to its preferred habitat rather than expand into the other one. Therefore, when resources are limiting, species will specialize, contract their niches, and reduce overlap. When resources are abundant, they should use either one or both habitats, depending on their intake rates in the two. Further, in the shared-preference case we have to assume that one of the species (1) is dominant and can exclude the other through its behavior or by other means such as scent marking. If species 2, the subordinate, is to coexist with species 1 then it must be more efficient at using the less preferred habitat B than is species 1. If the dominant species is more efficient in both habitats then it will exclude species 2. Therefore, when resources are limiting, one species – the dominant – will not change its habitat choice. The subordinate will change its choice from A to B: the competitive effect is asymmetrical, with the dominant having a large effect on the subordinate and the subordinate only a small effect on the dominant.

Fig. 6.15 Demonstration of distinct and shared preferences in habitat selection by three species of honeycreepers in Hawaii. The two main flowering trees are (a) *Sophora* and (b) *Metrosideros*. At low flower numbers *Loxops* (○) fed on *Sophora*, and *Himatione* (■) fed on *Metrosideros*, showing "distinct preference." *Vestaria* (Δ) fed on both trees, excluding the other species, but only at high flower numbers, indicating "shared preference." Note the reverse scale on the *x*-axis. (After Pimm and Pimm 1982.)

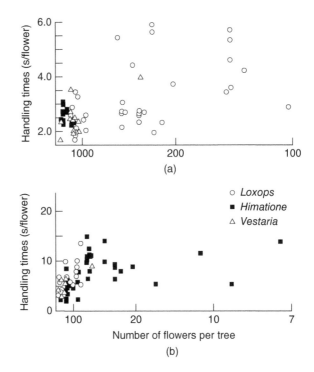

In a test of these hypotheses, Pimm and Pimm (1982) recorded the feeding choices of three nectar-feeding bird species (*Himatione, Loxops, Vestiaria*) on the island of Hawaii. There were two main tree species, *Metrosideros* and *Sophora*, which came into flower at different times of the year. The evidence for the distinct preference case is seen in Fig. 6.15. When the number of flowers is high, all three species feed on both trees. When flowers per tree are low (and assuming that this indicates a limiting resource), only *Loxops* feed on *Sophora*, and only *Himatione* feed on *Metrosideros*. Thus, both species reduce their niche width and specialize. There is also evidence of shared preference. *Vestaria* feed on both tree species, but only at high flower numbers, and physically exclude the other species by visual and vocal displays. In contrast, both *Himatione* and *Loxops* spend much of their feeding time on trees with few flowers. Thus, these two species are confined to poorer feeding areas during times when resources are low, as predicted by the theory.

Rosenzweig's theory predicts that niches contract when resources are limiting and there is interspecific competition. We have seen that the Hawaiian honeycreepers may conform to predictions, but what about other species? Information from wildlife both agrees and disagrees with the predictions. The overlap in diet of sympatric mountain goats (*Oreamnos americanus*) and bighorn sheep (*Ovis canadensis*) is high in summer but reduced in winter (Dailey *et al.* 1984), as predicted by the theory. In ducks, we have already seen that during winter there is a decrease in overlap (Fig. 6.9). Burning of grasslands increases the nutrient content of regenerating plants and may produce locally abundant food. Under these conditions, mountain goats and mule deer (*Odocoileus hemionus*) actually increase dietary overlap (Spowart and Thompson Hobbs 1985). In contrast, elk and deer in natural forests increase dietary overlap in winter, when resources are assumed to be least available (Leslie *et al.* 1984).

We should note that we do not have actual measures of the food supply in these examples so we cannot be sure that we are seeing competition. In the Serengeti, Tanzania, wildebeest are regulated by lack of food in the dry season (Mduma *et al.* 1999), so overlap with this species at this time should result in competition. However, overlap in both diet and habitat between wildebeest and several other ungulate species increases or does not change between wet and dry seasons (Hansen *et al.* 1985; Sinclair 1985). One interpretation could be that interspecific competition is asymmetrical, with the impact of the rarer ungulates on the numerous wildebeest being real but very slight, and the reverse not occurring due to these other ungulates being kept at low density by predation (Sinclair *et al.* 2003).

6.7 Competition in variable environments

So far we have discussed the patterns of occupancy and utilization of habitats as if they were constants for a species, or could only be changed seasonally. However, longer-term studies are now showing that species densities vary in the same habitat and change over a longer time scale, measured in years. Thus, populations may go through periods of abundant resources, during which although there is overlap with other species, there is no competition (Fig. 6.16). Occasionally there are periods of resource restriction and it is only at these times that one sees competition and niche separation (Wiens 1977).

6.8 Apparent competition

6.8.1 *Shared predators*

Some of the predicted outcomes from interspecific competition include the reduction of populations, the contraction of niches, and exclusion of species from communities. However, these predictions are also to be expected when species have non-overlapping resource requirements but share predators, especially when predators can increase their numbers fast.

Let us suppose there is a predator that is food-limited and which feeds on two prey species. The prey are both limited by predation and not by their own food supplies. If species 1 increases in number then this should lead to more predators, which in turn will depress species 2 numbers. This result is called *apparent competition* because it

Fig. 6.16 Changes over time in the mean (thick line) and variance (shaded area) of a selective constraint such as resource availability. At times A and B there are "bottlenecks" when competition is more likely. (After Wiens 1977.)

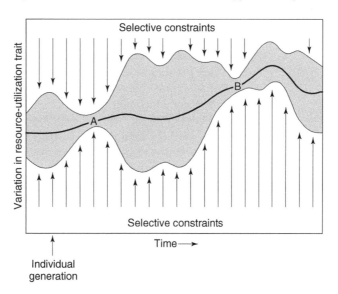

produces the same changes in prey populations as would be predicted from interspecific competition (Holt 1977, 1984). Examples of apparent competition are given in Section 6.10.2 and Chapter 22, where predators are causing the demise of secondary prey, the rare Roan antelope in Kruger National Park, South Africa (Harrington *et al.* 1999; McLoughlin and Owen-Smith 2003) and the wildebeest in Manyara National Park, Tanzania, as a result of a high abundance of buffalo, the primary prey.

If two prey species live in the same habitat, as in the wildebeest and buffalo example in Manyara, then at high intensities of predation coexistence is unlikely. On the other hand, coexistence is promoted if the two species select different habitats; that is, if niche partitioning occurs.

Another version of apparent competition can occur through shared parasites. One species can be a superior competitor if it supports a parasite that it transmits to a more vulnerable species. For example, when gray squirrels (*Sciurus carolinensis*) were introduced to Britain, they brought a parapox virus that reduced the competitive ability of the indigenous squirrel (*S. vulgaris*) (Hudson and Greenman 1998). The latter has largely been displaced, occurring now in only a few small locations of its former range. Gray squirrels are displacing red squirrels through competition in Italy and may be spreading through Europe (Wauters and Gurnell 1999).

6.8.2   *Implications*

Since the observed responses of prey populations to changes in predator numbers are similar to those from interspecific competition, we cannot infer such competition simply from observations or even experiments that show either changes in species population size or niche shifts. We need to know (i) whether resources are limiting and (ii) the predation rates and predator numbers.

6.9   **Facilitation**

6.9.1   *Examples of facilitation*

Facilitation is the process whereby one species benefits from the activities of another. In some cases, the relationship is *obligatory*, as in the classic example of the Nereid worm (*Nereis fucata*), which lives only in the shell of hermit crabs (*Eupagurus bernhardus*). The crabs are messy feeders; when they feed upon a carcass, scraps float away to be filtered out of the water by the worm. While the worm benefits, the crab appears not to suffer any disadvantage (Brightwell 1951). In other cases, the relationship may be *facultative*, by which we mean that the dependent species does not have to associate with the other one in order to survive, but does so if the opportunity arises. Thus, cattle egrets (*Ardea ibis*) often follow grazing cattle in order to catch insects disturbed by these large herbivores. Although the birds increase their prey capture rate by feeding with cattle, as they probably do by following water buffalo (*Bubalus bubalus*) in Asia and elephants and other large ungulates in Africa, they are quite capable of surviving without them (McKilligan 1984). The European starling (*Sturnus vulgaris*) also follows cattle on occasions. In contrast, its relative in Africa, the wattled starling (*Creatophora cinerea*), seems always to follow large mammals, and in the Serengeti migrates with the wildebeest like a camp follower.

Vesey-Fitzgerald (1960) suggested that there was grazing facilitation among African large mammals. Lake Rukwa in Tanzania is shallow and has extensive reedbeds around the edges. Grasses, sedges, and rushes there can grow to several meters in height, in which state only elephants can feed upon them. As the elephants feed and trample the tall grass, they create openings in which there is lush, regenerating vegetation. This provides a habitat for African buffalo, which in turn provide short grass patches that can be used by the smaller antelopes such as topi. In this case, elephants are creating a habitat for buffalo and topi that would not otherwise be able to live there. Therefore, the presence of elephants increases the number of herbivores that can live in the Lake Rukwa ecosystem. Vesey-Fitzgerald called this sequence of habitat change in the grasslands a *grazing succession*.

Bell (1971) has described a similar grazing succession among the large mammals of the Serengeti. In certain areas there is a series of low ridges bounded by shallow drainage lines; these ridges have sandy, thin soils and support short, palatable grasses. The drainage lines have fine silt or clay soils that retain water longer than those on the ridges and so support dense but coarse grasses, which remain green long into the dry season. Between these two extremes there are intermediate soil types on the slopes. The whole soil sequence from top to bottom is called a *catena*. In the wet season, when all areas are green, all five nonmigratory species (wildebeest, zebra, buffalo, topi, and Thomson's gazelle) feed on the ridge tops. Once the dry season starts, the different species move down the soil catena into the longer grass in sequence, with the larger species going first (Fig. 6.17). Thus, zebra is one of the first species to move, because it can eat the tough tall grass stems. By removing the stems, the zebra make the basal leaves in these tussock grasses more available to wildebeest and topi, and these in turn prepare the grass sward for the small Thomson's gazelle. Thus, there is a grazing succession.

Zebra, wildebeest, and Thomson's gazelle also have much larger migratory populations in the Serengeti, separate from the smaller resident populations discussed before. It is tempting to think that the movements of these migrants follow the same pattern as those of the resident populations. Indeed, McNaughton (1976) has shown that migrating Thomson's gazelle prefer to feed in areas already grazed by wildebeest because these produce young green regrowth not found in ungrazed areas. The gazelle take advantage of this growth, which was stimulated by the grazing, and so benefit from the wildebeest.

The relationship between the migrating zebra and wildebeest is more complex. While zebra usually move first from the short-grass plains to the long-grass dry-season areas, the wildebeest population (1.3 million), which is much larger than that of the zebra (200 000), often does not follow them but takes its own route, eating the

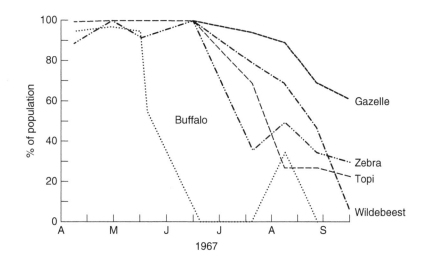

Fig. 6.17 The proportion of the population of different ungulate species using short grass areas on ridge tops (upper catena) in Serengeti. The larger species leave before the smaller at the start of the dry season. (After Bell 1970.)

long, dry grass. Therefore, most migrant wildebeest obtain no benefit from the zebra. In contrast, zebra may be benefiting from the wildebeest for a completely different reason. In the wet season, when there is abundant food, many zebra graze very close to the wildebeest, and by doing so they can avoid predation, because most predators (lions and spotted hyenas, *Crocuta crocuta*) prefer to eat wildebeest. Only if there are no wildebeest within range will predators turn their attention to zebra. Therefore, it pays zebra to make sure there are wildebeest nearby. In the dry season, however, zebra compete with wildebeest for food. Zebra, therefore, show habitat partitioning and avoid the wildebeest at this time. By doing so, they probably make themselves more vulnerable to predators (Sinclair 1985). Thus, zebra have to balance the disadvantages of predation attached to leaving the wildebeest with those of competition if they stay with them. We can see a seasonal change from facilitation in the wet season, when there is abundant food, to competition in the dry season, when food is regulating the wildebeest.

An example of facilitation has been recorded on the Isle of Rum in Scotland. Cattle were removed in 1957 but were reintroduced in 1970 to part of the island, where they grazed areas used by red deer. Pasture used by cattle in winter results in a greater biomass of green grass in spring compared to ungrazed areas. Gordon (1988) found that deer preferentially grazed areas in spring that had previously been used by cattle (Fig. 6.18), and subsequently there were more calves per female deer.

On the North American prairies, both black-tailed prairie dogs (*Cynomys ludovicianus*) and jackrabbits (*Lepus californicus*) benefit from grazing by cattle. If grazing is prevented then the long grass causes prairie dogs to abandon their burrows. At a site in South Dakota where cattle were fenced out there were half as many burrows as on adjacent areas where grazing was continued. Snell and Hlavachick (1980) showed that a large prairie dog site of 44 ha could be reduced to a mere 5 ha by the elimination of cattle grazing in summer to allow the grass to grow. Presumably under natural conditions when American bison grazed the prairies, facilitation by bison allowed the prairie dogs to live there. Facilitation may have been mutual, as both pronghorn and bison respond to the vegetation changes caused by prairie dogs, both species using prairie dog sites (Coppock *et al.* 1983; Wydeven and Dahlgren 1985; Huntly and Inouye 1988; Miller *et al.* 1994).

This example illustrates two management points which follow from the understanding of the interaction (facilitation) between large mammal grazers and prairie dogs: (i) a simple management program (through grazing manipulation) could be devised

Fig. 6.18 Facilitation of deer grazing by cattle is demonstrated by deer fecal-pellet groups on cattle-grazed plots during each of the 3 months of study. Deer preferred to graze plots used by cattle the previous winter. (After Gordon 1988.)

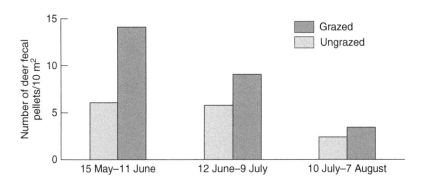

to control the prairie dogs, without the use of harmful poisons that might affect other species; and (ii) in many areas prairie dogs are becoming very scarce and their colonies need to be protected. In addition, the black-footed ferret (*Mustela nigripes*), one of the rarest mammals in the world, depends entirely on prairie dogs, and it is thought that its very low population has resulted from the decline in prairie dog populations. The conservation of both species would benefit from the manipulation of grazing practices.

In another example of facilitation improving management for wildlife, Anderson and Scherzinger (1975) showed that ungrazed grassland resulted in tall, low-quality winter food for elk. Cattle grazing in spring maintained the grass in a growing state for longer. If cattle were removed before the end of the growing season, the grasses could regrow sufficiently to produce a shorter, high-quality stand for elk; their population increased from 320 to 1190 after grazing management was introduced.

In Australia, rabbits prefer very short grass. Rangelands that have been overgrazed by sheep benefit rabbits through facilitation, and rabbit numbers increase. When sheep are removed and long grass returns, rabbit numbers decline. In Inner Mongolia, China, the substantial increase in livestock numbers since 1950 has produced a grass height and growth rate that favors Brandt's vole (*Microtus brandti*) populations, so that there has been an increase in the frequency of population outbreaks of this species (Zhang *et al.* 2003).

The saltmarsh pastures of Hudson Bay in northern Canada are grazed by lesser snow geese (*Chen caerulescens*) during their summer breeding (Bazely and Jefferies 1989; Hik and Jefferies 1990; Wilson and Jefferies 1996). The marshes are dominated by the stoloniferous grass *Puccinellia phryganodes* and the rhizomatous sedge *Carex subspathacea*. At La Perouse Bay, some 7000 adults and 15 000 juvenile geese graze the marsh, taking 95% of *Puccinellia* leaves. These are nutritious, with high amounts of soluble amino acids. From exclosure plots it was found that natural grazing by geese increased productivity by a factor of 1.3–2.0. Experimental plots with different levels of grazing by captive goslings showed that aboveground productivity of *Puccinellia* was 30–100% greater than that of ungrazed marsh. In addition, the biomass (*standing crop*) of the grass was higher if it was allowed to regrow for more than 35 days following clipping. Immediately after the experimental grazing, the biomass was less than the ungrazed plots, so that at some point between then and the eventual measurements the biomass on the treated and the untreated plots was the same. Even so, the production rate of shoots was higher on the grazed sites. Other experimental plots where grazing was allowed but from which goose feces were removed showed that biomass returned to the level of control plots, but no further. Thus, it appears that goose droppings, which are nitrogen-rich and easily decomposed by bacteria, stimulate growth of *Puccinellia*. Geese, therefore, benefit each other through their grazing by fertilizing the grass, a form of intraspecific facilitation.

In summary, facilitation occurs when one species alters a habitat or creates a new habitat in such a way as to allow the same or another species to benefit. We have discussed grazing systems in particular, but the concept applies in many other cases. For example, many hole-nesting birds and mammals in North America, such as wood ducks (*Aix sponsa*) and flying squirrels (*Glaucomys sabrinus*), depend on woodpeckers to excavate the holes, a form of facilitation. Knowledge of such interactions is important for the proper management and conservation of ecosystems.

6.9.2 *Do grasses benefit from grazing?*

If a species such as Thomson's gazelle benefits from the grazing effects of wildebeest due to the increased productivity of the plants, then do the plants themselves benefit? In other words, what benefits do the plants receive from being grazed and growing more? In evolutionary terms (see Chapter 20), we have to rephrase this as, "Does herbivory increase the fitness of individual plants?" In ecological terms one may ask, "Does the plant grow more after herbivory?"

The studies of lesser snow geese on the saltmarshes of Hudson Bay, which we discussed earlier, are now showing that the grass *Puccinellia* comes in different genotypes (Jefferies and Gottlieb 1983). Nineteen grazing experiments have shown that under-grazing there is selection for those genotypes that are fast-growing. These types have the ability to take up the extra nitrogen from the goose feces and seem to out-compete slower-growing genotypes. This is, therefore, an evolutionary benefit from grazing. After 5 years, grasses in plots where grazing is prevented are only just beginning to change to slower-growing genotypes. The more immediate ecological benefit from grazing again comes from the addition of nutrients, resulting in a 30–50% increase in biomass.

In general, there are few studies that show plants increasing their fitness as a result of herbivory (Belsky 1987). In contrast, we can look at communities of plants and see that if the majority of them, such as grasses, can tolerate grazing (i.e. survive despite herbivory) then a few intolerant species in the community may not survive due to nadvertent feeding or trampling by large mammals (i.e. apparent competition rather than true competition between plants). This may be simply a consequence of grazing and not necessarily an evolutionary advantage for the grass species. Nevertheless, McNaughton (1986) has argued in opposition to Belsky that grasses and grasslands have evolved in conjunction with their large-mammal herbivores, especially in Africa. From an evolutionary point of view, a grass individual that by chance has evolved an anti-herbivore strategy (such as the production of distasteful chemicals) should be able to spread through the grassland. We have to surmise at this stage that anti-herbivore adaptations are constrained in some way; it could be that production of distasteful chemicals results in the plant being less successful, in root competition, or in the uptake of nutrients, as in the example of lesser snow goose grazing.

On the surface, it appears disadvantageous for a grass to grow more as a response to grazing, as this will provide more food and invite further grazing. However, growth could also be viewed as a damage repair mechanism that is making the best of a bad situation (i.e. the grass may lose less fitness by growing than by not doing so).

In summary, we know too little about both the ecological and evolutionary consequences of herbivory on plants. We are left with many questions and opposing views, and more work is needed.

6.9.3 *Complex interactions*

Competition, parasitism, and predation are all processes that have negative effects on a species. However, when they act together, they may end up having a beneficial effect. For example, acorns of English oak (*Quercus robur*) are parasitized by weevils and gall wasps and are eaten by small mammals. Very high mortality rates are imposed on healthy acorns by small mammals, but parasitized acorns are left alone. While most of the parasitized acorns also die, some survive and are avoided by the small mammals. Thus, higher survivorship and, hence, fitness occurs when the plant is parasitized (Crawley 1987; Semel and Andersen 1988).

6.10 Applied aspects of competition

6.10.1 Applications

It is important that we should understand the underlying concepts of interspecific competition if we are to comprehend how species might or do actually interact in the field. There are several applications where we need to be aware of potential competition: (i) in conservation, where we might have to protect an endangered species from competition with another dominant species; (ii) in managed systems such as rangelands and forests, where there could be competition between domestic species and wildlife – for example, an increase in livestock or the expansion of rangeland might cause the extinction of wildlife species, or wildlife might eat food set aside for domestic animals; and (iii) if we want to introduce a new species to a system, such as a new gamebird for hunting, where there might be competition from other resident species.

6.10.2 Conservation

Let us imagine a situation where we want to conserve a rare species but we are concerned about possible competition from a common one. For example, roan antelope (*Hippotragus equinus*; a fairly rare species) were released in Kruger National Park, South Africa, as part of a conservation program. There were concerns that the numerous wildebeest would exclude the roan antelope. In this case, the management response was to cull the wildebeest (Smuts 1978). More recent evidence indicates that predators, supported by an abundant zebra population, are limiting and even excluding this rare species (Harrington *et al.* 1999; McLoughlin and Owen-Smith 2003). Thus, apparent competition is the dominant process here. A similar example involves the extremely rare Arabian oryx (*Oryx leucoryx*). A few of the last remaining individuals of this species were captured in Arabia in the early 1960s and taken to the San Diego Zoo. There, numbers increased, and some have been successfully reintroduced to Oman (Stanley Price 1989).

In both of these examples it would be important to detect whether there was competition with resident species. We have seen that simple measures of overlap or even changes of overlap with season may not be good indicators of competition. Similarly, observations that increase in a common species is correlated with a decrease in the rare species does not mean that competition is the cause, because of the problem of apparent competition. These measures are necessary but not sufficient. In addition, we would need a measure of: (i) resource requirement; (ii) availability of limiting resources and demonstration that one is in short supply; and (iii) the predation rates on both the target species and alternative prey.

A second kind of problem comes from changes in habitat. Assume there is coexistence and habitat partitioning between two species along the lines of Rosenzweig's shared preference hypothesis. Since studies of diet and habitat selection would show that both species prefer the same habitat, one might be tempted to manage an area by increasing the preferred habitat at the expense of the other ones. In this case, however, only one species, the dominant, would benefit, and the other would decline. The breeding habitats of the yellow-headed and red-winged blackbirds (Orians and Wilson 1964) may be a case in point. Both species prefer deeper-water marshes, but one may predict that increasing the depth of a marsh where both occur, thereby leaving little shallow water, could well result in the exclusion of the red-winged blackbird.

In Lake Manyara National Park, Tanzania, there is a habitat consisting of open grassland on the lake shore that is used by wildebeest. Adjacent to this shoreline is savanna consisting of longer grass with scattered trees and shrubs, preferred by African buffalo. In 1961, heavy rains caused the lake levels to rise and flood the open grasslands, a situation which remained for the rest of the 1960s. The wildebeest were forced to use the savanna habitat, which they did not prefer, and after 4 years the population went extinct. On the surface, this appeared to be due to competition with the buffalo. Closer inspection showed that lions, whose densities were high because of the high buffalo population, had eradicated the wildebeest. Wildebeest normally escape predation by running, which they can do in habitats with short grass and little cover for ambush by predators. Once wildebeest were confined to the savanna, however, they were less capable of avoiding lions. Buffalo, on the other hand, avoid predators by hiding in thickets and defending themselves with their horns, and this they could do in savanna but not on the open grassland. Thus, each prey species had its own specialized anti-predator habitat that allowed coexistence between them, as predicted by Holt's (1977) "apparent competition" hypothesis. Once this habitat partitioning broke down, the predator was able to eliminate one of the species. The process of apparent competition explains these observations better than true competition.

6.10.3 *Competition between domestic species and wildlife*

There are a number of studies designed to detect whether there is competition between livestock and wildlife. Thill (1984) recorded the seasonal diets of cattle and white-tailed deer in three forested and two clear-cut sites in Louisiana pine forests. Woody plants made up > 85% of the diet of deer on the forest sites throughout the year (Fig. 6.19). For cattle diets, these plants made up < 16% in summer and autumn but rose to 60% in winter and to 48% in spring. The overlap between the two species in overall diet was highest in winter at 46% and lowest in summer at 12%. In contrast, on cleared sites deer continued to eat mainly woody plants but cattle ate > 80% grass year round. Diet overlap was only 17% in summer and fell to 10% in winter. Since the two species were in the same habitat and there did not appear to be predators, there could be a real possibility for interspecific competition if cattle were confined to forest sites; in fact, most of them stayed on the open sites. It is possible that because cattle and deer have not evolved together, we do not see the expected decrease in overlap in winter, so that competition is increased rather than avoided at this time.

Thill (1984) points out the advantages for multiple-use management derived from the diet partitioning. As forest practices intensify, forest ages decrease and the young stands become impenetrable without artificial clearing. They are also poor areas for

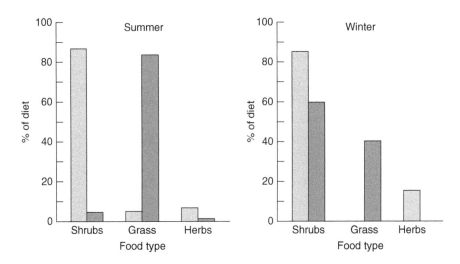

Fig. 6.19 The percentage of the diet of white-tailed deer (open bars) and cattle (shaded bars) made up of shrubs, grasses, and herbs. Diet overlap increased in winter when food was limiting. (Data from Thill 1984.)

deer forage. If cattle were used to graze these sites they could be kept open and so bene-fit deer by improving accessibility and increasing production of the second-growth deer food plants. This presumes that increasing deer numbers is the management objective. We should recognize that deer can have negative impacts, particularly on rare plants and birds (McShea *et al.* 1997), so management of deer needs to be carefully evaluated.

Hobbs *et al.* (1996) manipulated elk densities in randomized block experiments in sagebrush grassland to study the effects of competition with cattle. The effects of elk on cattle exhibited a threshold where low densities of elk had no effect but above a certain density there were both competition and facilitation effects. Food intake declined in direct proportion to elk density, because elk reduced the biomass of standing dead grass in winter. There were some weak facilitatory effects of elk grazing through an increase in digestibility and nitrogen content of the remaining grass available to cattle.

Cattle can also have indirect competitive effects by altering habitat structure. In a study of bird communities using the riverine shrub willow habitats in Colorado, Knopf *et al.* (1988) found that cattle grazing altered the structure of the shrubs but not the plant composition. Areas with only summer grazing contained larger bushes that were widely spaced and had few lower branches when compared with those areas that expe-rienced only winter grazing. The difference in structure affected migratory bird species according to how specific their habitat preferences were. Densities of those with wide habitat preference (e.g. yellow warbler (*Dendroica petechia*), song sparrow (*Melospiza melodia*)) did not change between the sites. Those with moderate niche width (Ameri-can robin (*Turdus migratorius*), red-winged blackbird (*Agelaius phoeniceus*)) were three times more numerous on the winter-grazed sites, while those with narrow niches (wil-low flycatcher (*Empidonax traillii*), white-crowned sparrow (*Zonotrichia leucophrys*)) occurred only on the winter-grazed sites.

Hayward *et al* (1997) found from a 10-year exclosure of cattle in riparian habitats of arid zones in New Mexico that small mammals were 50% more abundant in areas where cattle were excluded. Similarly, kangaroo rats (*Dipodomys merriami*) were more abun-dant in semi-desert shrubland where cattle grazing was reduced (Heske and Campbell 1991), and reptiles were also more abundant there (Bock *et al.* 1990).

6.10.4 *Introduction of exotic species*

Exotic species – those that do not normally live in a country – are introduced for a variety of reasons, and very often they become competitors with the native wildlife. Rabbits in Australia are perhaps the most conspicuous example of this, for they have been implicated in the decline of native herbivores through either direct or appar-ent competition by supporting exotic predators such as foxes (Short and Smith 1994; Short *et al.* 2000; Robley *et al.* 2001). Dawson and Ellis (1979) measured the dietary overlap between the rare yellow-footed rock-wallaby (*Petrogale xanthopus*), the euro (*Macropus robustus*; a type of kangaroo), and two introduced feral species, the domes-tic goat and the European rabbit. During periods of high rainfall, the rock-wallaby's diet was mostly of forbs (42–52%), but the proportion of forbs in the ground cover was only 14%. Under drought conditions, they still preferred forbs (13% forbs in the diet when forbs were hardly detectable in the vegetation), but trees and shrubs formed the largest dietary component (44% browse). In this season, major components for the other species were: euros, 83% grass; goats, 65% browse; rabbits, 25% browse. The rock-wallabies' overall diet overlap was 75% with goats, 53% with rabbits, and 39% with euros. In good conditions, dietary overlap was still substantial, but lower than when drought prevailed. At that time, the overlap was 47 and 45% with goats and euros, respectively. Thus, potential competition was greatest with goats and rabbits and least with the indigenous euro.

In North America, the introduced starlings (*Sturnus vulgaris*) and house sparrows (*Passer domesticus*) have competed with the native bluebird (*Sialia sialis*) for nesting sites, with the result that bluebird numbers have declined considerably (Zeleny 1976).

Not all introductions result in competition. Chukar partridge (*Alectoris chukar*) have been introduced to North America as a game bird. Their habitat includes semi-arid mountainous terrain with a mixture of grasses, forbs, and shrubs. In particular, they like the exotic cheatgrass (*Bromus tectorum*). Chukar introductions succeeded only where cheatgrass occurred. These habitat requirements are unlike those of native game-birds such as sage grouse (*Centrocercus urophasianus*), and thus little competition has taken place (Gullion 1965). Robley *et al.* (2001) showed that the endangered bur-rowing bettong (*Bettongia lesurur*) in Australia was able to cope with drought stress much better than rabbits because it could eat a variety of herbs and shrubs that rabbits could not. If anything, bettongs could out-compete rabbits. Thus, the decline of these marsupials was due to apparent competition from foxes (Short *et al.* 2000).

6.11 Summary

Interaction between species can be competitive or beneficial. Competition occurs when two species use a resource that is in short supply, but a perceived shortage in itself should not be used as unsupported evidence of competition. Instead, the relationships must be determined by manipulative experiments, reducing the density of one to deter-mine whether this leads to an increase of the other. Care should be taken to eliminate other factors such as predation that might cause the response.

Facilitation is the process by which one species benefits from the activities of the other. It often takes the form of one species modifying a less suitable food supply to make it more suitable for another species, or one species modifying a habitat to make it more favorable for another.

These two effects – competition and facilitation – can often be manipulated by man-agement to increase the density of a favored species.

7 Predation

7.1 Introduction

We start this chapter by describing the behavior of predators with respect to prey. With this knowledge, we explore some theoretical models for predator–prey interactions. Finally, we examine how the behavior of prey can influence the rate of predation. This chapter complements the approach given in Chapter 9 for analyzing interactions between trophic levels.

7.2 Predation and management

The previous two chapters dealt with interactions between individuals on the same trophic level. Predation usually involves interactions between trophic levels where one species negatively affects another. With respect to our three issues of management – conservation, control, and harvesting – predators and predation are of great interest. For rare prey species, the presence of a predator can make the difference between survival and extinction of the prey, especially if the predator is an introduced (exotic) species. This type of problem is particularly important on small islands, but also on isolated larger land areas such as New Zealand and Australia. In contrast, where prey are pests, predators may be useful as biological control agents. Ironically, it was for just this purpose that the small Indian mongoose (*Herpestes auropunctatus*) was introduced to Hawaii and the stoat (*Mustela erminea*) to New Zealand; unfortunately, in these cases the predators found the indigenous birds and small marsupial mammals easier to catch, so that the predators themselves became the problem. Finally, where harvesting of a prey species (by sport hunters, for example) is the objective, the offtake by natural predators must be taken into account, or else one runs the risk of overharvesting and causing a collapse of the prey population.

7.3 Definitions

Predation can be defined as occurring when individuals eat all or part of other live individuals. This excludes detritivores and scavengers that eat dead material.

There are four types of predation:

1 *Herbivory* This occurs when animals prey on green plants (grazing, defoliation) or their seeds and fruits. It is not necessary that the plants are killed; in most cases they are not. Seed predators (granivores) and fruit eaters (frugivores) often kill the seed, but some seeds require digestion to germinate. We discuss herbivory in Chapter 9.

2 *Parasitism* This is similar to herbivory in that one species, the parasite, feeds on another, the host, and often does not kill the host. It differs from herbivory in that the parasite is usually much smaller than the host and is usually confined to a single individual host. The behavior of nomadic herders in Africa, who live entirely on the

Wildlife Ecology, Conservation, and Management, Third Edition.
John M. Fryxell, Anthony R. E. Sinclair and Graeme Caughley.
© 2014 John Wiley & Sons, Ltd. Published 2014 by John Wiley & Sons, Ltd.
Companion Website: www.wiley.com/go/Fryxell/Wildlife

blood and milk of their cattle, would also fit the definition of parasitism. Insect parasites (parasitoids) lay their eggs on or near their host insects, which are later killed and eaten by the next generation. We discuss parasitism further in Chapter 8.

3 *Carnivory*   This is the classical concept of predation, in which the predator kills and eats the animal prey.

4 *Cannibalism*   This is a special case of predation in which the predator and prey are of the same species.

7.4 The effect of predators on prey density

Table 7.1 compares caribou and reindeer populations in areas with different levels of wolf predation (Seip 1991). Densities vary by two orders of magnitude, the highest densities being in areas with few or no predators and the lowest in areas subject to high and constant predation. Conversely, Fig. 7.1 shows, first, that wolf densities are positively related to moose densities in Alaska and the Yukon (i.e. the highest wolf densities are in areas with the highest moose density), suggesting that wolves are regulated by their food supply, and second, that when wolves are removed, moose densities go up (Gasaway *et al.* 1992). Other studies in Alaska and the Yukon have repeated these wolf removals and show similar increasing moose and caribou populations (Boertje *et al.* 1996; Hayes and Harestad 2000). In a natural experiment in which red foxes were removed for several years by an epizootic of sarcoptic mange, prey numbers increased, particularly those of hares and several grouse species (Lindstrom *et al.* 1994). In general, predator-removal experiments show that the prey population increases or that some index such as calf or fledgling survival increases.

The observations in Table 7.1 and Fig. 7.1 appear to go in opposite directions. No interpretation of cause and effect is possible because they represent correlations. We cannot tell, for example, whether the predators are truly regulating the prey at levels well below that allowed by the food supply or just catching those prey that are suffering from malnutrition (so that they are not regulating), or whether both processes are occurring together. Prey availability is influenced by a number of factors: (i) whether there are alternative, more preferred prey in the area; (ii) the size of this and other prey populations; (iii) the vulnerability of different age and sex classes; (iv) whether the predators specialize on particular prey; and (v) how the environment affects the

Table 7.1 Density of caribou and reindeer populations in relation to level of predation. (After Seip 1991.)

Category	Location	Density/km^2
Major predators rare or absent	Slate islands	4–8
	Norway	3–4
	Newfoundland (winter range)	8–9
	South Georgia	2.0
Migratory Arctic herds	George River	1.1
	Porcupine	0.6
	Northwest Territories	0.6
Mountain-dwelling herds	Finlayson	0.15
	Little Rancheria	0.1
	Central Alaska	0.2
Forest-dwelling herds	Quesnel Lake	0.03
	Ontario	0.03
	Saskatchewan	0.03

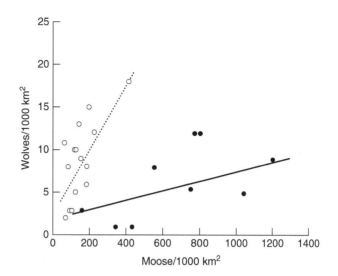

Fig. 7.1 Wolf density is related to moose density in Alaska and Yukon. In areas where wolves are culled (•) moose can reach higher densities than in areas where there is little culling of predators (○). (After Gasaway *et al.* 1992.)

efficiency of the predators in catching prey. To understand these processes we need to understand the behavior of predators.

7.5 The behavior of predators

We must first understand how predators respond to their prey if we are to interpret predator–prey interactions. We ask three questions. How do predators respond to: (i) changes in prey density; (ii) changes in predator density; and (iii) differences in the degree of clumping of prey? We look at these in the following three Sections.

7.5.1 The functional response of predators to prey density

The response of predators to different prey densities depends on: (i) the feeding behavior of individual predators, which is called the *functional response* (see also Sections 9.4 and 9.5), and (ii) the response of the predator population through reproduction, immigration, and emigration, which is called the *numerical response* (see also Section 9.5) (Solomon, 1949). We deal with the functional response first.

Understanding of the functional response was developed by Holling (1959). If we imagine a predator that (i) searches randomly for its prey, (ii) has an unlimited appetite, and (iii) spends a constant amount of time searching for its prey then the number of prey found will increase directly with prey density, as shown in Fig. 7.2a. This is called a *Type I* response. For the lower range of prey densities, some predators may show an approximation to a Type I response, such as reindeer feeding on lichens (Fig. 7.3), but for the larger range of densities these assumptions are unrealistic. For one thing, no animal has an unlimited appetite. Furthermore, a constant search time is also unlikely. Each time a prey is encountered, time is taken to subdue, kill, eat, and digest it (handling time, h). The more prey that are eaten per unit time (N_a), the more total time (T_t) is taken up with handling time (T_h) and the less time there is available for searching (T_s) (i.e. search time declines with prey density N).

Thus, handling time is given by:

$$T_h = h N_a$$

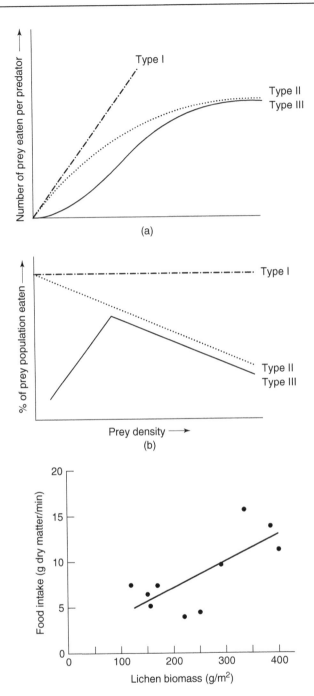

Fig. 7.2 (a) Types of functional response shown as the number of prey eaten per predator per unit time relative to prey density. (b) As for (a) but plotted as the percentage of the prey population eaten.

Fig. 7.3 The linear (Type I) functional response of reindeer feeding (dry matter intake) on lichen. (After White *et al.* 1981.)

and total time is:

$$T_t = T_h + T_s$$

The searching efficiency or attack rate of the predator a depends on the area searched per unit time a' and the probability of successful attack p_c, so that:

$$a = a' p_c$$

The number of prey eaten per predator per unit time N_a increases with search time, search efficiency, and prey density, so that:

$$N_a = a\, T_s N$$

Substituting the first two equations into this, we get:

$$N_a = a\, (T_t - h\, N_a)N$$

or:

$$N_a = (a\, T_t N)/(1 + a\, h\, N)$$

This is Holling's (1959) "disk equation," which describes a *Type II* functional response, where N_a increases to an asymptote as prey density increases (Fig. 7.2a).

When there are several prey types (species, sex, or age classes), the multispecies disk equation for prey type i eaten per predator is then:

$$N_{a\,i} = (a_i\, T_t\, N_i)/\, (1 + \Sigma_j\, a_j\, h_j\, N_j)$$

where the sum is across all prey types eaten.

The Type II functional response can be constructed from the parameters of the disk equation estimated from observations. Searching efficiency is the product of p_c and a'. The probability of capture is usually low, about 0.1–0.3 in most wildlife cases (Walters 1986). The area of search a' can be approximated from [distance moved × width of reaction field or detection distance]. Handling time per prey item h can be obtained from direct observation or from maximum feeding rates, because maximum rate $= 1/h$. Examples of such calculations are given in Clark *et al.* (1979) and Walters (1986).

The important effect of the Type II response is seen when numbers eaten per predator are re-expressed as a proportion of the living prey population (Fig. 7.2b). The Type II curve shows a decreasing proportion of prey eaten as prey density rises. Fig. 7.4a illustrates the Type II response of European kestrels (*Falco tinnunculus*) feeding on voles (*Microtus* spp.) in Finland (Korpimäki and Norrdahl 1991). The functional responses of herbivores are not as well known as those of carnivores, but where measured they appear to be Type II, as in Fig. 7.4b for bank voles (*Clethrionomys glareolus*) feeding on willow shoots (Lundberg 1988). Deer and elk show Type II functional responses to their food supply. Dale *et al.* (1994) report Type II responses of wolves preying on caribou.

Holling found a third type of functional response (*Type III*, Fig. 7.2). The number of prey caught per predator per unit time increases slowly at low prey densities but quickly at intermediate densities, before leveling off at high densities, producing an S-shaped curve. When those eaten are expressed as a proportion of the live population, the proportion consumed first increases, then declines. Hen harriers (*Circus cyaneus*) in the UK show a Type III functional response (Fig. 7.5) to changes in red grouse (*Lagopus l. scoticus*) populations (Redpath and Thirgood 1999).

The S shape of this curve is attributed to a behavioral characteristic of predators called *switching*. If there are two prey types, A being rare and B being common, predators will concentrate on B and ignore A. Predators may switch their search from B to

Fig. 7.4 The Type II
functional response of:
(a) European kestrel
feeding at different
densities of voles
(Microtus species).
"Kill" rate is voles eaten
per predator per
breeding season. (b)
Bank voles feeding on
willow shoots. (After
Lundberg 1988.)

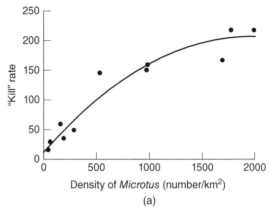

(a)

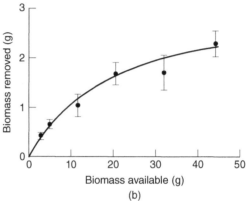

(b)

Fig. 7.5 The Type III
functional response of
hen harriers feeding on
red grouse chicks in UK.
(Source: After Redpath
and Thirgood, 1999.
Reproduced with
permission of John
Wiley & Sons Ltd.)

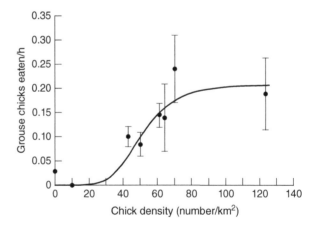

A, producing an upswing in the number of A killed, when A becomes more common.
There is often a sudden switch at a characteristic density of A. Birds have a *search image*
of a prey species such that they concentrate on one prey type while ignoring another.
As the rare prey (A) becomes more common, birds (such as chickadees (*Parus* spp.)
searching for insects in conifers) will accidentally come across A often enough to learn
a new search image and switch their searching to this species.

In practice, it is often difficult to determine whether there is a Type II or III response because the differences occur at low densities of prey and measurements are usually imprecise. The most robust evidence comes from determining whether predators ignore prey until there is a sizeable prey density available: that would indicate a Type III response.

7.5.2 *Predator searching*

The rate at which predators encounter prey always depends upon the density of the prey population. In some circumstances, interference among predators can also influence predator search efficiency, as we discuss further in Chapters 5 and 10. Such interference can take many forms. Mammal and bird predators are usually territorial and evict other individuals once a space has been fully occupied. If territory formation reduces access by some predators to prey, this will tend to reduce the average rate of attack.

Social predators can also experience reduced search efficiency if the group stays together while searching for prey, unless there is a compensatory increase in attack success such that each doubling in group size leads to a doubling in the probability of successful attack (Cosner *et al.* 1999; Fryxell *et al.* 2007). Such social interference is particularly likely in predators that form large groups, such as wild dogs and lions, simply because the probability of success cannot exceed 1. For example, large wolf packs in the Yukon have markedly lower rates of food intake per individual at a given prey density than do single wolves or pairs (Hayes *et al.* 2000).

Direct interference progressively reduces the searching efficiency of the predator as predator density increases, simply because of time wasted in aggressive encounters and avoidance of other predators (Beddington 1975). This can result in functional responses that are both prey- and predator-dependent or are dependent on the ratio of prey to predators (Arditi and Ginzburg 2012). The drop in searching efficiency caused by crowding lowers the asymptote of the functional response curve, which tends to stabilize numbers of predators and prey. Interference can also have a stabilizing influence on predator numbers if it causes dispersal once predators become too numerous.

7.5.3 *Predator searching and prey distribution*

Prey usually live in small patches of high density, with larger areas of low density in between; in short, prey normally have a clumped distribution. This can be seen in the patchiness of krill preyed upon by whales, of insects in conifers searched for by chickadees, of seeds on the floor of a forest eaten by deermice, of caribou herds preyed upon by wolves, and of impala herds hunted by leopards (*Panthera pardus*).

The searching behavior of predators is such that they concentrate on the patches of high density. By concentrating on these patches, predators have a regulating effect on the prey because of the numerical increase of predators by immigration, as will be discussed in the next Section.

7.6 **Numerical response of predators to prey density**

We define the *numerical response of predators* as *the trend of predator numbers against prey density* (see also Section 9.5 for other ways of looking at this). As prey density increases, more predators survive and reproduce. These two effects, survival and fecundity, result in an increase of the predator population, which in turn eats more prey. An example of this is Buckner and Turnock's (1965) study of birds preying on larch sawfly (*Pristiphora erichsonii*) (Table 7.2): as prey populations increased, the number of birds eating them also increased by reproduction and immigration; when plotted against prey density, predator numbers increased to an asymptote determined by interference

Table 7.2 Predation rate on larch sawfly in areas of tamarack (*Larix laricina*) (high density) and mixed conifers (low density). Bird predators include New World warblers and sparrows, cedar waxwing (*Bombycilla cedrorum*), and American robin (*Turdus migratorius*). (After Buckner and Turnock 1965.)

	High density (N/km^2)	Low density (N/km^2)
Sawfly larvae	528×10^4	9.88×10^4
Sawfly adults	50.75×10^4	1.16×10^4
Birds	58.1	31.1
%Predation larvae	0.5	5.9
%Predation adults	5.6	64.9

Fig. 7.6 The numerical response may be depicted as the trend of predator numbers against prey density.

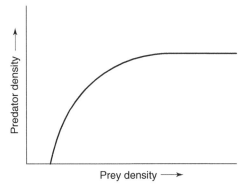

behavior such as territoriality (Fig. 7.6). Territoriality results in dispersal, meaning resident numbers stabilize. Wolves at high density have high dispersal rates: around 20% for adults and 50% for juveniles (Ballard *et al.* 1987; Fuller 1989). In New Zealand, the response of feral ferrets (*Mustela furo*) and cats to an experimental reduction of their primary prey (European rabbits) was a rapid, long-distance dispersal (Norbury *et al.* 1998). Extreme long-distance dispersal (800 km) of lynx has been observed in northern Canada when numbers of their primary prey, snowshoe hares, collapse.

The initial increase in numerical response may or may not be density-dependent. However, because of the asymptote, the numerical response at higher prey densities can only be depensatory (inversely density dependent). This means it has a destabilizing effect on the prey population, either driving the prey to extinction or allowing it to erupt. This is an important characteristic of populations and is illustrated in Buckner and Turnock's (1965) study: the proportion of sawfly eaten by birds in the high-prey-density area was lower than that in the low-density area (i.e. predation was depensatory and therefore could not keep the sawfly population down). The conditions under which regulation can or cannot occur are discussed in the Section 7.7.

7.7 The total response

We can now multiply the number of prey eaten by one predator (N_a, the functional response) by the number of predators (P, the numerical response) to give a total mortality M, where:

$$M = N_a P$$

The instantaneous change in prey numbers is:

$$dN/dt = -N_a P$$

Fig. 7.7 Theoretical total response curves (solid lines) of a predator population, measured as the percentage of prey population eaten, in relation to prey at different densities, (a) when there is density dependence in either the functional or numerical response, and (b) when there is no density dependence. The broken line represents the per capita net recruitment of prey $(dN/dt)(1/N)$ assuming logistic growth (i.e. after effects of competition for food have been accounted for). K is a stable point with no predators; A, C, and C1 are stable with predators; and B is an unstable boundary point.

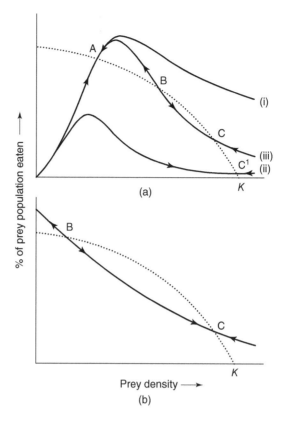

An approximation of changes in prey number over short intervals when prey populations do not change too much (< 50%) is given by:

$$N_{t-1} = N_t + N_t e^{-N_a P/N_t}$$

where $N_t = N$ in Holling's disk equation (Walters 1986).

If we express this total mortality M as a proportion of the living prey population N, we can get a family of curves as shown in Fig. 7.7, which depend on whether or not there is density dependence in the functional and numerical responses. If there is density dependence (e.g. from a Type III functional response) then we have a curve with an increasing (regulatory) part followed by a decreasing (depensatory) part. This is called a *total response* curve; examples of total response curves are shown for some of Holling's small mammals in Fig. 7.8a and for wolves eating moose in Fig. 7.8b (Boutin 1992).

7.7.1 *Regulatory effects of predation*

In Fig. 7.7 these total response curves have been superimposed on the per capita net recruitment rate of prey (dN/dt) $(1/N)$. For the case where we have density dependence (Fig. 7.7a), there are several stable equilibria (A, C, C^1) in which prey net recruitment is balanced by total predation mortality. Point B is an unstable equilibrium at which any perturbation to the system (from weather, for example) will result in the prey declining to A or increasing to C. In practice, B is never seen and is regarded as a boundary between domains of attraction towards A or C.

Fig. 7.8 Total response curves of predators at different prey densities. (a) Two shrews (Blarina, Sorex) and the deermouse (Peromyscus) eating European sawfly cocoons. (b) The proportion of moose populations killed by wolves in different areas of North America. (After Boutin 1992.)

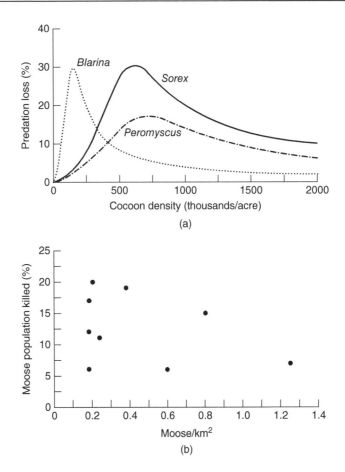

Curve (i) in Fig. 7.7 illustrates the case where predators can regulate the prey population under the complete range of prey densities and hold the prey at a low density A. One possible example of this occurs where both wolves and grizzly bears (*Ursus horribilis*) prey upon moose in Alaska (Ballard *et al.* 1987; Gasaway *et al.* 1992). Wolves appear to keep moose densities at low levels (< 0.4/km²). When wolves were removed in a culling operation, the mortality of juvenile moose caused by bears increased, so that moose numbers remained at the low level. Moose are kept at similar low levels by density-dependent predation by wolves in Quebec (Messier and Crete 1985). Red foxes can regulate some small marsupials in desert regions of Australia and some medium-sized marsupials in large eucalypt forests of Western Australia (Sinclair *et al.* 1998). The combined predation of two raptor species, the hen harrier (*Circus cyaneus*) and the peregrine falcon (*Falco peregrinus*), on red grouse in Scotland and England was found to be density-dependent in winter and probably regulated the prey (Thirgood *et al.* 2000). Regulation resulted at least partly from the Type III functional response referred to earlier (Redpath and Thirgood 1999).

Curve (ii) can occur when prey are regulated by intraspecific competition for food. Predators then kill malnourished animals and the effect on the prey population is depensatory rather than regulatory. This may be occurring on Isle Royale in Lake

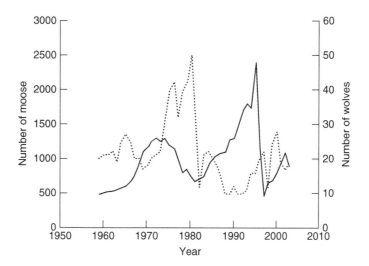

Fig. 7.9 Wolf (broken line) and moose (solid line) numbers on Isle Royale, during 1959–2003, show that the wolf population follows the fluctuations of moose, which are limited by food. (Source: After Peterson and Vucetich, 2003. Reproduced with permission of Michigan Technological University.)

Superior, where wolves cannot increase sufficiently to regulate the moose population (Fig. 7.9). Moose appear to be regulated by food (Peterson and Vucetich 2003), and wolf predation is merely depensatory (Fig. 7.9). Similar depensatory predation is exhibited in the total response of wolves depredating moose in the Findlayson Valley (data in Hayes and Harestad 2000).

Curve (iii) is the special case in which both A and C are present and there are multiple stable states. This situation has been suggested for a few predator–prey systems. One example is that of foxes feeding on rabbits in Australia (Pech *et al.* 1992). Foxes were experimentally removed from two areas and the rabbit populations increased in both, as would be expected from any of the curves in Fig. 7.7, so by itself the increase in prey tells us little about the nature of predation. However, when foxes were allowed to return to the removal areas there was some evidence that rabbits continued to stay in high numbers rather than return to their original low densities. This result suggests that we have curve (iii) and not (i) or (ii): the interpretation is that rabbit populations, originally at A, were allowed to increase above the boundary density, B, so that when foxes reinvaded the experimental area rabbit numbers continued towards C.

The "forty mile" caribou herd of the Yukon may have exhibited behavior characteristic of multiple stable states (Urquhart and Farnell 1986). Traditionally, this herd, whose range is on the Yukon–Alaska boundary, numbered in the hundreds of thousands – one estimate by O. Murie in 1920 being 568 000. In the 1920s and 1930s, goldminers and hunters killed tens of thousands. After the Second World War, when the Alaska Highway and associated roads were built, hunting increased further. By 1953 numbers were estimated at 55 000 and by 1973 there were only 5000 animals left. Although wolf numbers declined along with their prey, as one might expect, the proportional effect of predation was thought to be high. After 1973, hunting of caribou was restricted, and during 1981–83 wolf numbers were reduced from 125 to 60. Thereafter, wolf numbers returned to pre-reduction levels. Although caribou numbers increased marginally to 14 000 during the wolf reductions, they have remained at approximately this level since the early 1980s. Despite the lack of accurate population estimates, the density changes shown by the "forty mile" herd are so great (almost two orders of magnitude) that it is reasonably clear there has been a change in state from a high level determined by food to a low level determined by predators. The wolves may have been able to take over regulation because hunting reduced the caribou population size below the boundary level, B.

Another example of two states may be seen in the wildebeest of Kruger National Park, South Africa (Smuts 1978; Walker *et al.* 1981). In this case, high numbers of wildebeest were reduced by culling. When the culling was stopped, numbers continued to decline through lion predation, suggesting the system had been reduced below point B. A herbivore–plant interaction with two stable states is seen in Serengeti woodlands (Dublin *et al.* 1990; Sinclair and Krebs 2002). Woodland changed from high to low density in the 1950s and 1960s through severe disturbance from fires. In the 1970s, elephant browsing was able to hold woodlands at low density despite a low incidence of fires. Then poaching removed the elephants in the 1980s and the trees regenerated in the 1990s. Elephant numbers have been rebounding since 2000, but they cannot reduce the tree density.

7.7.2 *Destabilizing effect of predation*

Fig. 7.7b shows the case where predators have no regulatory effect but can cause the extinction of the prey species if prey numbers are allowed to drop below B. Predation mortality is greater than prey net recruitment below B, so that the prey population will decline to extinction. The conditions for this situation occur when there is no switching by predators (i.e. there is a Type II functional response), there is no refuge for the prey at low densities, and predators have an alternative prey source (their primary prey) to maintain their population when this (secondary) prey species is in low numbers.

Various mechanisms have been modeled by Gascoigne and Lipcius (2004). The inverse density-dependent effect of predation on secondary prey (i.e. greater proportional predation as numbers decline) has been shown experimentally using hen eggs in sooty shearwater (*Puffinus griseus*) nests in New Zealand. Smaller colonies of nests experienced higher proportional egg predation from rats and mustelids that were dependent on other primary prey, such as European rabbits (Jones 2003).

Low densities of the secondary prey could be produced by reduction of habitat (as has occurred with many endangered bird species) or hunting. For example, wolves prey upon mountain caribou in Wells Gray Park, British Columbia during winter, but not in summer when caribou migrate beyond the range of the wolves (Seip 1992). Recruitment for this herd in March is 24–39 calves per 100 females. In contrast, caribou in the adjacent Quesnel Lake area experience predation year-round and the average recruitment is 6.9 calves per 100 females. This population suffers an adult mortality of 29% (most of which is caused by wolves), well above the recruitment rate, and so the population is declining. Wittmer *et al.* (2005) have shown that the predation rate increases as caribou density declines, causing the populations to decline even faster (Fig. 7.10a), as predicted in Fig. 7.7b. Moose are now the primary prey in this system and maintain the wolf population. However, moose have only recently entered this ecosystem, having spread through British Columbia since the 1900s as a result of logging practices, so that previously wolves would not have had this species to maintain their populations at low caribou numbers. One interpretation, therefore, is that before the arrival of moose, caribou were probably the primary prey of wolves and the system was stable at either A or C; moose have now become the primary prey, caribou the secondary, and so the caribou may be vulnerable to local extinction (Hayes

Fig. 7.10 Depensatory total responses. (a) Wolf predation on different woodland caribou herds in British Columbia. Predation rate increases as caribou density declines, causing the populations to decline even faster. (b) Various mammal and bird predators on passerine bird nests as a function of forest patch size. These patches are an index of prey population size. (After Wilcove 1985.)

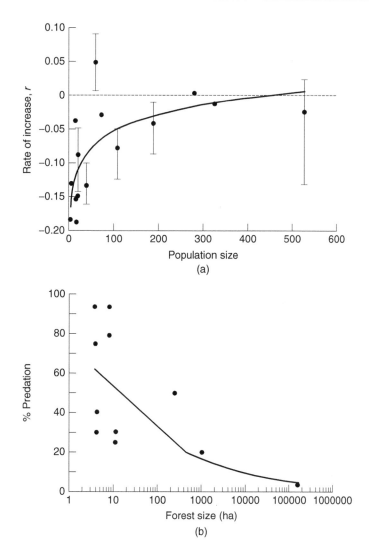

et al. 2000). Similar caribou declines have been recorded in central Canada (Rettie and Messier 1998).

Habitat fragmentation for passerine birds breeding in deciduous forests of North America is thought to be the primary reason for the major decline in their populations (Wilcove 1985; Terborgh 1989, 1992). The interiors of large patches of forest provide a refuge against nest predation from raccoons (*Procyon lotor*), opossums (*Didelphis virginiana*), and striped skunks (*Mephitis mephitis*) and against parasitism from brown-headed cowbirds (*Molothrus ater*). Fragmentation of the forests reduces this refuge, because nests are now closer to the edge of the forest where there are more predators and nest parasites. Predation rates are inversely related to forest patch size, which must be related to total prey population (Fig. 7.10b). In large forest tracts, nest predation is only 2%; in small suburban patches it is close to 100% and well above the recruitment rate. Since small fragmented forest patches are the norm in much of North

America, many populations of bird species may be in the situation shown in Fig. 7.7b where the density is left of the boundary B and declining to extinction.

A similar result was observed in southern Sweden with forest fragments embedded in an agricultural landscape. Andren (1992) recorded the impact of various species of the crow family as predators of artificial nests placed in the forest. Two species, the European jay (*Garrulus glandarius*) and the raven (*Corvus corax*), were confined to forest and were absent from small fragments, so their impact declined with fragmentation. Jackdaws (*C. monedula*) and black-billed magpies (*Pica pica*) were largely in agriculture. The hooded crow (*C. corone*) lived in agriculture but invaded forest patches, causing increased predation along forest edge and within small fragments.

In Kruger National Park, the expansion of zebra (*Equus burchelli*) populations into dry habitats when water holes were constructed in the middle of the 20th century allowed lions to move into those areas. Consequently, rare secondary prey such as roan antelope (*Hippotraus equinus*) and tsessebe (*Damaliscus lunatus*) have been driven towards extinction (Harrington *et al.* 1999).

Cougar (*Puma concolor*) appear to be having an inverse density-dependent effect, destabilizing bighorn sheep (*Ovis canadensis*) populations in the Sierra Ladron of New Mexico. These effects occur because cougar prey primarily on domestic cattle, which therefore subsidize the cougar population in this area (Rominger *et al.* 2004). The introduction of exotic predators and their exotic primary prey in Australia and New Zealand has caused declines and extinctions of endemic marsupials and birds. Thus, red foxes, which depend on European rabbits and sheep carrion, are able to drive black-footed rock wallabies (*Petrogale penicillata*) and other marsupials to extinction in Australia (Kinnear *et al.* 1998; Sinclair *et al.* 1998). In New Zealand, stoats (*Mustela erminea*), black rats (*Rattus rattus*), and brush-tailed possums (*Trichosurus vulpecula*), which depend upon exotic house mice (*Mus domesticus*), a variety of exotic passerine birds, and fruits, are driving endemic birds such as kokako (*Callaeas cinerea*) and yellowheads (*Mohoua ochrocephala*) to extinction (King 1983; Murphy and Dowding 1995); experimental reductions of these predators have allowed an increase in the endemic birds (Elliott 1996; Innes *et al.* 1999).

7.8 Behavior of the prey

We have seen how the behavior of predators can influence the nature and degree of predation. We will now examine how the behavior of prey affects predation rates.

7.8.1 *Migration*

If a prey species can migrate beyond the range of its main predators, its populations can escape predator regulation (Fryxell and Sinclair 1988a). This has been shown theoretically (Fryxell *et al.* 1988b), and there are some examples supporting this idea. The explanation for this escape from predator regulation is that predators, with slow-growing, nonprecocial young, are obliged to stay within a small area to breed. In contrast, ungulate prey, with precocial young, do not need to stay in one place as the young can follow the mother within an hour or so of birth. Thus, the prey can follow a changing food supply, while the predators cannot. For example, the wildebeest migrations in the Serengeti can follow seasonal changes in food and are regulated by food abundance; meanwhile their lion and hyaena predators, although commuting up to 50 km from their territories, cannot move nearly as far. Other examples from Africa are reported for wildebeest migrations in Kruger Park, South Africa (Smuts 1978) and white-eared kob (*Kobus kob*) in Sudan (Fryxell and Sinclair 1988b). In North America, a similar escape from predation is suggested for migrating caribou herds: the George

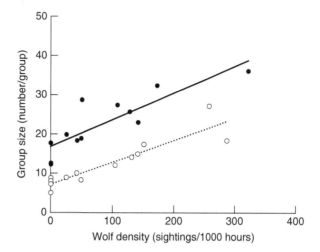

Fig. 7.11 Muskox group size on arctic islands in Northwest Territories, Canada, is related to wolf density in both summer (●) and winter (○). (After Heard 1992.)

River herd in Quebec (Messier *et al.* 1988); the barren-ground caribou (Heard and Williams 1991); the Wells Gray Park mountain caribou, through altitudinal migration (Seip 1992); and possibly the "forty mile" caribou, before hunting reduced the herd (Urquhart and Farnell 1986).

7.8.2 Herding and spacing

Theoretical studies propose that animals can reduce their risk of predation by forming groups, herds, or flocks (Hamilton 1971; Fryxell *et al.* 2007) and that group sizes should increase with increasing predator densities (Alexander 1974). The benefit of avoiding predators is counteracted by the cost of intraspecific competition within the group, however. There should be some group size at which the benefit–cost ratio is optimized (Terborgh and Janson 1986).

The effect of predators on herding behavior is illustrated in Fig. 7.11. The relationship between muskox (*Ovibos moschatus*) group size and wolf density suggests that predators are the most likely explanation for differences in group size in different populations (Heard 1992).

The opposite behavior to herding is shown by many female ungulates when they give birth. At that time they leave the herd and become solitary. This is seen in impala and other antelopes in Africa and in cervids, mountain sheep, and forest caribou in North America (Bergerud and Page 1987). This behavior relies on predators spending most of their search time in areas of high prey density, so that solitary prey at low density experience a partial refuge and hence lower predation rates.

Another form of refuge is used by deer in winter when they congregate in loose groups in the small areas between wolf home ranges (Rogers *et al.* 1980; Nelson and Mech 1981). These areas appear to be unused by the wolves, a sort of "no-man's-land," and hence they act as a refuge for the prey.

7.8.3 Birth synchrony

Some prey species synchronize their reproduction in order to lower predation rate on their young, a behavior called *predator swamping*. This synchrony is over and above that imposed by the seasons. For example, moose and caribou have highly synchronized birth periods (Leader-Williams 1988), as do many ungulates in Africa (Sinclair *et al.* 2000). Other examples of breeding synchrony are seen in lesser snow geese (Findlay and Cooke 1982) and colonial seabirds (Gochfeld 1980).

Experimental studies of foxes feeding in breeding colonies of black-headed gulls (*Larus ridibundus*) in England have shown that those gulls breeding outside the main nesting period are more likely to lose their nests to predators (Patterson 1965). However, synchrony may or may not be adaptive, depending on the abilities of the predator and the type of synchrony (Ims *et al.* 1988). Thus, if prey form groups and all groups are synchronized together, predator swamping can occur. However, if reproduction is synchronized within groups but not between them, predation rate on juveniles can be increased rather than decreased by this behavior, depending on the type of predator (Ims 1990). In general, breeding synchrony should not just be evaluated in terms of predation but should take into account other aspects such as seasonality of the environment. These aspects are important for conservation because species that rely on grouping behavior and synchrony of breeding will be vulnerable to excessive predation if human disturbances alter either aspect.

7.9 Summary

Some of the important points for conservation and management that we can derive from this discussion of predation are as follows:

1 Predator and prey populations usually coexist. Prey may be held at low density by predator regulation or at high density by intraspecific competition for food or other resources, and here predators are depensatory.

2 It is possible that both systems will operate in the same area, leading to multiple stable states. This may be generated by a Type III functional response or by a density-dependent numerical response at low prey densities. The system can move from one state to another as a result of disturbance. Such dynamics may occasionally underlie the outbreak of pest species and the decline of species subject to hunting.

3 Conversely, there are situations in which a prey population can go extinct, particularly with a Type II predator functional response, a lack of refuge for the prey, and alternative food sources for the predator. This is important in conservation, where habitat changes might reduce refuges, introduced pests such as rats might provide alternative prey for predators of rare endemic species, or invading prey such as moose or white-tail deer might assume the role of primary prey and so cause the original species to become vulnerable to extinction as secondary prey.

4 Which of these possibilities occurs depends on the ability of the predator to catch prey and the ability of the prey to escape, either by using a refuge or by reproducing fast enough to make up the losses. A very efficient predator defined by a high predator/prey ratio will hold the prey at low density.

8 Parasites and pathogens

8.1 Introduction and definitions

This chapter introduces parasitism and disease within wildlife populations. It addresses how an infection affects a population's dynamics and how it spreads through a population. The veterinary aspects of infection, special to each parasite and host, are not dealt with here. Instead we look at examples of how parasites and disease regulate populations, structure communities and affect conservation of endangered species, reduce the potential yield of harvested populations, and are of use in controlling pests.

Parasites feed on living hosts and (unlike predators) do not always kill them. Some parasites have many hosts, others are species-specific. Parasites and pathogens can best be divided into two classes: *microparasites*, which include viruses, fungi, and bacteria, and *macroparasites*, such as arthropods (e.g. fleas, ticks) nematodes, and cestodes (e.g. tapeworms). Microparasites and macroparasites have a roughly equivalent kind of effect upon their hosts and so can be lumped together as *parasites*. The debilitating effect of the parasite upon the host is termed *disease*. At the end of the book we include a glossary of terms often used in parasitology and epidemiology.

8.2 Effects of parasites

All animals support many species of parasites. For example, the American robin (*Turdus migratorius*) has at least 62 macroparasite species, the European starling (*Sturnus vulgaris*) 126 helminth species alone, the African buffalo over 60 species, and we ourselves (*Homo sapiens*) as many as 149 species (Windsor 1998). Many of these species live with their hosts through a substantial portion of the host lifespan, causing some minor debilitation. These parasite species are adapted to their hosts, and the hosts are adapted to the presence of the parasites. Such parasites are said to be *endemic*. The disease caused by this type of parasite is called *enzootic*. (Note the special use of the term "endemic" in this context. In another context, a species is endemic when it is confined naturally to one location, such as an island or a habitat).

Endemic parasites cause *chronic* impacts on a host; that is, low-level, persistent, nonlethal debilities or diseases. Other parasites cause *epizootic* disease (in animals) or *epidemic* disease (in humans). These cause relatively short-term, major, and often fatal debilities. As a result of human impacts and global climate change on ecosystems, we are experiencing the appearance of new diseases, sometimes termed *emerging infectious diseases* (EIDs). Enzootic and epizootic diseases have different effects on ecosystems, endangered species, and introduced pests. Parasites may lower the standing biomass

Wildlife Ecology, Conservation, and Management, Third Edition.
John M. Fryxell, Anthony R. E. Sinclair and Graeme Caughley.
© 2014 John Wiley & Sons, Ltd. Published 2014 by John Wiley & Sons, Ltd.
Companion Website: www.wiley.com/go/Fryxell/Wildlife

of a host population. Hence, they are disadvantageous if the host population is to be conserved or harvested, and advantageous if the host population is to be controlled.

The role of disease in mammals can be generalized to all vertebrates (Yuill 1987). Parasites can be expected in all wildlife species in every ecosystem. Death of the host is unusual and occurs only if: (i) serious illness facilitates transmission, as in rabies; (ii) the parasite does not depend on the infected host for survival and can complete its life cycle after the host dies; and (iii) the pathogen moves through host populations over a wide geographic area and over a long period of time. Disease may have a drastic effect on the survival of wildlife, but more commonly its effects are subtle. It can adversely affect natality or normal movement. Brucellosis in caribou has both effects. A caribou cow infected with brucellosis may abort her fetus, and the same disease can also cause lameness from degenerative arthritis in the leg joints. Infective agents can also affect the host's energy balance by reducing energy intake or increasing energy costs through higher body temperature and metabolic rate.

8.3 The basic parameters of epidemiology

Simple models for describing the way a disease establishes and spreads through a population start by assuming a constant host population size. This assumption allows us to understand transmission processes over short time intervals. More complex models can also account for changes in parasite and host populations.

8.3.1 Simple compartment models of parasite–host interactions

For directly transmitted infections of microparasites such as rinderpest, we can divide the host population (N) into three groups: *susceptibles* (S), *infected* (I), and *recovered* (R). The dynamic relationships between these groups are illustrated in a simple compartment model called the SIR model (Fig. 8.1; Anderson and May 1979). Host population size is determined by birth and death rates. Death rates arise from disease and other causes. The effects of disease are described by: (i) the per capita rate of mortality (α); (ii) the per capita rate of recovery (γ); (iii) the transmission rate or coefficient (β); and (iv) the per capita rate of loss of immunity.

The rate of change of the susceptible population is given by the rate of transmission of disease from infected to susceptibles. Thus:

$$\frac{dS}{dt} = -\frac{\beta SI}{N}$$

where N and β are assumed to be constant.

Fig. 8.1 The SIR model, showing the relationships between susceptibles, infected, and recovered components of the population.

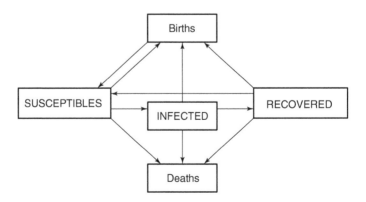

The rate of change of the infected population is given by the rate of transmission from infected to susceptibles minus the rate of recovery of infected animals. Thus:

$$\frac{dI}{dt} = \frac{\beta SI}{N} - \gamma I$$

where we assume γ is a constant and that transmission is directly related to the proportion of infected individuals in the population (I/N) times the size of the susceptible population (S).

The rate of change of the recovered population is given by the rate of recovered infected individuals. Thus:

$$\frac{dR}{dt} = \gamma I$$

8.3.2 Thresholds of infection and the transmission coefficient

When does a disease become epidemic; that is, start to spread through a population? The answer to this question depends on the net reproductive rate R_0 of the pathogen. For microparsites, R_0 is the average number of secondary infections produced by one infected individual, and for macroparasites it is the average number of offspring per parasite that grows to maturity. If the parasite has two sexes, it can also be defined as the average number of daughters reaching maturity per adult female.

If R_0 is less than unity, the initial inoculum of parasites will decay to extinction. R_0 is not a constant for a parasitic species but is determined by the varying characteristics of both the parasite and the host populations, particularly the density of the host. The conditions leading to persistence of the infection are given by Anderson and May (1986) and Anderson (1991) as the ratio of the rate at which new infectives are generated (β) to the rate at which they are lost:

$$R_0 = \frac{\beta}{\gamma}$$

An epidemic occurs if $R_0 > 1$, meaning that more infectives are generated than are lost. An epidemic stops when $R_0 = 1$. We can stop an epidemic by vaccinating a proportion (C) of the susceptible individuals. We can then reduce R_0 by:

$$R_0 = \frac{(1 - C)\beta}{\gamma}$$

The proportion that must be vaccinated in order to prevent an epidemic (i.e. to keep $R_0 = 1$) is:

$$C > 1 - \frac{\gamma}{\beta} = 1 - \frac{1}{R_0}$$

Thus, the proportion to be vaccinated is critically dependent on R_0. If $R_0 = 2$ then 50% of susceptibles must be vaccinated; if $R_0 = 10$ then 90% must be vaccinated (Krebs 2001).

The relationship can also be expressed in terms of a threshold host density N_T below which the infection will die out:

$$N_T = (\alpha + b + \gamma)/\beta > N$$

where b is the mortality rate of uninfected hosts. This makes the point that R_0 is dependent upon host density. Note that if the parasite is highly virulent (large α), or if recovery is rapid (large γ), or if the parasite transmits poorly between hosts (small β), then a dense population (large N_T) is needed to stop the infection dying out. This equation can be elaborated to take into account the effect of an incubation period and post-infection immunity (both of which increase N_T) and "vertical" transmission of the infection, whereby a fraction of the offspring of an infected female is born infected (which lowers N_T).

These equations encapsulate two important concepts of epidemiology:

1 Persistence or extinction of an infection is determined by only a few traits of the host and parasite.
2 The density of the hosts must exceed some critical threshold to allow the infection to persist and spread.

We next examine two examples of disease persistence in wildlife populations.

Swine fever

An example of the study of epidemiology involves classical swine fever (CSF) in wild pigs in Pakistan (Hone *et al.* 1992). This is a viral disease of pigs spread primarily by close proximity of hosts. It is widespread in Europe, Asia, and Central and South America. Understanding of its epidemiology is relevant to efforts to keep it out of North America and Australia.

CSF was introduced to a population from wild boar (*Sus scrofa*) in a 45 km^2 forest plantation in Pakistan. The known starting population (all of which were susceptibles) was 465. One infected animal was released into this population. After 69 days, 77 deaths had been recorded and it was assumed there were no deaths of uninfected animals. The regression of cumulative mortality over time provided a deterministic estimate of the transmission variable β as 0.00099/day. The threshold population of pigs (N_T) below which the disease cannot persist was estimated by:

$$NT = \frac{\alpha + \gamma}{\beta}$$

where α is the mortality rate from infection and γ is the recovery rate. Animals were infective for 15 days over this period. The mortality rate was 0.2/day and the recovery rate was 1/15 or 0.067/day. Thus, N_T was $(0.2 + 0.067)/0.00099 = 270$ animals.

The number of secondary infections (R_D) is the ratio of the number of susceptibles (S; in this case the starting population of 465) to the threshold population N_T. Thus:

$$R_D = S/N_T = 465/270 = 1.7$$

A disease establishes when R_D is unity or greater, but this is valid only for the initial population and not a prediction for persistence.

In general, six pieces of information are required from an epizootic to make predictions about the transmission of a disease:

1 the initial abundance of hosts;
2 the number of infectives initially involved;
3 the number of deaths during the epizootic;

Fig. 8.2 The threshold for establishment of brucellosis in bison of Yellowstone National Park is around 200 animals. (Data from Dobson and Meagher 1996.)

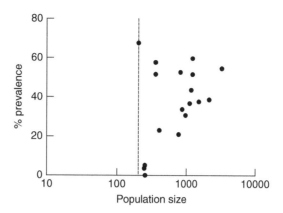

4 the incubation period;

5 the recovery rate;

6 the disease-induced mortality rate.

Brucellosis in Yellowstone National Park

Brucella abortus is a bacterium of the reproductive tract. It causes abortions and is transmitted by animals licking aborted fetuses and grazing contaminated forage. It is common in many ungulates of Africa and has been present in the elk and bison of Yellowstone National Park since the introduction of domestic stock to North America. There are species-specific differences in the effects of the disease on hosts. In elk, over 50% of females abort their first fetus, whereas in bison few if any do so (Thorne *et al.* 1978; Meyer and Meagher 1995).

Bison can acquire brucellosis from elk where the two species mix. Initially healthy bison in Grand Teton National Park acquired the disease from elk on the adjacent National Elk Refuge when the two species fed together in winter at Jackson Hole (Boyce 1989). Modeling of the epidemiology (Dobson and Meagher 1996) shows a threshold population for establishment in bison around 200 animals (Fig. 8.2), and the proportion of the host population infected increases directly with population density. The threshold population, however, is so low that it is very difficult to eradicate the disease – the population would need to be reduced below 200, a cull deemed to be unacceptable in a National Park.

8.4 Determinants of spread

Rate of spread *c* of an infection is determined, as is persistence, by traits of both the parasite and the host, particularly the rate of mortality α caused by the disease and the net reproductive rate R_0 of the pathogen. Källén *et al.* (1985) give the relationship as:

$$c = 2(D\alpha(R_0-1))^{1/2}$$

where *D* is a diffusion coefficient more or less measuring the area covered by the wandering of an infected animal over a given period of time. Dobson and May (1986a) calculated the constants of that equation for rinderpest in Africa from the observed radial spread of 1.4 km/day. Pech and McIlroy (1990) used a more elaborate version the other way around, estimating from knowledge of the equation's constants a potential spread of foot-and-mouth disease of 2.8 km/day through a population of feral pigs in Australia.

8.5 Endemic pathogens

8.5.1 Birth rates

Parasites usually take some of the energy and protein eaten by the host, and so the host suffers some loss. Such losses, if severe enough, can affect the reproductive ability of the host. When the nematode *Capillaria hepatica* was experimentally introduced to laboratory mice it resulted in reduced number of live young born and higher mortality of young before weaning. Such a reduction of natality and early survival might prevent the plagues of mice that are a feature of Australian wheatlands (Singleton and Spratt 1986; Spratt and Singleton 1986).

The bacterium *Brucella abortus* can reduce both conceptions and births in some ungulate species. In birds, parasites can reduce reproduction through forced desertion of nest sites, as in cliff swallows (*Hirundo pyrrhonota*) and many seabirds, or can reduce clutch size (barn swallow, *H. rustica*), cause delays in mating (great tit, *Parus major*), and lower body condition (house wren, *Troglodytes aedon*) (references in Loye and Carroll 1995). Red grouse (*Lagopus lagopus*) in northern England produced larger clutches of eggs and showed higher hatching success when the nematode *Trichostrongylus tenuis* was reduced with anthelmintic drugs (Hudson 1986). In general, there are still few data on the effect of parasites on host birth rates.

8.5.2 Mortality rates

Laboratory mice infected with the nematode *Heligmosomoides polygyrus* exhibited mortality rates in proportion to the intensity of infection (Scott and Lewis 1987). Soay sheep (*Ovis aries*) on the North Atlantic island group of St. Kilda exhibit population crashes every 3 or 4 years. Mortality is highest towards the end of winter, and dead animals have high nematode worm burdens. Live animals that were experimentally treated with anthelmintic drugs had higher survival rates (Gulland 1992). Other studies of rodents and hares show that mortality is often associated with high parasite burdens (e.g. helminths in snowshoe hares, Keith *et al.* 1984; bot flies in *Microtus* voles, Boonstra *et al.* 1980).

8.6 Endemic pathogens: synergistic interactions with food and predators

The great majority of parasites and diseases coexist with their hosts over long periods and do not exhibit wide fluctuations in prevalence over time. Direct mortality from these parasites is usually low. In contrast, they can have important indirect effects by (i) responding to the nutritional state of the host and becoming pathogenic or otherwise increasing vulnerability to predation and (ii) altering the behavior of the host.

8.6.1 Interactions of parasites with food supply

There is much evidence that the pathogenicity of parasites is influenced by the nutritional status of the host. In one experimental study, Keymer and Dobson (1987) repeatedly infected mice every 2 weeks for 12 weeks with larvae of the helminth *Heligmosomoides polygyrus*. Mice on a low-protein diet accumulated parasites in direct proportion to the infective dose. In contrast, those on high-protein diets had worm burdens that reached a plateau and even declined over time, and overall the worm burdens were lower for the same dose.

In a field study of snowshoe hares in Manitoba, Murray *et al.* (1997) reduced the natural burdens of sublethal nematodes using antihelmintic drugs. On three of six study areas, hares were provided with extra high-quality food during the winter, when food is normally limiting. The researchers found that survivorship of hares depended on a synergistic interaction of food and parasites. Overwinter survival was 56% in control animals (unfed and normal worm burdens). In unfed but parasite-reduced animals, survival was 60%, while in untreated but fed animals it was 73%. However, in fed and parasite-treated animals survival reached 90%.

Field experimental studies of the effects of parasites are rare, and most information comes from descriptive studies, where animals dying in poor nutritional condition also have high parasite burdens. Studies of the periodic mortality of Soay sheep on St. Kilda indicate that animals were emaciated and malnutrition was the cause of death. However, dead animals also had high nematode counts, indicating an interaction between food and parasites (Gulland 1992).

8.6.2 *Interaction of parasites with predators*

Parasites and pathogens can increase a host's vulnerability to predation by changing its ability to escape the predator. Snowshoe hares with high nematode burdens in spring were more likely to be caught in live traps than those with lower worm burdens (Murray *et al.* 1997). (Wood bison (*Bison bison*) populations may be held at low densities by predators only when there is a high prevalence of diseases such as tuberculosis (TB) and brucellosis (Joly and Messier 2004)).

In the red grouse (*Lagopus lagopus*), there is a complex interaction between the nematode *Trichostrongylus tenuis* and predators such as red fox (*Vulpes vulpes*). These game birds, being ground nesters, are vulnerable to predation while incubating eggs. Normally, grouse emit scent in the feces that can be detected by trained dogs (and presumably by foxes) up to 50 m away. However, during incubation female grouse stop producing caecal feces and dogs cannot locate the birds more than 0.5 m away. The parasite *T. tenuis* burrows into the caecal mucosa and disrupts its function so that the bird cannot control its scent (Dobson and Hudson 1994). Hudson *et al.* (1992a) demonstrated experimentally the effect of these worms on the detectability of incubating red grouse by dogs, treating some birds with anthelmintic drugs to reduce their worm burdens. Trained dogs found many fewer treated birds than untreated birds with naturally high worm burdens (Table 8.1). Thus, parasites increased the susceptibility of grouse to predation.

Parasites may also increase predation on hosts by altering the behavior of the host either as an incidental consequence of debility or as a specific adaptation to enhance transmission; the latter occurs when the predator is the final host in the life cycle of the parasite. In the former case, a disease that causes debility of the host makes it more conspicuous to predators through abnormal behavior, and especially flight behaviors. Other general responses are unusual levels of activity, disorientation, and altered responses to stimuli.

Three lines of evidence support the hypothesis that modified behavior of hosts is a strategy adapted to increase transmission (Lafferty and Morris 1996). First, hosts infected by transmissible stages of parasites often behave differently. Second, experimentally infected prey are more readily eaten by predators in laboratory experiments. Third, infected prey are eaten by predators more frequently than is expected in field

Table 8.1 Red grouse nests found by dogs (scent) and random search (researchers) with respect to treatment of the female with an anthelmintic to reduce burdens. (Source: Hudson *et al.*, 1992. Reproduced with permission of John Wiley & Sons Ltd.)

Year	Treatment	Number found	
		Dog scenting	Human search
1983			
Treated	Low worm burden	6	7
Untreated	High worm burden	37	10
1984			
Treated	Low worm burden	9	7
Untreated	High worm burden	29	7

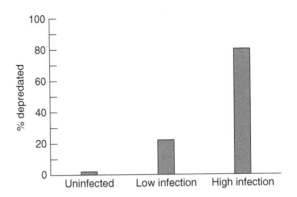

Fig. 8.3 Parasitized killifish are more heavily depredated by birds than are unparasitized fish. (Adapted from Lafferty and Morris 1996.)

studies. We have mentioned that parasitized snowshoe hares were more likely to be eaten by predators in spring (Murray *et al.* 1997). Conspicuous behaviors exhibited by killifish (*Fundulus parvipinnis*) were linked to parasitism by larval trematodes. Field experiments showed that parasitized fish were heavily depredated by birds, the final hosts (Fig. 8.3).

8.7 Epizootic diseases

Unlike enzootic diseases, epizootic ones have intermittent effects on host populations. They have outbreak phases, with rapid spread and high mortality, followed by periods of quiescence, when they lie dormant in host species. The great majority of the pathogens that cause epizootics are microparasites such as viruses and bacteria. Some case studies illustrate their behavior.

8.7.1 *Rinderpest*

Rinderpest is a virus from the Morbillivirus group (genus *Morbillivirus*, family Paramyxoviridae) that produces measles in humans and canine distemper in dogs, cats, and hyena. It is probably the oldest member of the group, from which others evolved. Predators can develop cross-immunity to distemper by feeding on herbivores infected with rinderpest (Rossiter 2001). Its natural host is cattle and it is endemic to Asia. It is highly contagious through droplet infection via licking and sneezing. It causes high fever and inflammation and lesions of the alimentary and respiratory passages.

From 1961 to 1976 an Africa-wide cattle vaccination campaign (called JP-15) aimed to eradicate the disease from cattle and thereby from wildlife. The latter, being unnatural hosts, could not maintain the disease by themselves; they obtained it by contact with cattle (Sinclair 1977, 1979a, 1995; Plowright 1982). Although the campaign largely succeeded, a few foci of infection remained in the remote regions of southern Sudan and Mali. By 1979, new outbreaks appeared in Mali, Mauritania, and Senegal,

Rinderpest, as far as evidence goes, was absent from Africa until it was introduced during the 1880s via Egypt to southern Sudan and Ethiopia. By 1889 it was causing epidemics in East Africa, where it killed 95% of the cattle and similar proportions of closely related wildlife, especially African buffalo but also wildebeest, and of the less closely related giraffe, warthog, greater and lesser kudu, and other antelope species (Rossiter 2001). By 1896 the epidemic had reached the tip of South Africa and the West African coast, causing similar mortality. Thereafter, rinderpest reappeared at roughly 20-year intervals, producing slightly less virulent epizootics. Mortality of susceptible animals was at least 50%.

and in 1981 it appeared in East Africa, dying out in 1984. Lax vaccination programs contributed to the spread, and it now appears that vaccination of cattle is required indefinitely (Walsh 1987; Rossiter 2001).

8.7.2 *Myxomatosis*

The *Myxoma* virus is endemic to rabbit species in South America. It was deliberately introduced to Australia in 1950 as a biological control agent for the European rabbit, which had become a serious exotic pest. The initial spread from the source infection on the upper Murray River was via mosquito vectors, which had increased as a result of recent floods. The first wave of the epidemic took 6 months to cross Australia and mortality was 99%. The virus has remained in the population since 1950 and every 2 years or so outbreaks occur, although mortality has declined to about 87%. Initial abundance of rabbits dropped considerably, but over the decades rabbit numbers have increased as they have become resistant to the virus and virulence has declined. Rabbit fleas (*Spilopsyllus cuniculus*) were introduced to augment the spread of the disease in wetter regions of Australia, and they now act as major vectors of the virus (Fenner and Fantini 1999).

8.7.3 *Rabbit hemorrhagic disease*

Rabbit hemorrhagic disease (RHD) is caused by a virus (*Lagovirus*, family Caliciviridae) that first appeared in domestic rabbits in China during the 1980s. It has since caused heavy mortality of wild rabbits throughout Europe. It is closely related to a disease killing European hares (*Lepus europaeus*).

It was being tested on Wardang Island, South Australia as a possible biological control agent for rabbits in Australia when it escaped and established in wild rabbits on the mainland in 1996 (Mutze *et al.* 1998). Although mechanisms of transmission and spread are not fully understood, blowflies (*Calliphora* spp.), a psychodid fly, the rabbit flea (*S. cuniculus*), and culicine mosquitoes are all carriers. Initial mortality of rabbits was high (about 90%) and rabbit numbers have remained depressed since its introduction (Bowen and Read 1998; Fenner and Fantini 1999).

8.8 **Emerging infectious diseases of wildlife**

The rinderpest in Africa is an early example of a number of diseases that have recently appeared in wildlife and human populations. The phocine distemper virus of grey seals (*Halichoerus grypus*) has spread along the coast of Europe (Kennedy 1990), while in Australia two orbiviruses cause blindness of eastern grey kangaroos (*Macropus giganteus*) (Hooper *et al.* 1999) and the *Chlamydia* bacterium causes blindness and urogenital disease in koala (*Plascolarctos cinereus*). These are wildlife cases, while acquired immunodeficiency syndrome (AIDS), ebola virus, tickborne spirochetal bacteria causing Lyme disease (*Borrelia burgdorferi*), and the virus causing severe acute respiratory syndrome (SARS) are human examples.

These EIDs are associated with a range of underlying causal factors. They can be classified on the basis of three main pathways of infection:

1 EIDs associated with a jump from domestic to wildlife populations living nearby.
2 EIDs connected with direct human intervention through translocation of host or parasite.
3 EIDs with no human or domestic animal associations.

The rinderpest is a clear case of the transfer of a virus from cattle to susceptible wildlife hosts that had not before met the disease. Similarly, canine distemper has spread into wild dog (*Lycaon pictus*) populations of the Serengeti, causing the local extinction of that species (Ginsberg *et al.* 1995); into lions, causing a 40% mortality; and also into hyena (Roelke-Parker *et al.* 1996). Most likely the rapidly expanding human population surrounding the Serengeti ecosystem, with its associated domestic dogs, which carry the disease, is the source of these new outbreaks. Another example is brucellosis (Meagher and Meyer 1994). This was introduced to North America with the import of cattle and then jumped to elk and bison in Yellowstone in the United States and Wood Buffalo National Park in Canada.

The translocation of wildlife for agriculture, hunting, and conservation has increased exposure to novel diseases. Translocation of fish and amphibians may have caused the ranavirus epizootics now threatening many amphibian populations (Daszak *et al.* 2000). Rabies epizootics in the eastern United States developed from translocations of infected racoons (*Procyon lotor*) from the southern part of the country, where the disease was enzootic (Rupprecht *et al.* 1995).

Zoos and captive feeding programs may inadvertently expose animals to novel diseases due to the close proximity of neighboring hosts. Asian elephants (*Elephas maximus*) in zoos have contracted a lethal herpes virus from neighboring African elephants (*Loxodonta africana*) (Richman *et al.* 1999). There is considerable concern that the agent for bovine spongiform encephalopathy (BSE) could be transferred to zoo-held wildlife through contaminated food, and thereby to free-living wildlife (Daszak *et al.* 2000).

A classic case of epidemic outbreak triggered inadvertently through translocation caused by humans is the catastrophic disease known as white-nose syndrome (Fenton 2012). Massive die-offs of small brown bats were discovered at a number of bat hibernacula in New York State in 2006. Many of the dead individuals had a white powdery infection of their nasal region. Subsequent laboratory analyses confirmed that the fungus *Geomyces destructans* was present in all of the dead specimens. The cause of death is still somewhat unclear, but is currently thought to stem from increased frequency of arousal from hibernation during the fall and winter. As each arousal event burns up a significant fraction of fat reserves, the disease causes death through increased risk of starvation.

Genetic data suggest that the fungus originated in Europe, where it occurs at low, nonpathogenic levels. It is speculated that infected European bats found their way across the Atlantic in a container or some other form of shipment. Subsequent laboratory trials verified that North American bats infected with the European fungal strain developed the full set of symptoms associated with white-nose syndrome (Warnecke *et al.* 2012), confirming that this is a novel disease for North American bats, with characteristically pathogenic properties.

Die-offs rapidly spread to neighboring states (Fig. 8.4), probably aided by bat dispersal during mating swarms in the late summer and fall, with similarly disastrous consequences. The disease has decimated whole bat colonies, with an average mortality rate of 73% per year (Frick *et al.* 2010). While bat survival rates tend to improve in subsequent years following introduction, the population growth rate can still remain negative for a number of years. Population viability analysis (Chapter 16) based on a Leslie matrix model with disease-mediated survival rates suggests that white-nose syndrome may be capable of causing regional if not complete extinction of the small brown bat, once one of the most common bat species in North America (Frick *et al.* 2010). The ecological consequences could be enormous, given the important role that bats play in feeding on aerial insects.

Fig. 8.4 Spatial spread of white-nose syndrome in small brown bat colonies in the northeastern United States during 2006–2010. (Source: Frick *et al.*, 2010. Reproduced with permission of the American Association for the Advancement of Science.)

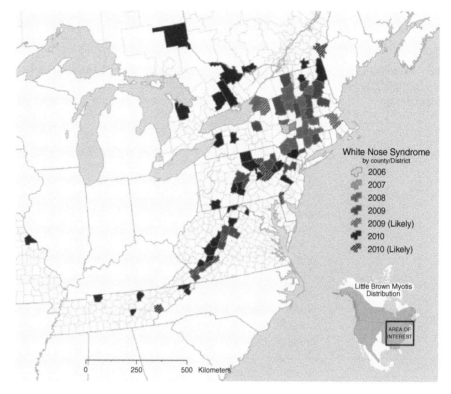

Climate change may be having an effect on the emergence, frequency, and intensity of epizootics. The chytrid fungus (*Batrachochytrium dendrobatidis*) causes mortality and decline in amphibian populations in many parts of the world via the fungal disease cutaneous chytridiomycosis (Berger *et al.* 1998; Pounds *et al.* 2006; Wake and Vredenburg 2008). The synchronous emergence of this novel disease, only discovered in the 1990s, in widely spaced sites (North, Central, and South America and Australia) could stem from global climate change (Pounds *et al.* 1999, 2006). Chytridiomycosis has been linked to massive die-offs of multiple frog species, through disruption of gas exchange at the moist skin surface of amphibians. It usually infects tadpoles, but does not kill them. Unfortunately, juveniles or adults are often reinfected following metamorphosis, with deadly consequences (Wake and Vredenburg 2008). Movement of infected adults from pond to pond is thought to aid in disease spread (Wake and Vredenburg 2008). Close correspondence between unusually warm years, chytridiomycosis outbreaks, and disappearances of multiple frog populations in montane study sites in the Central American tropics suggest that climate warming is a trigger for deadly disease outbreaks. Climate change has been similarly implicated in outbreaks of African horse sickness in South Africa (Bayliss *et al.* 1999) and various aquatic diseases (Marcogliese 2001).

In general, the causes of EIDs are largely ecological. These are: (i) movement and migration of hosts and pathogens to new environments; (ii) the change of environment *in situ* through global climate change; and (iii) a change in agricultural and forestry practices that bring species into contact. Changes in genetic characteristics of the pathogens play little if any part in EIDs, except perhaps in their ability to jump to new hosts (Krause 1992; Schrag and Wiener 1995).

8.9 Parasites and the regulation of host populations

As we have seen, most endemic parasites interact with other factors such as food and predators to reduce host population numbers. There are few examples where parasites on their own regulate a host population; that is, act in a density-dependent way. One clear example comes not from an endemic parasite but from an emerging epizootic disease. The poultry pathogen *Mycoplasma gallisepticum* has entered a previously unknown host, the house finch (*Carpodacus mexicanus*), in North America. The decline in finch population caused by the disease has been proportional to the initial density of finches, with the result that 3 years after the start of the epizootic most finch populations had stabilized at similar densities (Fig. 8.5). Thus, the mortality is density-dependent and the disease has regulated the finch population (Hochachka and Dhondt 2000).

We have already mentioned the emerging epizootic RHD, which was released in Australia in 1996 and has caused major declines in European rabbit numbers (Mutze *et al.* 1998). The disease now appears to be keeping rabbit numbers at very low levels. Bighorn sheep (*Ovis canadensis*) populations regularly experience pneumonia outbreaks caused by the bacterium *Pasteurella*. This has caused declines of bighorn sheep throughout western North America. Pneumonia can regulate bighorn sheep numbers, particularly in Idaho, keeping populations well below those determined by food resources. The source of the disease is domestic sheep, which are less susceptible to mortality from the pathogen than are the bighorns (Monello *et al.* 2001). More anecdotally, the rinderpest virus was probably regulating the African buffalo and wildebeest populations of the Serengeti, Tanzania before its removal through vaccination of cattle in 1963 (Sinclair 1977).

The role of endemic pathogens, particularly macroparasites, in regulating hosts is not clear. The nematode *H. polygyrus* regulated laboratory mouse populations (Scott 1987; Scott and Lewis 1987; Scott and Dobson 1989). We have yet to find examples from the field. However, recent studies suggest that macroparasites may at least be causing population cycles. For example, red grouse populations in Britain exhibit 7-year cycles, and it appears that these might be produced by the nematode *T. tenius*. Winter mortality

Fig. 8.5 The rate of change of a house finch population due to mortality by the pathogen Mycoplasma gallisepticum is density-dependent. (Data from Hochachka and Dhondt 2000.)

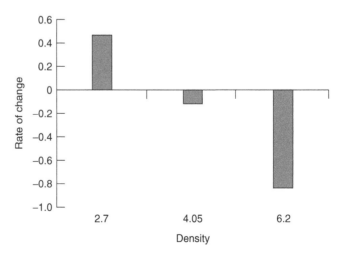

is the major factor determining changes in grouse numbers, although breeding losses are also important. Both winter loss and breeding loss are correlated with the intensity of parasite infection. Cycles may be resulting from time delays in the recruitment of parasites, so that they are partly out of phase with host numbers (Hudson *et al.* 1992b; Dobson and Hudson 1992). This idea was tested experimentally by reducing parasite burdens with anthelmintic drugs. Treatment of the grouse population prevented the normal decline in numbers, demonstrating that parasites were the cause of the decline phase of the cycle (Hudson *et al.* 1998).

8.10 Parasites and host communities

As more research on parasites is carried out, we are becoming aware of the role they play in structuring the diversity and abundance of host communities. This is a new area and much remains to be done (Minchella and Scott 1991; Poulin 1999). Most parasites have shorter life cycles and much faster rates of increase than their hosts. These features are the opposite to those of predators and, therefore, parasites can have different impacts on the structure of host communities.

8.10.1 *Altering species interaction*

Parasites can have three types of impact on host communities (Poulin 1999):

1 *Competition*   They can affect competitive interactions between two species by having a greater effect on one of the pair. A superior competitor may become an inferior competitor in the presence of the parasite. The northward spread of white-tailed deer in the hardwood forests of North America was accompanied by its meningeal nematode parasite *Parelaphostrongylus tenuis*. This worm is lethal to both moose and caribou, the original inhabitants of the forest, and populations of these species have declined (Anderson 1972; Nudds 1990; Schmitz and Nudds 1994). Thus, the parasite has altered the relative abundance of the three host species by affecting one less than the others. Schall (1992) shows that competition in *Anolis* lizards is altered by the presence of the malaria parasite (*Plasmodium azurophilum*). On the Caribbean island of St. Maarten, the normally dominant *A. gingivinus* excludes the subordinate *A. wattsi*, which is found only in the central hills. However, the parasite is common in *A. gingivinus* and rarely so in *A. wattsi*. In the presence of the malaria, the two coexist.

2 *Reducing predation*   They can reduce the efficiency of predators or herbivores in obtaining prey so that the prey increase at the expense of their competitors. In other words, parasites can alter the effect of "apparent competition" (see Chapter 9). Little has been documented at the carnivore trophic level. In herbivores, reduced food intake in reindeer is induced by gastrointestinal nematodes (Arneberg *et al.* 1996), allowing heavily grazed, palatable plants to increase in abundance. The presence of rinderpest in the Serengeti ecosystem (Section 8.7.1) reduced the dominant herbivore, wildebeest, by some 80%. One consequence of this reduction of wildebeest was to increase the biomass of grasses on the Serengeti plains and decrease both the diversity and the abundance of small dicot species, which are overshadowed and outcompeted by the grasses.

3 *Increasing prey susceptibility*   They can increase the availability of prey for a predator and so alter the competitive relationships between predators. We have already mentioned that parasites alter prey behavior to the benefit of their predators (Section 8.6.2). There are no data on how altered prey behavior affects the community of predators.

8.10.2 *Complex ecosystem effects*

Red grouse in northern England have declined in numbers due to an increase in prevalence of the tickborne virus louping ill, which affects the central nervous system.

The increase of the disease was produced by a change in the relative abundance of two plant species of the heath communities inhabited by grouse: heather (*Calluna vulgaris*), which is the major food for grouse, and bracken (*Pteridium aquilinum*), which produces a humid mat layer, the habitat for ticks. Bracken is increasing at the expense of heather because it can invade when heather is burned. Sheep ticks (*Ixodes ricinus*) are maintained by domestic sheep and mountain hares (*Lepus timidus*). The spread of bracken has increased the exposure of grouse to ticks and hence to louping ill virus (Dobson and Hudson 1986; Hudson *et al.* 1995).

Myxomatosis was introduced to rabbits in England in 1953. It caused severe mortality and resulted in several indirect effects on the ecosystem (Ross 1982). The normally closely-grazed grass lawns on the chalk downs changed to tussock grassland, with *Festuca rubra* and heather invading. Indeed, rabbits, which were introduced to Britain from Europe a thousand years ago, had maintained the species composition of these grasslands for so long that no one knew of any alternative state. There was an initial increase in diversity of flowering herbs, followed by dominance of tussock grasses, and eventually some areas turned to woodlands. Plant succession affected animal diversity: European hares (*L. europaeus*), voles (*Microtus agrestis*), and ants increased, while the sand lizard (*Lacerta agilis*) decreased. Predators that depended on rabbits, such as ermine (stoats, *Mustela erminea*) and buzzards (*Buteo buteo*), also declined. Similar changes were recorded in South Australia after RHD reduced rabbits there in 1996.

Dutch elm disease, caused by the fungus *Ceratocystis ulmi*, decimated elm trees (*Ulmus* spp.) in Britain in the early 1970s. These were among the most abundant trees in agricultural areas and their removal changed the physical structure of the habitats for birds. Death of the trees had less effect than their removal, because dead trees provided nesting and feeding sites. Bird diversity was reduced by eight species (from 36 to 28) as a consequence. Later, as the dead trees disappeared, increased light levels changed the herbaceous plant community (Osborne 1985).

These examples illustrate how the presence or absence of a disease can have complex indirect effects that filter down even to the plant community.

8.11 Parasites and conservation

Parasites and pathogens can be important in all three components of wildlife management. They can cause conservation problems by reducing the densities of species targeted for conservation, they can reduce the potential yield of harvested populations, but, on the positive side, they can be used to control pest species. This section provides examples in each of these areas to give a feel for the range of effects.

The long period of natural selection over which a parasite and its obligate host sort out an accommodation with each other ensures that a persistent infection has little influence on the density of the host. If, however, the specific characters of the host and parasite are such that usually $R_0 < 1$ then the infection is likely to be sporadic and may have a large but temporary depressing effect on the density of the host. Bubonic plague and (until recently) smallpox acted in this way against humans.

As we have already mentioned, parasites can reduce both birth and survival rates and, hence, affect population size. Therefore, they are relevant to the conservation of small populations and can be a cause of population decline (see Chapters 17 and 20). There are several ways in which threatened species may be exposed to parasites.

8.11.1 *Introduction of domestic or exotic species*

Microparasitic diseases are now implicated in the decline and extinction of several wildlife species, particularly carnivores (Ginsberg *et al.* 1995; Kat *et al.* 1995; Tompkins and Wilson 1998; Murray *et al.* 1999). Thus, African wild dog (*Lycaon pictus*), Ethiopian wolf (*Canis simensis*), and Blanford's fox (*Vulpes cana*) in Israel have all been decimated by rabies or canine distemper contracted from domestic dogs. The arctic fox (*Alopex lagopus semenovi*) on the Aleutian Islands has likewise contracted mange from dogs.

Parasites, especially microparasites, have their greatest effect when they jump from one species of host to another. This process is also a major source of evolutionary opportunities for parasites. For example, the knee worm (*Pelecitus roemeri*) has been "captured" by wallabies and kangaroos from the parrot family. The effect of the worm is unknown in parrots, but in macropods it induces a fibrous capsule up to the size of a cricket ball on the knee. Trans-specifics are the parasites and pathogens to watch out for. They can cause significant conservation problems, but they can also sometimes be used to control pest species. Other trans-species parasites must be guarded against because they cause considerable additional mortality. We have already mentioned the myxoviral rinderpest epidemic that swept the length of Africa in the 1890s and killed large numbers of wild ungulates, particularly African buffalo. Asian cattle were its original host, but it jumped across to wild ungulates when it reached Africa. The decline of moose populations in Nova Scotia and New Brunswick is associated with infestation by the nematode brainworm *Parelaphostrongylus tenuis*, which jumped from its original host, white-tailed deer. The infestation in moose is fatal, but there is little evidence that the parasite can maintain itself in the moose except by reinfection from white-tailed deer (Anderson 1972). However, the relationship between meningeal worm, white-tailed deer, and moose has not been studied experimentally and not all the circumstantial evidence is consistent (Samuel *et al.* 1992).

The translocation of domestic or exotic wildlife may lead to parasites and pathogens jumping to a new suite of species. In Australia, native animals such as kangaroos and wombats became infected with common liver flukes (*Fasciola hepatica*) acquired from sheep and cattle. The liver flukes cause severe lesions in the liver of the wombat (Spratt and Presidente 1981).

We have already seen how the presence of sheep brought the sheep tick to English moorlands and the louping ill virus to red grouse, causing their populations to decline. Similarly, high mortality of monkeys (*Presbytis entellies*, *Macaca radiata*) occurred after cattle were introduced to India, increasing the numbers of the tick *Haemophysalis spingera*. This tick carried the flavivirus causing Kysanur Forest Disease (Hudson and Dobson 1991).

Perhaps the most well known example of parasites introduced by wild exotic species is that of avian malaria in the Hawaiian Islands (Dobson and Hudson 1986; van Riper and van Riper 1986; McCallum and Dobson 1995; Cann and Douglas 1999). Early extinction of lowland native bird species (12th–19th centuries) resulted from agricultural clearing of forests, and later the introduction of rat and Indian mongoose (*Herpestes auropunctatus*), which depredated their highly vulnerable nests. The mosquito *Culex quinquefasciatus* was introduced in 1826. It spread across all of the islands but it did not carry avian malaria – no cases were detected from blood samples in the early 1900s.

Many species of exotic bird were introduced to the islands between 1900 and 1930 as a response to the damage caused by the clearing of forests for agriculture. As in New

Zealand and Australia previously, there was an organization (the Hui Manu) committed to introducing exotics. It is now clear that these exotics were responsible for bringing in the avian malaria (*Plasmodium relictum capistranoae*). Native birds were highly susceptible to the parasite and many species became extinct because of it. Now native species are restricted to habitats above 600 m, where both the mosquito and exotic birds are at low density. Reintroduction of native species in lowlands is not feasible in the presence of the parasite.

The species-jumping process may also operate the other way, with wildlife acting as reservoirs of parasites and pathogens transmitted to domestic stock. The controversies over brucellosis in bison (and its transmission to domestic stock in both the United States and Canada) and the transfer of TB (*Mycobacterium bovis*) from European badgers to cattle are obvious examples (Peterson *et al.* 1991; Clifton-Hadley *et al.* 2001). The appearance of SARS in humans in 2002 is thought to have arisen because people in southern China keep civets (*Viverra zibetha*) in captivity and eat them.

8.11.2 *The alteration of habitat*

We explore in Chapter 21 the general consequences of degraded habitats and fragmentation for conservation. One particular effect of habitat fragmentation is increased exposure to parasites. In birds, and perhaps in other animal groups, Loye and Carroll (1995) suggest three mechanisms: (i) increased edge habitat due to fragmentation increases the contact rate between species in adjacent habitats and exposes those in fragments to new vectors and new parasites to which they are more susceptible; (ii) loss of habitat could force birds to reuse old nests, exposing them to higher numbers of fleas, ticks, and other nest-living parasites; (iii) as a special case in birds, fragments expose them not only to predators commuting from the surrounding agriculture but also to brood-parasite birds such as the brown-headed cowbird (*Molothrus ater*) of North America.

Some of these mechanisms are illustrated by Loye and Carroll (1995) for the Puerto Rican parrot (*Amazona vittata*). This species is restricted to a single fragment of high-elevation forest. Habitat degradation, harvesting for the pet trade, and novel parasites have been factors in its decline. In particular, fatal parasitism of nestlings by muscid botflies became a problem after the pearly-eyed thrasher (*Margarops fuscatus*) invaded the forest fragment in about 1950. The thrasher nestlings are host to abundant blood-feeding botfly maggots. The introduced native thrasher therefore brought in its endemic parasite, which then spread to a new but rare host, the parrot.

8.11.3 *Captive breeding and reintroductions*

Parasites and pathogens can be a factor driving the decline of an endangered species and can become an issue in the recovery of endangered species generally. Parasites and pathogens can hinder or thwart attempts to establish captive-breeding populations. Thorne and Williams (1988) review the well-known example of the first attempts to establish a captive-breeding colony of black-footed ferrets (*Mustela nigripes*) in the United States. A previous attempt in the early 1970s failed because canine distemper virus (CDV) killed the only two litters. The source colony also disappeared. The extreme susceptibility of the black-footed ferrets to CDV became apparent when four of six died after being vaccinated for CDV in the 1970s. The vaccine had previously been shown to be safe in domestic ferrets. In 1981 the species was rediscovered in Wyoming, United States and the colony's vulnerability to disease was quickly realized. Precautions were taken to minimize human introductions of disease, especially CDV and influenza. The population declined from an estimated peak of 128 in 1984 to only

16 in 1985. The decline spurred an attempt to start a captive-breeding colony, but the first six ferrets captured rapidly succumbed to DCV. Despite all precautions, CDV infected the colony and most of its members died from the virus. The few survivors eventually formed a breeding population. Nonetheless, as Thorne and Williams (1988) note, "The captive breeding program went from a carefully planned approach with ideally selected, unrelated founder animals to a crisis situation with related animals, a poor sex ratio, and few mature, experienced breeder males."

Captive-bred animals released into the wild may spread disease or pick up parasites and pathogens from endemic wildlife. A potential example of the former is Jones's (1982) report of the release of Arabian oryx captive-raised in the United States for a national park in Oman, which was delayed when the animals tested positive for antibodies to bluetongue disease. The failure of the reintroduction of woodland caribou to an island within their historic range in Ontario, Canada is an example of the problems that can be encountered when reintroduced animals become infected with a disease from the endemic wildlife. The area had been colonized by white-tailed deer and the caribou became infected with meningeal worm via a gastropod secondary host (Anderson 1972).

Another example comes from the captive breeding of whooping cranes (*Grus americana*). An eastern equine encephalitis (EEE) virus fatally infected 7 of the 39 captive-bred population at the Patuxent Wildlife Research Centre in Maryland, United States. At that time in 1985, the captive population accounted for about 25% of the world's population. EEE virus causes sporadic outbreaks of disease in mammals and birds in the eastern United States and is spread by mosquitoes. No deaths are usually seen in endemic hosts, but introduced game birds such as ringneck pheasants (*Phasianus colchicus*) are vulnerable. Among the approximately 200 sandhill cranes in pens neighboring the whooping cranes, some were serum positive for EEE virus, but no clinical signs were found. The discovery of the vulnerability of the whooping cranes to a common pathogen was seen as an unrecognized risk and an obstacle to the species' recovery (Carpenter *et al.* 1989).

8.12 Parasites and control of pests

Spratt (1990), in reviewing the possible use of helminths in controlling vertebrate pest species, pointed out the marked contrast between the numerous successes in biological control of insects and the almost universal failure of such methods in controlling vertebrates. The one unequivocal success has been the use of the myxoma virus to control European rabbits in Australia (Fig. 8.6):

> Myxomatosis is a benign disease in *Syvilagus* (cottontail) rabbits in South America which is transmitted mechanically by mosquitoes. In the European rabbit (*Oryctolagus*), which is a pest in Australia and England, the virus from *Sylvilagus* produced a generalized disease that is almost always lethal. Myxomatosis was deliberately introduced into Australia in 1950 and into Europe in 1952. It was first spectacularly successful in controlling the rabbit pest, but biological adjustments occurred in the virulence of the virus and the genetic resistances of the rabbits. After 30 years of interaction, natural selection has resulted in a balance at a fairly high level of viral virulence. Fenner (1983)

The initial annual mortality rates were very high in Australia – over 95% – but they dropped progressively over the next few years. There is a widespread perception that

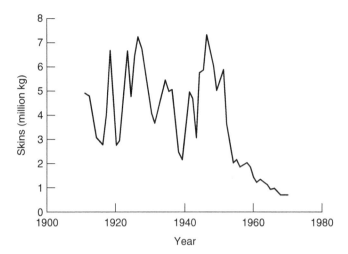

Fig. 8.6 The number of rabbits trapped in Australia (million kilograms of skins) shows a rapid decline after 1950, when the myxoma virus was introduced to control them. (Data from Fenner 1983.)

the rabbits and the disease accommodated to each other and, therefore, that myxomatosis provided only a temporary respite. This is not so. The rabbit density at equilibrium with the disease is considerably lower than the mean density in the absence of it.

Parer *et al.* (1985) demonstrated the controlling effect of this virus. They used a relatively benign strain of myxoma to immunize rabbit populations against the more virulent field strains that swept through the study area in most years. Rabbit densities increased by a factor of 10 under this treatment. Even after the rabbits and the virus had reached an accommodation with each other, the disease was apparently holding the mean density of rabbits to about 10% of that prevalent before its introduction.

8.13   Summary

Most parasites and pathogens have little effect on their hosts. When a parasite jumps from one host species to another, it is the "naivety" of the new host to the parasite or pathogen that is responsible for the reaction of the new host individual or population to the parasite. In both meningeal worm in cervids (other than white-tail deer) and liver fluke infection in wombats, it is the dramatic host immunologic response to the new parasite that is responsible for the debilitation in the animals. Such a response has been dampened over time as the meningeal worm has evolved in white-tail deer and liver fluke in sheep, so that we no longer see the same level of debilitation in the "normal" host species.

The key points from the epidemiology of parasites and pathogens are that the fate of an infection is determined by only a few traits of the host and parasite and that there is a critical density of the host that allows the infection to persist and spread. Efforts to reduce the effects of parasites and pathogens can be at their most important in the management of small populations of endangered species, be they in the wild or in captivity. Diseases of harvested wildlife are more rarely controlled, unless they present a potential hazard to people. Few attempts to use parasites and pathogens to control pest wildlife have been successful.

9 Consumer–resource dynamics

9.1 **Introduction**	In this chapter we explore those things an animal needs to eat so as to survive and reproduce: resources. This leads to a description of the structure and dynamics of consumer–resource systems, in which the consumers and their resources can interact in complex ways. We show how to analyze such systems by breaking them down into their dynamic components. This approach is used to compare several different systems: kangaroos and plants in Australia; trees, moose, and wolves in North America; small mammals in northern Europe; and snowshoe hares and lynx in Canada.

9.2 **Quality and quantity of a resource**	A *resource* is defined as something that an animal needs whose consumption by one individual makes it unavailable to another. The most obvious example is food, and to that may be added shelter, water, and nesting sites. By definition, a resource is beneficial. As the availability of resources rises, an individual's fecundity and probability of survival are enhanced.

Food resources are often characterized by two attributes: the amount of food available to an animal and the suitability of that food to the animal's requirements. For example, quality may be described as the percentage of digestible protein in the food, whereas quantity may be measured as dry mass of food per hectare. This often leads to a discussion on whether the quality or the quantity of the food is the most important to the animal. In most cases the distinction is meaningless. It indicates that the resource is being measured in the wrong units. If the resource is in fact digestible protein then that is what should be measured. The availability of the resource should be expressed as dry weight of digestible protein per hectare. Its measurement may entail measuring dry weight of herbage as an intermediate step, but that does not make herbage the resource.

9.3 **Kinds of resource**	It is necessary at this stage to give a classification of resources, because the interaction between a resource and the animals that depend upon it can take several forms. These in turn influence the dynamics of the population in different ways.

The use of a resource may be *preemptive*. An example is the use of nesting holes by parrots: individuals are either winners or losers. On the other hand, the use of a resource may be *consumptive*. All individuals have access to the resource and each individual's use of it reduces the level available to others. An example is the use of plants by herbivores. We see that both preemptive and consumptive uses of a resource remove a component of that resource from use by other individuals. Consumptive use removes the component permanently, whereas preemptive use removes it temporarily.

Wildlife Ecology, Conservation, and Management, Third Edition.
John M. Fryxell, Anthony R. E. Sinclair and Graeme Caughley.
© 2014 John Wiley & Sons, Ltd. Published 2014 by John Wiley & Sons, Ltd.
Companion Website: www.wiley.com/go/Fryxell/Wildlife

To complete the classification, there may be an *interactive relationship* between the population and the resource in that the level of the resource influences the rate of increase of the population, and reciprocally the level of the population's density influences the rate of increase of the resource. The dynamics of the animals interact with the dynamics of the resource. Examples are the relationship between a herbivore and its food supply and the relationship between a predator and its prey. In a *reactive relationship*, however, the rate of increase of the animal population reacts to the level of the resource (as before) but the density of the animals has no reciprocal influence on the rate of renewal of the resource. The relationship between a scavenger and its food supply or between a herbivore and salt licks are examples of reactive relationships.

We start by developing a general theoretical framework that applies in principle to all consumer–resource relationships, regardless of whether they focus on plants and herbivores, carnivores and their herbivorous prey, or all three.

9.4 Consumer–resource dynamics: general theory

The origin of consumer–resource theory can be traced directly to the contributions of two early ecologists: Alfred Lotka and Vito Volterra (Kingsland 1985). Starting from very different backgrounds, these two men simultaneously developed a similar framework for thinking about interactions between consumers and their resources (Lotka 1925; Volterra 1926b), a general framework that is still in common use, albeit with considerable change in biological details. The framework is a set of mathematical expressions for simultaneous changes in the density of consumers (symbolized here by N) and their resources (symbolized by V):

$$\frac{dV}{dt} = growth\ of\ resource - mortality\ due\ to\ consumption$$
$$\frac{dN}{dt} = growth\ of\ consumer\ due\ to\ consumption - mortality$$

In the case of resources, mortality is largely due to consumption, whereas for consumers there is a constant background level. For example, one might model reproduction and mortality by the following equations (Rosenzweig and MacArthur 1963):

$$\frac{dV}{dt} = r_{max}V\left(1 - \frac{V}{K}\right) - \frac{aVN}{1 + ahV}$$
$$\frac{dN}{dt} = \frac{acVN}{1 + ahV} - dN$$

where r_{max} is the maximum per capita rate of resource recruitment, K is the resource carrying capacity in the absence of consumers, a is the area searched per unit time by consumers, h is the handling time for each resource item, c is a coefficient for converting resource consumption into offspring, and d is the consumer per capita mortality rate. The particular form of the equations used in this example is based on the most commonly observed patterns. In the absence of consumption (e.g. when $N = 0$), the resource population has a logistic pattern of growth (see Chapter 5). In other words, the resource population is self-regulating. Consumption rates and per capita rates of growth by predators are presumed to increase and level off according to Holling's (1959) Type II functional response (see Chapter 7).

Depending on the parameter values that one uses, this model is capable of a variety of dynamics. In Fig. 9.1, we demonstrate one possible outcome: cyclic fluctuations of both consumers and resources over time.

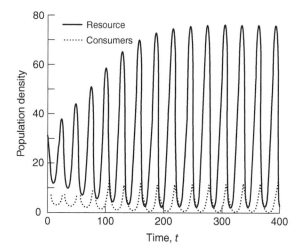

Fig. 9.1 Cyclic dynamics over time for the general consumer–resource model with the following parameter values: $a = 0.1, h = 0.2,$ $c = 0.2, d = 0.3, r_{max} =$ 0.4, and $K = 120$.

Rather than plot densities of both resource and consumer populations against time, ecologists often plot the density of consumers against that of resources. This is known as a phase-plane diagram. For example, in Fig.9.2 we replot the data shown in Fig. 9.1 as such a phase-plane diagram. The phase-plane trajectory in this figure displays a pattern spiraling outward from the starting point until it converges on a repetitive pattern known as a *stable limit cycle*. For most realistic consumer–resource models, this is a common outcome (Rosenzweig 1971; May 1972, 1973). If we had started at values outside the stable limit cycle, we would observe a spiral inward until the trajectory converged once again on the stable limit cycle.

There are some useful additional lines displayed in Fig. 9.2: the null isoclines (sometimes termed "nullclines") for consumers (the vertical line) and resources (the hump-shaped curve). Each of the null isoclines identifies a combination of consumer and resource densities at which $\frac{dV}{dt} = 0$ or $\frac{dN}{dt} = 0$ (i.e. one or the other population is unchanging). By rearranging our equations for simultaneous change in consumer and

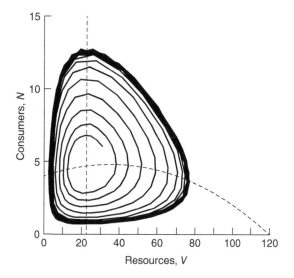

Fig. 9.2 Data from Fig. 9.1 replotted as a phase-plane diagram of consumers versus resources.

resource density, we can show that this will be the case when:

$$N = \frac{r_{max}V\left(1 - \frac{V}{K}\right)}{\frac{aV}{1+ahV}}$$

In other words, at any of the consumer and resource combinations lying on the hump-shaped isocline, consumption exactly matches the rate of resource production, so resource density will be unchanging.

Similarly, at the resource density shown by the vertical dotted line, the consumer population acquires just enough resources to allow it to balance mortality with off-spring production. This occurs when:

$$V = \frac{d}{a(c - dh)}$$

In this case, there is only one sustainable combination of consumer and resource densities at which both are unchanging: the point of intersection of the two null isoclines. If we somehow set both populations to this equilibrium point, they would stay there. Slight deviation from the equilibrium leads to spiraling outwards of the consumer–resource trajectory until the stable limit cycle is reached (Fig. 9.2). Hence, the coexistent equilibrium is dynamically unstable, at least for these parameter values. Other trivial equilibria are also present: when both N and $V = 0$ and when $N = 0$ and $V = K$. These equilibria are also unstable for the parameter combination shown in Fig. 9.2.

For other parameter values, a second sustainable outcome is possible: a stable equilibrium for both consumers and resources (Fig. 9.3). The only difference between the models plotted in Fig. 9.2 and Fig. 9.3 is the carrying capacity of resources. A decrease in the resource carrying capacity tends to be stabilizing, while its enrichment tends to be destabilizing. This has been termed the "paradox of enrichment," whereby provision of a better resource environment leads only to destabilization of consumers (Rosenzweig 1971).

Fig. 9.3 Phase-plane diagram of the dynamics over time for a stable form of the consumer–resource model with the following parameter values: $a = 0.1, h = 0.2, c = 0.1, d = 0.3, r_{max} = 0.4$, and $K = 70$.

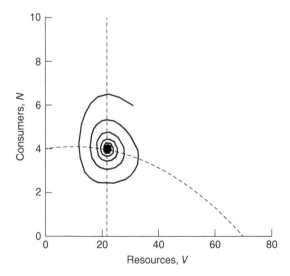

Although the complete explanation for this phenomenon is complex, the system can be usefully viewed as reflecting a dynamic tension between stabilizing influences (such as self-regulation by resources) and destabilizing influences (such as consumption of resources). The reason that consumption tends to be destabilizing is that the per capita risk of resource mortality for a given consumer density is inversely related to resource density (see Chapter 5). Hence, an increase in resource levels leads to diminished risk of death, which imparts a positive, rather than negative, feedback on population dynamics. When the carrying capacity is small, the consumer null isocline lies close to the resource carrying capacity, to the right of the hump in the resource null isocline. This is a region where stabilizing influences are stronger than destabilizing ones. In contrast, when the carrying capacity is large, the consumer null isocline lies far from the carrying capacity, where destabilizing influences hold sway.

With other parameter combinations, consumers would be unable to persist, simply because the intake of resources at any resource level would be unable to compensate for mortality. The possible outcomes of this consumer–resource model depend entirely on the parameter values. Predicting the outcome of even this highly simplified representation requires detailed knowledge of the magnitude of ecological parameters. We now go on to illustrate how this approach can be applied to a well-studied system: red kangaroos and their food plants in Australia.

9.5 Kangaroos and their food plants in semi-arid Australian savannas

9.5.1 Plant dynamics

The dynamics of a renewable resource can be quite complicated, containing elements of seasonality, intrinsic growth pattern, and the modification of the two by the animals using the resource. To clarify some general issues, we shall consider in detail a well-studied example: the growth of the herbage layer fed upon by kangaroos in the arid zone of Australia.

Fig. 9.4 shows Robertson's (1987) estimate of the *plant growth response*: growth by ungrazed herbacious plants in response to rainfall. He sampled growth rates on a kilometer grid over 440 km^2 of the arid zone of Australia. The measurements were repeated every 3 months for 3.5 years and rainfall was recorded for each 3-month interval. The curve labeled 60 mm indicates that the higher the biomass at the start of the 3-month period, the lower the increment of further biomass added over the next 3 months. This is to be expected as plants compete for space, water, light, and nutrients.

Fig. 9.4 Plant growth rate as a function of plant biomass at the beginning of the interval and rainfall during the interval, for pastures in Kinchega National Park. (After Robertson 1987.)

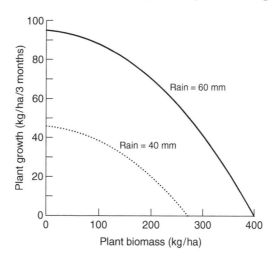

Table 9.1 100 years of rainfall statistics at central Australian study site.

	Mean (mm)	Standard deviation (mm)
Dec–Feb	62	59
Mar–May	57	47
Jun–Aug	59	34
Sep–Nov	61	44
Annual	239	107

The 60 and 40 mm curves shown in Fig. 9.4 are part of a family of curves each representing that trend for a given rainfall over 3 months. We can summarize the figure by saying that the higher the rainfall, the higher the growth increment, but that for a given rainfall, the higher the starting biomass, the lower the growth increment. Hence, the rate of plant growth is influenced by both rainfall and plant biomass at the beginning of the period.

Fig. 9.4 is a graphical representation of a regression analysis that estimated the function $\alpha(V, R)$, showing how incremental plant growth (in kg/ha) measured over the next 3 months is related to vegetation biomass (V, in kg/m^2) and rainfall (R, in mm):

$$\alpha(V, R) = -55.12 - 0.01535V - 0.00056V^2 + 2.5R$$

Unlike the logistic model, plant growth in the Australian study was highest at low levels of abundance, rather than at intermediate levels (Chapter 5). This is probably due to there being an ungrazeable plant reserve belowground. At low levels of plant abundance, rapid regrowth is enabled by translocation from these belowground tissues. Such an ungrazable refuge tends to lend a stabilizing influence to the interaction, as we shall shortly see.

Table 9.1 summarizes 100 years of rainfall statistics for the central Australian study site. There was no significant correlation of rainfall from one quarter to the next, nor between consecutive years.

9.5.2 The functional response of kangaroos to plant abundance

Having established how fast the resource grows in the absence of grazing and browsing, we now need to know what happens to it when a herbivore is present. The amount a herbivore eats per unit time is a constant only when it is faced by an *ad libitum* supply. Herbivores are seldom so lucky. The trend of intake against food availability is therefore curved, being zero when the level of food is zero and rising with increasing food to a plateau of intake. From there on, no increase in food supply has any effect on the rate of intake, because the animal is already eating at its maximum rate. Such a curve is called a *functional response* or *feeding response*: the trend of intake per individual against the level of the resource (see also Chapter 7). It can be represented symbolically by an equation such as:

$$\beta(V) = c(1 - \exp(-bV))$$

where $\beta(V)$ is plant consumption as a function of plant biomass, c is the maximum (satiating) intake, V is the plant biomass, and b is the slope of the curve, a measure of grazing efficiency. The last has another meaning: its reciprocal $1/b$ is the level of the resource V at which 0.63 (i.e. $1 - e^{-1}$) of the satiating intake is consumed.

Fig. 9.5 Food intake per individual red kangaroo per day at varying levels of food availability. (After Short 1987.)

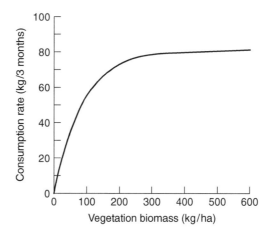

Fig. 9.5 shows the dry-weight food intake (I) of a red kangaroo at various levels of pasture biomass when it is grazing annual grasses and forbs interspersed with scattered shrubs (Short 1987). The equation for a 35 kg kangaroo is:

$$\beta(V) = 86(1 - \exp(-0.029V)$$

where I is the intake of food per red kangaroo measured in kilograms dry weight over 3 months, assuming no shrubs in the pasture layer (Short 1987). The satiating intake is 86 kg/3 months, occurring when pasture biomass exceeds 300 kg/ha.

Short (1987) estimated these two functional responses by allowing high densities of kangaroos and rabbits to graze down pasture in enclosures, the offtake per day being estimated as the difference between successive daily estimates of vegetation biomass corrected for trampling. Daily intake could be estimated for progressively lower levels of standing biomass because the vegetation was progressively defoliated during the experiment. We scale up this daily intake rate to intake per 3 months in order to maintain a similar time frame as for the plant growth data.

Although the functional response has been discussed here in the context of a plant–herbivore system, all of this discussion carries over to prey–predator systems. They are exactly analogous. The only difference lies in the difficulty of measuring a predator's food intake. The ability to measure intake by way of radioactive tracers has greatly simplified this problem. A good example is Green's (1978) use of radio-sodium to estimate how much meat a dingo eats in a day.

9.5.3 The numerical response of kangaroos to plant abundance

The functional response gives the effect of the animal upon a consumable resource. In contrast, the *numerical response* gives the effect of the resource on the change in animal numbers. If the resource is used in a preemptive rather than a consumptive way (e.g. nesting holes used by parrots) then it may be adequate to represent the numerical response by consumer density of the animals against the level of the resource (e.g. nesting holes per hectare). If the animals' use of the resource is consumptive, however, then the relationship between the animals and the resource is best portrayed as the instantaneous rate of population increase against the level of the resource.

Fig. 9.6 shows the numerical response relationship between the rate of increase of red kangaroos and the biomass of pasture. Bayliss (1987) estimated rates of increase

Fig. 9.6 Rate of increase
of red kangaroos on a 3-
monthly basis in relation
to food availability.
(After Bayliss 1987.)

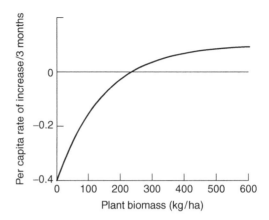

Fig. 9.6 Rate of increase of red kangaroos on a 3-monthly basis in relation to food availability. (After Bayliss 1987.)

from successive aerial surveys and pasture biomass from ground surveys. As with the functional response, the numerical response has an asymptote: there is an upper limit to how fast a population can increase and no extra ration of a resource will force that rate higher. The numerical response differs from the functional response in that negative values are both possible and logically necessary. If they were not, the population would increase to infinity.

The numerical response can usually be described by an equation of the form:

$$r(V) = -d + a(1 - \exp(-fV))$$

where r is the exponential rate of increase of the animals, d is the maximum rate of decrease, and a is the maximum extent to which that rate of decrease can be alleviated. Hence $a - d = r_{max}$ is the maximum rate of increase. Demographic efficiency, the ability of the population to increase when resources are in short supply, is indexed by f. For the present example, the constants were estimated (Bayliss 1987, modified by Caughley 1987) as:

$$r(V) = -0.4 + 0.5\,(1 - \exp(-0.007V\,))$$

The maximum rate of increase (i.e. when vegetation abundance is maximal) on a 3-monthly basis is $0.5 - 0.4 = 0.1$. Note that we have calculated the parameters for growth over 3 months, in order to remain consistent with the timeframe for other parameters used in the model. On an annual basis, r_{max} can be scaled up by simply multiplying by the four quarters in the year: $r_{max} = 0.4$. Hence, the population's maximum finite rate of increase over a year $\lambda = \exp(0.4) = 1.49$, a 49% increase per year.

9.5.4
Plant–kangaroo
dynamics

So far we have taken a plant–herbivore system and dissected it into its component processes: plant growth, the herbivore functional response to changes in plant biomass, and the numerical response of the herbivore, in terms of its rate of increase, to the biomass of the plants.

The evaluation of these component influences upon a population's dynamics provides two bonuses. First, it furnishes a tight summary of the dynamic ecology of the system. Second, it provides that summary in terms of causal relationships rather than correlations.

The numerical response of the herbivore allows us to calculate the equilibrium level to which plant biomass will converge in a constant environment under the influence of an unrestrained population of herbivores; that is, the null isocline. It is the *x* -intercept of the regression of rate of increase of the herbivores against plant biomass or, put another way, the plant biomass at which rate of increase of the herbivore is zero (see Fig. 9.6). The level of plant biomass *V* at which this occurs is:

$$V = (1/0.007)\ln(0.5/(0.5 - 0.4))$$

which equals 230 kg/ha dry-weight plant biomass.

This value is immensely important ecologically. It is the equilibrium level of plant biomass imposed by grazing in a constant environment. This is of some theoretical interest but of limited practical importance, because environments are not constant. However, it is also the level of plant biomass above which the herbivore population will increase and below which it will decrease (the *critical threshold*), and that is true whether the environment is constant or variable and whether the density of herbivores is high or low.

Using similar logic, we can calculate the combination of kangaroo and plant densities at which consumption exactly matches regrowth by plants. This will occur when $N\beta(V) = \alpha(V, R)$. We can rearrange the terms to isolate kangaroo density on the left-hand side of the equation, $N = \alpha(V, R)/\beta(V)$. Both null isoclines for the kangaroo–plant system are plotted as dotted lines in Fig. 9.7.

We can now reassemble the response functions of the system in their proper relationships to examine dynamics in the absence of stochastic variability in rainfall. We shall see what these in combination reveal about the system's dynamic behavior:

$$\frac{d}{dt}V(t) = \alpha(V, R) - N(t)\beta(V)$$
$$\frac{d}{dt}N(t) = N(t)r[V(t)]$$

Under a constant-rainfall regime (in this case, 60 mm per 3-month period), the system converges on the equilibrium: the point of intersection of both null isoclines (Fig. 9.7).

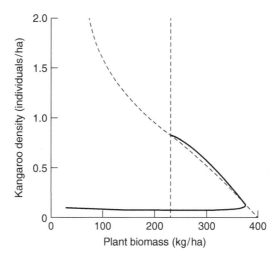

Fig. 9.7 Dynamics of kangaroos and plants over time, based on the Australian model discussed in the text.

Fig. 9.8 A typical
stochastic series of
rainfall amounts per
3-month period drawn
from a normal
distribution with mean
and variance equivalent
to the Australian data.
(After Caughley 1987.)

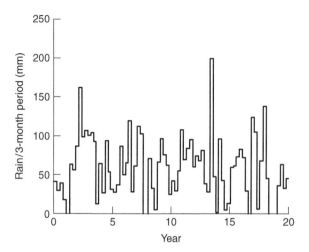

This shows that the equilibrium is stable, as one might have guessed based on the negative slope of the plant null isocline. Convergence on the equilibrium is circuitous, involving a burst of plant growth followed by plant decline as the kangaroo population stabilizes.

Rainfall can also be simulated as a sequence of random events from a normal distribution with a mean and standard deviation identical to the Australian data (Fig. 9.8), and the consequent changes in plant biomass and kangaroo numbers can be calculated accordingly.

Fig. 9.9 demonstrates a typical time trend for kangaroos, as generated by the equations describing the unpredictable rainfall (Fig.9.8) and the responses to it of the plants and herbivores. The only external input other than starting conditions is the random values from the 3-monthly rainfall distributions whose observed means and standard deviations are given above. The kangaroo population trajectory is a mathematical consequence of that rainfall, as its effect feeds through to plant growth, herbivore population growth, and grazing pressure.

The rainfall of this region takes the form of high-amplitude, high-frequency fluctuations (Fig. 9.8). The herb layer, whether grazed or ungrazed, generates a similar trace of high-amplitude, high-frequency fluctuations as it reacts speedily to rainfall or the lack

Fig. 9.9 A typical
stochastic time series for
kangaroos, using the
model discussed in the
text, for the rainfall
sequence shown in Fig.
12.8.

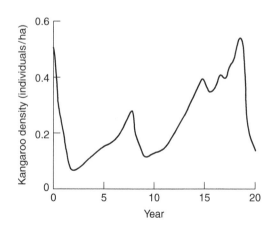

thereof. The fluctuations are paralleled by similar but more constrained fluctuations in the kangaroos' rate of increase as the population reacts dynamically to variations in food supply. The trend of kangaroo density differs from the rainfall regime, comprising fluctuations of high amplitude but of low frequency. This result might have been predictable from first principles: present density is an integration of past rates of increase, not of present conditions. Initial conditions are not highly influential: the system remembers previous plant biomass for only 3 years, but the memory of kangaroo density can linger for 10. As a consequence of the slow tracking of resources by kangaroos, there is a substantial time lag in their response to changing climatic conditions. This lag imparts an irregular fluctuation over time, rather than constancy in abundance, despite the stability of the system under deterministic (constant climatic) conditions. Caughley (1987) has christened systems that show slow convergence on stochastically shifting equilibria as "centripetal."

9.6 Wolf–moose–woody plant dynamics in the boreal forest

9.6.1 Models of the tri-trophic system

Few natural systems have been studied in sufficient detail to supply all the necessary parameters that we observed in the Australian kangaroo and plant system. Fortunately, it is often possible to estimate plausible parameter values from allometric reasoning or historical data from a variety of sources, allowing us to make educated guesses about system dynamics in a generic sense (Yodzis and Innes 1988; Turchin 2003). We shall demonstrate this approach for moose, wolves, and woody plants in the boreal forests of North America. This is an important system to understand, because it occurs across much of the extensive boreal forest biome spanning North America. We use Turchin's (2003) parameter estimates for the interactive model.

First, we recognize from the outset that this is fundamentally a tri-trophic system, meaning that there are three trophic levels that interact in the food chain. The framework we shall use simply expands the consumer–resource model outlined at the beginning of the chapter to a third trophic level (P, for wolves), which feeds on the second (N, for moose), which itself feeds on self-regulating plant resources (V). In all cases, we shall measure density in biomass (plants) or numerical abundance (for animals) per square kilometer. Mathematically, we can represent this interaction with the following system of equations:

$$\frac{dV}{dt} = r_{max}\left(1 - \frac{V}{K}\right) - \frac{aVN}{b + V}$$
$$\frac{dN}{dt} = \frac{aeVN}{b + V} - dN - \frac{ANP}{B + N}$$
$$\frac{dP}{dt} = \frac{AENP}{B + N} - DP$$

where a is the maximum rate of plant consumption by a single moose, b is the plant biomass at which plant consumption is half of the maximum, d is the rate of plant consumption at which moose just sustain themselves, e is the efficiency of conversion of food intake into new moose, and $A, B, D,$ and E represent the same set of parameters with respect to wolves.

We should note the similarity between the tri-trophic equations and the simpler consumer–resource model outlined at the beginning of the chapter. Resources have a self-regulating growth term, where the density-dependent term $(1 - V/K)$ reduces the growth rate proportionately with plant biomass. Plant consumption by moose is balanced against this positive contributor to resource abundance, with plant consumption expressed as the Michaelis–Menten form of the Type II functional response. Moose have a per capita growth function that depends on their intake of plants. Balanced against this is moose consumption by wolves. Finally, wolves have a per capita growth function that depends on their intake of moose, balanced against a constant per capita rate of mortality (presumably due to things like accidents, disease, and old age).

9.6.2 *Parameter estimation for the wolf–moose–woody plant system*

For a large part of the year, moose browse on the leaves and twigs of woody plants. Many species of plant contribute to the food supply of moose (Belovsky 1988). However, we know little about the web of ecological interactions within this plant guild, so we shall consider woody browse during winter (the period of the year when food is most often limiting to moose) as a single category. Edible biomass (measured in Mg/km^2) is symbolized by V. Field data suggest that it is rare to observe higher browse availability than $100\,g/m^2$, which is equivalent to $K = 100\,Mg/km^2$. Maximum moose density is thought to be 2 moose/km^2 (Messier 1994) and woody plant r_{max} is estimated as $3.33\,Mg/km^2/year$ (Turchin 2003). We see that the growth term for the edible plant biomass is maximized at low biomass, not at intermediate biomass as it would be for a logistic growth function. The rationale for low edible biomass is that moose have access only to regrowing tissues, such as twigs and leaves, so that the rest of the plant functions as an ungrazeable reserve. Regrowth capacity should not be inversely affected by herbivory so long as it does not jeopardize plant survival. The indigestible component is the same kind of refuse demonstrated in Robertson's study of food plants fed upon by kangaroos in semi-arid Australian grasslands.

The maximum rate of plant consumption by moose is set at 2 Mg/individual/year, based on maximum values quoted in the feeding studies literature (Crête and Bédard 1975). Fitting various curves to Vivås and Sæther's (1987) studies of moose foraging in Norway suggests a foraging efficiency of $b = 40\,Mg/km^2$. Moose can just meet their metabolic requirements at a level of intake of half the maximum, providing an estimate of $d = 1\,Mg/ind/km^2$. Given a maximum exponential rate of increase of 0.2 for moose (Fryxell *et al.* 1988a) and values for all the other parameters, one can solve for e using the following relationship:

$$e = \frac{r_0}{\frac{aK}{b+K} - d}$$

yielding $e = 0.467$.

Rates of wolf consumption of moose are modeled as a Type II functional response, based on Messier's (1994) review of several moose–wolf studies throughout North America. Each of these studies provides one or more estimates of rate of consumption by wolves at a given moose density. By combining all of the recorded data together in a single graph (Fig. 9.10), Messier was able to illustrate one of the most difficult kinds of ecological relationship, the functional responses of large organisms under free-living conditions. Such patterns are essential to our understanding of consumer–resource interactions, yet are prohibitively costly to gather in a single study. Use of aggregate data is a very useful way of solving this problem.

Scaling the wolf consumption rate to a yearly timeframe yields estimates of $A = 12.3$ moose/wolf/year and $B = 0.47$ moose/km^2. According to Fuller and Keith (1980), each wolf needs to eat 0.06 kg of meat per day to meet maintenance requirements, whereas a population whose individuals eat 0.13 kg each day can grow at the maximal rate. This yields estimates of $D = 0.6$ and $E = 0.1$.

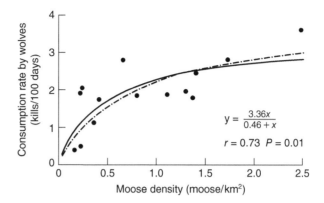

Fig. 9.10 The consumption rate by wolves of moose in relation to moose density. (After Messier 1994.)

$$y = \frac{3.36x}{0.46 + x}$$

$r = 0.73 \quad P = 0.01$

9.6.3 *Dynamics of the wolf–moose–woody plant system*

Combining these parameter values together, the outcome is a complex series of oscillations in moose and wolf abundance, which never quite repeat themselves (Fig. 9.11). This is a mild form of deterministic chaos, common in tri-trophic systems (Hastings and Powell 1991; McCann and Yodzis 1994). Even though the fluctuations are

Fig. 9.11 Predicted population dynamics of moose (top) and wolves (bottom) based on Turchin's (2003) tri-trophic model, described in the text.

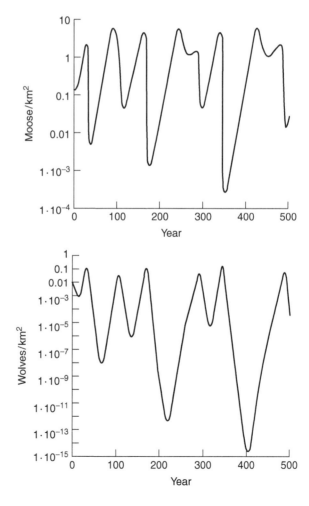

nonrepetitive, the time between successive peaks tends to be several decades – a very protracted pattern of fluctuation.

The manner by which parameters for the wolf–moose–woody plant model were derived, using a set of observations gathered around the globe, makes it fairly unlikely that we can predict the dynamics of any given system. It does suggest, nonetheless, that this system should exhibit an inherent tendency towards protracted fluctuations that recur over a decade-long timescale. Moreover, the model suggests that these fluctuations will not necessarily converge on a stable limit cycle, as do consumer–resource models with only two trophic levels. Rather, we may expect to see inconsistency as each population progresses from peak to peak.

One obvious objection to this model is that it ignores the role of wolf territoriality. In most landscapes, wolves form communal packs that partition the available habitat among themselves. Territorial strife among wolf packs can be intense, leading to substantial levels of mortality (Peterson *et al.* 1998). At least we should expect that the risk of this mortality should climb with wolf density, if only because of increasing frequency of encounters between members of different packs. One way to incorporate this effect is to make wolf mortality explicitly density-dependent:

$$\frac{dV}{dt} = r_{max}\left(1 - \frac{V}{K}\right) - \frac{aVN}{b+V}$$
$$\frac{dN}{dt} = \frac{aeVN}{b+V} - dN - \frac{ANP}{B+N}$$
$$\frac{dP}{dt} = \frac{AENP}{B+N} - DP - \frac{s_0 P^2}{\gamma}$$

where the maximum density of wolves (recorded from field studies) $\gamma = 0.1$ and the maximum per capita rate of wolves $s_0 = 0.4$. This modification imposes an additional per capita mortality term that increases by s_0/γ with each unit increase in wolf density P.

Territorial effects of this sort often have a stabilizing influence. Such is the case with the wolf–moose–woody plant model: the addition of density-dependent mortality due to territorial strife changes the dynamics of the system from deterministic chaos to a stable limit cycle (Fig. 9.12). The level of strife is insufficient, however, to completely stabilize the system.

The best long-term data set available on both moose and wolves is from Isle Royale, a small island 40 km off the coast of Canada in Lake Superior that supports a mix of deciduous and coniferous vegetation species typical of the boreal forest on the mainland. Moose apparently invaded Isle Royale a century ago, while wolves arrived by ice in the 1940s. Estimated patterns of abundance on Isle Royale certainly suggest protracted fluctuations over time (McLaren and Peterson 1994; Peterson 1999; Post *et al.* 1999), with moose populations fluctuating slowly with 25 years between successive peaks (Fig. 9.13).

It is difficult to conclusively tell from the Isle Royale time series data whether the system is cyclic or chaotic, because they are simply insufficient to evaluate even such a well-studied system. Such will nearly always be the case in slow-changing wildlife species. Nonetheless, the tri-trophic model seems to capture the fluctuating tendency of the Isle Royale system.

There are many other factors that could also contribute to the apparent instability of the Isle Royale populations. For example, complex changes over time in the age

Fig. 9.12 Predicted dynamics of the moose (top) and wolf (bottom) system with woody plants when wolves have additional density-dependent mortality due to territorial aggression, as described in the text.

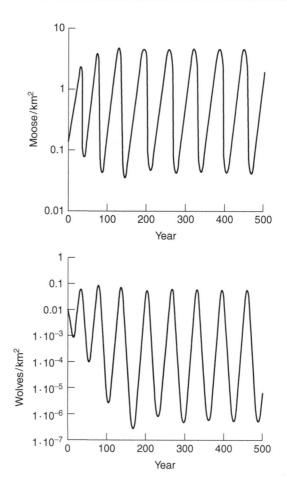

structure of moose could contribute to the propensity for fluctuations (Peterson and Vucetich 2003). Wolves are highly selective for specific age classes of prey, so changes in age distribution could translate into substantial changes in predation risk. As we discuss in Chapter 13, it can take many years for age distributions to stabilize in long-lived organisms. When age distributions are shaped by dynamic interactions with predators, this can be even more destabilizing. We also know that the wolf population on Isle Royale has much lower levels of genetic variability than do populations on the mainland (Wayne et al. 1991). This could influence wolf demographic parameters in unknown ways. Finally, there is evidence that complex interactions among climatic conditions, social grouping patterns of wolves, and predation risk of moose could contribute to instability. In years of deep snow, wolves form larger packs, which leads to increased rates of mortality on moose (Post et al. 1999). Nonetheless, the instability of this system seems to be intrinsic to the basic consumer–resource interactions (moose–vegetation and wolf–moose).

Truly long-term data for temperate-zone carnivores (wolves and coyotes) and ungulates (moose or white-tailed deer, depending on location) are scarce. Data from the Hudson's Bay Company probably represent the lengthiest data set. They suggest very slow oscillations in the abundance of wolves and coyotes during 1750–1900,

Fig. 9.13 Population dynamics of wolves (a) and moose (b), as well as annual growth of balsam fir trees (c), on Isle Royale. The solid lines represent observed values, the dotted lines polynomial regressions. The densities represent total abundance of each species recorded over the entirety of Isle Royale. (After Post *et al.* 1999.)

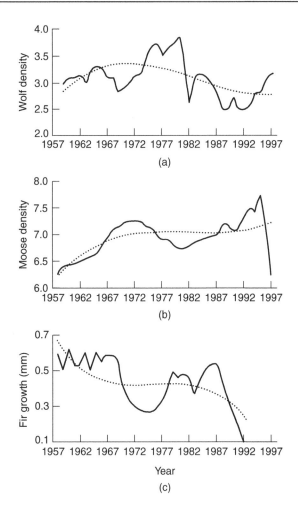

with roughly two cycles per century (Turchin 2003). Although the Hudson's Bay data on deer skins are more fragmentary, they too suggest long-term cycles in abundance (Turchin 2003). Slow oscillations by white-tailed deer in Canada (Fryxell *et al.* 1991) and moose in Finland (Lehtonen 1998) suggest that long-term oscillations are an important feature of some large mammal species.

9.7 Other population cycles

Long-term data for a number of other wildlife populations show pronounced cycles, first identified by Charles Elton (1924). Such cycles are sometimes regular, such as the 10-year cycle of snowshoe hares (Sinclair *et al.* 1993), sometimes erratic, such as the 3–6-year cycle of voles (*Microtus agrestis* and *Clethrionomys rufocanus*) in northern Europe (Turchin and Hanski 1997). They can be explained in many ways: unstable behavioral polymorphisms in cyclic populations (Chitty 1967; Krebs and Myers 1974); maternal effects transferred to offspring, imparting lagged density dependence (Inchausti and Ginzburg 1998); coupled interactions between plants, herbivores, and/or carnivores (Hansson 1987; Turchin and Hanski 1997; Turchin and Ellner 2000; Turchin and Batzli 2001). We shall review the northern European vole and

North American snowshoe hare populations here and consider the logic underlying consumer–resource explanations for population cycles.

Some of the longest continuous studies of vole populations come from sites in Scandinavia, Finland, and Russia (Turchin 2003). These data point to a fascinating geographical pattern: populations at southern latitudes show little evidence of repetitive, cyclic dynamics, whereas populations from more northerly latitudes exhibit repetitive cycles or perhaps even chaotic dynamics over time (Fig. 9.14).

Many ecological variables change as one progresses from the Arctic Circle to more southerly latitudes, including temperature maxima and minima, precipitation, vegetation cover and composition, primary productivity, mammalian and avian community diversity, and human population density. Most important among these variables, however, is the transition from a suite of generalist predators (red foxes, feral cats, badgers, and various owls, hawks, and other raptors) in the south to a narrow range of specialist predator species (primarily the least weasel, *Mustela nivalis*) in the most northerly areas. The abundance of generalist predators declines in northern latitudes because of the duration and depth of snow cover (Hansson and Henttonen 1985).

Least weasels exhibit a Type II functional response to changes in vole density (Erlinge 1975). As we have already discussed, this pattern of foraging tends to destabilize prey populations, because the per capita risk of mortality due to predators is inversely related to prey density. However, generalist predators switch feeding preferences to favor voles when they reach high density but ignore them when they collapse to low density (Korpimäki and Norrdahl 1991; Erlinge *et al.* 1983). As we show in Chapter 10, switching behavior can stabilize prey numbers, because the per capita risk of mortality for prey due to predation increases with prey density, at least over some prey densities. Because generalist predators can feed on a wide variety of other species, they may not be dependent on vole numbers (Turchin and Hanski 1997). In the absence of predators, vole population growth is self-regulating, due to density-dependent resource limitation and territorial spacing among individual voles.

Turchin and Hanski (1997, 2001) linked specialist predation by weasels, generalist predation, and the self-regulating population dynamics of voles with seasonal changes in vole logistic growth. In keeping with the empirical data, their model predicts (better than alternative models) complex cycles or chaos when generalist predators are rare but much more stable dynamics when they are common.

Data on the cyclical variation in abundance of snowshoe hares comes from fur records of the Hudson's Bay Company in Canada (see Fig. 5.7). These data show a regular oscillation of numbers, with a period of 10 years. Like the other examples we have discussed, snowshoe hares interact not only with their food supplies but also with a suite of carnivores that feed upon them (Krebs *et al.* 2001b). Some of these carnivores, especially the lynx, which is a specialist predator on hares, also display a 10-year cycle in abundance, slightly lagged behind that of the snowshoe hare. These characteristics suggest that the tri-trophic model might be a useful starting point in modeling the dynamics of hare and lynx populations. King and Schaffer (2001) estimated parameters in order to model dynamics of the woody plant–hare–lynx interaction. They found that realistic parameter values generated cycles in hare and lynx abundance of 8–12 years, consistent with the historical data.

Unlike in the other examples we have discussed, however, there is an inherent environmental cycle, the 11-year sunspot cycle, that apparently plays a crucial role in generating the hare–lynx cycle (Sinclair *et al.* 1993). Snow depth is strongly

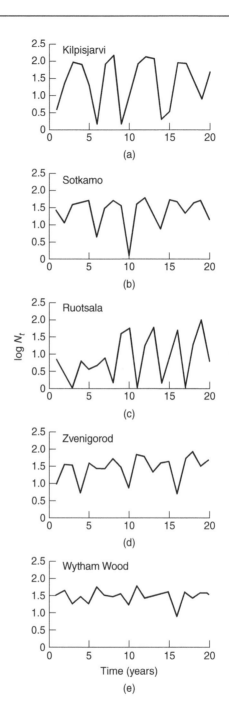

Fig. 9.14 The latitudinal gradient in vole dynamics across northern Europe, with the most northerly sites at the top of the figure. (After Turchin and Hanski 1997.)

influenced by the sunspot cycle, as evidenced by ice cores taken from glaciers. Disentangling the effect of the sunspot cycle from the endogenous rhythm of the tri-trophic consumer–resource interaction presents a sizeable challenge.

King and Schaffer (2001) also used the tri-trophic model to explain the outcome of a series of large-scale field experiments conducted in Kluane National Park, Canada, during the 1980s and 90s (Krebs *et al.* 1995, 2001b). The Kluane study involved experimental manipulations of food availability, predation risk, and both factors combined to tease apart bottom-up versus top-down trophic mechanisms. The Kluane team found that each of the manipulations had a considerable effect on hare densities and hare demographic rates. Food addition doubled hare densities and predator exclusion trebled them. Both had an 11-fold effect relative to controls. The clear implication is that both bottom-up and top-down processes are important to the natural regulation of snowshoe hares. Despite these results, however, none of the treatments dismantled the hare cycle. This may be because of the use of semi-permeable fencing in the experimental treatments, allowing hare populations within the exclosures to be driven by dynamics generated outside, via immigration.

The best interpretation of the existing information is that the snowshoe hare–lynx cycle is a complex tri-trophic interaction synchronized to some degree by the exogenous environmental rhythm of the sunspot cycle. These results suggest that coupled consumer–resource models can be a vital step in understanding the complex patterns of population dynamics that occur in natural ecosystems.

9.8 Summary

A resource is something an animal needs and whose consumption diminishes its availability to other consumers. Consumers and their resources often form a system in which the rate of increase of the resources is determined by the density of the animals eating them, and the rate of increase of the animals is determined by the density of the resources. Such a complex system can be studied only by breaking it down into its dynamic components, of which three dominate. First, there is the functional response of the animal, the rate of resource intake by a single consumer as a function of resource abundance. Second, there is the numerical response of the consumer, the rate at which its population increases as a function of the resource abundance. Finally, we require supplementary information on the growth rate of resources in relation to resource abundance. On the basis of these functional relationships, the full dynamic behavior of the system can be described. We illustrate this approach with two well-studied consumer–resource systems: kangaroos and their plants in Australia, and wolves, moose, and woody plants in North America. Interactive systems with these components can be deterministically stable (such as the Australian plant–kangaroo system) or unstable (such as the wolf–moose–woody plant system). Deterministic instability is evident in the repetitive population fluctuations (stable limit cycles) and in the nonrepetitive ones (deterministic chaos). Even stable food-chain models can show pronounced long-term fluctuations in response to stochastic environmental variability (centripetal systems). Two well-documented cyclical populations (voles in northern Europe and snowshoe hares in North America) have dynamics consistent with predictions of coupled consumer–resource models.

10 The ecology of behavior

10.1 **Introduction**

In this chapter we consider how ecological constraints shape the behavior of individual organisms and, conversely, the effect of individual behavior on the dynamics of populations and communities. This is part of the field known as behavioral ecology. We concentrate on foraging and social interactions because these characteristics often have important ecological ramifications that affect wildlife conservation and management. We start with a consideration of the many ways that organisms can choose what and where to eat, then move on to consider how ecological constraints affect social organization.

10.2 **Diet selection**

The range of mechanisms by which animals choose their diet is diverse. Some animals have a narrow range of preferences, sometimes even for a single species of plant. One of the best examples of this is the giant panda, which has evolved a special set of adaptations allowing it to feed largely on bamboo plants growing on the steep mountainsides of southern China (Schaller *et al.* 1985). Such species are termed "feeding specialists." Other species tend to the opposite extreme – feeding relatively indiscriminately from a wide range of items. A good example of this would be the moose, which feeds from an array of plants, including grasses, woody plants, herbs, forbs, and even aquatic plants (Belovsky 1978). Such species are termed "feeding generalists." Most wildlife species would fall between these extremes.

One can see an immediate advantage in having a broad diet: there is a much better chance of finding something to eat, no matter where the individual might find itself. There is also a disadvantage, however, to being a generalist: many of the possible items in the environment may be so nutritionally poor that they are scarcely worth pursuing, as we discuss in Chapter 2. For herbivores, it may be impossible for an individual to survive on poor-quality items, even when supplies are unlimited. For carnivores, variation among prey derives less from differences in nutritional quality than from differences in size, visibility, ease of capture, or the associated risk of injury during capture. In both cases, choosing wisely (becoming a specialist) among a wide range of food items might prove advantageous. Much foraging theory relates to this question: how does an animal choose a diet that yields the highest rate of energy gain over time, energy that can be devoted to enhancing survival and reproduction? This question forms the core of optimal foraging theory, a set of mathematical models predicting the patterns of animal behavior that might be favored by natural selection (Stephens and Krebs 1986).

10.2.1 *Optimal diet selection: the contingency model*

The simplest way to consider optimal diet is to start with the functional response: the rate of consumption in relation to food availability (we discuss this further in Chapters 7 and 9). Most organisms have a decelerating functional response to increasing food

Wildlife Ecology, Conservation, and Management, Third Edition.
John M. Fryxell, Anthony R. E. Sinclair and Graeme Caughley.
© 2014 John Wiley & Sons, Ltd. Published 2014 by John Wiley & Sons, Ltd.
Companion Website: www.wiley.com/go/Fryxell/Wildlife

Fig. 10.1 Comparison of the expected rate of energy gain by a forager specializing on the most profitable prey (prey 1, solid curve) relative to the profitability of alternative prey 2 (broken line). An optimal forager will expand its diet whenever the abundance of prey 1 drops below the density at which these two curves intersect (slightly more than 20 per unit area in this hypothetical example).

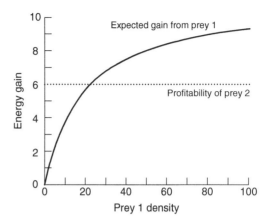

availability, often termed a *Type II* response (Holling 1959). The rate of energy gain $f(N_1)$ that an animal will experience as a consequence of the Type II functional response can be calculated as follows:

$$f(N_1) = \frac{e_1 a N_1}{1 + ahN_{11}}$$

where e_1 is the energy content of each item of the more profitable food type 1, a is the area searched per unit time by the consumer, h_1 is the time required to consume each item of food type 1, and N_1 is the population density of food type 1. The energy gain function $f(N_1)$ grows with increasing abundance of prey type 1, but there are diminishing returns to this relationship (Fig. 10.1). Indeed, there is an upper limit e_1/h_1 to the rate of energy gain, even when food is superabundant. This upper limit is set by the limited capacity of the animal to handle prey.

   If a forager specialized by feeding only on prey type 1, it would realize a rate of energy return equivalent to $f(N_1)$. How would this compare to the energy gain if the forager generalized by feeding on both prey types 1 and 2? If both prey types were mixed indiscriminately over the landscape traveled by our hypothetical forager, then the energy gain by a generalist would be calculated as follows:

$$g(N_1, N_2) = \frac{a(e_1 N_1 + e_2 N_2)}{1 + a(h_1 N_1 + h_2 N_2)}$$

This equation raises the following question: when does it pay to be a specialist and when to be a generalist? The answer is to specialize when $f(N_1) > g(N_1, N_2)$ but act like a generalist when $f(N_1) < g(N_1, N_2)$. Both strategies are equivalent when $f(N_1) = g(N_1, N_2)$. This special case occurs when $f(N_1) = e_2/h_2$. So, a forager that changed from being a specialist to a generalist whenever $f(N_1)$ fell below e_2/h_2 would do better than one that acted as either a specialist or a generalist all the time (Fig. 10.1). Such a foraging strategy is termed the "optimal" strategy, meaning that it yields the highest energetic returns over time. It is straightforward to extend this logic to any number of resource types. First, we rank prey in terms of profitability. Then we add prey to the diet so long as their profitability exceeds the expected rate of long-term intake obtained by specializing on all of the more profitable prey types.

Assuming that maximal acquisition of energy can improve the fitness of a forager (improve reproduction or survival), we might expect something like the optimal strategy to by favored by natural selection. Such selection does not necessarily mean that we would expect every animal to act like a little computer, perfectly assessing the implications of each behavioral decision it might make. Rather, a pattern of behavior that approximates the optimal strategy should be more successful at producing offspring than alternatives.

The optimal diet model makes a number of predictions that are testable by observation or, better still, experimental manipulation:

1 Foragers should rank food types in terms of their energetic profitability (energy content divided by handling time).
2 Foragers should always include the most profitable prey, then expand their diet to include less profitable prey when the expected rate of gain by specializing on the more profitable one matches the profitability of the poorer prey.
3 The decision to specialize or generalize should depend on the abundance of highly profitable prey, but not on the abundance of less profitable prey.
4 An optimal forager should have an all-or-nothing response. By this we mean that the perfect forager should either always accept alternative prey or never accept them, depending on whether $f(N_1) > e_2/h_2$.

The optimal diet model has been tested in a variety of settings since it was first proposed (MacArthur and Pianka 1966; Schoener 1971; Pulliam 1974; Charnov 1976a). Despite its simplicity, the optimal diet model has proved remarkably successful in predicting foraging behavior (Stephens and Krebs 1986). Sih and Christensen (2001) reviewed the outcome of over 130 diet choice studies and found that two-thirds of the species studied, ranging from invertebrate herbivores to mammalian carnivores, showed foraging patterns qualitatively consistent with the optimal diet choice model. Optimal diet models tend to perform particularly well in situations where all of the relevant parameters have been accurately measured, enabling precise quantitative predictions.

A classic example is captive great tits (*Parus major*) trained to pick mealworms off a conveyer belt as it passes in front of them (Krebs *et al.* 1977). The birds became adept at choosing whether to specialize on one prey or to accept both preys indiscriminately, in accordance with predictions **1**, **2**, and **3** of the model (Fig. 10.2). However, the birds

Fig. 10.2 Krebs *et al.*'s (1977) laboratory experiment on diet selection by Parus major. Two sizes of mealworms were delivered via conveyer belt in random order to the forager. The horizontal axis shows the rate of encounter with big prey. As the rate of encounter with big prey declined, the birds expanded their diet to include small prey.

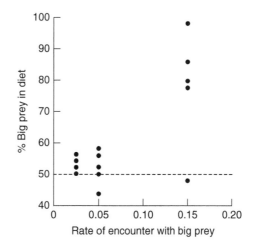

never mastered the all-or-nothing behavior that would be perfectly optimal. Instead, the foragers sometimes ate both prey types and sometimes only the more profitable prey, a pattern termed "partial preference." Such partial preferences are almost always observed, even in the most successful experiments, perhaps because foragers cannot discriminate perfectly among prey, or perhaps because foragers need continually to "test" alternative prey in order to assess their relative profitability.

Initial doubts that optimal diet choice theory could be applied to complex field situations with many prey species (Schluter 1981; Pierce and Ollason 1987) have been allayed by successful field studies (Sih and Christensen 2001), but the theory does require much effort in estimating many parameters. The optimal diet model has been less successful with predators that utilize mobile prey (e.g. weasels feeding on rodents) compared to those that utilize stationary prey (e.g. starfish feeding on mussels). There are a variety of reasons for this difference (Sih and Christensen 2001). In nature, food items are rarely mixed homogeneously across the landscape visited by the forager. This is particularly true of mobile prey, which may be attempting to avoid predators. If alternative prey species tend to occur in different microhabitats, differ in their activity patterns, or have a different capacity for avoidance of predators then simple frequencies of abundance may be a poor predictor of diet composition. In these cases, more complex models may be needed to predict optimal diet patterns.

In some cases, particularly with herbivores, foragers need to maintain a balanced intake of particular nutrients, rather than simply maximizing energy gain in whatever way possible (see Chapter 2). For example, howler monkeys tend to choose a diet more heavily laden with leaves than should be optimal, perhaps because they must balance nutrients (Milton 1979). Similarly, the marine gastropod *Dolabella auricularia* grows several times faster on a mix of algal species than when fed on a single species of algae (Penning *et al.* 1993).

The optimal diet model represents a simplistic view of foraging. Most species have biological features that introduce additional elements into the decision-making process. For example, many species forage outwards from a central place, whether that place is a den, perch, or resting site. If the forager sallies forth, retrieves one or more prey items, and then returns to the central place before feeding on the items then this additional travel time and energetic expenditure must be accommodated if we are to make useful predictions (Orians and Pearson 1979). Thus, beavers forage on a variety of woody plants on the shore surrounding the ponds or streams where they build their lodges. Several studies have shown that beavers feed more selectively the farther out food items occur from the lodge, as predicted by central-place foraging models (Jenkins 1980; McGinley and Witham 1985; Fryxell and Doucet 1993). The greater energetic cost of travel to more distant food requires that animals be more selective. As a consequence, the handling time of potential food items is dependent on distance, implying different patterns of diet selectivity.

10.2.2 *Optimal diet selection: effects on feeding rates*

Optimal patterns of foraging can have important effects on the rate of attack by consumers. The rate of consumption of prey type 1 ($f(N_1)$) by an optimal forager is predicted by the following multispecies functional response:

$$g(N_1, N_2) = \frac{aN_1}{1 + ahN_1 + \beta(N_1)ah_2N_{21}}$$

Fig. 10.3 Predicted effect of optimal diet choice on the rate of intake of the preferred prey.

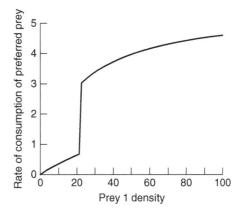

where $\beta(N_1)$ is the probability of foraging on the poorer prey, which is a function of the density of preferred prey, calculated according to the optimal diet choice rule (Section 10.2.1). This equation predicts that there will be a sharp drop in the consumption of the more profitable prey at the point at which the forager expands its diet to include less profitable types (Fig. 10.3). Such a drop implies that mortality risk for the more profitable species will decline accordingly. This process has been observed with beavers in large enclosures with either a single type of prey or a mix of prey species (Fryxell and Doucet 1993). When presented with a single species of prey, beavers had a Type II functional response, smoothly climbing with prey density, but at a decelerating rate (Fig. 10.4). When presented with a mix of prey, however, beavers expanded their diet as preferred prey declined in abundance, in accordance with an optimal diet model (Fryxell 1999). This led to a pronounced decline in predation risk to preferred prey as the diet expanded (Fig. 10.4).

As a natural consequence of these multispecies effects on feeding rates, optimal patterns of diet choice can have important implications for predator–prey interactions. An adaptive predator would shift its attention away from some species as they became more and more rare. Such behavior can have a stabilizing influence on predator–prey

Fig. 10.4 Functional response of beavers presented with a preferred species of prey (trembling aspen saplings) or a mix of prey species (trembling aspen, red maple, and speckled alder saplings) in two separate trials. The rate of consumption of preferred prey (aspens) dropped precipitously as the animals expanded their diet to include less profitable prey (maples and alders) in the multispecies trial, causing the pronounced deviation between the two curves at low densities of aspen saplings. (After Fryxell 1999.)

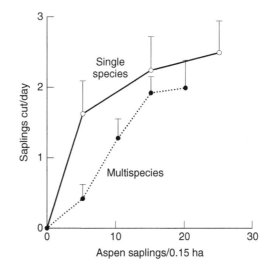

dynamics, reducing the degree of variability of cyclical predator–prey systems over time (Gleeson and Wilson 1986; Fryxell and Lundberg 1994; Krivan 1996). This is especially likely when the growth rate of the predator is poor on alternative prey, when the forager exhibits partial preferences, or when alternative prey do not have overlapping spatial distributions (Fryxell and Lundberg 1997).

10.2.3 Optimal diet selection: linear programming

An alternative way of modeling diet choice employs a technique called linear programming to identify the optimal solution to a requirement influenced by several constraints (Belovsky 1978; Belovsky *et al.* 1989). When applied to optimal foraging, this allows researchers to explore more subtle hypotheses. Linear programming can be used to predict the optimal diet for a forager that is trying to maximize its intake of energy and at the same time ensure that it obtains sufficient intake of a scarce nutrient to meet its metabolic requirements. When conducted for pairwise combinations of alternative foods, linear programming can be readily understood from simple graphs (Fig. 10.5).

Belovsky (1978) used linear programming to predict the optimal choice of aquatic versus terrestrial plants by moose, based on parameters for moose living on Isle Royale, a small island in Lake Superior. One constraint is that moose must obtain a daily intake of 2.57 g of sodium in order to meet their metabolic requirements. Terrestrial plants in this system are deficient in sodium, while aquatic plants have much higher concentrations. Like many other herbivores, moose have limits on the amount of food material they can process each day through their digestive tracts. The total daily consumption of aquatic and terrestrial plants eaten by a moose cannot exceed this processing rate, which we call the digestive constraint. Moose also have limits on the number of hours they can devote to cropping food items. Finally, each food type has a different profitability (ratio of digestible energy content to cropping time). Thus, time spent cropping energetically poor food items (such as aquatic plants) reduces the opportunity to look for energetically richer items (terrestrial plants). In other words, a moose might waste valuable time eating poor food that could be spent looking for better options.

All of these constraints vary linearly with the proportion of each food type in the diet (Fig. 10.5). The optimal solution will occur at one of the intersections of the linear constraint lines. By multiplying the energy content of each prey type by the daily intake of that item at each of the intersection points, we can assess which intersection point offers the greatest energy returns, while guaranteeing that moose maintain a minimum acceptable level of sodium intake. The optimal solution in this case is to have a diet dominated by terrestrial plants, with a small fraction of aquatic plants.

Fig. 10.5 Linear programming model of diet selection in moose, based on Belovsky (1978). The lines represent constraints that must be met. The range of diets that fall within these constraints is enclosed in the triangle in the middle of the diagram.

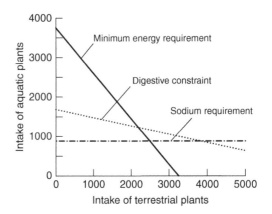

Linear programming has been successfully applied to predict simple dietary preferences (e.g. forbs versus grasses) in a wide variety of species (Belovsky 1986). It has proven less successful at predicting the actual mix of species in herbivore diets. Like the contingency model, linear programming models are ultimately limited by the reliability of parameter estimates and the degree to which proper constraints have been identified. Nonetheless, they remain a very useful means of incorporating multiple constraints into dietary predictions.

10.3 **Optimal patch or habitat use**

Many resources naturally have patchy patterns of spatial distribution. This presents a number of problems for foragers, such as how to decide which patches or habitats to exploit, how long to stay in each patch once chosen, and how to adjust habitat preferences in light of choices made by other foragers. Optimality principles can be usefully applied to each of these problems.

10.3.1 *Optimal patch residence time*

We start by considering how long an animal should stay in a given patch. Let us take for an example fig trees that are widely spaced throughout tropical rainforest. A toucan that wishes to eat figs is faced with deciding how long to feed at a particular tree before moving on to look for another. We have already seen that foragers must spend valuable time and energy looking for each food item that they might exploit. As a consequence, there are diminishing returns the longer the toucan stays at the tree, because most foragers have a functional response that declines as resource density declines. After an initial period of rapid energy gain, the rate of accumulation of further energy by the animal begins to slow as resource density drops lower and lower due to the animal's feeding (Fig. 10.6).

We can symbolize the cumulative energy gain with the function $G(t)$, meaning that cumulative gain depends on the time t spent in each patch. For simplicity, we assume that each patch is identical with respect to initial resource abundance and that the forager has no means of knowing exactly how long it will take to get to the next suitable patch, only how long it takes on average, based on its previous experience. The long-term rate of energy gain ($E(t)$) can be expressed as the total energy gained ($G[t]$) divided by the time spent within each patch (t) plus the average time it takes the forager to find a new patch ($1/\lambda$):

$$E(t) = \frac{G(t)}{\left(\frac{1}{\lambda}\right) + t} = \frac{\lambda G(t)}{1 + \lambda t}$$

Fig. 10.6 The cumulative gain of energy from a patch (solid curve) as a function of the time spent in the patch plus the average time spent traveling between patches. The broken line indicates the tangent to this gain curve that passes through the origin (0,0 on both axes). The slope of this tangent represents the long-term rate of energy gain relative to both travel and patch residence time. The point of intersection identifies the optimal patch residence time.

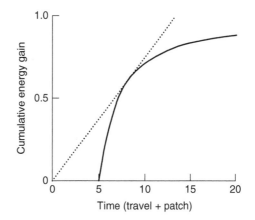

Long-term intake is usually maximized at an intermediate amount of time spent within each patch. The optimal residence time can be found graphically by drawing the tangent to the gain curve that passes through the origin (Fig. 10.6). This tangent is known as the "marginal value" in economic jargon, so the optimal patch use model has come to be known as the *marginal value theorem* (Charnov 1976b).

The marginal value theorem makes a number of useful predictions:

1 Foragers should leave all patches when the rate of intake in those patches reaches a threshold value. This will typically occur at a particular density of prey.
2 Foragers should leave resource-poor patches much sooner than resource-rich ones.
3 The average distance between patches should influence the optimal time to leave a patch, the *giving-up time* (and by analogy the optimal *giving-up density* of prey). The optimal decision would be to stay in each patch longer when the distance between them is long than when it is short.

Several studies have tested these predictions. Of 45 published, 70% show patterns of patch departure consistent with these predictions. In 25% of these studies, precise numerical predictions are upheld (Stephens and Krebs 1986). One of the most elegant examples is Cowie's (1977) study of patch use by great tits. Cowie built a series of perches in an aviary on which small containers with tight-fitting covers were attached. Several mealworms were placed in each container and covered with sawdust. Birds learned to pry the lid off each container before searching for mealworms within it, the container being the "patch." By changing the tightness of the lids, Cowie could control the time between the cessation of foraging in one patch and the initiation of a bout of foraging in a new one. He showed that birds were sensitive to travel time between patches, staying longer at patches when travel time was long than when it was short. Changes in departure time were well predicted by the marginal value theorem (Fig. 10.7).

10.3.2 *Patch use by herbivores*

For most large herbivores, food is continuously distributed across the landscape, rather than in definable patches. Nonetheless, local abundance of food still varies considerably from place to place. A slight modification to the marginal value theorem can

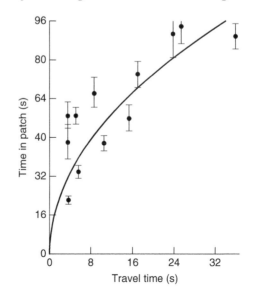

Fig. 10.7 Observed patch residence times by great tits foraging in an experimental aviary (shown by filled circles) versus the patch residence times predicted by the marginal value theorem (solid curve). (After Cowie 1977.)

readily accommodate this situation (Arditi and Dacorogna 1988). This model predicts that animals should feed whenever the cropping rate exceeds the average rate of cropping. A small herd of fallow deer (*Dama dama*) confined to a small pasture grazed according to the marginal value rule, concentrating its feeding in sites where food abundance was higher than average (Focardi and Marcellini 1996). However, a second deer herd that roamed over a much larger area showed little evidence of being sensitive to the marginal value of grazing. The marginal value rule seemed most applicable to the case where the deer had a much greater opportunity to develop detailed knowledge of the landscape. Similar patterns have been recorded in cattle (Laca *et al.* 1993; Distel *et al.* 1995) and dorcas gazelles (*Gazella dorcas*) (Ward and Salz 1994).

Large herbivores, particularly grazers, might also have good reason to avoid using patches of dense vegetation. The reason is that taller plants often have more cellulose and lignin than shorter plants, which provides support for their height and weight. Consequently, a herbivore that grazed tall plants would obtain less nutritious and less digestible food than one that concentrated on younger growth forms. However, at very low plant sizes, the rate of cropping is very low, and this can also compromise rates of food intake. As a result, grazers should benefit best by feeding on an intermediate height and biomass of grasses (Fig. 10.8). Several experimental studies have shown that large herbivores show grazing preference for swards of intermediate grass height and biomass, including cattle (de Vries *et al.* 1999), elk (Wilmshurst *et al.* 1995), bison (Bergman *et al.* 2001), red deer (Langvatn and Hanley 1993), and Thomson's gazelle (Fryxell *et al.* 2004). On the other hand, reindeer on the arctic island of Svalbard prefer patches of tall vegetation, even though it is nutritionally inferior (van der Wal *et al.* 2000), for reasons that are as yet unclear.

The marginal value theorem predicts that foragers should depart when the rate of food intake in a patch (i.e. the functional response) equals the average rate of food available elsewhere in the environment multiplied by a constant (proportional to the travel time between patches). This implies that foragers should concentrate in areas with above-average prey abundance, ignoring areas with lower levels. Incorporation of this behavior into models of predators and prey in patches has a stabilizing influence on metapopulation dynamics (see Chapter 21). Such behavior reduces the degree of variability in abundance over time of both predators and prey when averaged over all patches (Fryxell and Lundberg 1993; Krivan 1997). Average abundance in a collection of patches tends to be stable when abundance in a single patch at any given time tends

Fig. 10.8 Relative daily energy gain (O) and observed population density of Thomson's gazelles (●) in relation to the biomass of grass in particular patches in Serengeti National Park. (After Fryxell *et al.* 2004.)

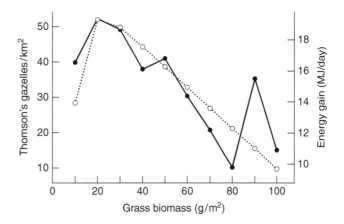

to be independent of that in other patches (de Roos *et al.* 1991; McCauley *et al.* 1993). This is more likely when predators abandon patches with low prey abundance than when movement in and out of patches is unrelated to resource abundance.

10.4 Risk-sensitive habitat use

Many foragers are themselves at risk of being attacked by predators. Frequently, such risk is highest when the forager is actively searching for food, rather than safely hidden away in a den or resting site. Incorporating predation risk is rarely straightforward in analyzing habitat use, yet we know from numerous empirical studies that it is important (Lima and Dill 1990). For example, risk-sensitive habitat use by the larvae of the aquatic insect *Notonecta* has been elegantly demonstrated in the laboratory (Sih 1980). Large *Notonecta* individuals often cannibalize smaller *Notonecta* individuals. Sih set up an experimental arena where individual *Notonecta* larvae could choose to feed in food-rich or food-poor patches. The larger *Notonecta* individuals selected food-rich patches, whereas smaller, more vulnerable individuals foraged in the poor patches. This seems to be a logical way of reducing the risk of predation, at the cost of reduced food intake.

One of the most elegant examples of the complex effects of risk-sensitive foraging is the series of experiments conducted by Schmitz and coworkers in small, caged populations of carnivorous spiders, herbivorous grasshoppers, and grasses and herbs (Schmitz *et al.* 1997). The grasshoppers suffer high rates of mortality from spiders under normal conditions. As a consequence, they tend to spend their time foraging on herbs, which are less nutritious than grasses but offer better cover from predators. At the spatial scale of a grasshopper, a single plant is a patch, so dietary preferences are in fact habitat preferences. By gluing shut the mouthparts of spiders, researchers were able to assess the demographic impact of perceived risk of predation versus real predation. Results showed that grasshoppers subject to perceived risk of predation (but not actual predation) suffered mortality levels similar to those of grasshoppers subject to real predation. Both treatment groups suffered considerably higher mortality than grasshoppers in cages without predators, which quickly learned to forage on the more nutritious grasses rather than the safer but less nutritious herbs.

Bluegill sunfish (*Lepomis macrochirus*) have been shown to balance the risk of predation against foraging benefits in choosing habitats. Nearshore habitats offer dense protective cover but relatively poor feeding. Open water offers better feeding but more exposure. When predators are present, young bluegills tend to concentrate in the habitat offering the greatest cover, whereas older, less vulnerable fish forage in the open, where energy gain is highest (Werner *et al.* 1983).

Sensitivity to predation risk also underlies patterns of habitat use by many large herbivores. For example, white-tailed deer in the boreal forests of Wisconsin and Minnesota tend to concentrate in the no-man's-land between wolf pack territories (Hoskinson and Mech 1976). Wolves tend to avoid going out of the area defended by their pack because of a pronounced risk of being attacked by hostile neighbors (Lewis and Murray 1993; Mech 1994). This effectively creates refuges in between territories in which individual deer are relatively safe.

One of the major difficulties in testing for risk-sensitive habitat use is finding a sensitive way of measuring risk, ideally from the animal's point of view. Brown (1988) suggested that the *giving-up density* at feeding trays could be used as a field measure of habitat attractiveness, which should be sensitive to both predation risk and alternative foraging opportunities in the surrounding habitat. This technique has been successfully applied in a large number of field studies. For example, different species of granivorous rodents in the Negev Desert have different assessments of risk in the same habitat, demonstrating interspecific differences in their perception of the risk of predation versus energetic gain (Brown *et al.* 1994).

A useful way of evaluating such decision-making, which balances trade-offs among competing risks and benefits to fitness, is known as *dynamic state variable modeling* (Mangel and Clark 1986; Clark and Mangel 2000). Although this approach is beyond what we can cover here, it offers a powerful means of evaluating the consequences of alternative behavioral activities that have complex trade-offs among energy gain, reproduction, and mortality risk. Indeed, it may be the only way to link complex sets of behaviors into a life-history framework. The monograph by Clark and Mangel (2000) offers an introduction to the techniques of dynamic state variable modeling and describes a wide set of applications.

10.5 Social behavior and foraging

10.5.1 *Interference among foragers*

The negative impact of other individuals on foraging success is sometimes termed *interference* by ecologists. This can result from direct aggression, stealing of food from other foragers, depletion of prey, or scattering or hiding of uneaten prey. If we presume that aggression is the main cause of interference, we can predict how it will affect feeding rates. While searching for prey, individuals should encounter other predators at random. If each encounter between predators results in a wastage of w time units then the foraging rate ($f(N, P)$) can be well approximated by a modified Type II functional response (Beddington 1975; Ruxton *et al.* 1992):

$$f(N, P) = \frac{aN}{1 + ahN + awP}$$

This formula predicts that interference should increase with predator forager density (Fig. 10.9). Similar logic can be used to develop an analogous model of interference arising from food thievery, also known as *kleptoparasitism* (Holmgren 1995).

Numerous field experiments have demonstrated such an increase in interference strength with forager abundance in organisms ranging from oystercatchers (*Haematopus ostralegus*; Goss-Custard and Durrell 1987) to caribou (Manseau 1996). It is conventional to measure interference from plots of log intake versus log forager abundance. Thus, Fig. 10.10 illustrates changes in intake of cockles by oystercatchers in the Netherlands as a function of forager density (Sutherland 1996).

Fig. 10.9 Changes in food intake as a function of predator (forager) density due to interference between predators.

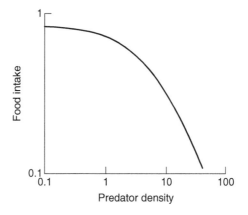

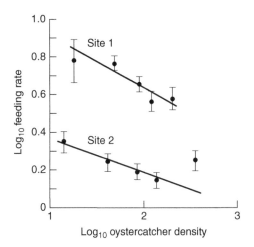

Fig. 10.10 Food intake by oystercatchers at two sites in the Netherlands declines as population density increases. (After Sutherland 1996.)

The ecological impact of interference can be profound (Beddington 1975; DeAngelis *et al.* 1975), adding a strong density-dependent effect to consumer–resource inter-actions that might otherwise be highly variable over time (see Chapter 9). Hence, interference can be a mechanism in the natural regulation of wildlife populations. An example of the effect of interference on carnivore population dynamics is the American marten, a mustelid carnivore in the forests of the United States and Canada (Fryxell *et al.* 1999).

10.5.2   *Territoriality*   Many wildlife species are territorial, meaning that they defend an area of (more or less) exclusive access from use by other members of the population. Males, females, or both sexes may be territorial, depending on the ecological circumstances that apply. Territories may be defended solely during the breeding season, as with many birds, or throughout the year, as with many vertebrate carnivores; and they may be defended by individuals, as with tigers, or by a pack of individuals, as with grey wolves. The mul-titude of territorial forms means that many different factors contribute to the adaptive significance of territoriality. Conversely, the consequences of territoriality at the pop-ulation and community levels can also vary considerably.

Central to most arguments about the ecological basis of territoriality is the notion of *economic defendability* (Brown 1964; Dill 1978; Kodric-Brown and Brown 1978; Schoener 1983; Stephens and Dunbar 1993; Fryxell and Lundberg 1997). It would make little sense to try to defend any territory offering trivial benefits or whose costs were astronomical. Let us assume that the purpose of a territory is to gain access to food supplies. The larger the territory, the greater the abundance of food. However, we have already seen that there are diminishing returns in terms of actual feeding rate as prey abundance increases. As a consequence, food benefits decelerate with increas-ing territory size. Similarly, the time and energy needed to patrol the perimeter also rise with territory size. Moreover, the larger the territory, the greater the risk that other individuals will intrude. As a result, costs continue to rise steadily while benefits show diminishing returns with increasing territory size. The profit margin is clearly greatest for individuals who hold territories of intermediate size (Fig. 10.11). Pro-vided that females are attracted to males that hold territories with sufficient resources to successfully rear offspring, the same sort of logic would predict that they favor

Fig. 10.11 Cost versus benefit of defending territories of different size. The optimal solution in this hypothetical case would be to defend an intermediate-sized territory of around 10.5 units, where there is maximum difference between the two curves.

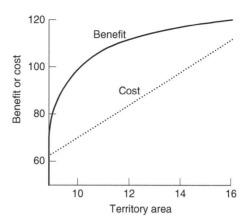

intermediate-sized territories. In short, territory formation can be viewed as an economic decision, like many of the other behavioral processes described earlier in this chapter. Natural selection should favor the evolution of territoriality, if it enhances survival and long-term reproductive success.

Economic models of territory formation predict that territory size should be negatively related to both prey abundance (because it affects diminishing benefits) and forager abundance (because it affects the cost of defending the area). These predictions have been borne out in several field studies. For example, the size of rufous hummingbird (*Selasphorus rufus*) territories is inversely proportional to the abundance of flowers per unit area (Gass *et al.* 1976; Kodric-Brown and Brown 1978). Hence, larger territories hold approximately the same resource abundance as smaller, richer ones. Similar patterns have been observed in many other species, ranging from shorebirds (Myers *et al.* 1979) to roe deer (*Capreolus capreolus*; Bobek 1977). Most convincing have been changes in territory size as a direct consequence of experimental alteration of either forager abundance (Bobek 1977) or resource levels (Myers *et al.* 1979). One difficulty with interpreting such experiments, however, is that experimental alteration of food levels often triggers changes in intruder pressure, so these factors tend to covary.

In many cases, individuals are faced with a choice of breeding (perhaps unsuccessfully) in a poor-quality territory or waiting for a vacancy in a better one. Such is the case for many passerine birds, with young birds often relegated to poor breeding habitats. We discussed one such example earlier in the chapter: low-ranking great tits occupy hedgerow territories rather than prime territories in woodland (Krebs 1971). Removal of prime territory holders leads to rapid replacement by members of the younger cohort. In other cases, younger individuals in poor habitat forgo breeding altogether, gambling instead on inheriting a good territory at a later date. A good example of this is the Seychelles brush warbler (*Bebrornis seychellensis*), with many youngsters choosing to stay at the nest and help their parents rear siblings rather than set off on their own (Komdeur 1992, 1993). Like all inheritances, this can be a risky proposition, because it depends on the probability that the helper will survive and the (hopefully lower) probability that the current occupant will not. Economic models predict that such helping behavior will be selected for when there are pronounced disparities in the quality of potential breeding sites and where the probability of long life is reasonably good. The Seychelles warblers proved this point by abandoning helping behavior as soon as openings for good territories were created.

Territoriality can play a stabilizing role in population dynamics (Fryxell and Lundberg 1997). If there is an upper limit to the number of territories that can be supported, this can effectively cap breeding by the predator population, preventing large-scale predator–prey cycles of the sort described in Chapter 9. Since many top carnivores are territorial (e.g. wolves, weasels, lions, hyenas, and tigers), this suggests that a deeper understanding of carnivore territory formation and dynamics in relation to changes in abundance of both predators and prey is essential to adequate conservation and management efforts.

10.6 Summary

Foraging success is strongly affected by the behavioral decisions of both predators and their prey. We consider a number of these decisions. First, foragers must decide which prey to attack and which to ignore. Optimal diet choice theory can be used to determine wise solutions to this problem, based on the opportunity cost of wasting time on poor prey in hand when better prey might yet be found. For herbivores, foraging decisions are shaped by multiple constraints, such as balancing the need to meet requirements for a scarce nutrient with the objective of maximizing energy intake. Such problems can be approached using linear programming. Optimal patch departure theory is useful in considering how long foragers should stay in a particular area before moving on. The best theoretical solution is that animals be sensitive to the opportunity cost of wasting time in a poor patch when a better patch lies close at hand. As for diet, herbivores choose patches in a manner that balances multiple constraints on their feeding.

Patch preferences can also be shaped by the need to trade off risk of predation against foraging gains. Risk-sensitive foraging demands complex approaches to decision evaluation, such as dynamic state variable modeling.

We also consider how social processes influence foraging decisions. Social interference plays an important role in this process, and we show how interference and territoriality can be viewed as adaptive responses to environmental conditions. All of these behaviors have important consequences for the dynamics of wildlife populations.

11 Climate change and wildlife

11.1 **Introduction** Evidence from studies of past climates shows that the world climate has always been changing continuously, and we can safely assume it will continue to do so. (A useful compendium on all aspects of climate change is presented in *The Britannica Guide to Climate Change* 2008). Presently the world is becoming warmer. The rate of warming is still within the range seen following the coldest point of the last ice age, from 20 000 to about 7000 years ago. However, the Intergovernmental Panel on Climate Change (IPCC 2007) stated in its fourth report that there is now unequivocal evidence that human causes are responsible for most of the current climate change; the impacts from human causes are over 10 times larger than those from all natural components combined. The main human source is the emission of greenhouse gases, in particular carbon dioxide, which comes from the burning of fossil fuels. Land use change from the destruction of forests for agriculture is a lesser cause. The IPCC has published a range of possible emission scenarios for the future that predict increases in global temperatures from 1.4 to 5.8 °C by the year 2100. Such warming will be more evident in higher latitudes (both northern and southern) and on land than at lower latitudes or over water. Precipitation (either snow or rain) will increase at higher latitudes and in the tropics but will decrease in the subtropics.

In this chapter, after looking at some of the evidence for changing climates we will explore the consequences for wildlife. We will look at how wildlife has been responding to environmental changes over the past 100 years and will review the mechanisms that produce these responses. Finally, we will examine how such changes will affect whole ecosystems.

11.2 Evidence for climate change

A synthesis of the earth's surface temperature from tree rings, ice cores, corals, and direct measurements over the past 2000 years (Fig. 11.1) shows a relatively steady temperature until the 20th century. The medieval warm period (AD 900–1200) was followed by the Little Ice Age (1450–1800) in Europe, which some suggest was due to the increase in solar reflectance (albedo) leading to cooling from the conversion of forests to agriculture. Warming started with the Industrial Revolution around 1800, reversing the previous cooling trend. The 20th century shows a gradual warming equivalent to the medieval period until about the middle of the century, when the rate of temperature increase accelerates to unprecedented levels. This is shown in more detail in Fig 11.2.

The foremost reason for such changes in temperature in the 20th century is the increase in atmospheric carbon dioxide, as shown by data collected at Mauna Loa on Hawaii (Fig 11.3). This gas, along with methane and sulfur dioxide, acts to trap irradiated heat. Most of the carbon dioxide has come from the burning of coal, oil, and natural gas by modern industrial societies.

Wildlife Ecology, Conservation, and Management, Third Edition.
John M. Fryxell, Anthony R. E. Sinclair and Graeme Caughley.
© 2014 John Wiley & Sons, Ltd. Published 2014 by John Wiley & Sons, Ltd.
Companion Website: www.wiley.com/go/Fryxell/Wildlife

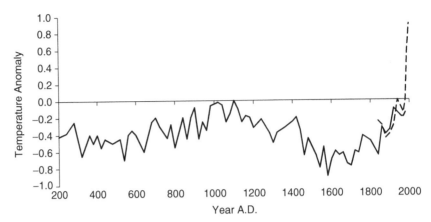

Fig. 11.1 Reconstructed surface temperature using proxy data (such as tree rings, ice cores, and corals), shown as deviation in degree C from long term average values (solid line, from data in Moberg *et al.* 2005). Instrumental surface-air temperature for 1850–2006 from the Climatic Research Unit, Norwich, UK(broken line). Figure redrawn from Mann *et al.* (2008).) The medieval warm period is seen 900–1200 A.D. and the little ice-age as 1400–1800 A.D.

Fig. 11.2 Global surface land temperature anomalies (1880–2012) as departures from the 20th century average (1901–2000). Data from the Climate Data Center, National Oceanic and Atmospheric Administration (NOAA) (see noaa.ncdc.gov).

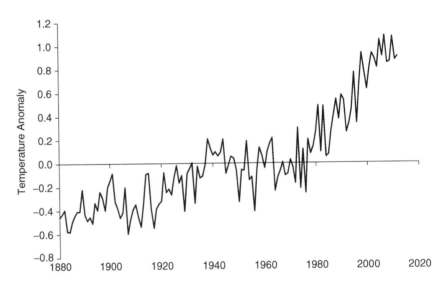

The physical consequences of atmospheric warming include the melting of the polar ice caps (Fig. 11.4), the melting of glaciers, the thawing of the permafrost, and a rise in sea level. In addition to the mean rise in temperature, greater variability in climate is predicted, producing more severe droughts and storms. There are biological consequences to these physical changes involving both oceanic and terrestrial systems; useful summaries can be found in Lovejoy and Hannah (2005), Boyce *et al.* (2006) and Pearson (2011). Here we look at some of the responses shown by wildlife.

11.3 Wildlife responses to climate change

One of the most well documented indicators of the effects of climate change is the change in species range (e.g. Root *et al.* 2003; Parmesan 2006). As atmospheric temperature changes with both latitude and elevation, we should expect to see geographical

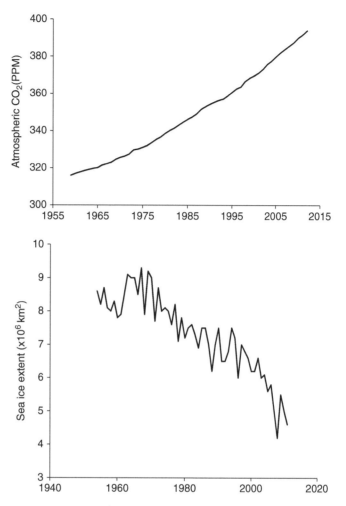

Fig. 11.3 Annual mean concentrations of atmospheric CO_2 at the Mauna Loa Observatory, Hawaii (PPM) 1959–2012. Scripps annual data from January 1959 – April 1974, and NOAA-ESRL data since May 1974. NOAA-ESRL annual data is posted at ftp://ftp.cmdl.noaa.gov/ccg/co2/trends/co2_annmean_mlo.txt

Fig. 11.4 Direct observations of minimum amount of sea ice remaining in the Arctic in September from satellite images (1954–2012). Redrawn from Stroeve *et al.* (2007) and Stirling and Derocher (2012), and updated with permission to 2012. Original data provided by J. Stroeve at the National Snow and Ice Data Center, USA.

shifts in range related to these two aspects. Other effects include changes in the timing of breeding and other life-history stages (phenology) and changes in demography (fitness consequences of changes in births and survival).

11.3.1 *Latitudinal changes in range*

If the range of animal species is determined by temperature then we could hypothesize that warming climates will result in a polewards shift in ranges. For example, in the northern hemisphere we would expect both the northern and southern limits of range to shift northwards. If range is determined by other factors, such as habitat, competition, predation, or human interference, then such shifts might not necessarily occur with changes in temperature.

British birds are well studied and their ranges have been measured since the 1960s (Thomas and Lennon 1999). A survey of some 100 species of birds covering a period of 30 years showed that there was a northwards movement of the northern edge of the range by an average of 1 km per year. A similar study in North America used data from the American Breeding Bird Survey, also collected since the 1960s. Of some 55 species that fitted the necessary criteria to test the hypothesis, most shifted their northern range edge polewards by some 2 km per year. However, in both studies the southern edge showed little change northwards. The authors attributed this to the method of data recording: it is far easier to detect when a species moves into a new area (in the north) than when it leaves a previously occupied area (in the south); there is a time lag in detecting this response (Hitch and Leberg 2007).

11.3.2 Elevational changes in range

Temperature changes with elevation over a far shorter distance than it does with latitude. Thus, mountain-living species can shift their range upwards even when they have limited dispersal ability. Of course, such species can shift polewards if the topography allows: north–south mountain ranges as in the Rockies or Andes would allow such shifts, but the east–west aligned Alps of Europe would prevent them. Climate change, therefore, predicts an upwards shift in elevation range. There is now considerable evidence that species are shifting their ranges upwards directly as a result of climate change rather than because of habitat modification. Expansions at the upper edge are constrained by maximum elevation available or suitable habitat (Wilson and Gutierrez 2012).

Madagascar has been cut off from the African mainland for a long time and a large majority of its vertebrates are endemic. This applies to the amphibians found in its rainforests. The elevational distribution of amphibians was recorded in a strip of forest on the Tsaratanana mountain range in northern Madagascar in 1993, and again in 2003 (Raxworthy et al. 2008). The midpoint of the elevational range was found to have shifted upwards by some 65 m, with some species moving up as much as 400 m. Air temperature increased over the same time period by about 0.10–0.37 °C, which would account for the observed 65m change.

A similar comparison of surveys over a much longer period was conducted by Craig Moritz et al. (2008), who made use of the early records of small mammals collected by Joseph Grinnell a century ago (1914–20) in the Yosemite transect of the Sierra Nevada Mountains. Moritz resurveyed the elevational distributions of 43 small mammal species in the 2000s and found that most had shifted upwards an average of 500 m. During this century-long interval the coldest monthly (mean minimum) temperature had increased by 3.7 °C. In general, upwards shifts in range occurred more frequently in high-elevation species; in addition, their lower range limit shifted up more frequently than their upper limit, resulting in a range contraction.

11.3.3 Effects of sea-level change

Sea-level changes are likely to affect shorelines and hence the species that use them: shorebirds that feed on mudflats, mangroves, estuaries, and swamps. Sea turtles that use beaches for egg laying will also be severely affected. Even on land there are consequences: in Florida, sea-level rise will reduce habitat and increase the probability of extinction in snowy plovers (*Charadrius alexandrinus*) (Aiello-Lammens et al. 2011).

11.3.4 Changes in phenology of reproduction and survival

One of the signals of climate change is the progressively early spring warming in the northern hemisphere. Starting in 1936, Aldo Leopold, and subsequently his family, recorded some 74 indicators of spring on the family farm in Wisconsin(USA) over 61 years (Bradley et al. 1999). Of these, 19 showed a spring advance of 1–2 days per decade. Not all indicators showed the spring advance because not all are determined by climate; some are determined by photoperiod, which is where problems arise with the breeding season of birds.

The pied flycatcher (*Ficedula hypoleuca*) is an insectivorous bird that breeds in northern Europe but migrates to Africa for the winter. Populations have been studied in

Holland for the past 30 years (Both and Visser 2001; Visser and Both 2005). One of the flycatcher's main foods is caterpillars on newly opening oak leaves. As a result of spring warming, the timing of peak abundance of caterpillars has advanced by 7.5 days per decade (Visser *et al.* 2006). Flycatchers return to Holland in spring to breed. Although the date of egg laying has advanced by 5 days per decade, they have not kept pace with the change in the peak of their food supply (Both and Visser 2001). Their laying date is constrained because their arrival date from Africa has not changed (Visser and Both 2005) – it is suggested that the birds are using photoperiod (increasing day length) as the cue for migration, which, of course, is not affected by climate change. Although the birds have a few days in which to bring forward laying, they now lag behind the peak of food and so chicks have less food available to them. The consequence is a decline in population, as much as 90%, in those areas where hatching of chicks is most out of synchrony with the food supply (Both *et al.* 2006). However, the mismatch of breeding with food supply depends on habitat. Caterpillars are strongly synchronized with leaf opening in oak forests but not in conifer forests. The mismatch is thus most severe in oak habitats and is not observed in conifer ones (Burger *et al.* 2012).

Mismatch of food and laying dates is also taking place in nonmigrant bird species, resulting in a loss of fitness for late-laying individuals; this has been measured in a 37-year study of great tits (*Parus major*) in Holland, where individuals that lay later have a lower breeding success (Reed *et al.* 2013).

In general, the arrival dates of migrants in the northern hemisphere have been getting earlier by as much as 4–6 days per decade (Parmesan and Yohe 2003; Root *et al.* 2003). Similar trends in phonological change are exhibited by southern-hemisphere birds. Mean maximum temperature in Australia has increased (although by a modest amount: 0.17 °C/decade). Breeding migrants arrive earlier and depart later, extending the breeding season; however, nonbreeding (overwintering) migrants arrive earlier but depart earlier too, so the timing has changed but not the duration (Beaumont *et al.* 2006).

Although day length triggers the onset of breeding, Wingfield *et al.* (2011) have shown that gonadal growth, once started, can be speeded up by increasing temperatures. The snowbunting (*Plectrophenax nivalis*) in Alaska is breeding earlier as spring temperatures warm, and as a consequence a greater proportion of birds are having two broods instead of the normal one.

Examinations of changes in the phenology of mammals have focused on the hibernation attributes of ground squirrels. These animals spend as much as 7 months underground over winter. They emerge in spring, and mate shortly afterwards. Dates of emergence have important fitness consequences, affecting opportunities for mating; dates of emergence and estrus are heritable and correlated in the Columbian ground squirrel (*Urocitellus columbianus*) (Lane *et al.* 2011). A 20-year study of a population in Alberta, Canada has shown a delay in emergence of 0.47 days per year, which has led to reduced reproduction and a decline in population. The reason for this was not a change in spring temperature, which has remained relatively constant, but rather an increase in the frequency of late winter storms, producing a delayed snowmelt, a feature of the increased variability of weather associated with climate change (Lane *et al.* 2012).

This increased variability of weather has also resulted in extreme events that reduce breeding success in birds. Seabirds that breed in the Antarctic, such as the snow petrel (*Pagodromas nivea*), are adapted to cold climates. Recent warm events, which have produced rain instead of snow while birds have been sitting on eggs, have resulted in the birds abandoning their nests – they are not adapted to these conditions (Wingfield *et al.* 2011).

Increasing climate variation also has effects on survival of the European beaver (*Castor fiber*), with higher variation leading to lower survival (Campbell *et al.* 2012). Furthermore, variable weather can have contrasting effects in different areas, and consequently opposite effects on vulnerable species: spotted owl (*Strix occidentalis*) populations in Arizona are predicted to decline rapidly due to increasing hot, dry, extreme weather, whereas those in California, which experience cold, wet conditions, are not so affected (Peery *et al.* 2012).

Another aspect of phenology is particularly important in migratory birds. These species have life-history requirements that partition their year: spring migration, breeding, molt, autumn migration, winter nonbreeding. Each stage takes place at a particular time of year and usually has a set duration. Changes in one period can affect others; for examples, a delay in the breeding season due to warmer weather could delay molt, and molt has to take place before the birds can migrate again. Phenological stages become backed-up, overlap, and are consequently out of phase with the environment (Wingfield 2008).

However, this issue is not confined to migrant species: the timing of nesting in duck species in the western boreal forests of North America is affected by earlier dates of snow melt. Earlier-breeding species, such as mallard (*Anas platyrhynchos*) and wigeon (*A. americana*), are more flexible and can accommodate these climatic changes better than late-breeding species such as scaup (*Aythya*) and scoter (*Melanitta*). The latter are experiencing reduced breeding success and declines in population numbers (Drever *et al.* 2012).

In marine turtles, sex of offspring is determined by incubation temperature, which is obviously prone to warming climate. Caribbean Columbian leatherback sea turtles (*Dermochelys coriacea*) now produce 92% female hatchlings; it is predicted that continued warming could result in complete feminization. Some cooler refuges between the main beaches may be their escape (Patino-Martinez *et al.* 2012).

11.4 Mechanisms of response to climate change

The range of a species is determined by environmental conditions that affect colonization–extinction dynamics: within the range, birth rates either exceed or at least equal death rates. Excess individuals colonize new areas. If suitable, this causes an expansion of range (in some areas, individuals can live but not breed; they use such areas while waiting to find a breeding area). In contrast, if the environment deteriorates (for that species) then we see three possible responses: (i) if a species has sufficient dispersal ability then one might expect to see a range shift; (ii) if there is substantial genetic variability and a strong selection gradient then one might predict rapid adaptation to the new conditions; (iii) if neither emigration nor adaptation is possible then the range will shrink and the population will decline and may even be extirpated. In this section we look at the mechanisms that produce these alternative responses.

11.4.1 Colonization

Colonization is most likely in species that have high dispersal ability, particularly birds that are long-distance migrants. We have already given several examples of this in Section 11.3.1. In addition, species with wide habitat preferences are less constrained when they move. Colonization is more difficult if species are preprogrammed to return to a specific site for breeding – such as sea turtles that return to particular beaches – and those sites become unsuitable. The least likely species to move are those with narrow habitat tolerances, high site fidelity, and low dispersal ability; many small mammals fall into this category.

Because all species have different dispersal abilities, the consequence of climate change is that different members of a community will move at different rates towards new and more suitable environments. We already know this from the movements of species north at the end of the last ice age (Davis and Shaw 2001). Studies of the effect of drought on North American birds show differential responses to drought, for example; species in montane areas increase in richness and abundance, whereas those on the semi-arid Great Plains decrease (Albright *et al.* 2010). Thus, a reshuffling of species within a community will likely take place. The problem is that new combinations of species may not be viable; in the extreme situation, predators to which prey are not adapted may arrive in the new community and eliminate the prey species. We have already experienced this outcome when prehistoric humans introduced predators into new ecosystems in Australia and New Zealand (see Chapter 17); such introductions in effect reshuffle the community.

11.4.2 *Adaptations of species to new environments*

A second result of climate change could be the adaptation of a species to the new conditions. Adaptation must have taken place during past climatic changes, and relatively quickly too. The change in temperature at the end of the last ice age took only 1000–2000 years, for example. However, the present human-induced change is likely to be even faster and more extreme. Can wildlife adapt to the coming fast rates? There is some evidence that evolutionary change is taking place (see Chapter 20). For example, increasing avian body size may be a response to increasing variability of climate or primary production (Goodman *et al.* 2012).

The eastern red-backed salamander (*Plethodon cinereus*) in North America comes in two morphs: an all-black variety and one with a red or yellow stripe down its back. In general, the black morph is associated with warmer environments than the red-striped one. The relative frequency of these morphs has been recorded since the early 1900s. Gibbs and Karraker (2006) have found that the frequency of the red-striped morph declined from 80 to 74% over the 20th century as the temperature within its range increased by 0.72 °C. This change in morph frequency illustrates an adaptation towards warmer conditions, which in this example probably resulted from human modification of habitat as well as climate change.

Although there are examples of adaptation, in general vertebrates, and especially large mammals, which have long generation times and low reproductive rates are less likely to be able to adapt genetically to fast-changing climates than are smaller species.

11.4.3 *Population declines*

With a changing environment, previously suitable conditions for reproduction and survival will decline. If a population cannot adapt or move then the area suitable for the species will be reduced to a fraction of the former range, or even disappear.

Polar bears (*Ursus maritimus*) have a circumpolar range with several subpopulations. In northern Canada they depend on sea ice, from which they hunt newly weaned ringed seals (*Pusa hispida*) in the spring. This abundant food source allows the bears to accumulate fat so that they can survive during the long summer fasting period when they are confined to the mainland and seals are not available (Stirling and Derocher 2012). Climate change is causing sea ice to break up earlier each year (Fig. 11.4), resulting in the bears having less time to put on fat, so they are of lower weight when they move on to the mainland (Fig. 11.5). Results include decreased access to denning areas, fewer

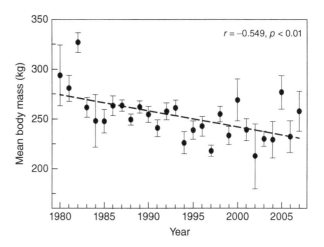

Fig. 11.5 Mean mass of female (and likely pregnant) polar bears in the western Arctic in autumn declined over the period 1980-2007. Broken line is the linear regression (Source: Stirling and Derocher, 2012. Reproduced with permission of John Wiley & Sons Ltd.). Weight loss is an indicator of lower hunting success as sea ice disappears.

and smaller cubs, and lower survival of both cubs and older age classes. Consequently, there will likely be population declines and perhaps eventual extirpation. This will begin in southern subpopulations such as those in the Hudson Bay or the southern Beaufort Sea.

The declining supply of seals has caused polar bears to find alternate foods. In the Hudson Bay region, bears have turned to eating lesser snow goose (*Chen caerulescens*) eggs (Rockwell and Gormezano 2009). Although this source may mitigate the loss of seals, the degree of predation is unlikely to be sustainable and geese are not available for all subpopulations of polar bears. Another consequence of these changes in food supply is that bears are now coming more frequently into conflict with human habitation. Stirling and Derocher believe that polar bears may persist in the northern Canadian Arctic Islands and northern Greenland for the foreseeable future, but they will have a much reduced and fragmented range on the islands, so their long-term viability is uncertain.

We have already seen that climatic warming causes species to shift their range towards higher elevations. This has the consequence of reducing range size, fragmenting and isolating the range on different peaks, and so reducing gene flow. In some cases, the habitat disappears if the elevation is insufficient. Monteverde is a tropical mountain in Costa Rica that supports a cloud forest. This forest obtains its moisture from water-laden air currents coming off the Caribbean; as the air hits the mountain, it lifts, and with the cooler temperature the moisture condenses as clouds. Since the 1970s, the warming temperature has caused an increase in the elevation at which clouds form, sometimes to above the mountain peak. The consequence is that the number of days that this montane forest is without cloud has risen, sometimes with periods of 16 days in a row mist-free (Pounds and Crump 1994; Pounds *et al*. 1999). The cloud forest supports a number of amphibian species, including the golden toad (*Bufo periglenes*). Monitoring of these amphibians since the 1970s showed relatively stable populations until the 1980s, when declines in population became evident, culminating in a catastrophic collapse in 1987. The golden toad never recovered and the last sighting before extinction was in 1989. Pounds *et al*. (1999) propose that at least one fundamental reason for this extinction was the change in the availability of moist clouds, reaching a threshold point at which the species could not exist. In essence, the livable range of the species had risen above the height of the mountain and so had disappeared.

This example illustrates the general phenomenon of decline in amphibians worldwide: in 2004 the Global Amphibian Assessment reported that 427 species were critically endangered, with 122 possibly extinct (Stuart *et al.* 2004). We will see later that these events are implicated in more complex ecological interactions.

11.5 **Complex ecosystem responses to climate change**

Direct effects of climate change on wildlife are relatively rare. The increased ferocity of storms that hit oceanic islands may cause extinctions, for example, but for the most part climate change has more important indirect effects. We consider three aspects of indirect effects in this section.

11.5.1 *Food chain effects*

We have focused so far on the direct responses of species to climate change through shifts in range and changes in phenology and demography. However, when one species changes its ecology it usually affects others, and so influences the whole ecosystem.

The marine food chain in the North Sea between Britain and Europe involves the normal links of phytoplankton that supports zooplankton, such as copepods, which in turn support fish larvae (Edwards and Richardson 2004). In spring, each of these trophic levels blooms at the appropriate water temperature, and each bloom normally coincides with the peak of its food supply. Over the past 50 years, the timing of the bloom in each trophic level has advanced with the warming of the water. The problem is that each level has advanced by a different number of days; thus, the phytoplankton peak has advanced by 23 days and the zooplankton peak by only 10, so that they are now out of synchronization with their food supply. Furthermore, the fish larvae have advanced by 27 days, so they are also out of synchronization. Thus, the productivity of this food chain is likely to decline, reducing fish survival and fish productivity.

Whitebark pine (*Pinus albicaulis*) in Yellowstone National Park supports a wide array of birds and rodents, as well as the grizzly bear, through its high production of nutritious seeds. Climate change is likely to reduce both the range of this tree and its seed production via increased incidence of disease, competition from lodgepole pine, and increased fire frequency. This will reduce a major food source for bears at a time when they need it to put on fat for winter; in years with low seed production, they turn to other components of the food chain, such as berries and ungulates, to compensate (Koteen 2002; Olliff *et al.* 2013).

We have already discussed changes in the timing of the insect food supply for the pied flycatcher (see Section 11.3.3). Among the major items in this bird's diet are the caterpillars of the winter moth (*Operophtera brumata*). Female winter moths climb oak trees in winter (hence the name) and lay their eggs on leaf buds. In spring, the eggs hatch to coincide with the early growth of nutritious leaves. As we have seen, the eggs of the winter moth are hatching earlier with climate warming (Visser *et al.* 2006), and this means that they are hatching too early for bud burst in the oaks. They thus have less food available, and numbers of caterpillars are declining (Visser and Holleman 2001). Thus, the sequence of oak leaves → caterpillar → pied flycatcher is disrupted at both links in the food chain.

11.5.2 *Interaction of climate change with other constraints*

We have seen that in mountainous regions, ranges can increase in elevation to counteract the increase in climate warming. However, if a mountain is not sufficiently high then the range is reduced or even disappears – as we saw with the golden toad (Section 11.4.3). In this sense, there is a physical constraint in response to climate change.

There are, however, other types of constraint that distort or limit a species' response to changing environments.

One major constraint is human modification of habitats to agriculture. Thus, suitable habitat may no longer be available in the new areas where climate has become suitable. In the pied flycatcher example, there were several original sites that experienced major declines in population or local extinctions because the populations were unable to move due to the constraint of agriculture. In general, the interaction of climate change and habitat loss is predicted to reduce the range of 71% of European bird species by 2050. Richness will decrease in southern Europe but increase in northern Europe as species' ranges shift (Barbet-Massin *et al.* 2012). Global terrestrial analysis shows that the interaction of climate change and habitat loss is exacerbated in areas with high maximum temperatures and ameliorated in areas with increasing rainfall (Mantyka-Pringle *et al.* 2012).

A second constraint involves the interaction of other biotic influences. Climate change could cause a species to become more vulnerable to interspecific competition, predation, or disease. We have already seen that the nests of lesser snow geese on the western shores of Hudson Bay may now be subject to increased predation from polar bears as the bears move their range to the mainland with the disappearance of sea ice (Rockwell and Gormezano 2009). Climate change is predicted to alter the competitive balance between three species of hare in Europe, allowing the common hare (*Lepus europeus*) to improve its competitiveness over the mountain hare (*L. timidus*) (Acevedo *et al.* 2012).

The major declines in amphibian populations around the world are now considered to be a result of the interaction of climate change with the exotic fungal disease chytridiomycosis (caused by *Batrachochytrium dendrobatidis*; Voyles *et al.* 2009) and habitat loss (Duarte *et al.* 2012). The fungus was originally endemic in South African *Xenopus* frogs; these were commonly used in medical studies around the world some decades ago. Somehow or other, they escaped and invaded their new continents. The fungus they carried lives in water and infects other species of frog through their skin – and most species are highly susceptible to it. Pounds *et al.* (2006) found that the decline in populations of a group of South American amphibians, the Harlequin toads (*Atelepus* spp.), was best explained by the interaction of a warmer climate and this fungus. The fungus grows best at 35–43 °C and without temperature extremes. The warming climate has increased cloudy days, reducing cold nights and hot days, which suits the fungus at midlevel elevations (many of the declines in Harlequin toad species have occurred at midlevel elevations, rather than at higher or lower ones, which are too cold or too hot for the fungus, respectively). Thus, the warming climate combined with the reduction in temperature extremes has promoted the viability of the fungus and the decline in amphibians. However, it should be mentioned that this disease is not the only cause of amphibian population declines; a number of other climate-related factors are also involved (Whitfield *et al.* 2007).

11.5.3 *Bioclimate envelopes*

By measuring the climate at sites where a species occurs, we can identify the environmental limits at which it can survive. Using sophisticated computer modeling, we can also predict the probability of occurrence within the range (not just the outer boundaries), thus providing *climate response surfaces*. This is an important advance over simple range maps because the models also incorporate abundances. These statistical descriptions are called *bioclimate envelopes* (see Huntley *et al.* 2012). This approach allows an analysis of whole communities in their response to climate change.

Huntley *et al.* (2012) fitted climate response surfaces to abundance and range data for 78 species of bird in southern Africa. Most of these were endemic to the region and of conservation concern. Their models projected the potential impacts of a series of climate change scenarios upon species abundance and range. The first important result was that abundance was more likely to change than range limits. Second, for most species (74%) both abundance and range were predicted to decrease by 2100. Third, impacts were particularly important if agriculture prevented species from dispersing, a point we have already discussed. If species are unable to disperse then abundance is predicted to decline by as much as 80% in 24% of the species. Thus, for most species climate change is likely to have negative consequences. Huntley and Barnard (2012) found similar results in their analysis of the biodiversity hotspot called "fynbos," a special vegetation zone at the southern tip of South Africa. As a result of climate change, species richness of fynbos and grassland bird assemblages is predicted to decrease by 30–40% by 2085. Areas of greatest richness will decrease and range shifts will be limited by agriculture. Species of conservation concern are projected to decrease in abundance more than is expected from range decrease alone, and two species are projected to go extinct. Similar predictions have been calculated for the world's freshwater turtles and tortoises (Ihlow *et al.* 2012). Some 86% will experience range contraction and 12% will have ranges completely outside their current distribution.

These are two specific examples of the bioclimate envelope approach. Others have examined wildlife in Mexico (Peterson *et al.* 2002), Australia (Williams *et al.* 2003), and Europe (Thomas *et al.* 2004). All come up with findings similar to those in southern Africa, namely a large number of species likely to experience declines in abundance and range, with a significant proportion likely to go extinct depending on the scenario used: whether species can disperse easily or not at all, and whether temperature increases by small or large amounts. Climate envelope models that incorporate other interacting factors may have more robust predictions: the rare South Australian glossy black cockatoo (*Calyptorhynchus latham*) is predicted to decline in numbers with climate warming, but the effect is exacerbated by disease, fire frequency, habitat change, and competition from other parrot species; management has to mitigate all of these factors (Harris *et al.* 2012).

One word of warning, however: these statistical approaches depend upon the assumption that species ranges and abundances are determined by climate. One can obtain correlations between abundance and climate, but this does not mean that climate is the determining factor. Beale *et al.* (2008) showed from simulations that such correlations can simply occur by chance, and so may not reflect climate as the main determinant of range. Thus, the bioclimate envelope approach needs to be backed up with more detailed analysis of the causes of population decline. In addition, the climate envelope approach does not account for possible evolutionary responses. As we show in Chapter 20, evolution can proceed at fast rates, and possibly at rates sufficient to respond to climate change: some species may be able to adapt to new conditions (Skelly *et al.* 2007).

11.6 Summary

The physical consequences of atmospheric warming include the melting of polar ice caps, the melting of glaciers, the thawing of the permafrost, and a rise in sea level. In addition to a mean rise in temperature, there is predicted to be greater variability in climate, producing more severe droughts and storms. These physical changes will have biological consequences, involving both oceanic and terrestrial systems. These include changes in the range of species with movement towards the poles and up mountains. Timing of breeding, especially in birds but also in small mammals, is affected by earlier spring warm-up and greater variability in weather.

The response of a given species to deteriorating environmental conditions can vary, ranging from emigration if the species has suitable dispersal ability, to evolutionary adaptation if there is sufficient genetic variability and selection gradients are sufficiently high to allow the species to adapt to the new conditions, to a reduction of range if neither emigration nor adaptation is possible. In the latter case, the range will shrink and the population will decline and perhaps be extirpated. Climate change has effects on whole communities, through alteration of the food chain. Nevertheless, responses to climate change may be impeded by other constraints, such as human modification of habitats to agriculture. New areas where climate has become more suitable may not be available if the habitat has been altered by humans. Other constraints involve the interaction of biotic influences with environmental change. A species might become more vulnerable to interspecific competition, predation, or disease.

For conservation, it is important to predict future changes in both local climate and the availability of suitable new sites. Models correlate present environmental parameters with a species distribution producing "climate envelopes." This approach has predicted many declines in abundance, contractions of range, and even extinctions of species. These models need to be tested, however, against future observations. Climate change models have their limitations and do not yet account for evolutionary change in a species' tolerance to new conditions or for the potentially complex effects of ecological interactions among community members. Many alternative climate models and potential biological outcomes are now being incorporated in "ensembles" of multiple models in order to overcome some of these uncertainties.

Part 2

Wildlife conservation and management

12 Counting animals

12.1 Introduction

Knowledge of the size or density of a population is often a vital prerequisite to managing it effectively. Is the population too small? Is it too large? Is the size changing, and if so in what direction? To answer these questions, we may have to count the individual animals, or we may be able to obtain adequate information by way of an indirect indication of abundance. In any event, we need to know when a census is necessary and how it might be done.

Although *census* is strictly the total enumeration of the animals in an area, we use the word here in its less restrictive sense of an *estimate of population size or density*. Such an estimate may come from a total count, from a sampled count, or by way of an indirect method such as mark–recapture.

Closely related to the census is the *index*, a number that is not itself an estimate of population size or density but which has a proportional relationship to it. The number of whales seen per cruising hour is an index of whale density. It does not tell us the true density but it does allow comparison of densities between areas and between years. *Indices provide measures of relative density* and are used only in comparisons. They are particularly useful in tracking changes in rates of increase and decrease.

Almost all decisions on how a population might best be managed require information on density, on trends in density, or on both. There are many methods to choose from and these differ by orders of magnitude in their accuracy and expense. Hence, before any censusing is attempted, the wildlife manager should ask a number of questions:

- Do I need any indication of density and what question will that information answer?
- Is absolute density required or will an index of density suffice?
- Will a rough estimate answer the question or is an accurate estimate required?
- What is the most appropriate method biologically and statistically?
- How much will it cost?
- Do we have that kind of money?
- Would that money be better spent on answering another question?

The trick in obtaining a usable estimate of abundance is to choose the right method. What works in some circumstances is useless in others. Here we present a wide range of options and indicate the conditions under which each is most effective.

12.2 Total counts

The idea of counting every animal in a population, or in a given area, has an attractive simplicity to it. It is the method used by farmers to keep track of the size of their flocks. No arithmetic beyond adding is called for and the results are easily interpreted.

Wildlife Ecology, Conservation, and Management, Third Edition.
John M. Fryxell, Anthony R. E. Sinclair and Graeme Caughley.
© 2014 John Wiley & Sons, Ltd. Published 2014 by John Wiley & Sons, Ltd.
Companion Website: www.wiley.com/go/Fryxell/Wildlife

This is why total counting was once very popular in wildlife management and why it is still the most popular method for censusing people.

Total counts have two serious drawbacks: they tend to be inaccurate and expensive. Nonetheless, they have a place. The hippopotami (*Hippopotamus amphibius*) in a clearwater stretch of river can be counted with reasonable facility from a low-flying aircraft. The number of large mammals in a 1 km² fenced reserve can be determined to a reasonable level of accuracy by a drive count. It takes much organization and many volunteers but it can be done. Every nesting bird can be counted in an adélie penguin (*Pygoscelis adeliae*) rookery, either from the ground or from an aerial photograph. That is an example of a "total count" providing an index of population size, because more than half the birds will be at sea on any given occasion.

Total counting of large mammals over extended areas was common in North America up to 1950. Gill *et al.* (1983) describe the system in Colorado:

> Biologists attempted to count total numbers of deer comprising the most important "herds" in the state. Crews of observers walked each drainage within winter range complexes and counted every deer they encountered. The sum of all counts over every drainage of a winter range was taken as the minimum population size of that herd (McCutchen 1938; Rasmussen and Doman 1943).

Total counting of large mammals from the air was a standard technique in Africa in the 1950s and early 1960s. Witness the total counts of large mammals on the 25 000 km² Serengeti–Mara plains (Talbot and Stewart 1964) and 20 000 km² Kruger National Park, South Africa:

> trends in population totals, spatial distribution, and social organization are obtained by means of surveys by fixed-wing aircraft. Due to the size of the Kruger National Park these (total count) surveys require three months to complete and are consequently undertaken only once annually (i.e. during the dry season from May to August (Joubert 1983)).

These massive exercises continued in Kruger until 1996, when they were abandoned due to cost. Similar methods are used to count pronghorn antelope in the United States (Gill *et al.* 1983). Total counts continue to be used on species that are highly clumped with wide spacing between clumps. For example, both African buffalo (*Syncerus caffer*) and African elephant (*Loxodonta africana*) live in widely dispersed large herds of several hundred animals in both Serengeti and Kruger National Parks, and total counting is still the best method of obtaining their population size. This is because the dispersion pattern of these species means that sample counts produce very high variances and hence wide confidence limits. A simulated transect sampling strategy for a known dispersion of buffalo showed that over 90% of the area had to be sampled before confidence limits were reduced to acceptable values (< 15% of the estimated total). Thus, total counting was more efficient because it was logistically easier than rigidly flown transects (Sinclair 1973). Similarly, the clumped distribution of pronghorn antelope (*Antilocapra americana*) in North Dakota produced such high variances from a variety of sampling strategies that Kraft *et al.* (1995) advised against using samples to estimate numbers.

12.3 Sampled counts: the logic

There are two important areas in which scientific thinking differs from everyday thinking: the selection of a random or unbiased sample and the choosing of an appropriate experimental control. Knowing how to sample and knowing how to design an experiment that gives an unambiguous answer are the two attributes distinguishing science from ideology. Sampling is the technique of drawing a subset of sampling units from the complete set and then making deductions about the whole from the part. It is used all the time in wildlife research and management, but often incorrectly.

The next section takes us through some of the mystery of sampling. It explores what actually happens when we sample a population in several different ways, making the point that the true estimate is independent of whatever mathematical calculations are applied to the data.

12.3.1 Precision and accuracy

If a large number of repeated estimates of density have a mean that does not differ significantly from the true density then each estimate is said to be *accurate* or *unbiased*. Accuracy is a measure of *bias error*. If that set of estimates has little scatter then the estimates are described as *precise* or *repeatable*. Precision is a measure of *sampling error*. A system of estimation may provide very precise estimates that are not accurate, just as a system may provide accurate but imprecise estimates. Ideally, both should be maximized, but often we must choose between one and the other according to what question is being asked. For example, is density below a critical threshold of one animal per square kilometer? Here we need an accurate measure of density and may be willing to trade off some precision to get it. But if we had asked whether present density was lower than that of last year then we would need two estimates, each of high precision. Their accuracy would be irrelevant so long as their bias was constant. Most questions require precision more than accuracy. Precision is obtained by rigid standardization of survey methods, by sampling in the most efficient manner, and by taking a large sample.

12.3.2 Bias errors

Bias errors derive from some systematic distortion in the counting technique, the observer's ability to detect animals, or the behavior of the animals. Often, but not always, the bias produces an undercount. Thus biases can accrue from sampling schemes that do not properly sample all habitats (e.g. using roads that avoid hills or riverine areas); from the observer missing animals on transects (because there are too many animals, because in counting one group the observer overlooks another, or simply because of observer fatigue); or from animals being hidden in thickets, under trees, or underwater.

The best way to measure bias error is to compare the census estimate with that from a known population. Pollock and Kendall (1987) review this method, along with the use of a subpopulation of marked animals, mapping with multiple observers, line transect sampling, and multiple counts on the same area. Visibility corrections have been calculated by comparing fixed-wing aerial surveys of waterfowl with ground counts (the known or unbiased population) (Arnold 1994; Bromley *et al.* 1995; Prenzlow and Lovvorn 1996). A similar approach was used to estimate bias in counts of wood stork (*Mycteria americana*) nests in Florida (Rodgers *et al.* 1995) and great blue heron (*Ardea herodias*) nests in south Carolina (Dodd and Murphy 1995). Moose usually live in dense habitats where they are difficult to see. Rivest *et al.* (1990) compared fixed-wing surveys of moose with the more accurate subsample surveys by helicopter to correct for visibility bias, an approach also used for counts of chicks in osprey (*Pandion haliaetes*) nests (Ewins and Miller 1995).

12.3.3 *Sampling frames*

Before an area is surveyed to estimate the number of animals on it, it must be divided into *sampling units*, which cover the whole area and are non-overlapping. The sampling units might comprise areas of land if we count deer, or trees if we count nests, or stretches of river if we count beavers or crocodiles. To allow us to sample from this *frame list* of sampling units, the list must be complete for the whole area. Hence, the frame of units contains all the animals whose numbers we wish to estimate.

For purposes of explanation, we use the first example: sampling units of land. The survey area may be divided up into units in any way the surveyor desires: into quadrats, transects, or irregular sections of land, perhaps delimited by fences. The choice is a compromise between what is most efficient statistically and what is most efficient operationally.

12.3.4 *Sampling strategies*

Suppose that we wished to estimate the number of kangaroos or antelopes in a large area by counting animals on a sample of that area. Several strategies are open to us. We could sample quadrats or transects, we could select these sampling units systematically or randomly, and, if the latter, we could ensure that each sampled unit occurred only once in the sample (sampling without replacement) or that the luck of the draw allowed units to be selected more than once (sampling with replacement). The efficiency of these systems will be demonstrated with the hypothetical data of Table 12.1, which may be thought of as the number of kangaroos standing on each square kilometer of an area totaling 144 km^2. In all cases 1/3 of the area will be surveyed. We can test the accuracy of the method by determining whether the mean of a set of repeated estimates is significantly different from the true total of 1737 kangaroos. The precision of a sampling system is indicated by the spread of those repeated and independent estimates, and that spread will be measured by the standard deviation of those estimates:

$$s = \sqrt{\left(\Sigma x^2 - \frac{(\Sigma x)^2}{N} \right) /(N - 1)}$$

where x is an independent estimate of total numbers and N is the number of such repeated estimates.

Table 12.1 Simulated dispersion of kangaroos on a 1 × 1 km grid of 144 cells. Marginal totals give numbers on 1 × 2 km transects oriented both across and down the region.

1	2	7	4	7	14	9	18	24	22	19	15	142
0	1	5	6	12	11	9	15	20	21	27	28	147
2	3	5	6	10	13	16	20	160	14	19	21	147
1	4	4	6	9	13	14	17	20	16	25	20	149
2	2	5	7	10	12	16	19	20	16	18	22	149
2	4	5	6	9	12	16	22	18	18	21	23	156
0	2	5	8	4	7	11	13	17	16	21	30	134
1	0	4	9	8	10	11	16	14	20	17	17	127
0	4	2	7	8	11	11	11	12	19	22	21	128
0	2	5	8	8	12	16	20	24	25	23	25	168
1	0	4	9	8	8	8	17	17	14	18	22	126
2	5	7	6	12	12	13	15	20	21	20	23	156
12	29	58	82	105	135	150	203	222	222	250	269	1737

We will first sample 1 km^2 quadrats randomly with replacement: *sampling with replacement (SWR)*. The quadrats are numbered from 1 to 144 and a sample of 48 of these is drawn randomly. Quadrats 27, 31, 50, and 53 are drawn twice and quadrat 7 three times, but since these are independent draws they are included in the sample as many times as they are randomly chosen. The quadrat is *replaced* in the frame list after each draw, allowing it the chance of being drawn again. The number of kangaroos in this sample of quadrats totals 523, and since we have sampled only a third of the quadrats we multiply the total by 3 to give an estimate of animals in the study area: 1569.

Note that this answer is wrong in the sense that it differs from the true total, known to be 1737 (i.e. it is not accurate). This disparity is called *sampling error*, which is quite distinct from *errors of measurement* resulting from failure to count all the animals on each sampled quadrat.

We now repeat the exercise by drawing a fresh sample of 48 units and get a sampled count of 493 kangaroos, which multiplies up to an estimate of 1479. The third and fourth surveys give estimates of 1836 and 1752. This exercise is repeated a thousand times with the help of a computer. The thousand independent estimates have a mean of $\bar{x} = 1741$, very close to the true total of 1737. We can be confident, therefore, that this sampling system produces accurate (i.e. unbiased) estimates. The thousand independent estimates have a standard deviation of $s = 153$, which tells us that there is a 95% chance that any one estimate will fall in the range $\bar{x} \pm 1.96s$ or 1741 ± 300, between 1441 and 2041. The standard deviation of a set of independent estimates is the measure of the efficacy of the sampling system and hence of the precision of any one of the independent estimates. It can be estimated from the quadrat counts of a single survey (see Section 12.5.1), and when estimated in this way it is called the *standard error of the estimate*. Hence the *standard error of an estimate is a calculation of what the standard deviation of a set of independent estimates is likely to be*.

With this background, we can now compare the efficiencies of several sampling systems.

12.3.5 *Sampling with or without replacement?*

When we use *sampling without replacement (SWOR)*, a quadrat may be drawn no more than once. This is in contrast to the previous system, which allowed a quadrat to be selected by the luck of the draw any number of times. We draw a unit, check whether it has been selected previously, and if so reject it and try again. Having drawn 48 distinct units, we calculate density. The sampling is again repeated a thousand times, yielding 1000 independent estimates – each based on a draw of 48 units – of the total number of animals. We know the true total to be 1737. These 1000 estimates have a mean of 1743 and a standard deviation of 131, which is appreciably lower than the $s = 153$ accruing from SWR.

The gain in precision with SWOR reflects the slightly greater information on density carried by the 48 distinct quadrats of each survey. SWOR is always more precise than SWR for the same sampling fraction, the relationship being:

$$s(\text{SWOR}) = s(\text{SWR}) \times \sqrt{1 - f}$$

where f is the sampling fraction, in this case 0.333. The $s(\text{SWR})$ from the 1000 repeated surveys is 153, and from this we can estimate, without needing to run the simulation, that the precision of the analogous SWOR system would be about:

$$s = 153 \times \sqrt{0.666} = 125$$

Our empirical s(SWOR) is 131, which is much the same as the $s = 125$ predicted theoretically.

However, it is not as simple as that. The quadrats chosen more than once in an SWR sample are not surveyed more than once, so the time taken for the survey is shorter. In the example, only about 41 of the 48 units drawn in an SWR sample would be distinct units, the other seven being repeats. To compare the precision of an SWOR sample with that of an SWR sample entailing the same groundwork, we would have to draw by SWR about 58 units. Ten are repeats, "free" units that do not need to be surveyed a second time. Intuitively, we would assume that the SWR sample of 48 distinct units and 10 repeats must give a more precise estimate than the SWOR sample with its 48 distinct units, none repeated. Not so. The smaller SWOR samples provide estimates that are more precise by a factor of $\sqrt{\left(1 - \frac{1}{2}f\right)}$. In all circumstances, SWOR is more precise than SWR (Raj and Khamis 1958). Precision is increased by rejecting the repeats and cutting the sample size back to that of the analogous SWOR sample.

Why then, if SWOR is always better, is SWR often used? First, when the sampling fraction is low ($< 15\%$), the precision of the two systems of sampling is similar. At $f = 0.1$ there is only a 5% difference in precision, reflecting the low likelihood of repeats at low sampling intensity. Most sampling intensity in wildlife management is of this order. Second, it is often convenient to sample with replacement when an area is traversed repeatedly by aerial-survey transects. There is not the same necessity to ensure that no transect crosses another or overlaps it. This is a useful flexibility for an aerial survey in a strong crosswind or for a ground survey in thick forest.

12.3.6 Transects or quadrats?

A frame of transects is a good or a bad sampling system according to how it is oriented with respect to trends in density. The dispersion of Table 12.1 has a marked increase in density from left to right. The precision of the estimate of total numbers will be relatively high if the transects are oriented along this cline but low if oriented at right angles to it. This can be demonstrated empirically by sampling the column totals at one-third sampling intensity. Each column represents a transect and each survey comprises four transects, randomly chosen. A thousand independent surveys produce a standard deviation of estimates of 512 for SWR and 427 for SWOR. If these transects were oriented at right angles, so that the rows rather than the columns formed the transects, the standard deviation of estimates of a thousand independent surveys would be approximately 80 for SWR and 69 for SWOR. In this case, precision is increased enormously by swinging the orientation of the transects through 90°.

Transects should go across the grain of the country rather than along it, should cross a river rather than parallel it, and should go up a slope rather than hug the contour. They should be oriented such that each samples as much as possible of the total variability of an area. In essence, we must *ensure that the variation between transects is minimized and therefore that the precision of the estimate is maximized*.

Much the same principle adjudicates between the use of quadrats as against transects. So long as the frame of transects is oriented appropriately, the resultant estimate will be more precise than that from a set of quadrats whose area sums to that of the transects. *The more clumped ithe distribution of the animals, the greater the gain in precision of transects over quadrats.* A quadrat is likely to land in a patch of either high or low density, whereas a transect is more likely to cut through areas of both. Table 12.2 shows that transects oriented along the cline in density of Table 12.1 provide estimates six times more precise than do quadrats of the same size and number.

12.3.7 *Random or nonrandom sampling?*

Sampling strategies grade from *strictly random* to *strictly systematic.* The region in between is described as *restricted random sampling.* One might decide, for example, to sample randomly but to reject a unit that abuts one previously drawn. Or one might break the area into zones (strata) and draw the same number of samples randomly from each zone. These two strategies depart from the requirement of strict random sampling, whereby each sampling unit has the same probability of selection. The extreme is systematic sampling, in which the choice of units is determined by the position of the first unit selected.

Systematic or restricted random sampling has several practical advantages over strict random sampling. First, it encourages or enforces SWOR, which, as we have seen, leads to a more precise estimate. Second, it reduces the disturbance of animals on a sampling unit caused by the surveying of an adjoining unit. This is particularly important in aerial survey, where the noise of the aircraft can move animals off one transect on to another. Third, any deviation from strictly random sampling tends to increase the precision of the estimate, because the sampled units together provide a more comprehensive coverage of total variability. Table 12.2 demonstrates this for our example. The standard deviation of a thousand independent surveys is lower for restricted random sampling than for random SWOR, and lower still for systematic sampling.

Statisticians do not like nonrandom sampling because the precision of the estimate cannot be calculated from a single survey. The formulae given in Section 12.5.1 for calculating the standard error of an estimate are correct only when sampling units are drawn at random, and they will tend to overestimate the true standard error when restricted random or systematic sampling is used. But not always. If a systematically drawn set of sampling units tends to align with systematically spaced highs and lows of density, the standard error calculated on the assumption of random sampling will be too low and the estimate of density will be biased.

Table 12.2 The effect of the sampling system on the precision of an estimate. All systems sample one-third of an area of 144 km² containing the dispersion of kangaroos simulated in Table 12.1. Each sampling system is run 1000 times to provide 1000 independent estimates of the true total of 1737.

Sampling system	Mean estimate	Standard deviation of 1000 estimates
Large quadrants, n = 4		
Random with replacement	1746	487
Random without replacement	1738	414
Small quadrants, n = 48		
Random with replacement	1741	153
Random without replacement	1743	131
Transects parallel to the density cline, n = 4		
Random with replacement	1732	80
Random without replacement	1734	69
Restricted random	1730	57
Systematic	1736	48

In practice, this tends not to happen. It is entirely appropriate to sample systematically or by some variant of restricted random sampling and to approximate the standard error of the estimate with the equation for random sampling. One can be confident that the estimate is unlikely to be biased and that the true standard error is unlikely to exceed that calculated.

12.3.8 *How not to sample*

There are a number of traps that sampling can lure one into, which can result in a biased estimate or an erroneous standard error. Suppose one decided to sample quadrats but, for logistical reasons, laid them out in lines, with the distance between the lines considerably greater than that between neighboring quadrats within the lines. The standard error of the estimate of density could not then be calculated by the usual formulae because the counts on those quadrats would not be independent. Density is correlated between neighboring quadrats, and this throws out the simple estimate of the standard error, which returns an erroneously low value. There are ways of dealing with the data from this design in order to yield an appropriate standard error (see Cochran 1977 for treatment of two-stage sampling and Norton-Griffiths 1973 for an example using the Serengeti wildebeest), but they are beyond the scope of this book. The simple remedy is to pool the data from all quadrats on each line, the line rather than the quadrat becoming the sampling unit. This procedure may appear to sacrifice information, but it does not (Caughley 1977a).

Another common mistake is to throw random points on to a map and to declare them centers of the units to be sampled, the boundary of each being defined by the position of the point. In this case, the requirement that sampling units cover the whole area and be non-overlapping is violated and the sampling design becomes a hybrid between SWR and SWOR, leading to difficulties in calculating a standard error. There is nothing wrong with choosing units to be sampled by throwing random points on a map so long as the frame of units is marked on the map first. The random points define the units to be selected. They do not determine where the boundaries of those units lie.

A third trap to watch for is a biased selection of units to be sampled. The most common source of this bias in wildlife management is the so-called "road count," in which animals are counted from a vehicle on either side of a road or track. Roads are not random samples of topography. They tend to run along the grain of the country rather than across it, they go around swamps rather than through them, they tend to run along vegetation ecotones, and they create their own environmental conditions, some of which attract animals and others of which repel them.

12.4 **Sampled counts: methods and arithmetic**

Sampled counts of animals fall easily into two categories. The first is the method of counting in sampling units with fixed boundaries. We might for example walk lines and count deer in the area within 100 m either side of the line of march, or count all the ducks in a sample of ponds, the shoreline of the pond providing a strict boundary to the sampling unit.

The alternative is unbounded sampling units (Buckland *et al.* 1993, 2001). Instead of restricting the counting to those animals within 100 m of a line of march, those outside the transect being ignored, we might count all the animals that we see. Since the observed density will fall away with distance from the observer, the raw counts are no longer an estimate of true density. They must therefore be corrected.

Of these two options (sampling units with boundaries and sampling units without boundaries), the first has immense advantages of simplicity and realism. If the transect width is appropriately chosen, what the observer sees is what the observer gets. The mathematics of such sampling are simple, elegant, and absolutely solid. In contrast, the accuracy of a corrected density estimated from unbounded transects depends heavily upon which model is chosen for the analysis. There are many to choose from and they give markedly different answers for the same data. The advantage of unbounded transects lies in all the sightings being used, with none being discarded. Since the precision of an estimate is related tightly to the number of animals actually counted, any sampling scheme that increases the number of sightings also tends to increase the precision of the estimate. That is an advantage if the increased precision is obtained without the sacrifice of too much accuracy.

The choice of one system or another is often determined by density. If the species is rare then we might be tempted to use all the data we can get. If it is common, we might be content to use the more dependable sampling units with fixed boundaries, knowing that fewer things can go wrong.

12.4.1 *Fixed boundaries to sampling units*

The appropriate analysis depends on whether the sampling units are of equal or unequal size, and how they are selected. Formulae were originally developed by Jolly (1969), based on Cochran (1977) (see also Norton-Griffiths 1978).

Notation

y = the number of animals on a given sampled unit;
a = the area of a given sampled unit;
A = the total area of the region being surveyed;
n = the number of units sampled;
D (or d) = the estimate of mean density;
$SE(D)$ = the standard error of estimated mean density;
Y = the estimate of total numbers in the region of size A;
$SE(Y)$ = the standard error of the estimate of total numbers.

The simple estimate (for equal-sized sampling units)

The simple estimate is used when sampling units are of constant size, as when the region being surveyed is a rectangle, which can be subdivided into quadrats or transects. It will provide an unbiased, although imprecise, estimate, even when sampling units differ in size – but more appropriate designs are available for that case. We will explore this design at some length because most of the principles are shared with the others.

The region to be surveyed, of area A, is divided on a map or in one's head into an exhaustive set of non-overlapping sampling units, each of constant area a. Let us assume, for illustration, that the region is as given in Table 12.1, and that this region of $A = 144$ km^2 is to be sampled by $n = 4$ transects each of area $a = 12$ km^2. Sampling intensity is hence $na/A = 4 \times 12/144 = 0.333$.

In Table 12.1 the rows represent transects and the marginal totals the number of animals on each. Numbering the transects from 1 to 12 and selecting at random with replacement from this set, we draw transects 4, 8, 1, and 4. On surveying these transects, we obtain counts of:

Transect:	1	4	4	8
Count:	142	149	149	127

Note that transect 4 has been drawn twice, so in practice we survey only three transects, although the count from transect 4 enters the calculation two times.

Density is estimated as the sum of the transect counts $(142 + 149 + 149 + 127)$ divided by the sum of the transect areas $(12 + 12 + 12 + 12)$. Thus:

$$D = \Sigma y / \Sigma a = 567/48 = 11.81/\text{km}^2.$$

The precision of this estimate is indexed by its standard error SE(D), which is itself an estimate of what the standard deviation of many independent estimates of density would be, each derived from four transects drawn at random with replacement:

$$SE(D) = 1/a \times \sqrt{(\Sigma y^2 - (\Sigma y)^2/n)/(n(n-1))}$$

This is a slight approximation. To be exactly unbiased, it should be multiplied by a further term $\sqrt{1 - (\Sigma a)/A}$, but that usually makes so little difference that it tends to be ignored.

The calculation tells us that this hypothetical distribution of estimates, each of them made in the same way as we made ours, with the same sampling frame and the same sampling intensity, only the draw of sampling units being different, will have a standard deviation in the vicinity of ±0.43. In fact, this is likely to be an underestimate because it is based on only four sampling units, three degrees of freedom. With samples above 30 sampling units we can form 95% confidence limits of the estimate by multiplying by 1.96, but for smaller samples we must choose a multiplier from a Student's t-table corresponding to a two-tailed probability of 0.05 and the degrees of freedom (d.f.) of our sample. In the case of d.f. = 3, the multiplier is 3.182 and so the 95% confidence limits of our estimate of density are $\pm3.18 \times 0.43 = \pm1.37$.

The number of animals Y in the surveyed region can now be calculated as the number of square kilometers in that region A multiplied by the estimated mean number per square kilometer D:

$$Y = AD = 144 \times 11.81 = 1701$$

This has a standard error of:

$$SE(Y) = \pm A \times SE(D) = \pm144 \times 0.43 = \pm62$$

Its 95% confidence limits are calculated as A multiplied by the 95% confidence limits of D:

$$\pm144 \times 1.37 = \pm197$$

We can check this against Table 12.2, which shows that the true total number Y is 1737, so the estimate with 95% confidence of $Y = 1701 \pm 197$ is entirely acceptable.

If the sampling is without replacement, the formula for SE(D) yields an overestimate. The standard error for SWOR is estimated by the formulation for the standard error with replacement multiplied by the square root of the proportion of the area not surveyed. This *finite population correction* (*FPC*) is:

$$FPC = \sqrt{1 - (\Sigma a)/A}$$

The simple estimate may validly be used even when sampling units are of unequal size. The constant a is then replaced by the mean area of sampling units. The precision of the estimate will be lower (i.e. the standard error will be higher) than that produced by the ratio method (see next subsection), but the estimate is unbiased and may be precise enough for many purposes.

The simple estimate, with minor modification, can be used when the total area A is unknown. One of us was forced to this exigency while surveying from the air a population of rusa deer (*Cervus timorensis*) in Papua New Guinea. The deer lived on a grassed plain, the area of which could not be gauged with any accuracy from the available map. The remedy was to measure the length of the plain by timing the aircraft along it at constant speed, and then to run transects from one side of the plain to the other at right angles to that measured baseline. The area of a sampling unit is entered as $a = 1$, even though they are of different and unknown areas. D then comes out as average numbers per transect, rather than per unit area. Total numbers Y on the plain can be estimated by replacing A with N, where N is the total number of transects that could have been fitted into the area. This is simply the length of the baseline divided by the width of a single transect. A similar approach was used for censusing of wildebeest in the Serengeti (Norton-Griffiths 1973, 1978).

The ratio estimate (for unequal-sized sampling units)

This is the best method for a frame of sampling units of unequal size, as might be provided by a faunal reserve of irregular shape sampled by transects. Statistical texts warn that the estimate is biased when the number of units sampled is less than 30 or so, but the bias is usually so slight as to be of little practical importance. The number of units may be as low as two without generating a bias of more than a few per cent.

The appropriate formulae are given in Table 12.3 and the notation at the beginning of Section 12.5.1. That for the standard error looks quite different from that for the simple estimate but they are mathematical identities when the sampling units are of equal size. The ratio estimate is general, the simple estimate being a special case of it. Hence, if these analyses are to be programmed into a calculator or computer, the ratio method is the only one needed.

Table 12.3 Estimates and their standard errors for animals counted on transects, quadrants, or sections. The models are described in the text.

Model	Density	Numbers
Simple		
Estimate	$D = \Sigma y / \Sigma a$	$Y = A \times D$
Standard error of estimate (SWR)	$SE(D)_1 = 1/a \times \sqrt{(\Sigma y^2 - (\Sigma y)2/n)/n(n-1)}$	$SE(Y) = A \times SE(D)_1$
Standard error of estimate (SWOR)	$SE(D)_2 = SE(D)_1 \times \sqrt{1 - (\Sigma a)/A}$	$SE(Y) = A \times SE(D)_2$
Ratio		
Estimate	$D = \Sigma y / \Sigma a$	$Y = A \times D$
Standard error of estimate (SWR)	$SE(D)_3 = n/\Sigma a \times \sqrt{(1/n(n-1))(\Sigma y^2 + D^2 \Sigma a^2 - 2D \Sigma ay)}$	$SE(Y) = A \times SE(D)_3$
Standard error of estimate (SWOR)	$SE(D)_4 = SE(D)_3 \times \sqrt{1 - (\Sigma a)/A}$	$SE(Y) = A \times SE(D)_4$
PPS		
Estimate	$d = 1/n \times \Sigma(y/a)$	$Y = A \times d$
Standard error of estimate (SWR)	$SE(D) = \sqrt{(\Sigma y/a)^2 - (\Sigma(y/a))^2/n)/n(n-1)}$	$SE(Y) = A \times SE(d)$

SWR, sampling with replacement; SWOR, sampling without replacement. Notation is given at the beginning of Section 12.5.1.

The PPS estimate

By the previous two methods, all sampling units in the frame have an equal chance of being selected. By the probability-proportional-to-size (PPS) method, the probability of selection is proportional to the size of the sampling unit. Suppose that the area to be surveyed is farmland. We might decide to declare the paddocks (or "pastures" or "fields," depending on which country you are in) as sampling units because the fences provide easily identified boundaries to those units.

If each sampling unit were assigned a number and the sample were chosen by lot, we would use the ratio method of analysis. However, we might decide instead to choose the sample by throwing random points on to a map. Each strike selects a unit to be sampled, the probability of selection increasing with the size of the unit.

The PPS estimate has the advantages that it is entirely unbiased and that the arithmetic (Table 12.3) is simple. Its disadvantage is that it can be used only when sampling with replacement and so it is not as precise as the ratio method used without replacement. Hence, this method should be restricted to surveys whose sampling intensity is below 15%. The PPS estimate is a mathematical identity of the simple estimate and the ratio estimate when units of equal size are sampled with replacement.

12.4.2 *Unbounded transects (line transects)*

The observer walks a line of specified length and counts all animals seen, measuring one or more subsidiary variables at each sighting (e.g. angle between the animal and the line of march; radial distance, the distance between the animal and the observer at the moment of sighting; the right-angle distance between the animal and the transect). If we know the shape of the sightability curve relating the probability of seeing an animal on the one hand to its right-angle distance from the line on the other, and if an animal standing on the line will be seen with certainty, it is fairly easy to derive an estimate of density from the number seen and their radial or right-angle distances. We seek a distance from the line at which the number of animals missed within that distance equals the number seen beyond it. True density is then the total seen divided by the product of twice that distance and the length of the line.

Therein lies the difficulty. That distance is determined by the shape of the sightability curve, which can seldom be judged from the data themselves. Consequently, the shape of the curve must be assumed to some extent, and the validity of the assumption determines the accuracy of the method.

We present here just two of the many models available, mainly to give some idea of their diversity. The first is the Hayne (1949) estimate, which is derived from the assumption that the surveyed animals have a fixed flushing distance and will be detected only when the observer crosses that threshold. If k is the number of animals detected and r the radial distance from a detected animal to the observer then:

$$D = (1/2L)\Sigma_k(1/r)$$

where L is the length of the line. Hence, density is the sum of the reciprocals of the radial sighting distances divided by twice the length of the line.

It is implicit in Hayne's model that $\sin\theta$, the sine of the sighting angle, is uniformly distributed between 0 and 1, and that the theoretically expected mean sighting angle is 32.7°. Hence, the reality of the model can be tested against the data. Eberhardt (1978) recommended tabulating the frequency of $\sin\theta$ in 10 intervals of 0.1 ($0.0 - 0.1, 0.1 - 0.2 \ldots 0.9 - 1.0$) and testing the uniformity of the frequencies by chi-square. He gave a worked example for a survey of the side-blotched lizard

(*Uta stansburiana*). Robinette *et al.* (1974) and Burnham *et al.* (1980) suggested that most mean sighting distances tended to be around 40° or more, the latter authors being convinced that the Hayne estimate is used far too uncritically in wildlife management. Robinette *et al.* (1974) compared the accuracy of the Hayne estimate with that of eight other line transect models, showing that when applied to inanimate objects or to elephants it tended to overestimate considerably. However, Pelletier and Krebs (1997) found both the Hayne estimate and line transect estimates provided relatively unbiased results when compared with a known population of ptarmigan (*Lagopus* species) in the Yukon. Buckland *et al.* (1993) provides a starting point for reading further about line transect methods.

Our second example is a nonparametric method developed by Eberhardt (1978) from work by Cox (1969). First, we choose arbitrarily a distance, Δ, perpendicular from the line. Eberhardt's estimate of density is:

$$D = (3k_1 - k_2)/4L\Delta$$

where k_1 and k_2 are the number of animals seen on either side of the line transect at distances that fall within the intervals $0 - \Delta$ and $\Delta - 2\Delta$, respectively. Eberhardt (1978) considered that the method is most useful as a cross-check on the results of other methods, because its estimate is likely to be imprecise. Precision is enhanced by choosing a large value of Δ but accuracy is enhanced by choosing a small one (Seber 1982).

Much of the present use of line transects in wildlife management stems from the belief that they are somehow more "scientific" than strip transects, just as there was once a belief that quadrats were statistically superior to transects. There are rare situations in which transect sampling will not work and where line transect methodology might (e.g. in very thick cover). The unbounded line transect method has advanced considerably with the use of the computer software Distance (http://www.ruwpa.st-and.ac.uk/distance), developed by Buckland *et al.* (1993, 2001). Although the use of the software is not easy, it is currently the most powerful tool for line censuses. In particular, it is most useful for rare observations, although it does require at least 30 records in order to be reliable. In addition, time must be allowed to make the necessary estimates of perpendicular distance from the line to the animal (or groups of animals). If there are insufficient observations of a particular species (or other category) in a station or habitat then one can repeat the line survey until a sufficient number have been accumulated. The only proviso is that animals distribute themselves randomly with respect to the line and there is no spatial correlation between surveys. The method is particularly suitable for rare species such as carnivores and rare ungulates and birds. It is less suitable where there are large numbers of animals (e.g. ungulates on the Serengeti plains).

Note that none of these unbounded methods can be used in aerial survey. They are all anchored by the assumption that all animals on the line of march (equivalent to the inner strip marker of aerial survey) are tallied by the observer. This assumption does not hold for aerial survey because the ground under the inner strip marker is at a distance from the observer, because an animal under a tree on that line may be missed, and because an observer cannot watch all parts of the strip at once and may therefore miss animals in full view on the line. In addition, the speed of the aircraft makes the measurement of distances from the observer unfeasible.

The assumption that all animals on the line are counted can be relaxed if the probability of detecting animals on the line can be estimated. This is particularly important for marine mammals, where only a fraction of a group or pod is on the surface at any one time. The probability of detection on the line for harbor porpoises (*Phocoena phocoena*) was estimated to be only 0.292, which illustrates just how many remain unseen. Furthermore, this estimate was made by experienced observers; for inexperienced observers, the sighting probability was only 0.079; that is, some 90% were missed. This shows the importance of training and experience (Laake *et al.* 1997).

The biologist must decide whether the statistical power of line transects justifies their practical application. Can the difficulty of measuring sighting distances and the unreliability of the resultant estimates be justified when an alternative with fewer problems is available? The line transect was originally introduced to circumvent the difficulty of counting all animals on a transect or quadrat. It cured that problem by replacing it with several others. Perhaps we should give some thought to ways of treating the original problem without introducing new ones. If animals are difficult to see on a transect of fixed width, why not walk two people abreast down the boundaries? If that does not work, put a third person between them. And so on.

12.4.3 *Stratification*

The precision of an estimate is determined by sampling intensity and by the variability of density among sampling units. Suppose there were two distinct habitats in the survey area and that from our knowledge of the species we could be sure that it would occur commonly in one and rarely in the other. If we surveyed those two sub-areas separately and estimated a separate total of animals for each, the combined estimate for the whole area would be appreciably more precise than if the area had been treated as an undifferentiated whole.

The process is called stratification and the sub-areas strata. By this strategy, we divide an area of uneven density into two or more strata within which density is much more even. The strata are treated as if they were each a total area of survey and the results are later combined. The estimate from each stratum will be called Y_h, which has a standard error of $SE(Y_h)$. Total numbers Y are estimated by $Y = \Sigma Y_h$. The standard error is the square root of the sum of the variances of the contributing stratal estimates. The variance of an estimate is the square of its standard error. Here it is designated Var(est) to distinguish it from the variance of a sample designated s^2. Calculate:

$$\text{Var}(Y_h) = (SE(Y_h))^2$$

for each stratum and then:

$$SE(Y) = \sqrt{\Sigma Var(Yh)}$$

to give the standard error of the combined estimate of total numbers.

Optimum allocation of sampling effort

If our aim is to get the most precise estimate of Y as opposed to a precise estimate of each Y_h, sampling intensity should be allocated between strata according to the expected standard deviation of sampled unit counts in each stratum. This requires a pilot survey or at least approximate knowledge of distribution and density gained on a previous survey. Often we have nothing more than aerial photographs or a vegetation map to give us some idea of the distribution of habitat, and only knowledge of the animal's ecology to guide us in predicting which habitats will hold many animals and which will hold few. This scant information is in fact sufficient to allow an allocation of sampling effort between strata that will not be too far off the optimum. The important point to understand is that for almost all populations the standard deviation of counts on sampling units rises linearly with density. From this can be derived the rule of thumb that the *number of sampling units put into a stratum should be directly proportional to what* Y_h *is likely to be.*

At first thought, this is a daunting challenge – to guess each Y_h before we have estimated it – but it is easier if we break it down into components. First, guess the density in each stratum. It does not matter too much if this is wrong, even badly wrong, because all we need to get roughly right is the ratios of densities between strata. Second, multiply each guessed density by the mapped area of its stratum to give a guess at numbers in the stratum. Third, divide each by the total area to give the proportion of total sampling effort that should be allocated to each stratum. Table 12.4 shows the calculation for a degree block that can be divided into three strata from a vegetation map and to which a total of 10 hours of aerial survey have been allocated.

12.4.4 Comparing estimates

If the sampling units are drawn independently of each other, the estimates of density from two surveys may be compared. The surveys may be of two areas, or of the same area in two different years, or of the same area surveyed in the same year by two teams or by different methods. A quick and dirty comparison is provided by the normal approximation, which is adequate if each survey covered more than 30 sampling units. The two estimates are significantly different when:

$$(est_1 - est_2)\sqrt{Var(est_1) + Var(est_2)} > 1.96$$

If sample sizes are too low, or if more than two surveys are being compared, the determination of significance should be made by one-factor analysis of variance. If the surveys are not independent, as when the same transects are run each year, a comparison may still be made by analysis of variance but with TRANSECTS now declared a factor in a two-factor analysis. Chapter 14 goes further into this and other uses of analysis of variance.

12.4.5 Merging estimates

If a comparison shows that two or more independent estimates of the same population are not significantly different we may wish to merge them to provide an estimate that is more precise than the originals. This procedure is quite distinct from stratification, where estimates from different populations are combined to give an overall estimate. Merging is restricted to the same population estimated more than once. We must make

Table 12.4 Allocation of $E = 10$ hours of aerial survey among strata to maximize the precision of the estimate of animals in the total area.

Stratum (h)	Area (A_h), km^2	Guessed density (D_h)	Guessed numbers ($Y_h = A_h D_h$)	Proportion of total effort ($P_h = Y_h/\Sigma Y_h$)	Hours allocated ($E_h = P_h E$)
1	2000	1	2000	0.03	0.3
2	7000	5	35 000	0.52	5.2
3	3000	10	30 000	0.45	4.5
	12 000		67 000	1.00	10.0

sure that environmental (e.g. different seasons) and biological (e.g. significant mortality or emigration) conditions do not differ between censuses. Merging is particularly powerful in obtaining a reduced confidence interval from a series of individual censuses, each with very wide confidence intervals. If we obtain a single estimate with a wide confidence interval (say, because too few samples were counted) then it will often pay to repeat the census as soon as possible and merge the two results.

There are two methods. The first is quick and dirty, to be used only when the individual estimates were made with about the same sampling intensity. The merged estimate $\hat{Y}$ can then be calculated as:

$$\hat{Y} = (Y_1 + Y_2 + Y_3 + \ldots + Y_N)/N$$

where there are N surveys. It has a variance of:

$$\mathrm{Var}(\hat{Y}) = (\mathrm{Var}(Y_1) + \mathrm{Var}(Y_2) + \mathrm{Var}(Y_3) + \ldots + \mathrm{Var}(Y_N))/N^2$$

Thus, the merged estimate is simply the mean of the individual estimates, and its variance is the mean of the individual estimate variances divided by their number.

$\mathrm{SE}(\hat{Y})$ is the square root of $\mathrm{Var}(\hat{Y})$. From these, the merged density estimate is $D = \hat{Y}/A$, which has a standard error of $\mathrm{SE}(D) = \mathrm{SE}(\hat{Y})/A$.

A second, more appropriate method, particularly for surveys utilizing markedly different intensities of sampling, is provided by Cochran (1954), who also considers more complex merging. Here the contribution of an individual estimate to the merged one is weighted according to its precision. Letting $w = 1/\mathrm{Var}(Y)$:

$$\hat{Y} = (w_1 Y_1 + w_2 Y_2 + w_3 Y_3 + \ldots + w_N Y_N)/(w_1 + w_2 + w_3 + \ldots + w_N)$$

with a variance of:

$$\mathrm{Var}(\hat{Y}) = 1/(w_1 + w_2 + w_3 + \ldots + w_N)$$

12.5 Indirect estimates of population size

This section outlines some of the methods available for calculating the size of a population using techniques that do not necessarily depend on accurate counts of animals. The line transect method could well come under this head but is placed in "Sampled Counts" because it requires accurate counting of animals on the line.

12.5.1 Index–manipulation–index method

If we obtain two indices of population size, I_1 and I_2, the former before and the latter after a known number of animals C was removed, the population's size can be estimated for the time of the first index by:

$$Y_1 = I_1 C/(I_1 - I_2)$$

The proportion removed is estimated as $p^* = (I_1 - I_2)/I_1$ and the proportion of those remaining as $q^* = 1 - p^*$. Following Eberhardt (1982), the variance of the estimate of Y can be approximated by:

$$\mathrm{Var}(Y_1) \approx Y_1^2 (q^*/p^*)^2 (1/I_1 + 1/I_2)$$

from which $SE(Y_1) = \sqrt{Var(Y1)}$. Eberhardt (1982) gives three examples from populations of feral horses. The data from his Cold Springs population were:

$I_1 = 301$;
$I_2 = 76$;
$C = 357$;
$p^* = 0.748$.

Thus, the population at the time of the first index is estimated as:

$$Y_1 = (301 \times 357)/(301 - 76) = 478$$

with the following estimated variance:

$$Var(Y_1) \approx 478^2 (0.252/0.748)^2 (1/301 + 1/76) = 428$$

from which $SE(Y_1) = \sqrt{428} = 21$.

The index–manipulation–index method assumes that the population is closed (no births, deaths, immigration, or emigration) between the estimation of the first and second indices. This assumption is approximated when the entire experiment is run over a short period.

12.5.2 Change-of-ratio method

If a population can be divided into two classes, say males and females or juveniles and adults, and one class is significantly reduced or increased by a known number of animals, the size of the population can be estimated from the change in ratio. Kelker (1940, 1944) introduced this method to estimate the size of deer populations manipulated by bucks-only hunting.

The two classes are designated x and y. Before the manipulation, there is a proportion p_1 of x individuals in the population; this becomes p_2 after the manipulation, which removes or adds C_x x individuals (additions are positive, removals negative) and C_y y individuals. $C = C_x + C_y$. The size of the population before the manipulation may be estimated as:

$$Y_1 = (C_x - p_2 C)/(p_2 - p_1)$$

As with the index–manipulation–index method, Kelker's method assumes that the population is closed. Hence, the two surveys to estimate the class proportions must be run close together. Additionally, all removals or additions must be recorded and the two classes must be equally amenable to survey.

Cooper et al. (2003) have extended this approach using likelihood estimates of the ratios. When harvesting is highly skewed towards a single sex or age class, the change in these ratios provides information about the exploitation rate, and when combined with absolute numbers removed also provides information on absolute abundance.

12.5.3 Mark–recapture

Mark–recapture is a special case of the change-of-ratio method. A sample of the population is marked and released and a subsequent sample is taken to estimate the ratio of marked to unmarked animals in the population. From data of this kind we can estimate the size of the population, and with further elaboration (individual markings, multiple recapturing occasions) the rate of gain and loss.

The huge number of mark–recapture models available has been reviewed adequately by Blower *et al.* (1981) and in detail by Seber (1982) and Krebs (1999). Bowden and Kufeld (1995) present some methods for estimating confidence limits for general mark–recapture calculations, using the example of Colorado moose (*A. alces*). Here we outline the range of methods, provide an introduction to the most simple cases, and emphasize their pitfalls, as well as mentioning some recent advances which might circumvent these pitfalls.

Petersen–Lincoln models

A sample of M animals is marked and released. A subsequent sample of n animals is captured, of which m are found to be marked. If Y is the unknown size of the population then clearly:

$$M/Y = m/n$$

within the limits of sampling variation. With rearranging, this allows an estimate of population size as:

$$Y = Mn/m$$

Intuitively obvious as this is, it is not quite right because of a statistical property of ratios that leads on average to a slight overestimation. This bias may be corrected by (Bailey 1951, 1952):

$$Y = (M(n + 1))/(m + 1)$$

which has a standard error of approximately:

$$SE(Y) = \sqrt{[(M^2(n + 1)(n - m))/((m + 1)^2(m + 2))]}$$

These formulae are for "direct sampling," where the number of animals to be recaptured is not decided upon prior to recapturing. There are further variants for SWR and for inverse sampling (see Seber 1982).

Except in the unlikely case of half or more of the population being marked, the distribution of repeated independent estimates of population size is always strongly skewed to the right – a positive skew. (The direction of skew is the direction of the longest tail.) Fig. 12.1 shows this effect from a computer simulation of 1000 estimates of a population of 500 animals containing 100 marked individuals. Each estimate is derived from a capturing of 50 animals. Apart from demonstrating the skew of estimates, this figure makes the point that only a limited number of estimated values is possible. With $Y = 500$ and $M = 100$, the probability of a given animal being marked is 0.2, so the expected number of marked animals in a sample of 50 is 10. This would give a population estimate of $Y = 464$. If nine were recaptured, the estimate would be $Y = 510$. No estimate between 464 and 510 is possible.

Since the estimates are skewed, the confidence limits of an estimate are also skewed and cannot easily be calculated from the standard error. Blower *et al.* (1981) recommended an approximating procedure. Let $a = m/n$. In a large sample, the 95%

Fig. 12.1 Simulated replications of estimates of a population of 500 individuals by mark–recapture where 100 are marked and 50 captured. Note the positive skew of estimates and the fact that only a limited number of estimated values are possible.

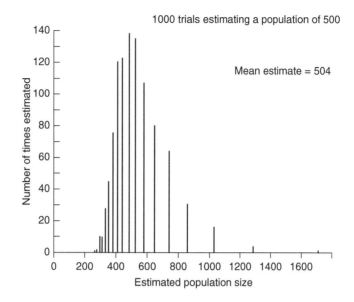

confidence limits of a are approximately $\pm 1.96\sqrt{(a(1-a)/n)}$. Since $Y = M/a$, the upper and lower 95% confidence limit of a can each be divided into M to give upper and lower 95% confidence limits of Y.

The Petersen estimate is the most simple of a family of estimation procedures. If animals are marked on more than one occasion and recaptured on more than one occasion then it is possible to estimate gains and losses from the population, in addition to its size. Seber (1982) describes most of the options.

The Petersen estimate depends on these assumptions:

1 All animals are equally catchable.
2 No animal is born or migrates into the population between marking and recapturing.
3 Marked and unmarked animals die or leave the area at the same rate.
4 No marks are lost.

Assumption **2** is not needed when marked animals are recaptured on more than one occasion, but the others are common to all elaborations of the Petersen estimate. The least realistic is the assumption of equal catchability, which is routinely violated by almost any population the wildlife manager is called upon to estimate (Eberhardt 1969). For this reason, the Petersen estimate and its elaborations (Bailey's triple catch, Schnabel's estimate, the Jolly–Seber estimate, and many others) are of limited utility in wildlife management.

Frequency-of-capture models

Petersen models work only when all animals in the population are equally catchable. Frequency-of-capture models are not constrained in this way but will work only if the population is closed: if there are no losses from or gains to the population over the interval of the experiment. This is easy enough to approximate by running the exercise over a short period.

Animals are captured on a number of occasions, usually on successive nights, and marked individually at the first capture. At the end of the experiment, each individual caught at least once can be scored according to the number of times it was captured. The data come in the form:

Number of times caught (i):	1	2	3	4	5	6	7	8	...	18
Number of animals (f_i):	43	16	8	6	0	2	1	0	...	0

which are from Edwards and Eberhardt (1967), who trapped a penned population of wild cottontail rabbits for 18 days. Of these, 43 were caught once only, 16 twice, 8 three times, and so on. $\Sigma f_i = 76$ gives the number of rabbits caught at least once and so the population must be at least that large. If we could estimate f_0, the number of rabbits never caught, we would have an estimate of population size:

$$Y = f_0 + 76$$

Traditionally, this has been attempted by fitting a zero-truncated statistical distribution (Poisson, geometric, negative binomial) to the data and thereby estimating the unknown zero frequency. Eberhardt (1969) exemplifies this approach. More complex mark−recapture models use sophisticated analytical techniques to cope with variation in the probability of capture due to time (seasonal trends, changes in weather), variation among individual animals (site fidelity, sex differences, dominance relationships), prior trapping history (capture-shyness and capture-proneness), and various combinations of these (Pollock 1974; Burnham and Overton 1978; Otis *et al.* 1978). The fit of each model can be tested against the data and an objective decision can be made as to which is the most appropriate, often using information theory (see Chapter 15). The computations are too lengthy to be attempted by hand, but several software programs are freely available on the Web: CAPTURE (White *et al.* 1982), SURGE (Lebreton *et al.* 1992), and MARK (White and Burnham 1999).

Estimation of density

All previously reviewed mark−recapture methods yield a population size Y that can be converted to a density D only when the area A relating to Y is known. In most studies, Y itself is meaningless because the "population" is not a population in the biological sense but the animals living on and drawn to a trap grid of arbitrary size.

Seber (1982) and Anderson *et al.* (1983) reviewed the methods currently used to estimate A as a prelude to determining density. Most rely on Dice's (1938) notion of a boundary strip around the trapping grid such that the *effective trapping area* A is the grid area plus the area of the boundary strip. Most of these methods are *ad hoc* and subject to numerous problems, or require large quantities of data to produce satisfactory estimates, or require supplementary trapping beyond the trapping grid.

Anderson *et al.* (1983) circumvented this problem with a method of mark−recapture that provides a direct estimate of density. The traps are laid out not in a grid but at equal intervals along the spokes of a wheel. Trap density therefore falls away progressively from the center of the web. The method pivots upon the assumption that the high density of traps at the center guarantees that all animals at the center will be captured. This is analogous to the assumption of line transect methodology that all animals are tallied on the line itself. The data collected as "distance of first capture from the center of the web" are analyzed almost exactly as if they were from a line transect (Buckland *et al.* 1993, 2001). This analysis can be run on the computer program Distance (Laake *et al.* 1993).

12.5.4 *Incomplete counts*

The problem of estimating the size of a population from "total counts" known to be inaccurate has been approached from three directions. One family of methods requires a set of replicate estimates, the second requires two estimates, and the third provides an estimate known with confidence to be below true population size.

Many counts

Hanson's (1967) method assumes that all animals have the same probability of being seen but that this probability is less than one. Hence, whether a given animal is seen or not on a given survey is a draw from a binomial distribution. It follows from the mathematics of the binomial distribution that $Y = \bar{x}x/(\bar{x} - s^2)$, where Y is the population size, $\bar{x}$ the mean of a set of (incomplete) counts, and s^2 the variance of those counts.

This method is not recommended, because of the restriction that all animals have the same sightability. In practice, sightability varies by individuals and between surveys. The variance of a set of replicate counts tends to be greater than their mean (a binomial variance is always less than the mean), indicating that the method is unworkable.

A modification of the method to circumvent this restriction was suggested by Caughley and Goddard (1972). It requires repeated counts made at two levels of survey efficiency (e.g. two sets of aerial surveys, one flown at 50 m and the other at 100 m altitude). However, Routledge (1981) showed by simulation that this method yields a very imprecise estimate unless the number of surveys is prohibitively large, and hence we do not recommend it.

The nonparametric *method of bounded counts* (Robson and Whitlock 1964) provides a population estimate from a set of replicate counts as twice the largest minus the second largest. Routledge (1982) dismissed this method also (as do we), because in most circumstances it greatly underestimates the true number.

Two counts

Caughley (1974) showed that if the counts of two observers of equivalent efficiency are divided into those animals (or groups of animals) seen by only one observer and those seen by both, the size of the population can be estimated. Henny *et al.* (1977) and Magnusson *et al.* (1978) extended the method to allow for the two observers being of disparate efficiency.

Essentially the method is a Petersen estimate, although animals are neither marked nor captured. Suppose that the entities being surveyed are stationary and that their individual positions can be mapped. Magnusson *et al.* (1978) surveyed crocodile nests and Henny *et al.* (1977) the nests of ospreys. If the area is surveyed independently twice, perhaps once from the ground and once from the air, the entities can be divided into four categories:

1 S_1 = the number seen on the first survey but missed on the second.
2 S_2 = the number seen on the second survey but missed on the first.
3 B = the number tallied by both surveys.
4 M = the number missed on both surveys.

This is equivalent to a mark–recapture exercise. The first survey maps (marks) a set of entities, each of which may or may not be seen (recaptured) on the second. But unlike a true mark–recapture exercise, the model is symmetrical and the first and second surveys are interchangeable.

If P_1 is the probability of an entity being seen on the first survey and P_2 the probability of its being seen on the second then:

$$P_1 = B/(B + S_2);$$
$$P_2 = B/(B + S_1);$$
$$M = S_1 S_2 / B;$$
$$Y = ((B + S_1)(B + S_2))/B.$$

where Y is an estimate of the size of the population.

The last equation may be corrected for statistical bias (Chapman 1951) to:

$$Y = [((B + S_1 + 1)(B + S_2 + 1))/(B + 1)] - 1$$

which has a variance given by Seber (1982) of:

$$\mathrm{Var}(Y) = (S_1 S_2 (B + S_1 + 1)(B + S_2 + 1))/((B + 1)^2 (B + 2))$$

Magnusson *et al.* (1978) reported that although the method is based on the assumptions that the two surveys are independent and that there is a constant probability of seeing an entity on a given survey (equal catchability), the second is not critical. The population estimate is close enough even when the probability of being seen varies greatly between individuals.

Caughley and Grice (1982) extended the method to moving targets, dropping the requirement that the position of stationary entities must be mapped so that they could be identified as seen or not seen at the two surveys. Groups of emus (*Dromaius novae-hollandiae*) were tallied simultaneously but independently by two observers seated in tandem on one side of an aircraft. Their counts of $S_1 = 7$, $S_2 = 3$, and $B = 10$ yielded $P_1 = 0.77$ and $P_2 = 0.59$, the population estimate being $Y = 22$ emu groups on the $843 \, \mathrm{km}^2$ of transects that they surveyed together, a density of 0.03 groups/km^2.

This method of simultaneous and fully independent tallying carries two dangers, one technical, the other statistical. First, the two observers must not unconsciously cue each other to the presence of animals in their field of view and ideally should be screened from each other. Second, the chances of "marking" and "recapturing" an entity should be uncorrelated, but they are not, because marking and recapturing occur at the same instant, the search images transmitted to each observer being nearly identical. Caughley and Grice (1982) showed by simulation that the effect of the close correlation was to underestimate density but that the underestimation became serious only when the mean of P_1 and P_2 was less than 0.5.

Known-to-be-alive

Most estimates of population size require that the manager makes a leap of faith. There is seldom any certainty whether the population fits the assumptions of the model, whether the estimate is wildly inaccurate, or whether the confidence limits have much to do with reality. The more complex the model, the greater the uncertainty. Many ecologists, particularly those working on small mammals, have decided that the work needed to achieve an unbiased estimate is not worth the effort. They would prefer an estimate that, although perhaps inaccurate, is inaccurate in a predictable direction and does not depend on a set of assumptions of dubious reliability. Hence the *known-to-be-alive* estimate, the number of animals that the researcher knows with certainty to be in the study area. These estimates for small-mammal populations are usually made by trapping an area at high intensity over a short period. Each animal is marked at first capture, the estimated population size being simply the number of first captures. Such estimates are acknowledged as underestimates but they have the advantage of yielding a real number, not an abstract concept, to work with.

Known-to-be-alive estimates are often the most appropriate in wildlife management. There are several problems of conservation and of harvesting for which an overestimate of density might lead to inappropriate management action. An underestimate, on the other hand, should simply produce inefficient but entirely safe management. The penalty for a poor estimate is often distributed asymmetrically around the true population size. It is not good to overestimate the number of individuals of an endangered species. It is not safe to apply a harvesting quota known to be safe for a given population size to one that is much smaller than you thought. Where the undesirable consequences of an overestimate are considerably greater than those accruing from an underestimate, the known-to-be-alive number is often the most appropriate estimate to work with.

12.6 Indices

An index of density is some attribute that changes in a predictable manner with changes in density. It may be the density of bird nests, the density of tracks of brown bears, or the number of minke whales (*Balaenoptera acutorostrata*) seen per cruising hour. A common index is the pellet or fecal dropping count (often used in studies of deer). This was used for endangered marsh rabbits (*Sylvilagus palustris*) in Florida, where pellet counts were closely correlated with radiotelemetry estimates (Forys and Humphrey 1997). Active burrow entrances were used for ground squirrel populations (van Horne *et al.* 1997) and call counts for mourning dove (*Zenaida macroura*) densities (Sauer *et al.* 1994). The North American Breeding Bird Survey is a standardized method in which some 2000 routes are sampled in June each year and the number of singing birds of each species is scored (Droege and Sauer 1989).

These indices reveal something about the density of birds, mammals, or whales. Without knowing anything about the proportional relationship between the index and the abundance of the animal, we could be confident that if the index halved or doubled it would reflect roughly a halving or doubling of animal density. Formally, this holds only when the relationship between index and density is a straight line that passes through the point of zero index and zero density.

Indices of density, if comparable, are useful for comparing the density of two populations or for tracking changes in the density of one population from year to year. Often a comparison is all we need. The relevant question may be not how large is the population but has it declined or increased under a particular regime of management. In such circumstances, the accuracy of an index is irrelevant; precision is paramount.

Let us compare an aerial survey designed to yield an estimate of absolute density with one designed to yield an index of density, as was conducted for pronghorn antelope in Colorado (Pojar *et al.* 1995). The first maximizes accuracy, the second precision. The "accurate" survey would probably inspect small quadrats by circling at a low but varying height above the ground. This is a good way to see animals but it is difficult to standardize between pilots. The "precise" survey would sample transects from a fixed height above ground at a constant speed. Since there is no requirement that all animals be counted on the sampled units, only a fixed proportion being sought, the survey variables are set according to how easily they may be standardized. Groundspeed is higher than for an "accurate" survey, to allow the pilot to maintain constant groundspeed safely even with a strong tailwind. Height above ground is set higher so that the inevitable variations in height will be proportionally less than at low level; $\pm 10\,\mathrm{m}$ around a height of 30 m results in large variations in search image. The same variation around 90 m has little effect. We might choose a transect width of 50 m per observer for an accurate survey but 200 m for a precise survey. The precision of the estimate is approximately proportional to the square root of the number of animals actually tallied (Eberhardt 1978) and so, although proportionally fewer will be seen on a 200 m strip, we choose the wider one in order to increase the absolute number that we see.

Consistency and rigid standardization of techniques are crucial when estimating an index. A good observer is not one who gets a high tally but one who has a consistent level of concentration and who produces results of high repeatability.

All the rules of sampling and of analysis hold as well for indices as for absolute counts of animals. Remember however that indices are useful only in comparisons and, therefore, the quantity to be estimated is the difference between two indices. The variance of an estimate of difference is the sum of the variances of the two estimated indices. As a rule of thumb, we should measure the two indices with a precision such that each standard error is less than a third of the difference we anticipate. Hence, an index must often be estimated much more precisely than is a one-off estimate of population size or density.

Errors in indices can be estimated by comparing results with a known population, similar to the way we estimate bias errors in counts (see Section 12.4.2; Eberhardt and Simmons 1987). For example, the number of sightings of fallow deer (*Dama dama*) in France along a transect (the index) was calibrated against a known population. The sighting index was found to be an effective standardized method for detecting trends in the population (Vincent *et al.* 1996).

12.6.1 *Known-to-be-alive used as an index*

Although known-to-be-alive is sometimes used as a one-off estimate of population size, it is more often used to track trends in population size. The operating rules governing these two uses are quite different. In the first exercise we seek the most accurate estimate we can get. In the second we seek consistency of method among several estimates, such that their bias is held constant. In the first case we put in as much work as possible. In the second we put in precisely the same amount of sampling on each surveying or capturing occasion. Otherwise, the trend in the estimates may reflect no more than variation in capturing effort.

A variant of this aberration of effort, very common in ecological research, is to boost the number known to be alive (because they were caught) on a given occasion by the number of individuals not caught on that occasion but which must have been there because they were caught on both previous and subsequent occasions. Although the accuracy of the estimate of absolute numbers is thereby enhanced, the consistency of the string of estimates is lowered. Estimates for the earlier occasions are inflated relative to those of later occasions, the rate of increase being underestimated if density is rising and the rate of decrease being overestimated if density is falling.

12.7 **Harvest-based population estimates**

In many cases, the only information that is available on wildlife or fishery abundance derives from harvest statistics. Indeed, many fisheries are managed almost exclusively on the basis of catch statistics, so the subject has attracted a great deal of attention from fisheries scientists. Harvest data vary widely in detail and in quality, depending largely on the time and resources that are used in collecting them, but they can be enormously useful if applied properly.

The simplest, but least reliable, index of abundance derived from harvests is a simple assumption that population size is proportionate to total harvest. This logic has been used to interpret long-term fur harvest records from the Hudson's Bay Company in order to study the well-known 10-year cycle of snowshoe hares and lynx, for example (Elton 1924; MacLulich 1937). There are several drawbacks to such an approach. Most importantly, in our example it assumes that the number of fur trappers and the efficiency with which they trap remains constant over time. This is highly unlikely given that economic climate varies over time, trapping techniques often evolve, and trapping is only possible in areas that haven't been turned over to other land use activities, such as agriculture. Nonetheless, such data represent our longest-standing records of crude abundance in the ecological record and have led to important scientific insights.

A much more robust method is to combine harvest totals with the effort expended to calculate *catch per unit effort* (*CPUE*). The logic for this approach is quite straightforward, being based on the same sort of behavioral processes that influence predation (see Chapter 10). Let us imagine that over the course of each full day of hunting a grouse hunter can effectively search 3 km^2. If we further assume that hunters are 100% successful in killing any animals that are encountered and that there is no limit on the number of animals hunted, then hunts would be expected to increase proportionately with grouse density. Following this logic, if the population density of grouse is 1 per km^2 then over the course of 5 days a single hunter would be expected to harvest 15 grouse. If the population density of grouse were to double, that number would be expected to jump to 30 grouse. In practice, harvesters cannot kill all animals encountered, and we don't know how efficiently they search the environment. Nonetheless, we would still expect harvest to scale proportionately with effort and abundance in the following manner:

$$H = qEN$$

where H is the harvest, E is effort, and q is the so-called *catchability* coefficient, which relates effort to the efficiency of search by hunters or fisherman. By rearranging this equation, we can show that:

$$N = \frac{H}{qE}$$

A minor modification (called the *random search equation*) is sometimes applied to accommodate the fact that a large number of hunters hunting over a large number of days will often tend to revisit sites that have already been well travelled. Under these circumstances, the relationship between harvest, abundance, and effort is better represented by a gentle curve than a straight line:

$$H = N(1 - e^{-qE})$$

or equivalently:

$$\frac{H}{N} = 1 - e^{-qE}$$

As an example, Fig. 12.2 shows that the proportion of white-tailed deer harvested from a game management area in Ontario varies in curvilinear fashion with the total

Fig. 12.2 Proportion of white-tailed deer harvested by hunters in Canada in relation to the number of days of hunting effort (males shown with filled symbols, females by open symbols). Lines indicate the best-fit models based on the random search formula (males solid line, females dotted line).

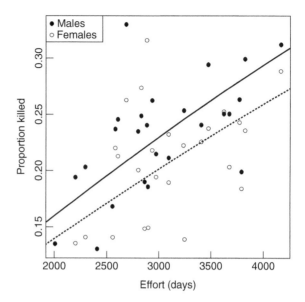

number of days of hunting (Fryxell *et al.* 1991), well approximated by the random search equation.

A more explicit calculation of abundance is possible when age-specific data are also available with respect to harvest. This methodology, known as *virtual population analysis* or *catch-at-age analysis*, is based on keeping track of the fates of specific cohorts (groups of individuals born in the same year) as they work their way through the population. For example, let $H_{i,t}$ represent the catch of age i individuals in year t. Let us further assume that all individuals have a constant probability of survival p and that no individuals live beyond $\tau = 5$ years of age. We start by recognizing that at minimum we know that $H_{i,t}$ individuals were alive before the harvest occurred at time t, hence age-specific abundance $N_{i,t} = H_{i,t}$. We similarly estimate the number of individuals that were alive the year prior to the harvest:

$$N_{i-1,t-1} = H_{i-1,t-1} + H_{i,t}/p$$

In other words, the number of individuals in the same cohort that were a year younger the previous year is determined by the number that show up in the harvests in both years, scaled by the proportion $(1/p)$ that survived long enough to be captured. Similar logic can be used to estimate abundance from age-specific harvest data using the generic formula:

$$N_{i,t} = H_{0,t} + \Sigma_i^\tau H_i/p^i$$

This procedure is sufficient to estimate abundance in completed cohorts (i.e. those that have worked their way to the maximum age τ). For incomplete cohorts, we need to estimate abundance in each age class for the final year, usually from CPUE:

$$N_{i,t} = \frac{H_{i,t}}{qE_t}$$

and then use these values in the last year of the dataset to calculate abundance in previous years:

$$N_{i-k,t-k} = N_{i,t} + \Sigma_{j=1}^{k} H_{i-j,t-j}/p^{j}$$

While the formulae look daunting because of all the subscripts, in practice this repetitive calculation is straightforward to apply within a spreadsheet.

12.8 Summary

Animal numbers can be estimated by total counts, sampled counts, mark–recapture, or various indirect methods. In each case, the usefulness of the method is determined by how closely its underlying assumptions are matched by the realities of what the animals do and how difficult they are to see, trap, or detect. The range of methods provided should allow wildlife managers to choose one that will be adequate in any given circumstance.

13 Age and stage structure

13.1 Introduction

In this chapter we deal with the internal workings of a population, often termed the *vital rates* or *demographic rates*, that result in a change of population size. The speed of this change is often measured as the exponential rate of increase. Any such change alerts us that the fecundity rate has changed, that the mortality rate has changed, that the age distribution has changed, or that more than one of these has changed. Each of those parameters will be considered in turn and the relationships between them explained.

We will then link the vital rates into an age-structured model to better understand how age structure influences patterns of change over time. A similar approach will be taken with populations whose demographic characteristics are better described in terms of growth stages, rather than organismal age. We demonstrate an important practical technique, elasticity analysis, which is often used by conservation biologists to evaluate the critical vital rates that may be holding back recovery of a threatened species. Finally, we use long-term demographic data from the Soay sheep population in Scotland to demonstrate how age structure, density dependence, and environmental stochasticity interact to influence population dynamics in long-lived wildlife species.

13.2 Demographic rates

A population's rate of increase is determined by its size, how many animals are born, and how many die during a year. Birth rate is thus an important component of population dynamics, and it can be measured in a number of ways. Of these, the most useful is fecundity rate, which we measure as the number of female live births per female per unit of time (usually one year). That figure is often broken down into age classes to give a fecundity schedule as in Table 13.1. Each value is symbolized m_x, female births per female in the age interval x to $x + 1$.

The number of animals that die over a year is another important determinant of rate of increase, and again it can be measured in a number of ways. We measure it as the mortality rate: the number of animals that die during a unit of time (usually one year) divided by the number alive at the beginning of the time unit. As with fecundity, the rate is often given for each interval of age.

The pattern of mortality with age is summarized as a life table, which has a number of columns as in Table 13.2. The first is the age interval, labeled by the age at the beginning of the interval and symbolized x. The second is survivorship l_x, the probability at birth of surviving to age x. The third is mortality d_x, the probability at birth of dying in the age interval $x, x + 1$. The fourth, and most useful, is mortality rate q_x, the probability of an animal age x dying before the age of $x + 1$. The fifth, age-specific survivorship p_x, is the probability at age x that an animal will still be alive on its next birthday.

Wildlife Ecology, Conservation, and Management, Third Edition.
John M. Fryxell, Anthony R. E. Sinclair and Graeme Caughley.
© 2014 John Wiley & Sons, Ltd. Published 2014 by John Wiley & Sons, Ltd.
Companion Website: www.wiley.com/go/Fryxell/Wildlife

Table 13.1 Fecundity schedule calculated for chamois. (Source: Caughley, 1970. Reproduced with permission of De Gruyter Publishers.)

Age in years, x	Sampled number, f_x	Number pregnant or lactating, B_x	Female births/ female $(B_x/2f_x)$, m_x
0	–	–	0.000
1	60	2	0.017
2	36	14	0.194
3	70	52	0.371
4	48	45	0.469
5	26	19	0.365
6	19	16	0.421
7	6	5	0.417
>7	10	7	0.350

Table 13.2 Construction of a partial life table.

Age, x	Survivors, f_x	Survivorship, l_x	Mortality, d_x	Mortality rate, q_x	Survival rate, p_x
0	1200	1.00	0.58	0.58	0.42
1	500	0.42	0.17	0.40	0.60
2	300	0.25	0.08	0.32	0.68
3	200	0.17	-	-	-
...	...	...	...	...	...

Probabilities are estimated from proportions. The probability of a bird surviving to age x can be estimated, for example, by banding 1200 fledglings and recording the number still alive 1 year later, 2 years later, 3 years later, and so on. Let us say those frequencies are 500, 300, 200. Survivorship at age 0 (i.e. at birth) is $1200/1200 = 1$; by age 1 year it has dropped to $500/1200 = 0.42$; at 2 years it is $300/1200 = 0.25$; and at 3 years it is $200/1200 = 0.17$.

No further data are needed to fill in the other columns corresponding to these values of l_x because each is a mathematical manipulation of the l_x column. Mortality d_x is calculated as $l_x - l_{x+1}(1 - 0.42 = 0.58$ for $x = 0$ and $0.42 - 0.25 = 0.17$ for $x = 1$). Mortality rate q_x is calculated as $(l_x - l_{x+1})/l_x$ or $d_x/l_x(0.58/1 = 0.58$ for $x = 1$ and $0.17/0.42 = 0.40$ for $x = 2$). Table 13.2 shows the table fully constructed up to age 2 years. The entry for age 3 years is partial because data for age 4 years are needed to complete it. The subsequent rows would be filled in each year as the data became available.

So, constructing a life table is straightforward when the appropriate data are available. Pause for a moment to contemplate the difficulty of obtaining those data. Banding 1200 fledglings, or whatever number, poses no more than a problem in logistics. The difficulty comes in estimating what proportion of those birds are still alive at the end of the year. Nonetheless, there have been a number of direct studies of vital rates in wildlife species, based on mark–recapture methods (Lebreton et al. 1992; Gaillard et al. 2000). Approximation methods are also available, based on age structure. If one can age a sample of the living population, or alternatively establish the ages at death of a sample of deaths from that population, an approximate life table can, in some circumstances, be constructed from those age frequencies.

13.3 Direct estimation of life table parameters

There are basically two different ways in which life table data can be directly estimated. The first, and rarest, is to monitor the fates of all individuals in a relatively small population that is carefully studied over a long time. For example, virtually every young lion born to the population inhabiting the Serengeti plains and adjacent woodlands has been carefully monitored over the past 3 decades (Packer *et al.* 2005). The unique combination of facial spots, scars, and other features make it possible to visually recognize every individual and keep track of their fate. By collating data for each specific cohort, one can readily calculate the probability that any member born to this group survives to age x (the l_x series), by simply dividing the number of survivors at age x by the initial group size.

Even in this ideal situation, however, there are thorny problems associated with the estimation of life table parameters. The difficult issue is that survival is like a game of chance: the outcome can vary considerably from one replicate to another (see Chapter 16). For example, a 0.5 probability of survival for an initial group of 4 individuals can lead to 0 survivors (expected 6.3% of the time), 1 survivor (expected 25% of the time), 2 survivors (expected 37.5% of the time), 3 survivors (expected 25% of the time), or even 4 survivors (expected 6.3% of the time). So the fact that 2 out of 4 individuals in a cohort survive over a given year does not conclusively demonstrate that the probability of survival really is 0.5; nor is the observation of no individuals surviving conclusive evidence against such a rate. As a result of the inherently variable nature of demographic processes, it is difficult to ascribe a particular risk of mortality with high likelihood, unless very large numbers of individuals are involved or such observations are repeated over many years.

The second way to estimate life table parameters directly is to mark a large number of individuals at time t (A_t), then recover some of those individuals (b_{t+1}) in a subsequent sampling session, say a year later, to estimate the probability of survival. Marked individuals might be equipped with leg bands (as in many bird studies), ear tags (as in many studies of small mammals), or even radio transmitters (as in many studies of large mammals). If the true number of survivors is B_{t+1} then the number of marked animals in the sample (b_{t+1}) depends on the detectability of individuals in each sample (c), typically under the presumption that $b_{t+1} = cB_{t+1}$. In this situation, not only is there stochastic variation to contend with but there is also sample variation associated with detectability of individuals in the population. By chance, we might detect a relatively large number of marked individuals in a subsequent sample, for reasons wholly unrelated to survival probability.

The confidence we ascribe to survival probabilities estimated using these mark–recapture techniques depends critically on sample size, probability of recapture if an animal is still alive, the mobility of marked animals and their loyalty to the site at which originally caught, the number of replicate sampling intervals, and whether or not newly marked animals have been repeatedly added to the population (Lebreton *et al.* 1992; Nichols 1992). Over the past 2 decades, there has been a revolution of sorts in the analysis of mark–recapture data, using sophisticated computer programs such as SURGE (Lebreton *et al.* 1992) and MARK (White and Burnham 1999). Many of these programs are available free of charge from their makers' Web sites. We point interested readers to the encyclopedic review of demographic methodology by Williams *et al.* (2002) for an insightful discussion of different mark–recapture approaches and their statistical analysis.

Sophisticated mark–recapture experiments aimed at estimating demographic parameters often involve comparisons among a large number of competing models (does survival vary among sexes, over time, across age groups, or between sites?). As we discuss in Chapter 15, such comparisons often require the use of information-theoretic approaches to identify the most parsimonious model by which to explain a given dataset. Recent versions of demographic analysis software, such as MARK, commonly include formal means of making choices among competing models (commonly either via AIC comparison (see Chapter 15) or likelihood-ratio testing).

13.4 Indirect estimation of life table parameters

If certain conditions (see the end of this section) are met, the age distribution of the living population can be used as a surrogate for the survival frequency f_x of Table 13.1 to produce an approximate life table. Many of the bovids can be aged from annual growth rings on the horns; some species of deer, seals, and possums produce growth layers in the teeth; and fish form growth lines on the scales. Unbiased samples of animals from the live population or which died from natural causes yield data that may be amenable to life table analysis.

It is sometimes possible to estimate a life table from a sample of individuals taken indiscriminately from the live population. This is most often done from hunting statistics, although the reliability of such measures is often questionable, given the tendency for most hunters to select bigger or older animals. It is better to rely on catastrophic events that indiscriminately sample a cross-section of individuals in the population.

Flash floods during the autumn of 1984 killed thousands of woodland caribou from the George River herd in northern Quebec as they were migrating to their winter range. A large number of carcasses from this freak event washed up on the banks of the Caniapiscau River, where wildlife biologists working with the Quebec government retrieved them (Messier *et al.* 1988). The resulting sample of 875 female caribou aged 1 year or older was assumed to reflect the standing age composition of the living population. The frequency of newborns was estimated from calf–mother counts on the calving grounds.

If any study population is unchanging (termed "stationary"), the standing age distribution reflects survival frequencies by age. In the case of the George River caribou herd, however, a series of censuses were available showing strong evidence of exponential increase over the previous 2 decades, with $r = 0.11$ (Fig. 13.1). This introduces a bias into life table parameter estimation, because older animals were born into a much smaller population than were younger individuals. The appropriate way to cope with this bias is to transform the age-frequency data before deriving the life table. Table 13.3

Fig. 13.1 Exponential population growth of the George River caribou herd, as discussed in the text. (After Messier *et al.* 1988.)

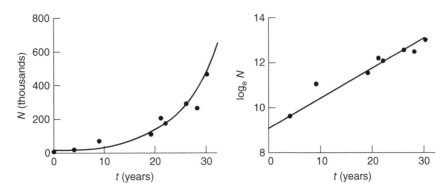

Table 13.3 Life table for female caribou in the George River herd (Source: Adapted from Messier *et al.*, 1988). Column 2 gives the original data from dead animals. Column 4 both corrects column 2 by multiplying by e^rx and smooths out column 3.

Age	Frequency	Corrected frequency	Smoothed frequency	l_x	d_x	p_x	q_x	m_x
0	236.1	236.1	236.1	1.000	0.286	0.286	0.714	0
1	138	154.0	168.5	0.714	0.007	0.010	0.990	0
2	156	194.4	167.0	0.707	0.017	0.024	0.976	0.06
3	113	157.2	163.0	0.690	0.027	0.039	0.961	0.35
4	94	145.9	156.6	0.663	0.037	0.056	0.944	0.4
5	83	143.9	147.9	0.626	0.044	0.070	0.930	0.4
6	65	125.8	137.3	0.582	0.053	0.091	0.909	0.4
7	63	136.1	125.0	0.529	0.057	0.108	0.892	0.4
8	57	137.4	111.4	0.472	0.063	0.133	0.867	0.4
9	40	107.6	96.6	0.409	0.065	0.159	0.841	0.4
10	24	72.1	81.2	0.344	0.067	0.195	0.805	0.4
11	18	60.4	65.4	0.277	0.067	0.242	0.758	0.4
12	12	44.9	49.5	0.210	0.066	0.314	0.686	0.4
13	7	29.2	33.9	0.144	0.064	0.444	0.556	0.4
14	1	4.7	18.8	0.080	0.061	0.763	0.238	0.4
15	4	20.8	4.4	0.019	0.019	1.000	0.000	0.4

demonstrates how to transform the age structure data, by multiplying the observed frequency at age x (f_x) by a coefficient (e^{rx}) that corrects for the bias in observed age frequencies caused by population growth.

One often needs to further smooth the age-frequency data, especially when the data come from a relatively small sample of animals, in order to guarantee a continual decline in frequency with each successive age group. This is usually done by fitting a quadratic or cubic curve to the age distribution, using the values derived from the curve in place of the actual observations, as demonstrated for the George River caribou in Table 13.3. The survivorship series is then constructed by dividing each age frequency by 236, the d_x series as $l_x - l_{x+1}$ and the q_x series as d_x/l_x. If the age-frequency data had not been smoothed, there would have been instances in which the observed frequency of an older age group exceeded that in the next youngest age group, implying survival rates exceeding 100%, an obvious impossibility.

An unbiased sample of ages at death (due to natural causes), as might be obtained by a picked-up collection of skulls, may in some circumstances be treated as a multiple of the d_x series. Table 13.4 gives an example from African buffalo (Sinclair 1977). Only those skulls aged 2 years or older were counted, because skulls from younger animals disintegrate quickly. These age frequencies are given in the second column of the table and total 183 skulls. The third column corrects for the missing younger frequencies: sample counts of juveniles in the field showed that the mortality rate over the first year of life was 48.5% and that 12.9% of the original cohort died in the second year. Hence, if the original cohort is taken as 1000, 485 of these would die in the first year of life and 129 in the second year. These values are tabled. They account for 614 of the original cohort, leaving 386 to die at older ages. The age frequencies of the 183 animals in the second column are thus each multiplied by 386/183 to complete the third column. The fourth column, d_x, is formed by dividing the fd_x frequencies by 1000 so that they sum to unity. Survivorship at age 0 (i.e. birth) is then set at one and the subsequent l_x values calculated by subtracting the corresponding d_x from each. Mortality rates q_x are calculated as before, as $q_x = d_x/l_x$.

Table 13.4 Construction of a life table from a pick-up sample of African buffalo skulls. Uncorrected for rate of increase.

Age, x	Mortality frequency, f_x	Mortality corrected, fd_x	Mortality, d_x	Survivorship, l_x	Mortality rate, q_x
0	-	485	0.485	1.000	0.485
1	-	129	0.129	0.515	0.250
2	2	4	0.004	0.387	0.010
3	5	11	0.011	0.383	0.029
4	5	11	0.011	0.372	0.030
5	6	13	0.013	0.361	0.036
6	18	38	0.038	0.348	0.109
7	17	36	0.036	0.310	0.116
8	20	42	0.042	0.274	0.153
9	17	36	0.036	0.232	0.155
10	15	32	0.032	0.196	0.163
11	16	34	0.034	0.164	0.207
12	18	38	0.038	0.130	0.292
13	15	32	0.032	0.092	0.348
14	14	29	0.029	0.060	0.483
15	8	17	0.017	0.031	0.548
16	5	10	0.010	0.014	0.714
17	1	2	0.002	0.004	0.500
18	0	0	0.000	0.002	0.000
19	1	2	0.002	0.002	1.000
Total	183	1001	1.001		

The reliability of any life table developed indirectly from a sample of either the live population or animals that died of natural causes depends on how closely the data meet the underlying assumptions of the analysis:

1 The sample is an unbiased representation of the living age distribution in the first case or of the true frequency of ages at death in the second. The exercise would have to control the usual biases implicit in hunting activities if the sample of the living age distribution were obtained by shooting. One would be unlikely to use a sample obtained by sporting hunters, for example. The first age class is usually underestimated in a picked-up sample of ages at death because the skulls of young animals disintegrate much faster than do those of adults, significantly biasing the table.

2 Age-specific fecundity and mortality must have remained essentially unchanged for a couple of generations.

3 Whether the sample is of the living population or of ages at death, the population from which it came must have a rate of increase very close to zero, or else the data must be transformed to accommodate the observed rate of population change over the past two generations. Major fluctuations in recent rates of population growth invalidate virtually all such indirect methods. This can limit the usefulness of such exercises in wildlife management.

13.5 Relationships among parameters

We restrict the following discussion to females for simplicity, but the points made apply equally to the male segment of the population. Remember that l_x is survivorship to age x, m_x is production of daughters per female at age x, and r is the exponential rate at which the population increases. These parameters are all interrelated according to the following formula (commonly termed Euler's equation):

$$\Sigma l_x m_x e^{-rx} = 1$$

If the survivorship and fecundity schedules hold constant, the population's age distribution will converge to the constant form of:

$$S_x = l_x e^{-rx}$$

which is called the *stable age distribution*. S_x is the number of females in a particular age class divided by the number of females in the first age class. The basic equation may thus be written $\Sigma S_x m_x = 1$. In the special case of rate of increase being zero, the stable age distribution, now called the *stationary age distribution*, is $S_x = l_x$ by virtue of $e^{-0x} = 1$. This is the justification for using such an age distribution to construct a life table. The *stationary age distribution* is the special case of the stable age distribution that results when $r = 0$. It has been argued that, since fecundity and mortality schedules seldom remain constant for long, the stable age distribution is little more than a mathematical abstraction, albeit a useful one. Although the stable distribution can be attained fairly quickly (roughly two generations) after mortality and fecundity patterns stabilize, most wildlife species that have been adequately studied have mortality and fecundity schedules that fluctuate, sometimes substantially, from year to year.

13.6 Age-specific population models

All wildlife populations have individuals of different ages. The vital rates (i.e. birth rates and probabilities of survival and mortality) often vary with age. Hence, a population composed of old individuals might well exhibit a different potential for growth than does a younger population. Assessing these kinds of process requires an age-specific model of population growth (Caswell 2001).

The standard technique is to use a Leslie matrix model, named in honor of the British ecologist who pioneered this approach (Leslie 1945, 1948). This involves multiplying age-specific population densities by a transition matrix (A). The top row of A reflects the probability of survival p from the previous age class multiplied by fecundity m at age x. The matrix elements along the subdiagonal and in the lower right-hand corner reflect the age-specific survival probabilities, p_0, p_1, and p_2.

In order to run the Leslie matrix model, we need to specify three other vectors: the initial age distribution of the population n, the age-specific rates of survival p, and the age-specific fecundity rates m. The demographic parameters we are using in this case come from the Serengeti population of cheetahs, based on a study of individually recognizable individuals that spans 3 decades (Kelly and Durant 2000; Durant *et al.* 2004). Typical of most long-lived organisms, survival rates of immature cheetahs (68%) are lower than those of adults (85%).

Provided that we have an initial age distribution, we can apply the Leslie matrix to estimate the abundance of individuals in each age group in subsequent years. This is done by multiplying the matrix A by the age vector n. A reminder of matrix algebra may be helpful here. Provided that an age vector has the same number of rows as the matrix has rows and columns, we can multiply them in the following manner. The first subscript refers to the row and the second to the column:

$$n_{0,t+1} = A_{0,0}n_{0,t} + A_{0,1}n_{1,t} + A_{0,2}n_{2,t}$$
$$n_{1,t+1} = A_{1,0}n_{0,t} + A_{1,1}n_{1,t} + A_{1,2}n_{2,t}$$
$$n_{2,t+1} = A_{2,0}n_{0,t} + A_{2,1}n_{1,t} + A_{2,2}n_{2,t}$$

This would obviously be rather cumbersome to calculate by hand for very long, especially for a species with more age groups. Fortunately, there is a simple way to automate the procedure, using matrix operations:

$$n^{\langle t+1 \rangle} = A \cdot n^{\langle t \rangle}$$

The "<>" notation used here refers to the age-specific abundance in year t. Hence, n $^{<2>}$ stands for the vector of age-specific abundances in year 2.

The age-specific totals predicted by the model are shown in the matrix termed "n." The first column of the matrix is the initial density of individuals in each age group, with newborns on the top row and old individuals on the bottom. Each column thereafter represents densities in successively later years.

$$n = \begin{pmatrix} 30 & 1.36 & 1.77 & 7.77 & 6.91 & 6.27 & 6.97 & 7.39 & 7.61 & 7.95 & 8.32 & 8.69 \\ 2 & 20.37 & 0.92 & 1.2 & 5.28 & 4.69 & 4.26 & 4.73 & 5.02 & 5.17 & 5.4 & 5.65 \\ 3 & 3.92 & 17.17 & 15.28 & 13.85 & 15.4 & 16.33 & 16.82 & 17.56 & 18.39 & 19.2 & 20.05 \end{pmatrix}$$

Model predictions of total population density, obtained by summing the densities of all age groups present at any point in time $\left(N_t = \sum n_{x,t}\right)$, are shown in Fig. 13.2.

After an initial period of adjustment, the population settles into a pattern of geometric (i.e. exponential) population growth, whose finite rate of increase depends on the integrated combination of age-specific parameters.

Each age-specific proportion $W_{x,t}$ is given by:

$$W_{x,t} = \frac{n_{x,t}}{N_t}$$

As the population settles into a geometrical growth pattern, the proportions of each age class in the population ($W_{x,t}$) converge on a stable age distribution, illustrated in Fig. 13.3. The time it takes for the age structure to stabilize is typically 2–3 generations, where generation time is the average time that elapses between the birth of a female and that of her daughters.

So, over time, we can be sure of two things: (i) the population will eventually grow geometrically and (ii) once this happens, the proportions of individuals in each age group will also become constant. The findings of geometric increase in N and stable age

Fig. 13.2 Predicted pattern of population change over time for the Leslie matrix model example discussed in the text.

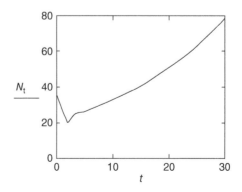

Fig. 13.3 Predicted age distribution over time for the Leslie matrix model example discussed in the text. After 6 years, the proportions of each age group have stabilized.

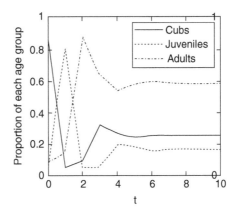

distribution imply that the following mathematical statements are equivalent: $N_{t+1} = \lambda N_t$ and $n^{<t+1>} = An^{<t>}$, where N and λ are *scalars* (i.e. single, countable numbers) and A and n are *matrices* or *vectors*. In other words, a simple model of geometric increase ($N_{t+1} = \lambda N_t$) yields the same results as the Leslie matrix model ($n^{<t+1>} = An^{<t>}$). This means that we can estimate λ (the finite annual rate of increase) from the transition matrix A, by finding something called the *dominant (largest) eigenvalue* of the transition matrix. The largest of the eigenvalues (1.04) is the finite rate of population increase (λ) once the population has reached a stable age distribution:

$$eigenvals(A) = \begin{pmatrix} -0.1 + 0.44i \\ -0.1 - 0.44i \\ 1.04 \end{pmatrix}$$

Hence, after the initial period of uncertainty, the total population will increase by 4% per year (because $\lambda = 1.04$).

Just as there is a simple means of estimating the eventual rate of population growth (dominant eigenvalue), there is an equally simple way of predicting the eventual proportion of individuals in each age group – the stable age distribution S alluded to earlier. We calculate the so-called "right eigenvector" corresponding to the "dominant eigenvalue" of the transition matrix A:

$$S_{raw} = eigenvec(A, 1.04) \qquad S_{raw} = \begin{pmatrix} 0.38 \\ 0.25 \\ 0.89 \end{pmatrix}$$

It is easier to interpret these values if we transform them into proportions:

$$S = \frac{S_{raw}}{\Sigma_i S_{raw_i}} \qquad S = \begin{pmatrix} 0.25 \\ 0.16 \\ 0.58 \end{pmatrix}$$

In other words, once the rate of growth has stabilized, newborns will comprise 25%, 1-year-olds 16%, and adults 58% of the population.

The discussion of right eigenvectors and eigenvalues can be unnerving for many biologists, even for hardened professionals, but do not worry too much. Although the terms sound mysterious, the meanings of "eigenvalue" and "eigenvector" are actually fairly simple. Once a population has converged on its stable age distribution, thereafter every year the total population N increases by a multiplicative factor λ (the dominant eigenvalue), meaning that each age group in the population also increases year-to-year by the same factor. So, an eigenvector is just a string of numbers (the stable age distribution) that produces exactly the same outcome when multiplied by the constant λ as by the transition matrix A. Fortunately, we can use this string of numbers in a very practical way, because it tells us the relative proportion of individuals we can expect eventually to see over time in each age category.

13.7 Elasticity of matrix models

The largest eigenvalue of the cheetah transition matrix equals 1.04, implying that the population is just able to sustain itself (a value of 1.0 is required for sustainability, i.e. a stationary population). Age- or stage-specific models also offer useful insights into possible remedies to counteract such population declines. By modifying vital rates, one can interpret the effectiveness of possible conservation actions that might be taken. In the case of cheetahs, for example, it might be possible to protect cubs or alternatively to improve adult survival. How can we evaluate these options?

There is an easy (but not very exact) way to do this, by substituting different values for each parameter in the transition matrix and seeing which one has the biggest effect. You should do so for yourself, to determine which parameter is most conducive to improving the growth rate λ. A more elegant answer can be obtained by using calculus to determine how *sensitive* λ is to a proportionate change in each parameter A_{ij} in the transition matrix, where i refers to row and j to column. This is termed by ecologists the *elasticity* of λ, and the derivation is presented in Box 13.1 (based on Caswell 1978; de Kroon *et al.* 1986). It is calculated in the following manner:

$$E_{i,j} = \frac{A_{i,j}}{\lambda} \cdot \left(\frac{V_i \cdot S_j}{V \cdot S} \right)$$

Box 13.1 Deriving the elasticity of matrix models (Caswell 1978).

For small changes in λ as a function of small changes in any element of the transition matrix (A_{ij}), elasticity is defined in the follow manner:

$$\text{elasticity} = \frac{d}{dlnA_{ij}} \ln \lambda$$

$$\frac{d}{dlnA_{ij}} \ln \lambda = \frac{A_{ij}}{\lambda} \cdot \frac{d}{dA_{ij}} \lambda$$

and:

$$\frac{A_{ij}}{\lambda} \cdot \left(\frac{d}{dA_{ij}} \lambda \right) = \frac{A_{ij}}{\lambda} \cdot \left(\frac{V_i \cdot W_j}{V \cdot W} \right)$$

So:

$$\text{elasticity} = \frac{A_{ij}}{\lambda} \cdot \left(\frac{V_i \cdot W_j}{V \cdot W} \right)$$

where $V \cdot W$ in the denominator refers to the scalar or dot product obtained by multiplying together the column vector representing the stable stage distribution W and the row vector of reproductive values V.

where W is the vector corresponding to the stable age distribution and V is the *vector of reproductive values*. It is the left eigenvector (as opposed to the stable age distribution, which is the right eigenvector) that corresponds to the dominant eigenvalue λ. The left eigenvector is a string of numbers that satisfies the following equality:

$$V^T \cdot A = \lambda \cdot V^T$$

where the T means that we are transposing the vector, so that it is now a row (rather than a column) vector. Stage-specific reproductive values for Serengeti cheetahs are calculated as follows:

$$V_{raw} = eigenvec(A^T, 1.040) \qquad V_{raw} = \begin{pmatrix} 0.33 \\ 0.51 \\ 0.79 \end{pmatrix}$$

It is traditional to normalize this vector by dividing each age-specific reproductive value by that of newborn individuals:

$$V = \frac{V_{raw}}{V_{raw_0}} \qquad V = \begin{pmatrix} 1 \\ 1.54 \\ 2.37 \end{pmatrix}$$

We now have all the data needed to calculate the elasticities of the elements of the cheetah transition matrix:

$$E = \begin{pmatrix} 0 & 0 & 0.13 \\ 0.13 & 0 & 0 \\ 0 & 0.13 & 0.6 \end{pmatrix}$$

If we look carefully through the elasticity matrix E for the Serengeti cheetah population we see that one value (0.6) is clearly larger than the rest. This value corresponds to the adult survival rate in the original transition matrix A. It implies that a change in adult survival would have more impact on the rate of population growth by cheetahs than a change of similar proportion of any other demographic parameter. This is a very useful statistic, allowing conservation biologists to identify which demographic parameters have the strongest influence on long-term viability.

13.8 Stage-specific models

In many organisms it makes more sense to think about different demographic stages or size classes, rather than specific age classes. This can also be a convenient means of approximating the dynamics of long-lived organisms, by lumping age groups into stages, because often we do not have information on exact ages. Such an approach is known as a Lefkovitch stage-class model (Lefkovitch 1965; Caswell 2001). This involves multiplying stage-specific population densities by a transition matrix (A). The top row of A reflects the probability of survival for stage class i multiplied by its fecundity f_i. The diagonal reflects the probability of surviving and remaining within stage i p_i, while the subdiagonal represents the probability of surviving and growing into the next stage g_i.

$$A = \begin{pmatrix} p_0 & f_1 & f_2 & f_3 & f_4 & f_5 & f_6 \\ g_0 & p_1 & 0 & 0 & 0 & 0 & 0 \\ 0 & g_1 & p_2 & 0 & 0 & 0 & 0 \\ 0 & 0 & g_2 & p_3 & 0 & 0 & 0 \\ 0 & 0 & 0 & g_3 & p_4 & 0 & 0 \\ 0 & 0 & 0 & 0 & g_4 & p_5 & 0 \\ 0 & 0 & 0 & 0 & 0 & g_5 & p_6 \end{pmatrix}$$

When we multiply this Lefkovitch matrix by a vector of stage groups, the use of the two sets of diagonals allows some individuals in each stage to mature into the next while others remain in the same stage as before. Just as we saw for simple age-structured models, the stage-class model predicts geometric increase (or decrease, depending on the magnitude of λ) after the age structure has equilibrated.

A good example of the application of this kind of model involves the loggerhead sea turtle (*Caretta careta*), a marine species that lays its eggs in sandy beaches of the southeastern United States. Demographic parameters are difficult to estimate for a long-lived species like the loggerhead, which roams widely across the Atlantic Ocean. Hence, accurate age-specific data are unavailable, as are age-specific estimates of fecundity. Crude data are available, however, on the relative survival rates and fecundities of different stages: eggs, hatchlings, and mature individuals. Additional predictable variation in fecundity stems from body size. Crouse *et al.* (1987) developed a Lefkovitch stage-class model (Box 13.2) to evaluate which of seven life stages would be most responsive to conservation efforts. This is a considerable simplification of the 54 age classes that would be needed for a full Leslie matrix model.

Box 13.2 The Lefkovitch stage-class model for the loggerhead sea turtle (Crouse *et al.* 1987).

There are seven stages in the model. The youngest (0) represents eggs and hatchlings. The next two (1 and 2) represent small and large juveniles. Sub-adults are represented by stage 3. All individuals beyond stage 3 are capable of breeding. Stage 4 represents novice breeders. The last two stages correspond to young (5) and older (6) adults. Crouse *et al.* (1987) estimated the proportion of individuals growing into the next class by assuming a stable age distribution – perhaps a risky assumption. On this basis, they derived the following Lefkovitch matrix model:

$$p = \begin{pmatrix} 0.0 \\ 0.737 \\ 0.661 \\ 0.691 \\ 0.0 \\ 0.0 \\ 0.809 \end{pmatrix} \quad g = \begin{pmatrix} 0.675 \\ 0.049 \\ 0.015 \\ 0.052 \\ 0.809 \\ 0.809 \\ 0 \end{pmatrix} \quad f = \begin{pmatrix} 0 \\ 0 \\ 0 \\ 0 \\ 127 \\ 4 \\ 80 \end{pmatrix}$$

$$A = \begin{pmatrix} p_0 & f_1 & f_2 & f_3 & f_4 & f_5 & f_6 \\ g_0 & p_1 & 0 & 0 & 0 & 0 & 0 \\ 0 & g_1 & p_2 & 0 & 0 & 0 & 0 \\ 0 & 0 & g_2 & p_3 & 0 & 0 & 0 \\ 0 & 0 & 0 & g_3 & p_4 & 0 & 0 \\ 0 & 0 & 0 & 0 & g_4 & p_5 & 0 \\ 0 & 0 & 0 & 0 & 0 & g_5 & p_6 \end{pmatrix} \quad n^{(0)} = \begin{pmatrix} 10 \\ 5 \\ 3 \\ 2 \\ 1 \\ 1 \\ 0 \end{pmatrix}$$

13.9 Elasticity of the loggerhead turtle model

The largest eigenvalue of the loggerhead sea turtle transition matrix equals 0.95, implying that the population cannot sustain itself (a value of 1.0 is required for sustainability, i.e. a stationary population). Age- or stage-specific models also offer useful insights into possible remedies to counteract such population declines. By modifying vital rates, one can interpret the effectiveness of possible conservation actions that might be taken. In the case of loggerheads, for example, one could protect nesting sites on beaches or alternatively devote larger effort to improving survival while the animals are out at sea. One way to evaluate these options is by calculating elasticities of the transition matrix. The elasticity calculations for the loggerhead sea turtle are shown in Box 13.3.

Changing some demographic parameters obviously has a stronger impact on turtle growth rates than does changing others. The largest elasticities correspond to the probabilities of survival and remaining within the adult and juvenile stage classes. If one had a finite amount of money, energy, and time to devote to conserving this species, it

Box 13.3 Calculation of elasticities of the Lefkovitch matrix for the loggerhead sea turtle. (Based on data from Crouse *et al.* 1987.)

The stable age distribution for the loggerhead turtle is calculated from the right eigenvector of transition matrix A in Box **13.2**, for which 0.947 is the dominant eigenvalue.

$$W = \begin{pmatrix} 0.21 \\ 0.67 \\ 0.11 \\ 0.01 \\ 3.69 \times 10^{-4} \\ 3.16 \times 10^{-4} \\ 0 \end{pmatrix}$$

In other words, after the loggerhead turtle model has proceeded for a number of years, we will expect to see 21% of the population being composed of eggs or hatchlings, 67% juveniles, and the rest (12%) sub-adults and adults.

For the loggerhead sea turtle, the vector V depicting stage-specific reproductive value is as follows:

$$V = \begin{pmatrix} 1 \\ 1.4 \\ 6 \\ 114.31 \\ 562.07 \\ 500.74 \\ 581.02 \end{pmatrix}$$

With these values we can estimate elasticities for every parameter in the transition matrix:

$$E_{i,j} = \frac{A_{i,j}}{\lambda} \cdot \left(\frac{V_i \cdot W_j}{V \cdot W} \right)$$

$$E = \begin{pmatrix} 0 & 0 & 0 & 0 & 0.01 & 3.3 \times 10^{-4} & 0.04 \\ 0.05 & 0.18 & 0 & 0 & 0 & 0 & 0 \\ 0 & 0.05 & 0.12 & 0 & 0 & 0 & 0 \\ 0 & 0 & 0.05 & 0.14 & 0 & 0 & 0 \\ 0 & 0 & 0 & 0.05 & 0 & 0 & 0 \\ 0 & 0 & 0 & 0 & 0.04 & 0 & 0 \\ 0 & 0 & 0 & 0 & 0 & 0.04 & 0.23 \end{pmatrix}$$

would be most effectively spent on improving adult or juvenile survival. Interestingly, most conservation effort prior to the paper of Crouse *et al.* (1987) had been devoted to enhancing breeding success on the beaches. Although this was no doubt useful, the elasticity calculations suggest that efforts would have been better spent on improving survival at sea.

Turtle excluder devices (TEDs) can greatly reduce mortality of turtles that get caught accidentally in fish and shrimp trawls. The elasticity calculations suggest that application of TEDs should be the most efficient means of improving the long-term viability of the population. Let us assume that such devices improve survival of old adults from 81% to 95%. Would this be sufficient to ensure long-term viability?

The demographic modified matrix obtained by changing adult survival to 95% leads to a finite rate of increase $\lambda = 1.01$, which is indeed just sufficient to allow sustainability. Hence, strenuous enforcement of TEDs would be sufficient to allow population recovery. Given the slender margin, however, other conservation practices are also called for, such as improved breeding success and enhanced hatchling survival to the juvenile stage. In practice, enhanced usage of TEDs has led to dramatic improvement in loggerhead turtle survival, a genuine conservation success story!

13.10 Short-term changes in structured populations

Although the Leslie matrix, Lefkovitch matrix, and geometric models make similar predictions in the long term, they definitely make different ones in the short term, before the age distribution has had a chance to stabilize. This may be particularly important in conservation and management of many wildlife species that tend to be long-lived. Many wildlife populations accordingly have age distributions that are far from stable (Owen-Smith 1990; Coulson *et al.* 2001a, 2003; Lande *et al.* 2002). Under these circumstances, it would be useful to have a reliable means of quantifying which demographic parameters have the greatest short-term impact on population growth. A promising approach has recently been outlined by Fox and Gurevitch (2000), based on use of the full set of eigenvalues and eigenvectors, rather than just the dominant set. Application of this new approach with an endangered cactus species (*Coryphanta robbinsorum*) demonstrated that the key demographic parameters for population growth in the short term differed considerably from those identified by standard elasticity assessment (Fox and Gurevitch 2000).

13.11 Environmental stochasticity and age-structured populations

When we link age structure with density dependence and environmental stochasticity, we begin to understand the multiple factors that can influence population dynamics. A well-known example comes from the feral Soay sheep on the St. Kilda archipelago off the coast of Scotland. These sheep, similar in many ways to the ancestral sheep first domesticated by humans, were initially introduced during the second millennium BC. They have roamed wild for decades on several of the St. Kilda islands, the best known of which is the small island of Hirta. A fraction of the Hirta population uses an open, grassy area once occupied by people. Here the pregnant females use abandoned stone huts for shelter during birth. As a consequence, sheep numbers can be counted accurately and a large fraction of the newly born lambs can be caught and marked. Tracking these known individuals over the subsequent years has allowed unusually detailed calculations of age-specific reproduction and survival (Clutton-Brock *et al.* 1997).

The demographic pattern that has emerged from these field studies shows pronounced threshold effects of population density on sheep survival (Fig. 13.4). When

Fig. 13.4 Survival of Soay sheep lambs on the island of Hirta in relation to adult population density: females (solid line), males (broken line). (After Clutton-Brock *et al.* 1997.)

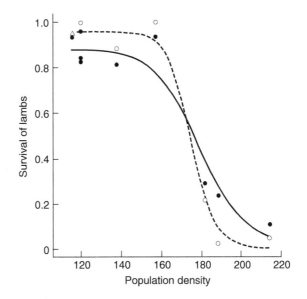

the population is less than 200 adult sheep, survival of lambs, yearlings, and adults tends to be high: typically > 90% in adults and yearlings and > 80% in lambs. Increase in sheep abundance beyond the threshold tends to be accompanied by a precipitous decline in survival to low levels, sometimes as low as 10%. In order to fully understand the impact of these threshold effects, we need to construct a model that combines density-dependent and age-dependent features (see Box 13.4).

Box 13.4 Model of the Soay sheep population on St. Kilda.

Threshold effects on mortality can be well described by a sigmoid function:

$$\Psi(i, N) = \frac{p_{max_i}}{1 + (\alpha_i \cdot N)^{\beta_i}}$$

where i refers to the age group (from 0 for newborns up to 2 for adults), N is the population density of yearlings and adults, p_{max} is the maximum survival rate, and α and β are parameters determining the shape of the sigmoid survival function (Ψ). Clutton-Brock *et al.* (1997) estimated the parameters of the Ψ function from several years of data as follows:

$$p_{max} = \begin{pmatrix} 0.88 \\ 0.94 \\ 0.96 \end{pmatrix} \alpha = \begin{pmatrix} 0.00562 \\ 0.00484 \\ 0.00467 \end{pmatrix} \beta = \begin{pmatrix} 15.3 \\ 9.46 \\ 8.93 \end{pmatrix}$$

Similar sigmoid functions (Ω) can be fitted to age-specific fecundity rates of females:

$$m_{max} = \begin{pmatrix} 0.335 \\ 0.643 \\ 0.643 \end{pmatrix} \gamma = \begin{pmatrix} 0.00629 \\ 0.00589 \\ 0.00589 \end{pmatrix} \varphi = \begin{pmatrix} 24.1 \\ 14.1 \\ 14.1 \end{pmatrix}$$

$$\Omega(i, N) = \frac{m_{max_i}}{1 + (\gamma_i \cdot N)^{\varphi_i}}$$

By applying these density-dependent survival and fecundity rates to specific age-classes, we can estimate changes in abundance over time:

$$\begin{pmatrix} n_{0,t+1} \\ n_{1,t+1} \\ n_{2,t+1} \end{pmatrix} = \begin{bmatrix} \sum_i \left(n_{1,t} \cdot \Omega \left(i, \sum_j n_{j,t} \right) \cdot \Psi \left(i, \sum_j n_{j,t} \right) \right) \\ n_{0,t} \cdot \Psi \left(0, \sum_j n_{j,t} \right) \\ n_{1,t} \cdot \Psi \left(1, \sum_j n_{j,t} \right) + n_{2,t} \cdot \Psi \left(2, \sum_j n_{j,t} \right) \end{bmatrix}$$

Simulation models constructed with threshold survival and fecundity effects (Fig. 13.5) generate regular fluctuations of Soay sheep at 6-year intervals (Fig. 13.6), qualitatively similar to the patterns seen by the real population (Fig 13.7) (Grenfell *et al.* 1992). The model does not capture all of the variability in sheep abundance observed on St. Kilda. Like all models, it leaves out many important features. It has

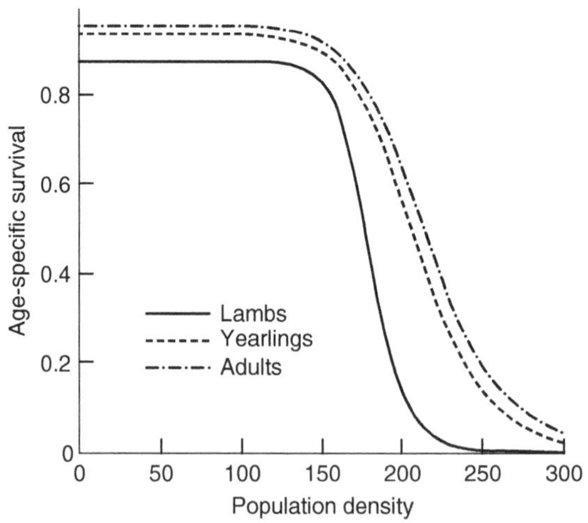

Fig. 13.5 Sigmoid survival functions in relation to the population density of adult + yearling Soay sheep, for adults (dashed-dotted line), yearlings (dotted line), and lambs (solid line), estimated by Clutton-Brock *et al.* (1997).

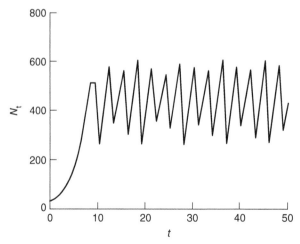

Fig. 13.6 Simulated dynamics of an age-structured population with sigmoid survival and fecundity functions with the same parameters as those of the Soay sheep population on the island of Hirta. (After Clutton-Brock *et al.* 1997.)

Fig. 13.7 Observed variation in the population of Soay sheep over the entire St. Kilda archipelago. (After Clutton-Brock *et al.* 1997.)

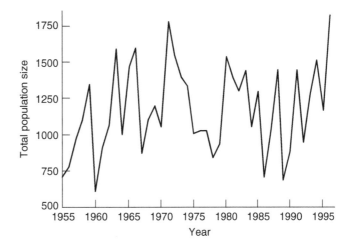

no direct link with food supply or disease, both of which are important in shaping dynamics. Catastrophic mortality is largely caused by starvation, and vulnerability to starvation is exacerbated by high nematode infestation in the intestinal tract of individual sheep (Gulland 1992; Clutton-Brock *et al.* 1997). Nonetheless, the model clearly demonstrates that the combination of age structure and threshold effects predispose the Soay sheep population to recurrent cycles in abundance.

Strong effects of weather variation can influence the population dynamics of Soay sheep (Grenfell *et al.* 1998; Coulson *et al.* 2001a). Populations of sheep on adjacent, isolated islands tend to be loosely synchronized, because they share a common climate (Grenfell *et al.* 1998). Although density-dependent processes regulate Soay sheep, the precise pattern of regulation is strongly affected by age structure. Different age groups vary in their degree of density dependence and sensitivity to weather conditions (Catchpole *et al.* 2000; Coulson *et al.* 2001a). As a result, a population dominated by young animals would have quite different population dynamics in the short term from one with a more equitable distribution of age groups. The mix of age groups on any of the islands is highly changeable and slight modifications in age structure alter the dynamic consequences of density-dependent processes (Coulson *et al.* 2001a). The Soay sheep example illustrates how populations can fluctuate through a combination of (i) stochastic environmental effects, (ii) nonlinear demographic responses, and (iii) delays that arise through a complex age structure.

13.12 Summary

The dynamic behavior of a population – whether it increases, decreases, or remains constant – is determined by its age- or stage-specific mortality and fecundity rates interacting with the underlying distribution of ages or stages in the population. A wide variety of techniques are available for estimating age-specific parameters, summarized in the life table. When age-specific rates of fecundity and survivorship remain constant, the population's age distribution assumes a stable form, even though its size may be changing. These demographic parameters determine the rate of population change over time, forming the logical basis for many conservation and management decisions.

The long-term rate of population change can be calculated through use of a transition matrix composed of these vital rates. Determination of the elasticity of the rate of population change due to slight modifications of the vital rates can be a useful means of evaluating alternate conservation and management options, as illustrated by the case studies with cheetahs and loggerhead turtles.

In most wildlife species, population dynamics over the long term will be strongly influenced by age structure, density dependence, and how each of these parameters is affected by environmental stochasticity. These synergistic effects are well demonstrated by the Soay sheep population off the coast of Scotland.

14 Experimental management

14.1 Introduction

In practical terms, there are two different modes of wildlife management: that in which management decisions flow from personal experience and received wisdom, and that in which they are based upon data and analysis. For want of better names, we will call these the "traditional" mode and the "experimental" mode of wildlife management. The value of the traditional mode should not be underestimated. Its dominant characteristic is conservatism, a lack of interest in trying out new ideas. This is sometimes a strength rather than a weakness, because most new ideas turn out to be wrong (Caughley 1985). However, some new ideas are useful, and these are best identified by the experimental approach. In this chapter we explain how a technical judgment can be evaluated, by posing it as a question (hypothesis).

One way to evaluate a hypothesis is through a standard statistical test of inference. We provide guidelines here for designing effective experiments and outline some standard statistical methods of analysis of such experiments. We describe the use of replication to sample the natural range of variability and the use of controls to render the conclusions unambiguous. However, standard statistical tests are sometimes inadequate for identifying which of many possible hypotheses provides the "best" explanation for the observations at hand and for deciding which course of management will be most effective. In order to pursue these issues, we need a different approach, namely model evaluation and inference. In Chapter 15 we will outline some of the principles of model evaluation and show how this method can provide a powerful tool in the resource manager's arsenal.

14.2 Differentiating success from failure

Wildlife management is not like civil engineering. The theory and practice of civil engineering is placed on public display every time a bridge is built. No expertise is needed to interpret that test. If the bridge remains operational for the length of its design life, the engineers got it right. If it collapses, they got it wrong, and we look forward to hearing the details of how and why at the subsequent court case.

Wildlife management differs from civil engineering in a number of respects. First, the managers are not erecting something new but acting as custodians of something already there. They are not responsible for the initial conditions, but these often constrain their options.

Second, in civil engineering the question is usually obvious. In ecology it is seldom so. Choosing the appropriate question is the most difficult task, much more so than answering that question. Good design does not correct an inadequate grasp of the problem.

Wildlife Ecology, Conservation, and Management, Third Edition.
John M. Fryxell, Anthony R. E. Sinclair and Graeme Caughley.
© 2014 John Wiley & Sons, Ltd. Published 2014 by John Wiley & Sons, Ltd.
Companion Website: www.wiley.com/go/Fryxell/Wildlife

Third, in wildlife management criteria for success and failure are seldom tight and often are not available to the public. Compare these two statements:

1 The provision of nest boxes for wood ducks will increase the size of the population.
2 The provision of nest boxes for waterfowl will benefit their overall ecology.

The first is a hypothesis testable against a predicted outcome. The second is of the type that covers everything and probably cannot be disproved. What is an "overall ecology"? How is it measured? What species should it be measured on? Wildlife management objectives that are framed in an unverifiable form (example **2**) are not very useful, whereas those in the form of a testable hypothesis (example **1**) allow us to learn more about a system.

Fourth, even when there are such criteria by which to judge success, the wildlife manager is seldom in complete control of the situation and hence cannot be held fully and personally responsible for the outcome. A failure is usually referable to the acts of many people, often interacting with changes in the environment (sometimes referred to as "acts of God").

Fifth, the wildlife and its habitat usually form a robust ecological system. Within rather wide limits, that system will absorb the most inappropriate or irrelevant of management and still look good.

Because the criteria for success are often fuzzy in wildlife management, the outcomes of different management systems are sometimes difficult to rank. For example, when managing deer populations do we shoot only bucks, shoot only does, shoot 70% bucks and 30% does, shoot 30% bucks and 70% does, or shoot neither. All these schemes have been tried and all have been reported as highly successful. "Highly successful" with what end in view? How "highly successful"? Perhaps we ask these questions less often than we should.

Agriculture made a major advance when R.A. Fisher invented the analysis of variance ("A of V" or "ANOVA") and a few agriculturalists recognized that here was a technique that could differentiate the effects of different management treatments. More importantly, they believed that differentiation was necessary. Wildlife management can learn from the history of agriculture by incorporating more statistical design in management programs.

14.3 Technical judgments can be tested

The wisdom of a *technical judgment* can be evaluated according to strict criteria. If a manager decides that supplementary feeding will increase the density of quail, this can be tested and the decision can be rated right or wrong. If a manager decides that elephants must be culled because otherwise they will eliminate *Acacia tortilis* trees from the area, this decision is right or wrong and can be demonstrated to be one or the other by an appropriate experiment. Note that the decision on whether the local survival of the acacia justifies the proposed reduction of elephants is a *value judgment* and hence not testable.

So, there are value judgments and there are technical judgments and these must not be confused one with the other. Technical judgments can be tested and should be tested. By this means we learn from our failures as well as from our successes. A recurring theme of this book is that wildlife management advances only when the efficacy of a management treatment is tested. For that to happen, the technical decision as to the appropriate treatment must be stated in a form that predicts a verifiable outcome.

14.3.1 *Hypotheses*   Research questions are usually phrased in positive form, such as, "Does the mean body weight of black bears (*Ursus americanus*) change as we move from the equator towards the pole?" That question is most easily tested statistically if we frame it in negative form, the so-called *null hypothesis* symbolized H_0: mean body weight of black bears does not change with latitude. If this hypothesis is falsified by data showing that mean weights are not the same at all sampled latitudes, we reject the null hypothesis in favor of an *alternative hypothesis* H_a. Whereas a question can generate only one null hypothesis, there may be a number of competing alternative hypotheses. In the bear example, the alternative to no change of weight with latitude might be an increase with latitude, a decrease with latitude, a peak in the middle latitudes, or a trough in the middle latitudes.

The procedures by which we test hypotheses make up the realm of statistical analysis. They come quite late in the research sequence, which proceeds in the following manner:

1 Pose a research question (usually our best guess or prediction as to what is going on).
2 Convert it to a null hypothesis.
3 Collect the data that will test the null hypothesis.
4 Run the appropriate statistical test.
5 Accept or reject the null hypothesis in the light of that testing.
6 Convert the statistical conclusion to a biological conclusion.

Most statistical tests estimate the probability that a null hypothesis is false. A probability of say 10% is often interpreted loosely as meaning that there is only a 10% chance of the null hypothesis being true. That is not quite right. Suppose our null hypothesis states that there is no difference in bill length between the females of two populations of a particular species. We draw a sample from each, perform the appropriate statistical test for a difference, and find that the test statistic has a probability of (say) 10% for the sample sizes that we used. That 10% is the estimated probability of drawing two samples as different in average bill length as or even more different than those that we drew *if the populations from which they were drawn did not differ in that estimated attribute*.

If there really is no difference between the two populations in average bill length then the probability returned by the statistical test will be in the region of 50%. This implies that the chance of drawing more disparate samples than those we actually drew is the same as the chance of drawing less disparate samples than those we actually drew. A probability greater than 50% means that the two samples are more similar than we would expect from random sampling of identical populations. If the probability approaches 95% or so, we should investigate whether the sampling procedure was biased.

Statistical tests deal in probabilities, not certainties. There is always a chance that we are wrong. Such errors come in two forms, the *Type 1 error* (also known as an alpha error), in which the null hypothesis is rejected even though it is true, and the *Type 2 error* (beta error), in which the null hypothesis is accepted even though false. Following Zar (1996), the relationship between the two kinds of error can be shown as a matrix:

	If H_0 is true	If H_0 is false
If H_0 is rejected	Type 1 error	No error
If H_0 is accepted	No error	Type 2 error

Obviously we are not keen to make either kind of error. The probability of committing a Type 1 error is simply the specified significance level. The probability of committing a Type 2 error is not immediately specifiable, except that we can say that it is related inversely to the significance level for rejecting the null hypothesis. The two kinds of error cannot be minimized simultaneously, except by increasing the sample size. Hence, we need a compromise level of significance that will provide an acceptably small chance of rejecting a factual null hypothesis, but which is not so small as to generate too large a chance of committing a Type 2 error. Experience has indicated that a 5% chance of rejecting the null hypothesis when it is true provides reasonable insurance against both kinds of error. We therefore conventionally specify the 5% probability as our significance level, although that level is essentially arbitrary and little more than a gentleman's agreement. What is not arbitrary is that the hypothesis to be tested and the level of significance at which the hypothesis is rejected must be decided upon *before* the data are examined and preferably before they are collected. Otherwise the whole logic of testing is violated.

Our standard statistical tests concentrate on minimizing Type 1 errors. The extent to which they minimize Type 2 errors is called *power*. Depending on context, avoidance of Type 2 errors may be more important than ensuring the warranted rejection of the null hypothesis.

14.3.2 *Asymmetry of risk*

Converting a statistical result back into a biological conclusion is not at all straightforward. The classical null hypothesis method is at its best when testing whether a treatment has an effect, the treatment representing a cost and the effect a benefit. An example might be supplementary feeding to increase the clutch size of a game bird. Here the feeding costs money and time, and we will use it operationally only if an adequate response is clearly demonstrated. First, the null hypothesis must be rejected, and then the difference in response between experimental control and treatment must be evaluated to determine whether the cost of the treatment is justified by the size of the response in fecundity. If the null hypothesis (no effect of treatment) is not rejected, we are simply back where we started and no harm is done. Type 1 and Type 2 errors are both possible; both are inconvenient but neither is catastrophic. A Type 1 error leads to unnecessary expense until the mistake is identified; a Type 2 error results in a small sacrifice in the potential fecundity of the game bird population.

Null hypothesis testing is less effective and efficient when the treatment itself is a benefit and the lack of treatment itself is a cost. Suppose a marine fish stock appears to be declining but there is considerable year-to-year variation in the index of abundance used: catch-per-unit-effort. Further, there are good reasons to suspect that the fishing itself is heavy enough to precipitate a decline. The null hypothesis is that fishing has no effect on population size. In this case, the failure to reject the "no effect" null hypothesis is not sufficient reason to operate on the assumption that the fishing is having no effect. At the very least, one would first want to know something about the power of the test. In this case, the cost of making a Type 2 mistake greatly outweighs the benefit of getting it right. The effect of continuing to fish when one should have stopped could be disastrous and irreversible, whereas unnecessary cessation of fishing results only in a temporary cost until fishing resumes. This is an *asymmetry of risk*. It is particularly prevalent in work on endangered species, where an error can result in extinction. Asymmetry of risk demands conservative interpretation of statistical results.

14.4 The nature of the evidence

14.4.1 *Experimental evidence*

A management treatment may be successful or it may be a failure. If the former, the manager needs to know whether the success flowed from the treatment itself or if it would have happened anyway. Otherwise an expensive and unnecessary management scheme might run indefinitely. If the latter, the manager must first establish that the management failed to achieve its stated aim without doubt and then find out why. Was the failure caused by some extraneous factor that formed no part of the treatment? Was the entire management treatment inappropriate or only a part of it? Would a higher intensity of the treatment have been successful where the lower intensity was not?

To find out, the management must be run as an experiment. There are rules to designing an experiment, which are there for one very important reason: if they are broken, the questions the experiment is designed to answer cannot be answered unambiguously.

Suppose a manager wishes to increase the density of quail in an area by supplementing their supply of food with wheat. How this is done determines whether anything will be learned from the exercise. There is a graded series of approaches ranging from useless, in that they yield no verification of the worth of the treatment, through suggestive, in that their results allow a cautious choice between alternative interpretations, to definitive, in that the results can be interpreted without error.

1 Grain is scattered once a month but density is not monitored. The manager assumes that since the treatment *should* increase the density of quail it *will* increase their density.

This is no test, because the outcome of the treatment is assumed rather than observed.

2 The manager measures density on two occasions separated by 1 year, the first before supplementary feeding is instituted. If the density is higher on the second occasion then the manager might assume that the rise resulted from the feeding.

This is the classic fallacy of "before" being taken as a control on "after." Interpretation of the result rests on the assumption that the density would have remained stable without supplementary feeding, and there is no guarantee of that. It may, for example, have been increasing steadily for several years in response to a progressive and general increase of cover.

3 The manager designates two areas, one on which the birds are fed (the treatment) and another on which they are not (the experimental control). The density of quail is measured before and after supplementary feeding is instituted. If the proportionate increase in density on the treatment area is greater than that on the control area then the difference is ascribed to the effect of feeding.

This design is a radical improvement but still yields ambiguous results: the difference in rate of increase may reflect a difference between the two sites rather than between the two treatments. We say that the effect of site and the effect of treatment are *confounded*. Perhaps the soil of one site was heavy and that of the second light, the vegetation on the two sites therefore reacting differently to heavy winter rainfall and the quail reacting to that difference in plant growth.

Flawed as it is, this design is often the only one available, particularly if the treated area is a national park. In such cases, the control must be chosen with great care to ensure that it is in all important respects similar to the treated area, and the response variable should be monitored on each area for some time before the treatment is instituted to establish that it behaves similarly in the two areas. Another way around this problem is to reverse the two treatments and see if the same result is obtained.

The effect of such local and extraneous influences on the results of the experiment is countered by replication.

4 The manager designates six sites, three treated by feeding and the other three left as controls. The category of each site is determined randomly. Before and after the experiment, measurements of density are made at the same time of year in all six areas. The biological question, "does supplementary feeding affect density?", is translated into a form reflecting the experimental design: "Is the difference in quail density between treatments (feeding versus not feeding) greater than the difference between sites (replicates) within treatments?"

This is an appropriate experimental design in that the outcome provides an unambiguous test of the hypothesis. Its efficiency and precision could be increased in various ways but its logic is right.

The form of an experimental design is dictated by logic rather than by the special requirements of the arithmetic subsequently performed on the data. This is an immensely important point. If the manager has no intention of applying powerful methods of analysis to the data, that in no way sanctions shortcuts in the basic experimental design.

Another common fallacy is the belief that although a logically designed experiment is necessary for publication in a scientific journal, the manager need not bother with all that rigmarole if the only aim is to find out what is going on. The manager might then simply run an "empirical test" like questions 2 or 3 without realizing that the measurements do not reveal what is going on.

14.4.2 *Why replicate the experiment?*

Suppose we wish to determine whether grazing by deer affects the density of a species of grass. The experimental treatment is grazing by a fixed density of deer and the experimental control is an absence of such grazing. We cannot simply apply the two treatments each to a single area, because no two areas are precisely the same. We would not know whether the measured difference in plant density was attributable to the difference in treatment or whether it reflected some intrinsic difference between the two areas. There will always be a measurable difference between areas in the density of any species whatever one does or does not do.

We can postulate that a difference between treated areas is caused by the disparate treatments applied to them only when the difference between the treatments is appreciably greater than the difference within treatments. To determine the scale of variation within the "population" of treatments, we must look at a sample of areas that have received the same treatment. The minimum size of a sample is two. Thus, we must designate at least two areas as grazing treatments and two as controls. A sample of three is better.

The density of a plant species is usually measured within small quadrats scattered over a treatment area. We might measure 50 in each. These 50 quadrats are not replicates: they are subsamples of a single treatment, and their invalid use as "replicates" is called "pseudoreplication." Sampling within a treatment is not treatment replication. Data from such subsamples could be fed into an ANOVA, which would then provide what might appear to be a rigorous test of the hypothesis, but that is an illusion: the arithmetic procedures have been fulfilled but the logic is not satisfied. The result is actually a test of whether the combination of the treatment and the intrinsic characteristics of a single area differ from another treatment combined with the intrinsic characteristics of another area. We say that area and treatment are *confounded*. Their individual effects cannot be disentangled. No strong test of the effect of the treatments themselves is possible unless those treatments are replicated.

Replicates are not meant to be similar. They are meant to sample the natural range of variability. Consequently, one does not look for six similar sites to provide three treatments and three controls: one picks six sites at random. A common excuse offered for a lack of replication in management experiments (and even in research experiments) is that sites similar enough to act as replicates could not be located. Such an excuse is not valid and points to a lack of understanding of the nature of evidence.

These principles carry over to all other forms of comparison. We cannot conclude from two specimens that parrots of a given species have a higher hemoglobin count near the tops of mountains than at lower altitudes. We get no further forward by taking a number of blood samples (subsampling) from the two individuals. Instead, we must test the blood of several parrots from each zone, look at the variation within each group of parrots, and then calculate whether the average difference between groups is greater than the difference within groups. Hence, we must replicate. The arithmetic of such a comparison can be extracted from any book on statistical methods. That is the easy part. The difficult part is getting the logic right.

14.5 Experimental and survey design

Experimental design has its own vocabulary. The thing that we monitor, in this case the density or rate of increase of the quail, is the *response variable*. That which affects the response variable, in this case WHEAT, is a *factor*. In our imaginary experiment, the factor we examined had two *levels*: no supplementary feeding of wheat and some supplementary feeding of wheat (Fig. 14.1). Its levels could have been set at 0, 30, 70, and 250 kg of grain per hectare per month, as in Fig. 14.2. The levels of a factor need not be numbers as in that example, however. The levels of factor HABITAT, for example, might be pine, oak, and grassland. The levels of factor ORDER might be first, second, third, and fourth. The levels of factor SPECIES might be mule deer, white-tailed deer, and elk.

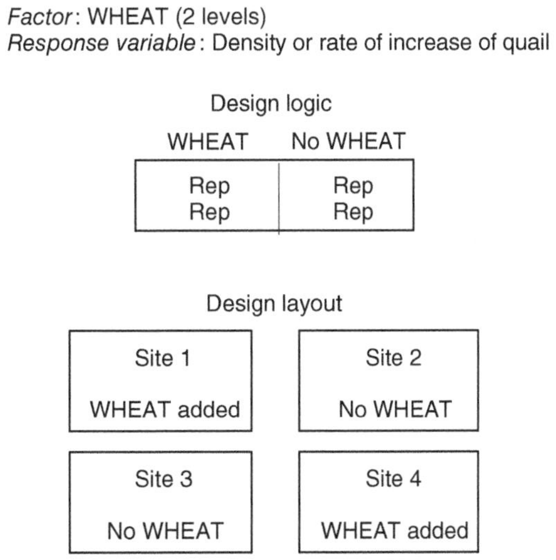

Fig. 14.1 Minimum one-factor experimental design.

Factor: WHEAT (2 levels)
Response variable: Density or rate of increase of quail

Design logic

WHEAT	No WHEAT
Rep Rep	Rep Rep

Design layout

Site 1 WHEAT added	Site 2 No WHEAT
Site 3 No WHEAT	Site 4 WHEAT added

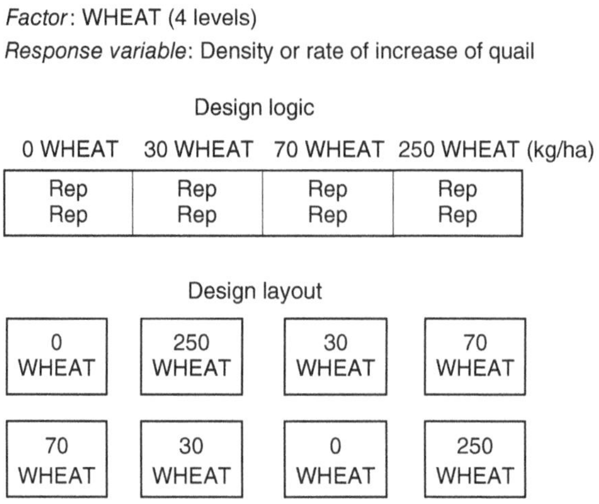

Fig. 14.2 One-factor experimental design where the factor has more than two levels.

Factor: WHEAT (4 levels)

Response variable: Density or rate of increase of quail

Design logic

0 WHEAT   30 WHEAT   70 WHEAT   250 WHEAT (kg/ha)

Rep Rep	Rep Rep	Rep Rep	Rep Rep

Design layout

0 WHEAT	250 WHEAT	30 WHEAT	70 WHEAT

70 WHEAT	30 WHEAT	0 WHEAT	250 WHEAT

Suppose we wished to examine the effects of two management treatments simultaneously. Instead of looking at just the effect of wheat on density of quail, we might also wish to examine the effect of supplying extra salt. There are now two factors: WHEAT and SALT. The questions become:

1 Does WHEAT affect density?
2 Does SALT affect density?
3 Is the effect of WHEAT on density influenced by the level of SALT, and *vice versa*?
In statistical language, question 3 deals with the interaction between the two factors: whether their individual effects on density are additive (i.e. independent of each other) or whether the effect of a level of one factor changes according to which level of the other is combined with it. Section 14.6.3 considers interactions in greater detail.

Fig. 14.3 gives an appropriate experimental design for such a two-factor experiment. Its main features are that each level of the first factor is combined with each level of the second, that there are therefore $2 \times 4 = 8$ cells or treatments, that each treatment is replicated, and that the number of replicates per treatment is the same for all treatments.

14.5.1   *Controls*

A control is that level of a factor subjected to zero treatment. That is not to say that it is necessarily left undisturbed: everything done to the other levels must also be done to the control, other than the manipulation that is formally the focus of the treatment. If vehicles are driven over the quail plots to distribute the grain, they must also be driven over the control areas.

Controls must obviously be appropriate, and often a good deal of thought is needed to ensure that they are. We have previously dealt with the mistake of declaring "before treatment" a control on "after treatment" (see Section 14.4), but there are more subtle traps to keep in mind. If the treatment is an insecticide dissolved in a solvent, the control plots must be sprayed with the solvent minus the insecticide. If treated birds are banded then the control birds must be banded. If animals are removed from the field to the laboratory for treatment and then released, the control animals must also be subjected to that disturbance. And so on.

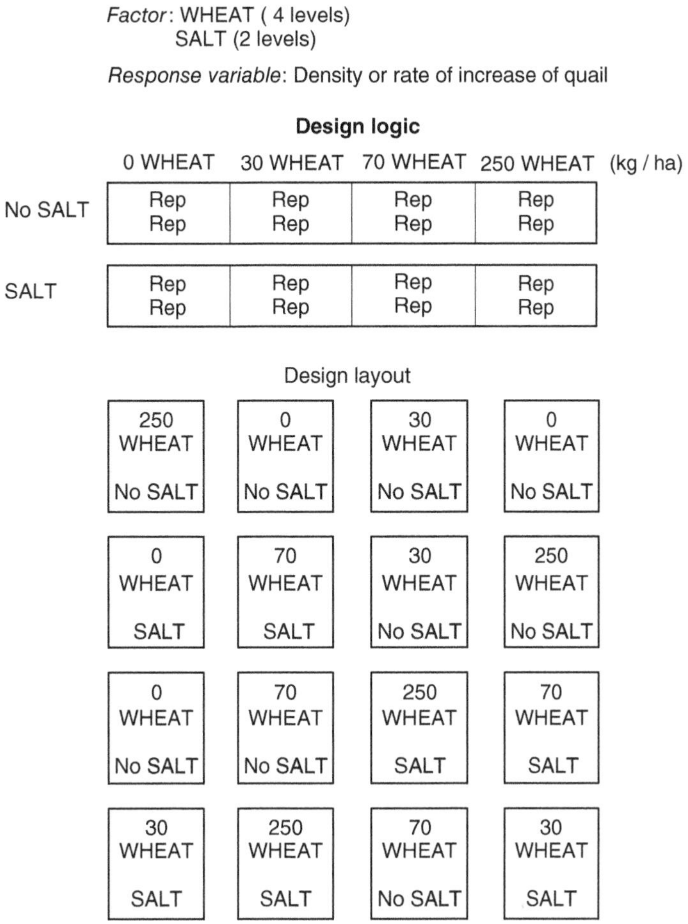

Fig. 14.3 Two-factor experimental design. Do the two factors act independently of each other or do they interact? If the latter, do they reinforce each other or oppose each other?

Factor: WHEAT (4 levels)
SALT (2 levels)

Response variable: Density or rate of increase of quail

Design logic

	0 WHEAT	30 WHEAT	70 WHEAT	250 WHEAT	(kg / ha)
No SALT	Rep Rep	Rep Rep	Rep Rep	Rep Rep	
SALT	Rep Rep	Rep Rep	Rep Rep	Rep Rep	

Design layout

250 WHEAT No SALT	0 WHEAT No SALT	30 WHEAT No SALT	0 WHEAT No SALT
0 WHEAT SALT	70 WHEAT SALT	30 WHEAT No SALT	250 WHEAT No SALT
0 WHEAT No SALT	70 WHEAT No SALT	250 WHEAT SALT	70 WHEAT SALT
30 WHEAT SALT	250 WHEAT SALT	70 WHEAT No SALT	30 WHEAT SALT

14.5.2 *Sample size*

There is no general answer to the question, "How many replicates are necessary?" other than the trite, "At least two per treatment." It depends upon the number of treatments to be compared, the average variance among replicates within treatments, and the magnitude of the differences one expects or is attempting to establish. These may be estimated from a pilot experiment or from a previous experiment in the same area.

As a general rule, however, the fewer the treatments, the more replicates are needed per treatment, but there is little to be gained from increasing replication beyond 30 degrees of freedom for the residual. Suppose the experiment has three factors, with i levels in the first, j in the second, and k in the third. There are thus ijk treatments and $ijk(n-1)$ degrees of freedom in the residual, where n is the number of replicates per treatment.

14.5.3 *Standard experimental designs*

Most questions on the effect of this or that management treatment have a similar logical structure, even if they deal with different animals in different conditions. The most common questions lead to standard experimental designs.

One factor, two levels

Fig. 14.1 represents the simplest design that will provide an answer that can be trusted. It evaluates the operational null hypothesis that supplementary feeding with wheat has no effect on the density (or rate of increase) of quail. What is tested, however, is the statistical null hypothesis that the difference between treatments is not significantly greater than the difference between replicates within treatments. If the experiment rejects that null hypothesis, we accept as highly likely the alternative hypothesis that supplementary feeding affects the dynamics of quail populations living in conditions similar to those of the populations being studied.

This design tests the effect of only one factor (WHEAT) and evaluates it at only two levels (no wheat and some wheat). Note that the diagram of design logic calls for two replicates at each level. The diagram of design layout shows that the treatments are interspersed: we do not have the zero treatments (i.e. controls) bunched together in one region and the wheat-added treatments in another.

One factor, several levels

This design (Fig. 14.2) is similar to the last, with the difference that the effect of supplementary feeding with wheat is evaluated at four levels: 0, 30, 70, and 250 kg/ha of wheat distributed each month. It allows an answer to two questions: whether supplementary feeding has any effect at all upon the density of the quail and whether that effect varies according to the level of supplementary feeding. An answer to the second question allows a cost–benefit analysis of the optimum level of supplementary feeding. Treatment replication and interspersion of treatments is maintained.

Two factors, two or several levels per factor

In this design (Fig. 14.3), the effect of supplementary feeding on quail density is evaluated in tandem with an evaluation of a second factor, the provision of rock salt. Although the two factors could be evaluated by two separate experiments, there are large advantages in combining them within the same one, which provides an answer to a question that might prove to be of considerable importance: do the two factors interact?

Hypothetically, additional salt in the diet of quail might affect their physiology and hence their dynamics, particularly in sodium-depleted areas, and the same may be true of supplementary feeding with wheat. However, suppose that supplementary feeding has an effect only when there is adequate salt in the diet. In such circumstances, two separate experiments would produce the fallacious conclusion that, whereas salt has an effect, wheat has none. The interactive relationship between the two factors would be missed and the resultant management would be inappropriate. One looks for an interaction by calculating whether the effect of the two factors in combination is greater or less than the addition of the two effects when the factors are evaluated separately. This is achieved by ensuring that each level of the first factor is run in combination with each level of the second. The factors are said to be mutually *orthogonal* (at right angles to one another).

The design logic (Fig. 14.3) is seen as a simple extension of the logic of the one-factor design and the design layout continues to adhere to replication and interspersion of treatments. Since there are now eight treatments, each with two replicates, the interspersion of treatments can best be achieved by laying them out either in a systematic manner, as with a Latin square, or, as in the example, by assigning their positions on the ground with random numbers.

14.5.4
Weak-inference designs

Very often a field experiment breaches one or more rules of experimental design and so no longer answers unambiguously the question being posed. Such an occurrence has two causes: an unfortunate mistake or a necessary choice.

Mistakes

Very often there may be no logistical or technical justification for using an inappropriate design. Such a flaw is simply a mistake. One of the most common in ecological and wildlife research is *pseudoreplication* (= subsampling), used under the misapprehension that it constitutes treatment replication (Hurlbert 1984). In this case, site and treatment are confounded (see Section 14.4.1).

A second common mistake is the unbalanced design. Fig. 14.4 illustrates an experiment to evaluate the effect of grazing by sheep and rabbits on the density of a species of grass. There are two factors (SHEEP and RABBITS), each with two levels (presence and absence). "Presence" is taken for sheep as the standard stocking rate and for rabbits as the prevailing density. Variation of rabbit density across the area is taken care of by replication.

The four treatments may be symbolized by a code in which 1 indicates presence and 0 absence. Most of the practical details of setting up such a trial are simple. A rabbit-proof fence around a quadrat excludes both sheep and rabbits (treatment R0 S0). A sheep-proof fence excludes sheep but allows rabbits in (R1 S0). A quadrat to measure the effect of sheep and rabbits together is simply an unfenced square marked by four pegs (R1 S1). Thus, three of the four treatments are easily arranged. They can be set up and then temporarily forgotten, the experimenter returning after several months or even years to harvest the data.

The final treatment (R0 S1) cannot be managed in this way. No one has yet invented a fence that acts as a barrier to rabbits but allows sheep free access to a quadrat. Hence R0 S1 must be handled differently. It requires erecting a rabbit-proof fence around the quadrat to exclude rabbits (as for R0 S0) but keeping sheep at standard stocking density *within* the enclosure. This treatment cannot be set up and then left untended: sheep

Fig. 14.4 Design logic for a two-factor experiment on the effect of sheep and of rabbits on the biomass of pasture.

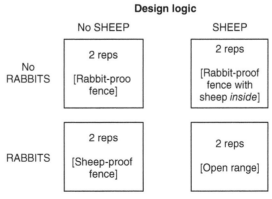

need water and husbandry. Hence, this treatment is often left out of the experiment. The result is a set of data in which the individual effects of sheep on vegetation cannot be disentangled from the effects of rabbits – the total justification for the experiment in the first place.

We have seen many such incomplete experiments set up, often at some expense. They provide estimates of the effect on vegetation of rabbits alone and of sheep and rabbits together, but not of sheep alone. The effect of sheep alone cannot be obtained indirectly by subtraction because that works only where the two effects are additive (i.e. there is no interaction). A significant interaction can quite safely be assumed because each blade of grass eaten by a sheep is no longer available to a rabbit, and *vice versa*.

Necessary compromises

There are a number of problems that involve passive observation of a pattern or process not under the researcher's manipulative control. In these circumstances, a tight experimental design is sometimes not possible, or alternatively the problem may not be open to the classical scientific method. In many fields, such as astronomy, geology, and economics, such problems are the rule rather than the exception. A common example from ecology is the environmental impact assessment (EIA). As Eberhardt and Thomas (1991) put it: "the basic problem in impact studies is that evaluation of the environmental impact of a single installation of, say, a nuclear power station on a river, cannot very well be formulated in the context of the classical agricultural experimental design, since there is only one 'treatment' – the particular power-generating station." In fact, the problem is even more intractable: EIA studies do not test hypotheses. However, EIAs are still necessary. That they generate only weak inference is no good argument against doing them.

Weak inference also results from a second class of problem: where tight experimental design is theoretically possible but not practical. In such circumstances we might have an unbalanced design, poor interspersion of treatments, insufficient replication, or even no replication. Again, the results are not useless but must be treated as what they are: possibilities, which may be confirmed by further research. This brings us into the realm of meta-analysis, where replication of complete studies is the answer (Johnson 2002).

Weak inference is seldom harmful and can be very useful so long as its unreliability is recognized. Weak inference mistaken for strong inference can be ruinously dangerous.

14.6 Some standard analyses

There are several possible analyses available for any given experimental or survey design. Sometimes they give much the same answer and sometimes they produce different ones. The former case reflects only that there is more than one way of doing things; the latter reflects differing assumptions underlying the analyses. Hence, it is important to know what a particular analysis can and cannot do, lest one choose the wrong one. For example, chi-square tests are used only on frequencies (i.e. counts that come as whole numbers); ANOVA can deal both with frequencies and continuous measurements. Student's "t" test is a special case of ANOVA and shares its underlying assumptions.

We will use the ANOVA to introduce a broad class of analyses appropriate for the majority of experimental and survey designs. Any statistical textbook will take this discussion further and present additional analytical options.

14.6.1 *One-factor ANOVA*

The one-factor ANOVA tests the hypothesis that the response variable does not vary with the level of the factor. The alternative hypothesis is that the response variable differs according to the level of the factor, generally increasing or decreasing with its level, going up then down or the reverse, or varying in an unsystematic manner.

Our example (Box 14.1) comprises counts of kangaroos on randomly placed east–west transects, each 90 km in length, on a single degree block in southwest Queensland, Australia. The question of particular concern is whether there is an order effect in terms of days of survey. Did the kangaroos become increasingly disturbed by the aircraft and, therefore, seek cover whenever one was heard? Or did they become progressively habituated to the noise, such that more were seen each day as the survey progressed? The null hypothesis is that the average seen per transect per day is independent of the day order.

Box 14.1 Red kangaroos counted on the Cunnamulla degree block 10 870 km^2 in August 1986. Each replicate is the number of kangaroos counted on a transect measuring 0.4 by 90 km.

Day 1	Day 2	Day 3
96	71	28
38	45	43
80	45	29
35	67	36
50	31	37
55	28	59
38	84	Lost
64	70	Lost

$n_1 = 8$	$n_2 = 8$	$n_3 = 6$
$T_1 = 456$	$T_2 = 441$	$T_3 = 232$
$\bar{x}_1 = 57.0$	$\bar{x}_2 = 55.1$	$\bar{x}_3 = 38.7$

k = number of classes = 3
N = number of samples = $n_1 + n_2 + n_3 = 22$
$\Sigma X_{ij} = 96 + 38 + \ldots + 37 + 59 = 1129$
$\Sigma X_{ij}^2 = 96^2 + 38^2 + \ldots + 37^2 + 59^2 = 66\ 251$
$\Sigma (T_i^2/n_i) = 456^2/8 + 441^2/8 + 232^2/6 = 59273$

Main effects sum of squares (SS):

$$\Sigma (T_i^2/n_i) - (\Sigma X_{ij})^2/N = 59\ 273 - 1129^2/22 = 1335$$

Residual SS:

$$\Sigma X_{ij}^2 - \Sigma(T_i^2/n_i) = 66\ 251 - 59\ 273 = 6978$$

Total SS:

$$\Sigma X_{ij}^2 - (\Sigma X_{ij})^2/N = 66\ 251 - 1129^2/22 = 8313$$

ANOVA

SOURCE	SS	d.f.	MS	F
Main Effect	1335	$k - 1 = 2$	667.5	$\dfrac{667.5}{367} = 1.8$
Residual	6978	$N - k = 19$	367	
Total	8313	$N - 1 = 21$		

$F = 1.8$ with 2 d.f. in the numerator and 19 in the denominator. The probability is 0.19, too high to argue for rejection of the null hypothesis that observable density does not differ by day of survey.

Note that factor DAY contains three levels: the first day, the second day, and the third day. The latter contains only six replicates, in contrast to the eight of the first two days. This makes the point that the arithmetic of one-factor ANOVA does not require that the design be balanced (i.e. that the number of replicates be the same for all levels). The analysis can be run without balance, although the result must be interpreted more cautiously. Balance should always be sought, but it need not necessarily always be attained.

The analysis of Box 14.1 leads to an F ratio testing the null hypothesis. Appendix A gives its critical values. The probability of 20% is too high to call the null hypothesis into serious question. That value is the probability of drawing by chance three daily samples as disparate as or more disparate than those that were drawn, *when there is no difference in density or sightability between days*. We would require a P around 10% before we became suspicious of the null hypothesis, and one below 5% before we rejected it in favor of some alternative explanation.

14.6.2 *Two-factor ANOVA*

A two-factor ANOVA tests simultaneously for an effect of two separate factors on a response variable and for an interaction between them. Even though the arithmetic is simply a generalization of the one-factor case, the two-factor ANOVA differs in kind from the one-factor because of the interaction term. There are also a number of other differences, but we will get to them after we have considered an example.

Data for a two-factor ANOVA are laid out as a two-dimensional matrix, with the rows representing the levels of one factor and the columns the levels of the other. These are interchangeable. Each cell of the matrix contains the replicate readings of the response variable, whatever it is. Table 14.1 outlines symbolically and formally the calculation of the sums of squares and degrees of freedom for the four components

Table 14.1 Calculations of sums of squares for two-factor ANOVA.

ROW effect	$(1/nc) \sum T_i^2 - (1/nrc)T^2$	d.f. $= r - 1$
COLUMN effect	$(1/nr) \sum T_j^2 - (1/nrc)T^2$	d.f. $= c - 1$
ROW × COLUMN effect	$(1/n) \sum T_{ij}^2 - (1/nc) \sum T_i^2 - (1/nr) \sum T_j^2 + (1/nrc)T^2$	d.f. $= (r - 1)(c - 1)$
Residual	$\sum X_{ijk}^2 - (1/n) \sum T_{ij}^2$	d.f. $= rc(n - 1)$
Total	$\sum X_{ijk}^2 - (1/nrc)T^2$	d.f. $= rcn - 1$

T_{ij}, total number of replicates in the cell at the ith row and jth column; T_i, total number of replicates in the ith row; T_j, total number of replicates in the jth column; T, grand total; r, number of rows; c, number of columns; n, number of replicates per cell; d.f., degrees of freedom.

into which the total sum of squares is split: the effect on the response variable of the factor (represented by the rows), the effect of the factor (represented by the columns), the effect of the interaction between them (of which more soon), and the remaining or residual sum of squares, which represents the average intrinsic variation within each treatment cell and which, therefore, is not ascribable to either the factors or their interaction.

Box 14.2 provides a set of data amenable to a two-factor ANOVA. As with the one-factor example, these are real data from an aerial survey whose purpose was to establish whether the counts obtained on a given day were influenced by the disturbance or habituation imparted by the survey flying of previous days. Two species were counted this time, the red kangaroo and the eastern gray kangaroo, and since they might well react in differing ways to the sound of a low-flying aircraft, their counts are kept separate for purposes of analysis. Red kangaroos and eastern gray kangaroos are now the two levels of the factor SPECIES.

Box 14.2 Red kangaroos and grey kangaroos counted on the Cunnamulla degree block $10\,870$ km^2 in June 1987. Each replicate is the number of kangaroos counted on a transect measuring 0.4 by 90 km.

	Day 1	Day 2	Day 3	T_i
Red kangaroos	45	19	18	
	17	51	44	
	8	8	61	
	28	11	35	
	26	72	65	
	48	34	76	
	53	67	52	
	62	27	30	
$T_{ij} =$	287	289	381	957
$\bar{x} =$	35.9	36.1	47.6	
Grey kangaroos	66	27	27	
	52	47	66	
	34	13	75	
	8	16	104	
	35	101	109	
	36	150	170	
	42	116	51	
	65	66	14	
$T_{ij} =$	338	536	616	1490
$\bar{x} =$	42.5	67.0	77.0	
$T_j =$	625	825	997	$T = 2447$

$r =$ number of rows $= 2$
$c =$ number of columns $= 4$
$n =$ number of replicates per cell $= 8$
$(1/nc)\Sigma T_i^2 = (1/24)(957^2 + 1490^2) = 130\ 665$
$(1/nr)\Sigma\ T_j^2 = (1/16)(625^2 + 825^2 + 997^2) = 129\ 079$
$(1/n)\ \Sigma\ T_{ij}^2 = (1/8)(287^2 + 289^2 + \ldots + 616^2) = 136\ 506$
$(1/nrc)T^2 = (1/48)(5\ 957\ 809) = 124\ 746$
$\Sigma X_{ijk}^2 = 45^2 + 17^2 + \ldots + 14^2 = 184\ 081$

Sum of squares

ROW (Species)	130 665 – 124 746	= 5919
COLUMN (Days)	129 079 – 124 746	= 4333
ROW × COLUMN	136 506 – 130 665 – 129 079 + 124 746	= 1508
RESIDUAL	184 081 – 136 506	= 47 575
TOTAL	184 081 – 124 746	= 59 335

ANOVA

SOURCE	SS	d.f.	MS	F
ROW (Species)	5919	$r - 1 = 1$	5919	5.22
COLUMN (Days)	4333	$c - 1 = 2$	2166	1.91
Species × Days	1508	$(r - 1)(c - 1) = 2$	754	0.67
RESIDUAL	47 575	$rc(n - 1) = 42$	1133	
TOTAL	59 335	$rcn - 1 = 47$		

In the ANOVA at the bottom of the box, the sum of squares of each source of variation is divided by the respective degrees of freedom to form a mean square. ("Mean square" is just another name for variance.) The three sources of variance of interest, those of the two factors and that of their interaction, are divided (in this case) by the residual mean square to form the F ratios (named for R.A. Fisher, who invented ANOVA) that are our test statistics. The one for SPECIES is 5.22, and we check that for significance by looking up the F table in Appendix A, which shows that an F with 1 degree of freedom in the numerator and 42 (say 40) in the denominator will have to exceed 4.08 if the magic 5% or lower probability is to be attained. We therefore conclude that the disparity in observed numbers between reds and grays, 957 versus 1490, is more than a quirk of sampling – that gray kangaroos really were more numerous than reds on the Cunnamulla block at the time of survey.

In like manner, we test for a day effect. The trend in day totals −625, 825, and 997 kangaroos – suggests that the animals are becoming habituated to aircraft noise and hence are becoming progressively more visible day by day. However, the F tables show that with degrees of freedom of 2 and 42, a one-tail probability of 5% or better would require F = 3.23. Ours reached only 1.91, equivalent to a probability of 16%, so we are not tempted to replace our null hypothesis (no day effect) with the alternative explanation suggested by a superficial look at the data.

The F ratio for interaction was less than 1, indicating that the mean square for interaction is less than the residual mean square. It cannot, therefore, be significant, and we do not even bother to look up the probability associated with it.

14.6.3 *What is an interaction?*

In the last section we tested an interaction, and found it nonsignificant, without really exploring what question we were answering. A nonsignificant interaction implies that the effect of one factor on the response variable is independent of any effect that might be exerted on it by the other factor: that the two factors are each operating alone.

The effect of the two factors acting together is exactly the addition of the effects of the two factors each acting in the absence of the other.

If an analysis produces a significant interaction, the relationship should be examined by graphing the response variable against the levels of the first factor. Fig. 14.5 shows the kinds of trend most commonly encountered. A significant interaction shows that no statement can be made as to the effect on the response variable of a particular level of the first factor unless we know the prevailing level of the second. The graph will make this clear.

It is entirely possible for an ANOVA to reveal no main effect of the first factor, no main effect of the second, but a massive interaction between them. A graphing of the response variable will reveal a crossing-over pattern, as in the last graph of Fig. 14.5.

Fig. 14.5 Common forms of interaction in two-factor ANOVAs. ns, not significant; sig, significant.

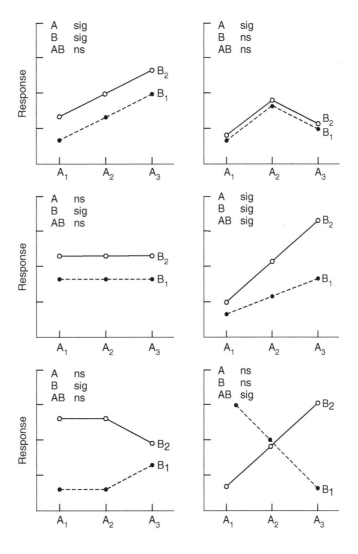

14.6.4
Heterogeneity of variance

The main assumption underlying ANOVA is that the variance of the response variable is constant across treatments. The means may differ (and that is in fact what we are testing to discover) but the variances remain the same. A violation of this assumption can seriously bias the test. Consequently, we need to test for heterogeneity of variance and, if we find it, either transform the data to render the variances homogeneous or use an alternative method such as analysis of deviance that does not employ the assumption of homogeneity.

The most common test for heterogeneity of variance is Bartlett's. It can be found in almost all statistical tests. Recent work has shown however that it is too sensitive. ANOVA is an immensely robust test that performs well even when the assumptions of the analysis are not met in full. It copes well with minor heterogeneity of variance and with deviations from normality. About the only thing that throws it out badly is bimodality of the response variable. A better test is Cochran's C, the test statistic of which is simply the largest variance in a cell divided by the sum of all cell variances. For the two-factor ANOVA given in Box 14.2, the largest cell variance is returned by the replicate counts of gray kangaroos on day 2: $s^2 = 2566$. The sum of all six variances is $\sum s^2 = 6796.6$, so Cochran's C = $2566/6796.6 = 0.378$. Looking up a table of the critical values of Cochran's C (Appendix B) reveals that the test statistic would have to exceed C = 0.398 (seven degrees of freedom (d.f.) per variance, with six variances) to represent a significant departure from homogeneity of variance. We can thus choose to analyze without transformation.

In many biological cases, the variance of the response variable rises with the mean. This is particularly true of counts of animals that tend to fit a negative binomial distribution. A transformation of the counts to logarithms after adding 0.1 (to knock out the zeros) will usually stabilize the variances. If the animals are solitary rather than gregarious, their counts are more likely to fit a Poisson distribution, which is characterized by the mean and the variance being identical. Transformation to square roots after adding three-eighths to each will homogenize the variances. The variances of almost all body measurements regress on their means, and the data require log transformation. An arcsine transformation should be used if the data come as percentages.

14.6.5 *Are the factors fixed or random?*

Before any data are collected, let alone analyzed, we must decide precisely what questions are to be asked of them. Take the example of comparing counts of kangaroos on successive days. We were asking whether the act of flying over the study area on one day influenced the counts obtained the next. The influence could conceivably be negative (disturbance forces the kangaroos into cover before the surveying aircraft) or positive (the kangaroos become progressively habituated to aircraft noise).

Now take another question: do viewing conditions differ between days of survey? This would be answered by counts obtained by days sampled at random. We would want those days to be spaced rather than consecutive, as they were in answering the question about the effect of day order. Otherwise the answers to the two questions would be confounded and we would not know which was being answered by a significant day effect. In the question concerning an order effect of days, the factor DAY is said to be fixed. No arbitrary selection of any three days will do. The days have to follow each other, without gaps between them.

Whether a factor is declared fixed or random determines both the question being asked and the denominator of the F ratio that answers it. Table 14.2 shows the

Table 14.2 Mean square providing the denominator for an F ratio testing the significance of a factor or interaction (i.e. source of variation).

Model	Source	Denominator of F
One-factor ANOVA		
A fixed	A	MS_e
A random	A	MS_e
Two-factor ANOVA		
A fixed, B fixed	A	MS_e
	B	MS_e
	AB	MS_e
A fixed, B random	A	MS_{AB}
	B	MS_e
	AB	MS_e
A random, B random	A	MS_{AB}
	B	MS_{AB}
	AB	MS_e
Three-factor ANOVA		
A fixed, B fixed, C fixed	A	MS_e
	B	MS_e
	C	MS_e
	AB	MS_e
	AC	MS_e
	BC	MS_e
	ABC	MS_e
A fixed, B fixed, C random	A	MS_{AC}
	B	MS_{BC}
	C	MS_{ABC}
	AB	MS_e
	AC	MS_e
	BC	MS_e
	ABC	MS_e
A fixed, B random, C random	A	XXX
	B	MS_{BC}
	C	MS_{BC}
	AB	MS_{ABC}
	AC	MS_{ABC}
	BC	MS_e
	ABC	MS_e
A random, B random, C random	A	XXX
	B	XXX
	C	XXX
	AB	MS_{ABC}
	AC	MS_{ABC}
	BC	MS_{ABC}
	ABC	MS_e

XXX, no explicit test possible; MS_e, mean square of residual.

appropriate choice of denominator. For two of the three-factor models there is no explicit test for the significance of some of the factors. Various messy approximations are available (see Zar 1996, appendix), but it is far better to rephrase the question to one logically answerable from consideration of the data.

Let us generalize the difference between fixed and random factors. A fixed factor is one whose levels cover exhaustively the range of interest. MONTHS therefore usually constitute a fixed factor, because they index seasons. YEARS may be fixed or random depending on context. REGIONS may be fixed if its levels are the only ones of interest; if they are simply a random sample of regions, and any other selection of regions would serve as well, the factor REGIONS is random.

Note that questions change according to whether the factor is fixed or random. Suppose that the response variable is the growth rate of a species of pine and we wish to test for a difference among three soil types (factor SOILS) covering the entire range of soil types of interest. The factor SOILS will thus be fixed. If the question concerns the best region in which to plant a plantation of that species, and there are four and only four regions that are possible candidates, the choice of regions is fixed and the appropriate denominator of the F testing a difference of growth rate among the soil types of those regions will be the residual mean square. However, if we ask the more general question of whether grow rate differs among soil types across regions in general, any random selection of a set of regions will suffice and the denominator of the F testing SOILS will be the mean square of the interaction between SOILS and REGIONS (Table 14.2).

14.6.6 Three-factor ANOVA

The three-factor ANOVA follows the general lines of the two-factor except that seven rather than three questions are addressed simultaneously. Table 14.3 gives the sums of squares and degrees of freedom.

Box 14.3 gives an example, again of kangaroos counted by aerial survey, but with factor YEARS added. YEARS is fixed because we wish to test specifically whether density changed between 1987 and 1988; we do not wish to test the more general question of whether kangaroo populations remained stable with time. The ANOVA shows that of the main effects, the three first-order interactions and the one second-order interaction, only the main effect of kangaroo species is significant. We conclude therefore that the two species certainly differed in density but that there was insufficient evidence to identify a day effect or a change in density between years. Neither did any factor appear to interact with any other.

Table 14.3 Calculation of sums of squares and degrees of freedom for three-factor ANOVA.

ROW effect	$(1/ncl) \sum T_i^2 - (1/nrcl)T^2$	d.f. $= r - 1$
COLUMN effect	$(1/nrl) \sum T_j^2 - (1/nrcl)T^2$	d.f. $= c-1$
LAYER effect	$(1/nrc) \sum T_k^2 - (1/nrcl)T^2$	d.f. $= l-1$
RC interaction	$(1/nl) \sum T_{ij}^2 - (1/ncl) \sum T_i^2 - (1/nrl) \sum T_j^2 + (1/nrcl)T^2$	d.f. $= (r-1)(c-1)$
CL interaction	$(1/nr) \sum T_{jk}^2 - (1/nrl) \sum T_j^2 - (1/nrc) \sum T_k^2 + (1/nrcl)T^2$	d.f. $= (c-1)(l-1)$
RL interaction	$(1/nc) \sum T_{ik}^2 - (1/ncl) \sum T_i^2 - (1/nrc) \sum T_k^2 + (1/nrcl)T^2$	d.f. $= (r-1)(l-1)$
RCL interaction	$(1/n) \sum T_{ijk}^2 - (1/nl) \sum T_{ij}^2 - (1/nr) \sum T_{jk}^2 - (1/nc) \sum T_{ik}^2 + (1/ncl) \sum T_i^2 + (1/nrl) \sum T_j^2 + (1/nrc) \sum T_k^2 - (1/nrcl)T^2$	d.f. $= (r-1)(c-1)(l-1)$
RESIDUAL	$\sum X_{ijkm}^2 - (1/n) \sum T_{ijk}^2$	d.f. $= (n-1)rcl$
TOTAL	$\sum X_{ijkm}^2 - (1/nrcl)T^2$	d.f. $= nrcl - 1$

d.f., degrees of freedom.

Box 14.3 Red kangaroos and grey kangaroos counted on the Cunnamulla degree block 10 870 km² in 1987 and 1988. Each replicate is the number of kangaroos counted on a transect measuring 0.4 by 90 km.

		June 1987			June 1988	
	Day 1	Day 2	Day 3	Day 1	Day 2	Day 3
Red kangaroos	45	19	18	72	31	28
	17	51	44	32	29	9
	8	8	61	94	34	48
	28	11	35	27	47	38
	26	72	65	66	21	91
	48	34	76	55	49	138
	53	67	52	102	29	67
	62	27	30	41	67	106
$T_{ijk} =$	287	289	381	489	307	525
$\bar{x} =$	35.9	36.1	47.6	61.1	38.4	65.6
Grey kangaroos	66	27	27	116	65	10
	52	47	66	81	57	22
	34	13	75	63	41	43
	8	16	104	25	33	63
	35	101	109	75	120	74
	36	150	170	77	28	101
	42	116	51	59	62	76
	65	66	14	59	17	82
$T_{ijk} =$	338	536	616	555	423	471
$\bar{x} =$	42.3	67.0	77.0	69.4	52.9	58.9

ANOVA

SOURCE	SS	d.f.	MS	F	$F_{0.05}$*
ROW (Species)	4551.3	1	4551.3	4.7	4.0
COLUMN (Days)	3227.3	2	1613.6	1.6	3.1
LAYER (Years)	1086.8	1	1086.8	1.1	4.0
RC interaction	1018.1	2	509.0	0.5	3.1
CL interaction	4681.6	2	2340.8	2.3	3.1
RL interaction	1708.6	1	1708.6	1.7	4.0
RCL interaction	1444.7	2	722.3	0.7	3.1
RESIDUAL	86 359.4	84	1028.1		
TOTAL	104 077.7	95			

*$F_{0.05}$ is the critical level that must be exceeded by the observed F to qualify for a probability equal to or less than 5%.

14.7 Summary

Testing of wildlife management treatments requires rigorous definition of the expected outcome of the treatments. Once a verifiable outcome is posed as a hypothesis, the data to test it can be collected by following the logic of experimental design. Insufficient replication of treatments to sample the range of natural variability is a common shortcut but nullifies the point of the exercise.

The principles illustrated in this chapter can be summarized as the basic rules of experimental design. There are exceptions to several of them, but until the manager or researcher learns how and in what circumstances they may safely be varied they should be followed in full.

1 To determine whether a factor affects the response variable under study, more than one level of that factor should be examined. The levels may be zero (control) and some nonzero amount, or they may be two or more categories (e.g. habitat types) or nonzero quantities (e.g. altitudinal bands).

2 "Before" is a poor control for "after," because subsequent trends can be caused by other influences unrelated to the treatment under study.

3 Treatments must be replicated, not subsampled. (See Hurlbert 1984 for an excellent exposition of the pitfalls of "pseudoreplication" in ecological research.)

4 The number of replications per treatment (including the control treatment) should be as close as possible to equal across treatments.

5 Treatments must be interspersed in time and space. Do not run the replications of Treatment A and then the replications of Treatment B. Mix up the order. Do not site the replications of Treatment A in the north of the study region and the replications of Treatment B in the south. Mix them up.

6 If the influence of more than one factor is of interest, each level of each factor should be examined in combination with each level of each other factor (factorial design).

7 If an extraneous influence (site in the quail example) is likely to be correlated with one of the designated factors, either it should be declared a factor in its own right and the design modified accordingly or its range should be covered at random by the replication. Its influence is factorized out in the first option and randomized across treatments in the second.

8 All these rules may be broken, but this degrades the design to one yielding neither strong inference nor an unambiguous conclusion. Such results are still useful so long as their dubious nature is fully appreciated and declared. EIAs are an example of this.

15 Model evaluation and adaptive management

15.1 Introduction

In ecology, a *model* is a hypothesis that is usually expressed mathematically. A mathematical description allows a more precise definition of the hypothesis than does a verbal description, and this precision is particularly important for complex, nonlinear processes. We can often construct more than one model to describe a process, and these alternatives are equivalent to alternative hypotheses. In this chapter, we explore the methods for choosing between such alternative models or hypotheses.

In Chapter 14 we introduced the concept of statistical inference, which uses standardized criteria for decision-making, to help ensure that decisions are not swayed by personal opinion or pressure brought to bear by politicians or the public. Despite its widespread use, however, statistical inference is not the only, nor even necessarily the best, way to choose among a wide variety of alternate hypotheses, whether these arise in the quest for "pure" or more "applied" knowledge (Johnson 1999; Anderson *et al.* 2000; Guthery *et al.* 2001; Johnson and Omland 2004). Statistical tests are effective at ruling out null hypotheses. The trouble is, the null hypothesis is sometimes an explanation that we need not seriously entertain, so rejecting it is not helpful in increasing our understanding of observations. For example, the null hypothesis in many wildlife habitat studies is that animals have no habitat preferences. We would be astounded if this ever proved true, so what progress do we make in rejecting it?

There are far fewer examples of hypothesis testing that are directed at evaluating a suite of alternative models or hypotheses that vary subtly from one another. It is hard enough to gather enough data to discriminate between random versus "significant" patterns of association, let alone tease apart subtle variants. More importantly, however, classic statistical methods are often impossible to use when alternative models are not special cases of more general ones. This situation is particularly common in the kind of nonlinear models that we find in ecology. Such "non-nested" models, in the jargon of professional statisticians, present special problems for finding a suitable statistical test.

Statistical inference is also plagued by "statistical" versus "biological" significance. Recall that in statistical hypothesis testing, a P value of less than 0.05 is taken to mean that there is a remote probability (1 in 20) that an observed pattern could have been produced by chance alone. This probability is quite sensitive, however, to the amount of data that go into the assessment. Endangered species are often plagued by a crucial lack of data. This can preclude sufficient sample sizes and replicated treatments needed for significance testing, leaving us with no reliable option for making management

Wildlife Ecology, Conservation, and Management, Third Edition.
John M. Fryxell, Anthony R. E. Sinclair and Graeme Caughley.
© 2014 John Wiley & Sons, Ltd. Published 2014 by John Wiley & Sons, Ltd.
Companion Website: www.wiley.com/go/Fryxell/Wildlife

decisions or informed scientific judgments if we rely on standard statistical approaches. Even when we have a large sample of data that yields a statistically significant result (i.e. a *P* value less than 0.05), this may be of trivial biological significance. Hence, slavish adherence to statistical significance alone can distract one from the real issues at hand. It is often much more important to decide which factor has the strongest effect on the pattern or variable of interest.

Recognizing these limitations, ecologists have developed an alternative branch of statistics that allows them to evaluate a series of alternative models of the same phenomena (Hilborn and Mangel 1997; Burnham and Anderson 1998; Anderson *et al.* 2000; Johnson and Omland 2004). The philosophical spirit of this approach, called *model evaluation*, is not so much to discriminate significant from insignificant factors, but rather to decide which of many competing explanations is *most consistent* with the facts at hand, so that one can make an informed judgment about the best course of management action.

15.2 Fitting models to data and estimation of parameters

While model fitting can be readily applied to experimental data (e.g. Hobbs *et al.* 2003), it is applied most frequently to the evaluation of observational data gathered routinely by many wildlife agencies, such as mark–recapture data (Jorgenson *et al.* 1997), spatial distribution data (Fryxell *et al.* 2004), or annual censuses of abundance (Hebblewhite *et al.* 2002; Taper and Gogan 2002). Early in the exercise of model fitting, one should ensure that no major changes have occurred over time in the way that data have been gathered, or if they have, that some method exists by which to directly compare data gathered at different points in time. Discrepancies in the way data are gathered are more frequent than one would like. Biologists are constantly tempted to modify observational techniques, to improve either the ease of observation or the repeatability of observations. Such changes in methodology can be a good idea, but they can make it difficult to compare data from different eras. Methodological changes are often neglected when someone (e.g. a consultant) decides to analyze the cumulative historical data. For this reason, any analysis team ought to include at least one person familiar with the original data, ideally someone who has gathered the data themselves.

As an example of a long-term dataset set we shall consider the census data on migratory wildebeest in Serengeti National Park, illustrated in Fig. 5.11. Population estimates in this system date back to the early 1960s, when the Serengeti Research Institute was first established. It was recognized early that aerial counting was perhaps the best way to monitor the broad expanses of savanna grasslands and broadleaf woodlands that make up the park. Counting methods were established early and there has been little deviation over the years, despite changes in observers and technological changes in aircraft and navigation equipment. Serengeti wildlife ecologists used a stratified aerial count design, with photographs taken at known altitude used to count individuals in the center of wildebeest aggregations and visual counts made in areas with lower numbers of animals.

The results of these censuses over the past 40 years show a rapid increase in wildebeest abundance over the first 20 years and a subsequent leveling off and erratic fluctuation around roughly 1.25 million individuals. There are numerous ways one could mathematically depict this general pattern, however, and they all have different management implications with respect to the long-term viability of the wildebeest population. Which model is most consistent with the data? We'll use a formal model evaluation, based on information criteria, to find out.

There are a great many ways that we can interpret the underlying causes of this population growth. Ecological interactions among wildebeest, other large mammals, and the rest of the environment could influence the patterns of wildebeest population growth and natural regulation. We can begin by determining how fast the population is growing. A useful way to do this is to convert the census data into estimates of the exponential rate of increase r between sequential population estimates. We recall from Chapter 5 that the growth rate for a population can be expressed either as finite growth rate ($\lambda = N_{t+1}/N_t$) or as its exponential equivalent ($r = \ln(N_{t+1}/N_t)$). The exponential growth rate is especially convenient when population censuses or estimates are timed irregularly, rather than occurring every year, because it can be readily translated into shorter or longer time intervals by simple multiplication or division operations. We calculate the natural log of the ratio of subsequent to initial population abundance for each time interval and divide this ratio by the number of years between successive population estimates τ:

$$r_t = \ln\left(\frac{N_{t+\tau}}{N_t}\right)/\tau$$

We encourage you to check the procedure by calculating the first two or three estimates of r by hand.

Why do we need to divide by τ? In many cases, we will not have annual data to work from. We can handle irregular timing of censuses by dividing by the number of years between them, τ. Table 15.1 shows the result of such calculations for Serengeti wildebeest. Once values of r have been calculated, we can readily translate back into values of λ by exponentiation: $\lambda = e^r$.

The next step is to fit a mathematical relationship to the multiple estimates of r recorded over time. The accepted convention in such matters is to find a mathematical model whose values best fit the existing data, where "best fit" means to minimize the sum of squared deviations between the model estimates and the observed data. The recorded estimates for Serengeti wildebeest certainly seem to decline with increasing wildebeest abundance, so one obvious model candidate would be a linear decline in r

Table 15.1 Instantaneous rate of increase calculated for the Serengeti wildebeest population.

Year	$N \times 1000$	Rate of increase r
1958	190	0.108835146
1961	263.262	0.130874700
1963	356.124	0.104751424
1965	439.124	0.047919596
1967	483.292	0.090021783
1971	629.777	0.109589002
1972	773.014	0.124420247
1977	1440.000	−0.142352726
1978	1248.934	0.03443494
1980	1337.979	−0.050802822
1982	1208.711	0.050754299
1984	1337.849	−0.077244421
1986	1146.340	0.012747402
1991	1221.783	−0.095578878
1994	917.204	0.070080128
1998	1165.908	0.110475091
1999	1302.096	−0.044661447

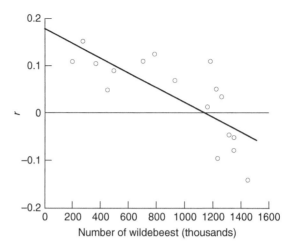

Fig. 15.1 Predicted (line) and observed (circles) exponential rates of increase shown by Serengeti wildebeest in relation to population density.

with N. In mathematical terms, this means we postulate the linear model $r(i) = a + b *$ $N(i)$, where a is the intercept of the straight line and b is the slope of the relationship between N and r in any particular census i. These values can be estimated from any linear regression package. We then convert the slope and intercept into the parameters of the Ricker logistic model, remembering that the intercept $= r_{max}$ and the slope $= -r_{max}/K$, yielding the following parameter estimates: $r_{max} = 0.18$ and $K = 1\,142\,000$.

We can plot the observed value of r for each time interval versus population abundance at the beginning of the interval and then overlay these values on those predicted by the linear model whose parameters we have just estimated (Fig. 15.1). The linear model seems to do a reasonable job of predicting the rate of population growth by Serengeti wildebeest. The consistent tendency for deviation below the regression line at either very low or very high wildebeest densities, coupled with deviation above the regression line at intermediate densities, suggests that a curve might fit these data even better.

15.3 Measuring the likelihood of the observed data

We now consider how we choose between several possible models that could represent our observations, in this case the trend in wildebeest numbers. We have illustrated one model of trend, the straight line, and observed that a curve might be a better model. However, there could be several types of curve. Therefore, we need to decide how "likely" each of several alternate mathematical models might be, based on their ability to explain the census data obtained for the Serengeti wildebeest. First, we describe the distribution of residual variability around the postulated relationships (i.e. models). In the case of the Serengeti data, there are 17 estimates of r_i (a sample size that we will call n). For these n estimates of r, we can calculate the residual variance around the regression line σ^2:

$$\sigma^2 = \frac{1}{n} \sum_{i=1}^{n} \left(r_i - r_{max} \left(1 - \frac{N_i}{K} \right) \right)^2$$

The standard deviation of these residuals ($\sigma = 0.055$) can then be calculated by taking the square root of the residual variance (see Chapter 5).

In our example (Fig. 15.2), the residuals seem to be reasonably well depicted by a normal distribution, although the sample size is too small to be totally sure. On the

Fig. 15.2 Frequency distribution of residuals around the Ricker logistic regression of r versus N (histogram) and a normal probability curve with the same mean and variance plotted against these data.

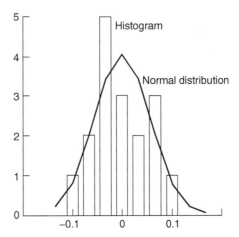

presumption that the residuals do follow a normal distribution with mean of zero and $\sigma = 0.055$, we can assess how well each model explains the existing data through use of the concept of likelihood (Λ). For example, we could use this approach to evaluate the likelihood that the carrying capacity of Serengeti wildebeest is a particular value of K:

$$\Lambda(K) = \prod_{i=1}^{n} \frac{1}{\sigma\sqrt{2\pi}} \exp\left(\frac{-\left(r_i - r_{max}\left(1 - \frac{N_i}{K}\right)\right)^2}{2\sigma^2}\right)$$

Likelihood is proportional to the probability of obtaining a particular set of observed data given that population dynamics are dictated by a given model. Likelihood is calculated from the probability function defining the residual variability that affects each estimate (in this case, the normal distribution) and the value predicted by the model. Hence, we can use the equation defining the normal distribution, which we obtain from any statistics textbook, with an expected value for each observation derived from the Ricker logistic model.

Because likelihoods are often very small or very large numbers, it is customary to evaluate their negative natural log-transformed values (termed the *negative log-likelihood*). This transforms the function from a "dome" to a "valley" shape, where the most likely parameter is the value that is at the bottom of the valley (Fig. 15.3).

$$L(K) = n\left(\ln(\sigma) + \frac{1}{2}\ln(2\pi)\right) + \sum_{i=1}^{n} \frac{\left(r_i - r_{max}\left(1 - \frac{N_i}{K}\right)\right)^2}{2\sigma^2}$$

The likelihood profile shown in Fig. 15.3 is useful for demonstrating how the negative log-likelihood will change with different parameter values. We see that the value of K that minimizes the negative log-likelihood is the value identified using least-squares regression. The minimum log-likelihood is the maximum likelihood, so we refer to the "best-fit" value of K as a maximum-likelihood estimate. Just as we can use likelihood to evaluate the most probable parameter values, in analogous fashion we can use it to assess the plausibility of alternative models in explaining observed data.

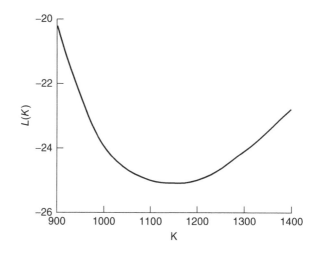

Fig. 15.3 Negative log-likelihood for a range of values of K in the Ricker logistic model applied to the wildebeest population in a Serengeti ecosystem.

15.4 Evaluating the likelihood of alternate models using AIC

We now want to decide which of numerous mathematical expressions best represents the data we have at hand. The most general approach that can be applied to non-nested models is Akaike's information criterion (AIC) or one if its variants (Akaike 1973; Burnham and Anderson 1998). An AIC score can be used to evaluate the plausibility of a given model for predicting a set of experimental or field data. For further details, the interested reader should consult the comprehensive treatise on AIC by Burnham and Anderson (1998). Perhaps the easiest way to introduce the procedure is to apply it to our data for Serengeti wildebeest.

15.4.1 Ricker logistic model

The Ricker logistic model that we have fitted to the Serengeti wildebeest data (which we will call model 1) has three parameters, as we have estimated r_{max}, K, and the standard deviation of the residuals around the linear regression line. We evaluate the plausibility of the first model by calculating the AIC score based on its negative log-likelihood L_1, sample size n, and number of parameters (in this case $p_1 = 3$):

$$AIC_1 = 2L_1 + 2p_1 \left(\frac{n}{n - p_1 - 1} \right)$$

Applying this formula to the Ricker logistic model fitted to the Serengeti wildebeest data, $AIC_1 = -42.35$. Note that the equation used here is appropriate for the small sample sizes that are all too typical in wildlife studies. For much larger data sets (where $n > 40$), one could use a simpler formula: $AIC = 2L + 2p$.

AIC values become smaller with increasing likelihood but larger with increased number of parameters. One would ordinarily expect that the likelihood of a model would go up when a larger number of parameters was available to model the data. On the other hand, the AIC approach applies an increasing penalty for each parameter in the model. A model with a low AIC score is considered plausible because it explains a great deal of the variation in the data but doesn't require a great many parameters to do so. In other words, a low AIC score signals a parsimonious tradeoff between explanatory power and the complexity of the mathematical model needed to capture the pattern in the data. By itself, the AIC score isn't very useful – it only has meaning relative to AIC scores arising from other plausible models. We now consider two such candidates.

15.4.2 *Theta logistic model*   The theta logistic model is appropriate for populations that have a threshold curvilinear relationship between r and N, with population growth predicted by the following model:

$$N_{t+1} = N_t \exp\left(r_{max}\left(1 - \frac{N_t}{K}\right)^{\theta}\right)$$

Predictions of the theta logistic model depend on the magnitude of θ. For $\theta > 1$, growth rates change little with density when N is modest, but the density-dependent response becomes much steeper as N becomes large. The opposite is true when $\theta > 1$. For $\theta > 1$, the intensity of density-dependent processes becomes disproportionately severe at high population densities. If there is a threshold effect, the theta logistic might be a plausible model. Exponential growth rates are a nonlinear function of N, with the precise pattern dictated by three parameters: r_{max}, K, and θ. Many statistical programs have a routine to estimate the parameters for nonlinear relationships such as this. In the case of Serengeti wildebeest, the best-fit parameter estimates are $r_{max} = 0.105$, $K = 1\,241\,000$, and $\theta = 5.946$.

With these values we can plot the theta logistic curve against the data (Fig. 15.4). This certainly seems to better capture the rapid decline in growth rates that occurred as the wildebeest population rapidly grew in the 1960s. To evaluate the theta logistic model's likelihood for Serengeti wildebeest, we first need to calculate the residual variance around the nonlinear growth function:

$$\sigma_2{}^2 = \frac{1}{n}\sum_{i=1}^{n}\left(r_i - r_{max}\left(1 - \frac{N_i}{K}\right)^{\theta}\right)^2$$

Taking the square root, $\sigma_2 = 0.04$. The negative log-likelihood function for the theta logistic model is calculated in a similar manner to that of the Ricker model, except we

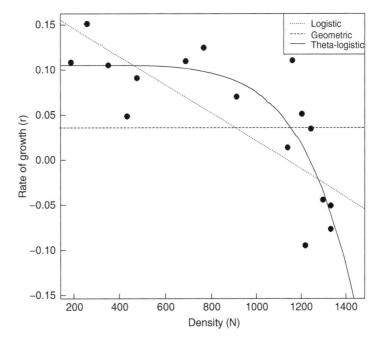

Fig. 15.4 Observed (symbols) versus predicted exponential rates of increase by Serengeti wildebeest in relation to population density, based on the geometric growth model (dashed line), Ricker logistic model (dotted line), and theta logistic model (solid line).

modify the expected value and the standard deviation of the residuals:

$$L_2 = n \left(\ln \left(\sigma_2 \right) + \frac{1}{2} \ln(2\pi) \right) + \sum_{i=1}^{n} \frac{\left(r_i - r_{max} \left(1 - \frac{N_i}{K} \right)^{\theta} \right)^2}{2\sigma_2^2}$$

Note that the theta logistic model has four parameters (r_{max}, K, θ, and the standard deviation of the residuals around the linear regression line), which are necessary for the more complex, nonlinear model. Its AIC score is calculated as -49.57, a bit smaller than the score for the Ricker logistic model, which was -42.35. This implies that the theta logistic is a more plausible model than the Ricker logistic, because it offers a more parsimonious explanation of the observed changes in wildebeest numbers over time.

15.4.3 *Geometric growth model*

As our final model, we consider an old standby, the geometric growth model (see Chapter 5). This model assumes that there is no change in growth rates as the population increases over time (Fig. 15.4). The residual variation around the mean value (r_{max}) is calculated as follows:

$$\sigma_3^2 = \frac{1}{n} \sum_{i=1}^{n} (r_i - r_{max})^2$$

Taking the square root of the residual variance, $\sigma_3 = 0.085$. The negative log-likelihood function for the geometric model is calculated in a similar manner as for the Ricker and theta logistic models, except we modify the expected value and the standard deviation of the residuals:

$$L_3 = n \left(\ln \left(\sigma_3 \right) + \frac{1}{2} \ln(2\pi) \right) + \sum_{i=1}^{n} \frac{(r_i - r_{max})^2}{2\sigma_3^2}$$

Note that the geometric model has the least number of parameters (r_{max} and the standard deviation of the residuals around the linear regression line, so $p_3 = 2$). Its AIC score is calculated as -30.69, considerably larger than the scores for the Ricker and the theta logistic models. This implies that the geometric model doesn't offer a very parsimonious explanation of the observed changes in wildebeest numbers over time, in keeping with its poor fit to the data in Fig. 15.4.

The model with the lowest AIC score is taken as the most plausible, because smaller scores imply a more parsimonious fit to the data. We can evaluate the merit of each model relative to the best model using the difference between their AIC scores ($\Delta AIC_i = $ AIC score of model $i - $ AIC score of the best model). Differences in AIC of less than 2 ($\Delta AIC < 2$) are considered trivial by statistical experts (Burnham and Anderson 1998).

On the basis of this model evaluation, one would conclude that the theta logistic model offers the most parsimonious explanation for the observed changes in Serengeti wildebeest abundance recorded over the past 40 years, followed by the Ricker logistic model ($\Delta AIC = 7.2$), and then the geometric growth model ($\Delta AIC = 18.9$).

We can more formally assess the likelihood of each of the competing models by calculating their Akaike weights. The Akaike weight for a given model i is calculated by dividing the likelihood of that model ($\exp(-\Delta AIC_i/2)$) by the summed likelihoods of all of the competing models $\left(\sum \exp \left(-\Delta AIC_j/2 \right) \right)$:

$$w_i = \frac{\exp(-\Delta AIC_i)}{\sum\limits_{j=1}^{3} \exp(-\Delta AIC_i)}$$

The Akaike weight for the theta logistic model is 0.99, whereas the Akaike weights of all other models are less than 0.01. This implies that there is a 99% probability that the theta logistic model is the most plausible model (i.e. the most parsimonious) among the set of models under consideration. That is not to say that the theta logistic model is correct – only that it offers the most parsimonious explanation of the data, balancing the need for accurate prediction with a tolerably small number of parameters. Akaike weights offer a practical means of assessing how seriously to take alternative predictions derived from each of the models.

The simplest biological interpretation of this pattern is that wildebeest require some resource, such as food of suitable quality, whose availability is strongly related to wildebeest abundance. This interpretation is further strengthened by the observation that wildebeest survival rates measured locally are positively related to the amount of green grass available per individual animal, which varies with monthly rainfall (Mduma *et al.* 1999).

To simplify our example, we evaluated only three alternative models. One can readily imagine other plausible models that might be even more useful in predicting changes in wildebeest abundance over time, such as age-dependent models or models incorporating interactions with predators. These new models should also be ranked in terms of their AIC scores. A wise researcher or wildlife manager would consult all of the models whose ΔAIC values were within a difference of 2 of the leading model, because all such models are consistent with the evidence. The robustness of future population projections can then be reevaluated as further evidence is accrued. This learning procedure is especially useful if comparisons can be made under conditions in which plausible models make different predictions.

15.5 Adaptive management

Nowhere is model evaluation more important than in the emerging concept of adaptive management (Walters 1986; Walters and Holling 1990). No ecosystem is completely understood. As a consequence, we can never predict with certainty how any ecosystem will respond to human intervention, such as conservation programs (Chapter 17) or harvesting (Chapter 18). What this means, of course, is that any management policy that we choose to adopt should be viewed as an experiment whose outcome is uncertain. Good wildlife managers have always recognized this, at least subconsciously, and accordingly gathered data to monitor the statuses of the species with which they are charged. Where "adaptive" management departs from simply "good" management is in formalizing a mechanism by which management policies can be improved over time, by reducing at least some of the uncertainties.

One way to go about this is to make the best possible use of historical data to judge which model out of many possible ones is most useful. We have already demonstrated such an analysis to explain the demographic patterns of Serengeti wildebeest. In sifting through possible candidate explanatory models, information theory (likelihood, AIC, and the like) provides a powerful set of tools for data analysis. This process is known as *passive adaptive management*, because it relies on natural variability to expose the underlying relationships. Without wide variation in wildebeest abundance, for example, we would have had great difficulty discriminating among the three subtly but dynamically different classes of density-dependent model that might apply.

The variation in numbers of wildebeest occurred through a fortuitous accident, namely the unintentional elimination of rinderpest. This event immediately suggests a radically different approach: manipulative experimentation with the express intention of clarifying our understanding of the system. Such intentional experimentation at the ecosystem level is known as *active adaptive management*. In principle, all of the attributes of good experimental design (see Chapter 14) should be incorporated: use of controls versus experimental treatments, replication, and a factorial design to identify possible interactions among processes. In practice, however, it is difficult to implement an ideal experimental design at the enormous spatial scale at which wildlife management policy is typically conducted. This argues against using conventional inferential statistics to evaluate the outcome of active adaptive management, such as the ANOVA designs described in Chapter 14. Information theory and Bayesian methods of analysis offer a more practical toolkit, based on the objective of finding the best explanatory model among many possible models, rather than definitively rejecting a null model. Active adaptive management policies trade off the short-term goal of maximizing some output, such as a harvest, against the long-term goal of gaining greater understanding of important ecological, physical, or social processes. In that sense, active adaptive management is rather like industrial research and development, reinvesting current revenue to enhance future profits.

One unfortunate aspect of this terminology is that "active" seems a great deal more attractive than "passive" adaptive management. It is not always clear that this is so. Experimentation implies consciously ignoring the current "best" model to explore alternatives. This can only come at a cost, perhaps a great cost, if the alternate models prove to be inferior. For example, Anderson (1975) suggested that higher harvests of mallard ducks could be sustained if the breeding stock was allowed to climb considerably above the then current levels. Choosing to explore this option would require considerable reduction in harvest quotas for hunting enthusiasts and resultant loss of revenue to agencies, tourist operators, and equipment suppliers. If we were to test Anderson's hypothesis and it proved incorrect, the cost of learning would be a reduction in harvests (and profits). Small wonder that this might not be an attractive option for everyone concerned! One way to evaluate the wisdom of embarking on an active adaptive experiment is to evaluate the costs and benefits of sticking with the untested status quo hypothesis versus adopting an experiment to test an alternate one. This can be symbolized by a decision matrix (Table 15.2).

For example, let us say that the current stocking rate of mallard ducks really is most productive. If we increase the stock to a less productive level, this will lead to a reduction in average harvest from 1.0 to 0.5 units. On the other hand, if mallards really are more productive at a higher stock size then increasing the stock may increase yields by a given amount, say 20%. If each model is equally plausible based on our current data then we can calculate the expected payoff by simply averaging the possible outcomes across each row.

In this example, the expected payoff is 1.0 for the status quo option (= $(1.0 + 1.0)/2$), whereas the experimental option has an expected payoff of 0.85

Table 15.2 Hypothetical matrix of net benefit from two alternative stocking models. One suggests the best harvest of mallard ducks is obtained from current stock values, the other that it would come from higher stock values.

| | Model (hypothesis) | |
Policy option	Current stock is optimal	Higher stock is optimal
Maintain status quo	1.0	1.0
Increase stock	0.5	1.2

Modified from Walters and Holling (1990).

$(= (0.5 + 1.2)/2)$. This suggests that the experiment is too costly and/or too unprofitable to warrant testing. In contrast, if the payoff in the lower right-hand cell of the matrix were 1.75 then the expected pay off would be 1.125 and the experiment would be justifiable. The point is that minor improvements in yield accompanied by major costs may not justify experimentation.

Even when the decision to adopt an experimental procedure is justifiable, many individuals may not value slight improvements in management efficiency if it interferes with their personal recreational values. This situation may explain why moose and trout populations close to population centers in North America are probably well below the level of maximal productivity. Alternatively, some resource users might not want to take any risk of losing income, simply because there are no other economic options. Hence, the payoff matrix will often vary among different special interest groups.

While passive adaptive management might be attractive to risk-averse decision-makers it also has its perils. For one thing, it can lead to lost opportunity for all resource users, should several models continue to make similar predictions. For another, it makes it difficult to discriminate between good management and good luck. High harvests could accrue by chance during a series of good years despite application of a wrong model.

As an example of the potential utility of active adaptive management, let us consider harvesting of waterfowl, specifically mallard ducks, in more detail. Harvest quotas for a variety of ducks are determined in part using a sophisticated system of stratified aerial surveys crisscrossing the extensive area of breeding habitat on the North American prairies (Nichols *et al.* 1995). Density levels and pond availability are used to predict stochastic variability in duck recruitment rates, and these recruitment rates are interpreted as a harvestable surplus (Anderson 1975). Much of the stochastic variability in demographic parameters stems from variation in rainfall on the prairies. Wet weather generates large numbers of small ponds and pothole lakes on the prairies, which in turn generate increased success in offspring recruitment. Banding records, obtained from the recovery of identification bands (in the hunting season), allow an estimation of mortality rates. This information on offspring recruitment and mortality is then used in quantitative population models to predict safe harvesting levels year to year. The remarkable consistency in duck numbers over time attests to the robustness of this program (Nichols *et al.* 1995).

There are indications, nonetheless, that the harvest allocation for some waterfowl species could be considerably improved. A key uncertainty in the harvest evaluation procedure is whether mortality is compensatory or not (Anderson 1975; Williams *et al.* 1996). In this context, perfect compensation means that increased duck mortality due to harvest has no effect on overall duck mortality, at least over some range of harvest rates, because survival in the wild adjusts perfectly to losses imposed by man (Fig. 15.5). The alternative hypothesis is that there is no compensation: hunting

Fig. 15.5 Schematic representation of the compensatory (broken line) and additive (solid line) models for survival in waterfowl in relation to harvest levels.

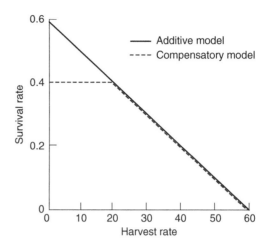

mortality is in addition to natural mortality, so total mortality is linearly related to harvest rates (Fig. 15.5). Current data are inadequate to discriminate between these two hypotheses, yet they have critical implications with respect to both the risk of overharvesting, particularly in poor years, and the optimal harvest policy (Anderson 1975). Simulation models have been used to show that by far the most efficient way to decide which of these alternative models is correct is active adaptive management (Nichols *et al.* 1995; Williams *et al.* 1996). Indeed, this may be the only realistic way to reduce the uncertainty in biological processes, at least within our lifetime.

Such an active adaptive management procedure has been implemented despite the inherent difficulty in coordinating agencies and resource users in a variety of jurisdictions (Nichols *et al.* 1995). If this kind of coordinated model evaluation can be conducted for waterfowl, it can be used even more readily for less mobile species. The key may be that there has been a long-standing tradition in waterfowl management of applying biomathematical models to the production of recruitment and harvest management. Such models have been applied rarely to wildlife species, for which harvesting policy is often developed in a more haphazard fashion. The adaptive approach demonstrates a more productive option.

15.6 Summary

Statistical hypothesis testing is not always the best way to make informed decisions about causal factors associated with wildlife population dynamics, due to a preoccupation with rejection of null hypotheses rather than evaluation of the merits of a suite of more plausible models. We outline an alternative approach to inference that is based on information-theoretic theory, allowing one to decide which model or suite of models offers the best explanation for existing patterns of data. Such an approach complements the practical need to make the best management decisions possible on the basis of incomplete scientific information. A cornerstone of all model evaluation procedures is some measure of goodness-of-fit between models and data. Such model evaluation is an essential component of adaptive management regimes, where alternate explanatory models are vigorously pursued using historical data or experimental perturbation. We show how adaptive management can be used to improve management of harvesting in migratory waterfowl populations in North America.

16 Population viability analysis

16.1 Introduction

In this chapter we deal with a theory that has been developed to account for why and how populations become extinct. Most of this theory deals with extinction as a consequence of chance demographic events affecting small populations or of random variation in the environment, such as that arising from year-to-year changes in weather. A second class of extinction processes – those caused by a permanent and deleterious change in a once-healthy population's environment leading to long-term declines – is less well served by theory (Caughley 1994), but promising new approaches are under development. These will be examined in turn, particularly in the context of evaluating the likelihood of a population's going extinct.

16.2 Environmental stochasticity

Year-to-year variation in environmental conditions often has a profound effect on the rate of population change, a process that is commonly termed "environmental stochasticity." The most important source of environmental stochasticity is variation in weather conditions. Temperature has a direct effect on the demography of plants, invertebrates, and cold-blooded vertebrates through its immediate effects on physiological rates, metabolic activity, and behavior. Rates of growth in these taxa are consequently a direct function of temperature as measured in degree days. Temperature effects are typically less pronounced on mammals, especially at temperatures within the so-called thermoneutral zone in which mammals can readily thermoregulate without investing additional energy expenditures in shivering (in cold weather) or panting and sweating (in hot weather). On their own, precipitation and humidity tend to have more subtle ecological effects that are often mediated indirectly through food supply. Wind speed, precipitation, and humidity can also interact with low temperature, however, to create stressful thermal conditions that can multiply several-fold the energy expenditures of mammals and other wildlife taxa. A classic example is Soay sheep off the coast of Scotland (Coulson *et al.* 2001a), in which winter severity interacts synergistically with population density and resource availability to determine survival rates (see Chapter 13).

We symbolize the variance in exponential rate of population change r caused by a fluctuating environment as $\mathrm{Var}(r_e)$. This can be measured as the actual year-to-year variance in r exhibited by a population whose size is large enough to swamp the effect of variance due to individual variation. Even so, $\mathrm{Var}(r_e)$ will typically be overestimated because its measurement contains a further component of variance introduced by sampling variation associated with population estimates used in year-to-year rates of increase. To make this more realistic, we shall look at a real example (Fig. 16.1): the

Wildlife Ecology, Conservation, and Management, Third Edition.
John M. Fryxell, Anthony R. E. Sinclair and Graeme Caughley.
© 2014 John Wiley & Sons, Ltd. Published 2014 by John Wiley & Sons, Ltd.
Companion Website: www.wiley.com/go/Fryxell/Wildlife

Fig. 16.1 Population
dynamics over time of
female grizzly bears in
Yellowstone National
Park during 1959–82.
(After Eberhardt *et al.*
1986.)

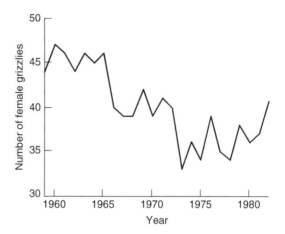

Fig. 16.1 Population
dynamics over time of
female grizzly bears in
Yellowstone National
Park during 1959–82.
(After Eberhardt *et al.*
1986.)

Yellowstone population of grizzlies, which is the largest remnant population left in the continental United States. For the period 1959–82, the Yellowstone grizzly population hovered around 35–40 female bears (Eberhardt *et al.* 1986). Such small population levels are often considered dangerously low, due to the risk of chance demographic events or Allee effects. This has raised conservation concerns for the long-term viability of grizzlies in Yellowstone. The average *r* recorded over this period was −0.00086. There was cause for concern because if *r* was negative for long enough it would have meant extinction.

In order to calculate how likely extinction of the Yellowstone grizzlies might be, we first need to calculate the residual variation in bear growth rates around the mean rate of growth from the Yellowstone data. This is done by calculating the average exponential rate of increase for each time interval in the census data, as described in Chapter 8. We then subtract the average exponential growth rate from the observed growth rate in each time interval that is available, square these deviations, and calculate the mean squared deviation by dividing by the sample size, yielding the following equation:

$$Var(r_e) = \sum_{t=0}^{32} \frac{(r_t - (-0.00086))^2}{16}$$

In the case of Yellowstone grizzlies, $Var(r_e) = 0.0064$. This estimate of environmental stochasticity can be used to evaluate long-term viability.

16.3 PVA based on the exponential growth model

Population viability analysis (PVA) is a procedure by which we can estimate the probability of persistence (or its converse, extinction) over a specified time interval (Boyce 1992). Depending on the biological facts known for the population in question, PVAs can be based on exponential, density-dependent, interactive predator–prey, metapopulation, or even age- or stage-dependent models (Boyce 1992; Beissinger and McCullough 2002; Morris and Doak 2002). Regardless of structural details, all PVAs use estimated variation in demographic parameters to add noise to simulated populations. By repeating such Monte Carlo models many times, one can assess the probability of falling below an arbitrary critical value (termed a *quasi-extinction threshold*). Why is this critical level not set at 0? First, many otherwise useful models do not have reliable behavior as density approaches zero (e.g. in the logistic model a population always increases even when it is near zero). Second, we know that demographic stochasticity

Fig. 16.2 Simulated dynamics of a grizzly bear population with mean and variance in opulation growth rate similar to that of the Yellowstone population during 1959–1982 and starting with the population size recorded in 1982. Note that the simulated population has just reached the critical "quasiextinction" threshold of 10 animals in year 97.

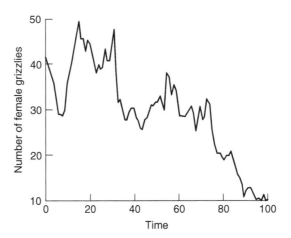

(discussed in Section 16.6) will often doom any population that spends too long at "too low" a density. Third, we might feel that crisis management is called for below this arbitrary threshold.

Some populations are so small that they are unlikely to experience any major changes in net recruitment due to increasing density. If so, we would expect them to grow according to an exponential population model, such as:

$$N_{t+1} = N_t e^r$$

We can use the long-term demographic data from Yellowstone, basic as they are, to conduct a simple stochastic simulation of the population dynamics of grizzly bears using the following modification of the geometric growth model:

$$N_{t+1} = N_t e^{\mu + \varepsilon_t}$$

where μ is the mean exponential growth rate recorded in the past ($\mu = -0.00086$) and ε_t is the magnitude of environmental variation simulated for year t, drawn from a normal distribution with a mean of zero and a residual standard deviation equivalent to that recorded in the past ($\mathrm{sd}(r_e) = 0.08$, calculated by taking the square root of $\mathrm{Var}(r_e)$).

Let us assume that a value of 10 bears is the lowest bear population that can be tolerated before emergency measures are called into play. This becomes our so-called quasi-extinction threshold. By setting the lowest value on the y-axis to 10, we can readily monitor when our simulated bear population falls below this critical threshold (Fig. 16.2). Repetition of this process many times suggests a roughly 8–10% risk of quasi-extinction within the next century (Fig. 16.3). In real life, of course, it seems highly unlikely that Yellowstone National Park authorities would allow the grizzly population to go extinct, but the quasi-extinction estimate gives us some idea of the risk of a serious population crash that might require remedial action.

16.4 PVA based on the diffusion model

There is a more mathematically elegant way to estimate extinction risk in populations with exponential population growth, using what is known as a "diffusion model" (Lande 1993). We follow the discussion of diffusion models in Morris and Doak (2002). A diffusion model can be visualized as if a vast number of beads are released at a single point near the bottom of an infinitely deep river. Over time, the cloud of beads will spread, due to individual beads bouncing up or down as a result of turbulence

Fig. 16.3 Results of 100 replicate
Monte Carlo simulations of the
Yellowstone grizzly population,
based on an exponential growth
model with $\mu = -0.00086$ and
$\sigma = 0.08$. The lower critical density
is arbitrarily set at 10 individuals.
Between 5 and 10% of the replicates
tend to reach the critical threshold
within a century.

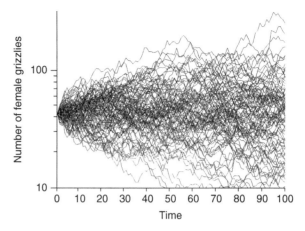

as they flow downstream, just as the population trajectories for the geometric growth model tend to spread over time (Fig. 16.3). A probability density formula known as the inverse Gaussian distribution is used to predict the analogous distribution of population densities over time, with residual variability in the exponential growth rate of the population generating the turbulence. This equation generates a bell-shaped distribution of population densities at each point in time, with the degree of spread of the bell-shaped curve growing wider with every year. The initial position at which the beads are released corresponds to the starting population density N_c. This is important, because more beads will strike the bottom when they are released low in the water column, just as the probability of extinction is highest for populations starting at low numbers. The quasi-extinction threshold is symbolized by N_x, μ is the average exponential rate of increase (which equals the arithmetic mean of $\ln(\lambda)$), and σ^2 is the Var(r_e). The probability of extinction in any given year t and the cumulative probability of extinction are calculated according to the equations shown in Box 16.1.

Box 16.1 Equations for calculating the probability of extinction in any given year t and the cumulative probability of extinction.

Diffusion model for predicting the probability of extinction in any given year, $P(t)$, and the cumulative probability for all years prior to and including t, $G(t)$.

$$P(t) = \frac{d}{\sqrt{2\pi\sigma^2 t^3}} \cdot \exp\left[\frac{-(d + \pi t)^2}{2\sigma^2 t}\right]$$

where:

$$d = \ln(N_c) - \ln(N_x)$$

where N_c is the initial density and N_x is the quasi-extinction threshold. By integrating the time-specific equation from minus infinity to t, we get the following cumulative function:

$$G(t) = \Phi\left(\frac{-d - \pi t}{\sqrt{\sigma^2 t}}\right) + \exp\left(\frac{-2\pi d}{\sigma^2}\right) \cdot \Phi\left(\frac{-d + \mu t}{\sqrt{\sigma^2 t}}\right)$$

where:

$$\Phi(z) = \left(\frac{1}{\sqrt{2\pi}}\right) \int_{-\infty}^{z} \exp(-y^2)\,dy$$

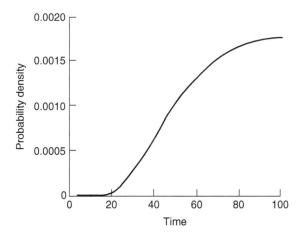

Fig. 16.4 The predicted probability of extinction in any future year, based on the diffusion equation approximation for an exponentially growing grizzly bear population with demographic parameters identical to the Yellowstone population studied during 1959–82.

We will illustrate the application of the diffusion equation, using population censuses of females from the Yellowstone grizzly bear population during 1959–82, to estimate mean r and the standard deviation in r. As before, we will assume a lower critical threshold $N_c = 10$ animals and use the 1982 census total as the initial density, $N_x = 41$.

The predicted probability of extinction increases over time, because it takes several poor years in sequence for a population of 41 grizzlies to crash to below 10 individuals. The diffusion model predicts that the cumulative risk of extinction for Yellowstone grizzlies should accordingly tend to increase over time, initially at an accelerating rate, then later at a diminishing one (Fig. 16.4). For a population whose mean exponential growth rate $\mu < 0$ like this one, eventual extinction is certain, provided one waits long enough. The question is: how long might this process take? By year 100, the diffusion model estimates a 9% probability of extinction (Fig. 16.4), just as we found from our earlier simulations.

In reality, the Yellowstone grizzly population managed to recover to much higher numbers in the 1990s (Fig. 16.5). Hence the risk of imminent extinction in the early 1980s proved to be a false alarm, at least in this case, as one might expect in the majority

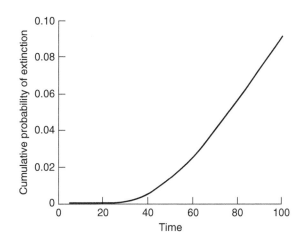

Fig. 16.5 The cumulative probability of extinction over a given span of time for a grizzly bear population with demographic parameters identical to the Yellowstone population studied during 1959–82. $\sigma = 0.08$.

of cases (> 90%). Nonetheless, the PVA approach gives us a formal starting point in evaluating extinction risk.

16.5   PVA based on logistic growth

All populations show some degree of variability in conditions from year to year. Such stochastic or random variation can have a strong influence on the dynamics of even tightly regulated species. We can explore this by applying the Ricker logistic model to some typical empirical data, which come once again from Yellowstone National Park. Fig. 5.18 shows records of elk censused on the northern range between 1968 and 1989 (Coughenour and Singer 1996). We see that there is the barest hint of a sigmoid pattern in these data. Nonetheless, exponential growth rates r_t calculated over this 2-decade period show a strong density-dependent decline in growth rates when the population is large (Fig. 16.6).

The scatter around the regression line (termed "residual" variation) in Fig. 16.6 shows that natural regulation explains only part of the demographic response by a wild population to changes in density. Even when the population is tightly regulated, as is obviously the case here, there can be considerable variation in growth rates from year to year, which is not explained by density dependence. Some of this variability is due to the stochastic climatic variation that characterizes every natural environment – some places more than others. In the case of Northern Yellowstone elk, for example, precipitation in the preceding 2 years is probably responsible for much of the residual variation shown in Fig. 16.6, judging from its effect on offspring production and survival rates (Coughenour and Singer 1996). This probably stems from a strong linkage between precipitation and forage availability to elk. Just as with the Yellowstone grizzly bear data, we must first calculate the observed exponential rates of elk growth for each time interval from the census data. In the case of the Yellowstone elk, 0.518 is the intercept (r_{max}) of the regression line drawn through the observed values of r_t versus N_t, and −0.000044 is the slope.

We can then use this density-dependent model of logistic growth to calculate the residual variation in growth rates around the regression line, using similar logic as for grizzlies. The only difference is that now the deviation is based on the difference between observed growth rates for elk and the values predicted by the regression equation, rather than the deviation between observed growth rates for bears and the

Fig. 16.6 Exponential growth rates for Northern Yellowstone elk between 1968 and 1989 in relation to population density at the beginning of each yearly interval. (Data from Coughenour and Singer 1996.)

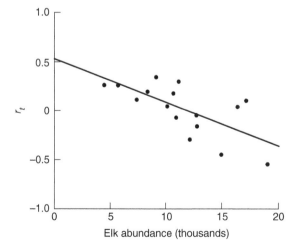

Elk abundance (thousands)

Fig. 16.7 Simulated dynamics of elk, based on the Yellowstone National Park population. (Data from Coughenour and Singer 1996.)

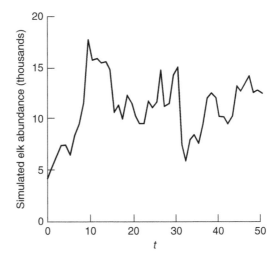

average growth rate. This is calculated in the following manner:

$$\text{Var}(r_e) = \sum_{i=1}^{16} \frac{(r_t - (0.518 - 0.000044N_t))^2}{16}$$

We calculate the deviation between each observation of r and the value predicted by the regression line at that population density, square each deviation to standardize positive versus negative values, sum the squared deviations, and divide by the sample size (16 in this case) to estimate the mean squared deviation. The square root of this value is the residual standard deviation. For the Northern Yellowstone elk, the residual variation in growth rates $\text{Var}(r_e) = 0.04$.

Once equipped with an estimate of the residual deviation in growth rates based on the observed data, we draw values of the random variable ε from a bell-shaped (i.e. normal) probability distribution with the same magnitude of residual variation. For the elk example $\text{mean}(r_e) = 0$ and $\text{sd}(r_e) = 0.19$. We then combine the random normal deviate at any point in time ε_t with the rate of increase predicted by the Ricker logistic equation ($r_{max}(1 - N_t/K)$) to predict changes in abundance.

$$N_{t+1} = N_t e^{r_{max}\left(1 - \frac{N_t}{K}\right) + \varepsilon_t}$$

We plot the simulated elk data in Fig. 16.7, where it can be seen that the trends in the simulated population are completely different from those of the real population, but the overall magnitude of variability is similar. This similarity occurs because we have included both of the important factors influencing population dynamics: environmental stochasticity, which tends to perturb the population away from its carrying capacity, and the natural regulatory processes, which tend to restore the perturbed population back towards its equilibrium. Both density dependence and environmental stochasticity are common in the natural world, and therefore we often need to accommodate them in our management planning.

16.6 Demographic stochasticity

Demography deals with the probability of individuals living or dying and the probability that they will reproduce. Those probabilities, accumulated over all individuals in a population, determine what the population as a whole will do next: whether it will increase, decrease, or remain at the same size. A population's expected rate of increase is determined by age-specific fecundity rates interacting with age-specific mortality rates (see Chapter 13), but that expectation is reliable only when there are substantial numbers of individuals. If numbers are low, the actual rate of increase may vary markedly in either direction from that predicted by the life and fecundity tables, because births and deaths are chance events (see Chapter 5). This effect is called *demographic stochasticity*.

The principle is familiar to anyone that has performed a coin toss, played a game of cards, or spun a roulette wheel. The outcome of such random processes can be readily predicted using the binomial probability distribution. For example, the odds of flipping heads three times in a row are $0.5 \times 0.5 \times 0.5 = 0.125$. The binomial distribution allows us to predict all the possible outcomes of such a set of trials in which each trial falls into one of two discrete classes, termed a success or a failure (e.g. heads versus tails, survival versus death). Let p = the probability of success in any single trial and $1 - p$ = the probability of failure. Over N trials, the probability of getting z successes is calculated by the following equation:

$$P(z) = \frac{N!}{z!(N-z)!} p^z (1-p)^{N-z}$$

It should be straightforward to see how this translates into an ecological setting. For a given probability of survival (say $p = 0.64$), we do not necessarily expect exactly 64 out of 100 individuals to survive, but rather anticipate that by chance sometimes a larger fraction will survive, sometimes a smaller fraction. For example, in a population of five individuals with such a survival rate, the probability that all five will survive $P(z = 5) = 0.107$. By substituting different values into the binomial probability density function, one can similarly calculate the probability of all the other possible outcomes: $P(z = 4) = 0.302$, $P(z = 3) = 0.340$, $P(z = 2) = 0.191$, $P(z = 1) = 0.054$, and $P(z = 0) = 0.006$. The final outcome is particularly important from a conservation point of view, because it corresponds to extinction. Concentrating solely on this outcome, the probability of extinction due to chance survival events is negatively related to population size by the following relationship:

$$P(z = 0) = (1 - p)^N$$

This formula shows that the risk of extinction due simply to chance events affecting survival will tend to decline with increasing population size, following a simple exponential decay curve.

Consider the hypothetical case of a population of 1000 large mammals whose intrinsic rate of increase $r_{max} = 0.28$, based on the following assumptions: a female can produce no more than one offspring per year, successful reproduction occurs with a probability $b = 0.95$, and the probability of survival of an adult $p = 0.9$. The population is at a low density of 0.01 animals per square kilometer, so there will be little competition for resources and consequently the rate of increase will be close to r_{max}. The beginning of the year is defined as immediately after the birth pulse, at which time the population contains 500 males and 500 females. By the end of the year the population will have been reduced by natural mortality to about 900 (i.e. 1000×0.9), and these animals will produce about 428 offspring (450×0.95) at the next birth pulse. The population therefore starts the next year with about 1328 individuals ($900 + 428$), having

Table 16.1 Probabilities for the population outcome over a year of a population comprising two individuals, one of each sex. The chance of an individual surviving the year is $p = 0.9$ and the chance of the female producing an offspring at the end of that year is $m = 0.95$.

N_{t+1}	What happens	Probability of outcome		r
		Symbolic	Numerical	
0	Both die	$(1-p)^2$	0.01	$-\infty$
1	One dies	$2p(1-p)$	0.18	-0.69
2	Both live, no offspring	$p^2(1-m)$	0.0405	0.0
3	Both live, one offspring	p^2m	0.7695	0.41
			1.0000	

registered a net increase over the year at about the rate $r = \log_e (1328/1000) = 0.28$, or 32%. The actual outcome will be very close to these figures because the differences in demographic behavior between individuals tend to cancel out.

Now consider a subset of this population restricted to a reserve of $200\,\text{km}^2$. The density of $0.01/\text{km}^2$ translates to a population size of two individuals. These two obviously cannot increase by 32% to 2.64 individuals as the large-population estimate would imply. They can only increase to 3, remain at 2, or decline to 1 or even to 0. Table 16.1 gives the probabilities of those outcomes.

Table 16.1 shows that the most likely outcome is 3 animals and a rate of increase of $r = 0.41$. But even though the population is "trying" to increase, the actual rate of increase may by chance vary between a low of minus infinity and a high of $r = 0.41$. Hence the dynamics of a small population are highly influenced by the luck and misfortune of individuals. It is a lottery. In contrast, the dynamics of a large population are ruled by the law of averages. We say that the outcome for a small population is highly stochastic and that for a large population is highly deterministic. In Section 16.7 we show how to estimate the magnitude of demographic and environmental stochasticity.

Table 16.2 shows that at a population size of $N = 50$ the effects of small numbers can result in a rate of increase varying (at 95% confidence) between $r = 0.48$ (i.e. $0.28 + 0.202$) and $r = 0.08$ (i.e. $0.28 - 0.202$). At $N = 10$ the possible outcomes vary between a high rate of increase and a steep decline. In this example the deterministic rate of increase might only be a reliable guide to the actual rate of increase once the population has attained a size of several hundred.

Although the details are special, the message is general: populations containing less than about 30 individuals can quite easily be walked to extinction by random demographic variation between individuals, even when those individuals are in the peak of health and the environment is entirely favorable.

Table 16.2 Deviation from expected rate of increase resulting from stochastic variation. The influence of one individual on variance in r is taken as $\text{Var}(r)_1 = 0.5$.

N	Expected r	Var(r)	SE(r)	95% confidence limits of r
10	0.28	0.05	0.224	±0.500
50	0.28	0.01	0.100	±0.202
100	0.28	0.005	0.071	±0.139
500	0.28	0.001	0.032	±0.063
1000	0.28	0.0005	0.022	±0.043

16.7 **Estimating both environmental and demographic stochasticity**

In small populations, variability in population growth rates will be substantially affected by both environmental and demographic stochasticity. Provided that data are available on the survival and reproductive success of specific individuals, it is possible to disentangle demographic from environmental sources of stochasticity (Sæther *et al.* 2000a). The variance in *r* that stems from demographic stochasticity can be estimated from the fates of an individually recognizable population with seasonal breeding (the norm for virtually all wildlife species) by the following relationship:

$$\text{Var}(r_{\text{d}}) = \frac{1}{n-1} \sum_{i=1}^{n} (\omega_i - \overline{\omega})^2$$

where ω_i is the sum of offspring produced by a given female plus 1 if she survives to the next breeding season. The extent to which actual *r* is affected by demographic stochasticity in a given year can then be estimated by dividing the demographic variance by population size ($\text{Var}(r_{\text{d}})/N$).

We will illustrate this using data for a small population of song sparrows (*Melospiza melodia*) on Mandarte Island, just off the coast of British Colombia, Canada. The Mandarte sparrow population has been closely monitored for several decades, allowing precise estimates of total abundance and allowing field biologists to follow the fates of all individuals in the population (Smith *et al.* 2006). Over 3 decades, the Mandarte song sparrow population underwent a series of population increases followed by severe crashes in abundance, with the number of adult females fluctuating between a low of 3 birds and a peak of > 70 (Fig. 16.8). Like most populations for which sufficient data have been gathered, the rate of population change in Mandarte sparrows was negatively related to population size (Arcese *et al.* 1992), but this density dependence was clearly insufficient to maintain stable population numbers. Several factors might have contributed to the population crashes, including severe winter weather (cold, rain, and high winds), disease, nest parasitism due to cowbirds, and predators such as crows, gulls, and deer mice (Smith *et al.* 2006). At low points, this population was small enough for demographic stochasticity to play an important role. Sæther *et al.* (2000a) used the survival and recruitment data gathered by the Mandarte field team for each year to estimate $\text{Var}(r_{\text{d}})$. There was considerable variation in fitness ω among females (Lande *et al.* 2003), with some contributing multiple offspring to the

Fig. 16.8 Number of adult sparrows alive on Mandarte Island from 1975–2002. (Source: Smith *et al.*, 2006. Reproduced with permission of Oxford University Press.)

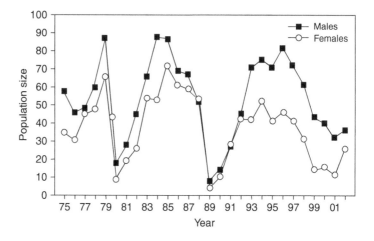

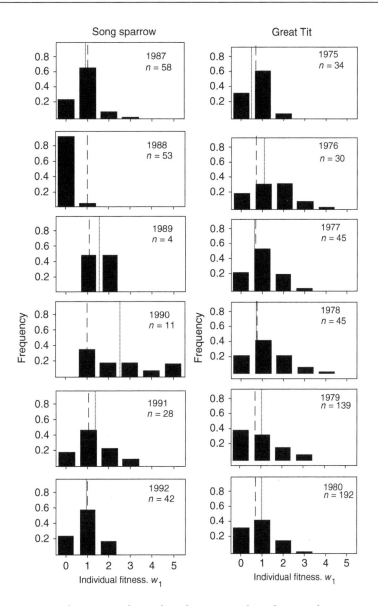

Fig. 16.9 Variation in fitness ω among female song sparrows on Mandarte Island and great tits in woodlands near Oxford, UK. (Source: Lande *et al.*, 2003. Reproduced with permission of Oxford University Press.)

next generation and surviving themselves, but many others dying without successfully breeding (Fig. 16.9).

Once the demographic variance has been estimated, the environmental variance can be teased out using the following relationship (Lande *et al.* 2003):

$$\text{Var}(r_e) = \left(\frac{\Delta N}{N} - g(N) + 1 \right)^2 - \frac{\text{Var}(r_d)}{N}$$

where N is the population abundance at the beginning of a given time interval, ΔN is the observed change in abundance over the time interval, and $g(N)$ is the annual growth rate ($\lambda = N_{t+1}/N_t$) predicted by the logistic growth model for that time interval. Averaging these estimates over all the years of the Mandarte study, $\text{Var}(r_d) = 0.66$

and $\text{Var}(r_e) = 0.41$. This is one of the highest estimates of environmental stochasticity recorded in a vertebrate species, reinforcing the notion that this population is unusually vulnerable to unfavorable changes in conditions from year to year, whatever the underlying cause.

16.8 PVA based on demographic and environmental stochasticity

If suitable data are available, it is relatively straightforward to include variability in growth rates due to environmental and demographic stochasticity as well as any density-dependent effects. We do this by simulating stochastic variation in the environment, affecting the population similarly at all levels of abundance, and adding this to the additional density-dependent variation in growth rates that stems from demographic stochasticity. Both of these sources of variation are used in combination with normal density-dependent effects on population growth rates to estimate the stochastic growth rate r_t that applies at any point in time. The Mandarte sparrow data provide a nice example. Sæther *et al.* (2000a) used the following equation to encompass all of the sources of variation in population growth rates (density, $\text{Var}(r_e)$, and $\text{Var}(r_d)$):

$$N_{t+1} = N_t e^{r_{max}\left(1-\left(\frac{N_t}{K}\right)^{\theta}\right)+\varepsilon_t}$$

where the stochastic term (ε_t) is drawn from a normal distribution whose standard deviation at any given time step σ_t is calculated by combining the demographic and environmental variance appropriate to a given level of population abundance:

$$\sigma_t = \sqrt{\text{Var}(r_e) - \frac{\text{Var}(r_d)}{N_t}}$$

Monte Carlo simulations based on this combination of environmental and demographic stochasticity suggest that quasi-extinction (in this case set at a threshold of one individual) is a meaningful possibility in the Mandarte song sparrow population. Out of 100 replicates shown in Fig. 16.10, five simulated populations went extinct over a 30-year timespan. The fact that such events haven't been seen yet is perhaps not surprising, given that the odds against them are $19:1$.

16.9 Strengths and weaknesses of PVA

PVA models are attractive because they supply hard numbers for the kind of uncertain (stochastic) processes that threaten small populations. This has led to their widespread proliferation, as discussed by several in-depth reviews (Boyce 1992; Beissinger and McCullough 2002; Morris and Doak 2002). However, we should be cautious about the reliability of projected extinction risks, for a number of reasons. First, we rarely have precise, reliable estimates of growth rates in any population, let alone one that is under threat. Minor errors in our estimates of demographic parameters (because the number of years of observation is short) multiply geometrically over time, leading to biased estimates of extinction risk (Ludwig 1999; Fieberg and Ellner 2000). Projection of extinction risk beyond a short time interval (10–20% of the length of the time series) is particularly unreliable (Fieberg and Ellner 2000). In other words, 40 years of data on grizzly bears might allow us to make a reliable assessment of extinction risk over the next 10 years, but certainly not over the next 100.

This problem is compounded when we have no idea about the reliability of population estimates in any given year. This is obviously less of an issue in the rare cases where every individual is recognizable. Most populations are known only from samples taken

Fig. 16.10 Monte Carlo simulation of a Mandarte song sparrow population, incorporating density dependence, environmental stochasticity, and demographic stochasticity.

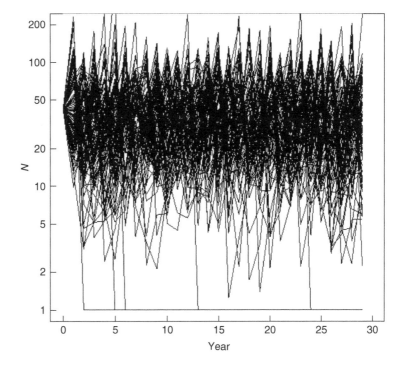

from a small fraction of the inhabited range, leading to considerable uncertainty in true abundance. Observation errors play a key role, because they convey a false impression about the true magnitude of environmental and demographic stochasticity, as well as biased estimates (usually downward) of the strength of density dependence. As a consequence, even well-studied populations may yield biased predictions of extinction risk (Ludwig 1999).

Data for well-studied populations illustrate that catastrophic climatic events play an important role in causing population collapse. Such catastrophes tend to be difficult to predict using Monte Carlo simulations, particularly when long-term census data are lacking (Coulson *et al.* 2001b). Most importantly, application of PVAs is founded on an underlying faith that conditions (e.g. climate, habitat availability, and human interference) will hold far into the future (Coulson *et al.* 2001b). For example, the Yellowstone grizzly bear data for 1959–82 provide a substantially higher risk of extinction than do the later demographic data, suggesting a change in environmental conditions or park management policy.

On the basis of these weaknesses, some critics have claimed that PVA is virtually meaningless (Ludwig 1999; Coulson *et al.* 2001b). Moreover, preoccupation with the stochastic dynamics of small populations ignores the ecological, physical, and anthropomorphic causes of population decline (Caughley 1994; Harcourt 1995; Walsh *et al.* 2003). Other conservation biologists argue that, while not infallible, PVAs are nonetheless quite useful as a means of comparing relative extinction risk among populations or in various subpopulations of a single species, or of assessing the relative risks associated with alternative management actions (Lindenmayer and Possingham 1996; Brook *et al.* 2000; Morris and Doak 2002).

To resolve these different views, Brook *et al.* (2000) gathered demographic data from 21 well-studied populations. They used the first half of the data for each time series to parameterize PVA models, then used the resulting PVA models to predict the outcome of the second half of each data set. They concluded that the risk of decline closely matched predictions and that there was no significant bias in the predictions. They also found few major differences in the quality of predictions of any of the most common models that are commercially available to decision-makers. Coulson *et al.* (2001b) countered that the 21 data sets considered by Brook *et al.* were unrepresentative of the endangered species most likely to be candidates for PVA. Rare organisms are, by their very nature, poorly understood. Nonetheless, PVA is an important tool for risk assessment by both field biologists and decision-makers.

16.10 Extinction caused by environmental change

Despite the preoccupation of most PVAs with stochastic extinction processes, the most common cause of extinction is a critical change in the organism's environment. This is distinct from year-to-year fluctuation due to either demographic or environmental stochasticity. We identify the new environment as the driving variable responsible for the population's decline, and the population may be driven to extinction by its action. A population seldom "dwindles to extinction." It is pushed. If we can identify the agent imparting the pressure and neutralize that pressure, we can save the population.

The three most common causes of driven extinctions, roughly in order of importance, are: (i) contraction and modification of habitat; (ii) unsustainable harvesting by humans; and (iii) introduction of a novel pathogen, predator, or competitor into the environment Hilton-Taylor 2001. Case studies are considered in some detail in Chapters 8, 17, 18, 21, and 22. Before we consider the historical record, however, we will briefly examine how theoretical models can provide useful insight into each of these processes.

16.11 Extinction threat due to introduction of exotic predators or competitors

The first major way that humans wreak havoc on threatened species is through modification of trophic relationships within a pre-existing community. Often this is via introduction of a competitor and/or predator for which an endemic species is poorly prepared. This is particularly common for endemic species on islands that have evolved for considerable periods without risk of predation. Such species are poorly equipped to cope with a novel predator – the typical traits for wariness, stealthy lifestyle, and inconspicuous coloration have not provided any selective advantage and may have disappeared. This sets the scene for a brief, but sadly inevitable, slide into oblivion once a novel predator has arrived.

The existing body of predator–prey theory is sometimes suitable to elucidate such processes. If a predator is particularly efficient (high rates of capture even at low prey densities, high efficiency of conversion of prey into new predators) then prey densities will be expected to plunge suddenly to dangerously low levels, at which there is a high probability of extinction due to demographic or environmental stochasticity. A particularly graphic example is the brown tree snake (*Boiga irregularis*), introduced on to the island of Guam in the 1950s (Savidge 1987). In the course of 2 decades, this generalist predator spread rapidly across the island, its range expansion coinciding with the rapid decline and (in some cases) disappearance of 11 native species of forest birds.

Special circumstances sometimes apply, however, and these can be far from obvious. For example, introduction of rabbits on to a number of islands in the South Pacific has apparently triggered the collapse of endemic bird species that were previously well able to withstand predation. A case in point is the loss of an endemic parakeet and a banded rail species from Macquarrie Island (Taylor 1979). This has been attributed to hyperpredation (see Chapter 22), a form of apparent competition by which an exotic prey that is capable of withstanding predation subsidizes population growth by a resident predator (Smith and Quin 1996). Endemic prey then decline because they are more vulnerable to predation than the exotic species.

This scenario of asymmetric apparent competition (Holt 1977) induced via subsidies to a common predator population was developed theoretically by Courchamp *et al.* (2000a). They represented the hyperpredation process with the following system of differential equations:

$$\frac{d}{dt}B(t) = r_{\max_B}B(t)\left(1 - \frac{B(t)}{K_B}\right) - C(t)\mu_B B(t)\frac{\alpha B(t)}{R(t) + \alpha B(t)}$$

$$\frac{d}{dt}R(t) = r_{\max_R}R(t)\left(1 - \frac{R(t)}{K_R}\right) - C(t)\mu_R R(t)\frac{R(t)}{R(t) + \alpha B(t)}$$

$$\frac{d}{dt}C(t) = \frac{(\lambda_B\mu_B\alpha B(t)^2 + \lambda_R\mu_R R(t)^2)C(t)}{\alpha B(t) + R(t)} - \nu C(t)$$

where B represents the density of the endemic prey species, R the density of exotic prey species, and C the density of the predator that both prey have in common. Note the similarity in structure to other consumer–resource models discussed in Chapter 9. Both prey species have logistic patterns of growth in the absence of predation, with maximum rate of increase dictated by $r_{\max}$ and carrying capacity dictated by K. The predator is assumed to have a Type I (linear) functional response to changes in prey density, with an attack rate of μ on each prey type. The predator is assumed to have a fixed preference for the endemic species, the magnitude of which is proportional to α (defined as the frequency of captures of preferred endemic prey to less-preferred exotic prey when both are equally abundant). Nonetheless, the actual fraction of each prey in the predator's diet changes proportionately with shifts in their relative abundance, according to the ratio $\alpha B/(R + \alpha B)$. The presumption used in the model is that exotic prey individuals are much less easy to catch than endemic prey (α is much greater than 1). Any prey individuals that are successfully attacked are converted into new predators with an efficiency of λ, but predators also have a constant per capita rate of mortality ν.

Depending on the magnitude of the key parameters, these equations can lead to several different outcomes of conservation interest: (i) extinction of endemic prey but perpetuation of exotics and predators; (ii) extinction of exotics but perpetuation of endemic prey and predators; (iii) extinction of predators but perpetuation of both prey; and (iv) coexistence of all three species. A common outcome is depicted in Fig. 16.11: introduction of the exotic leads to high predator density, collapse of endemic prey to dangerously low levels at which demographic or environmental stochasticity threatens extinction, and substantial numbers of exotics. This scenario tends to play out when the exotic species has much higher carrying capacity and intrinsic growth rate than the endemic species (both of which are often true of introduced pest species like rabbits) and the predator tends to have a stronger probability of encountering endemic prey

Fig. 16.11 Simulation of hyperpredation, leading to collapse of an endemic prey species when an exotic alternative prey species is translocated into the system. The following parameter values were used: $\alpha = 3, r_{\max_B} = 0.1,$ $r_{\max_R} = 2.0, K_B = 1000,$ $K_R = 5000, \mu_B = 0.1,$ $\mu_R = 0.1, \lambda_B = 0.01,$ $\lambda_R = 0.01,$ and $v = 0.5.$ (After Courchamp *et al.* 2000b.)

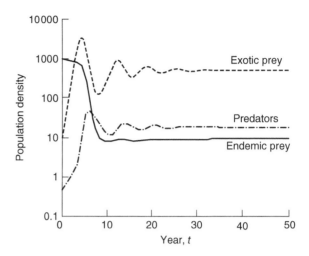

than exotic prey (which is also often true when the endemic has little or no prior experience with predation).

A superb example of this kind of process is seen on the Channel Islands off the coast of California. Several islands have had accidental translocations of an exotic herbivore, the feral pig (*Sus scrofa*). In response to exploding pig populations, golden eagles (*Aquila chrysaetos*) have recently immigrated into the Channel Islands from the mainland and begun to breed successfully. Rising eagle abundance has led to the rapid decline of an endemic prey species, the island fox (*Urocyon littoralis*; Fig. 16.12). Foxes have completely disappeared from two of the islands and have experienced a 10-fold decline in abundance on the largest island (Santa Cruz) following colonization by eagles in the early 1990s (Roemer *et al.* 2002; Courchamp *et al.* 2003). The decline in fox abundance has led to a parallel increase in a competitor, the island spotted skunk (*Spilogale gracilis amphiala*). Although there has been a successful effort to translocate some eagles away from the islands, several pairs still remain, thwarting fox recovery. This presents a conservation conundrum: do we harm or remove a protected species (the golden eagle) in order to save an endangered species (the island fox)? An obvious countermeasure is to cull feral pigs. Without timely reduction in eagle abundance, the model predicts that pig eradication could inadvertently lead to heightened predation pressure on foxes, perhaps even doom them rather than helping them (Courchamp *et al.* 2003). This is a clear demonstration of the utility of trophic models as a means of evaluating alternative conservation actions. In Chapter 9 we consider consumer–resource dynamics in much greater detail.

16.12 Extinction threat due to unsustainable harvesting

For many big game species, harvesting does not pose a conservation threat. Indeed, such species usually become entrenched as "game" because their life history attributes (high reproductive capacity, broad geographical distribution, ability to tolerate interference by hunting humans) make them relatively robust. Modern exceptions to this are species whose male ornaments (horn, tusks, antlers, or other body parts) make them particularly attractive to humans, regardless of the cost and energy required to kill them. There are several obvious examples: black rhinos, elephants, and big cats (lions, leopards, and tigers). When the profit from a rhino horn can exceed a rural African's expected income for a decade, it is not surprising that overharvesting occurs.

Fig. 16.12 (a) The (a) number of golden eagle sightings on the Channel Islands, California, increased as the feral pig population increased. (b) The abundance of the Channel Island fox (●), as indicated by trapping, declined due to depredation by eagles, and this allowed the increase of the endemic skunks (□). (Data from Roemer *et al.* 2002.)

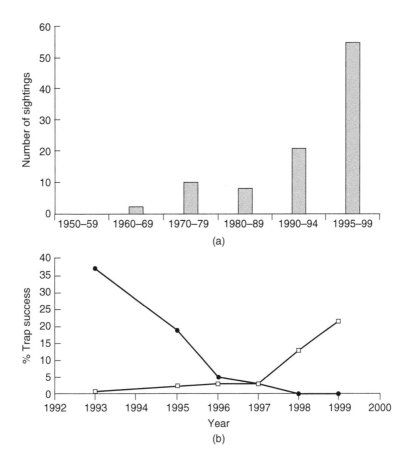

In many cases, however, the ornaments of interest appreciate in value as animals get older. When successful breeding depends on having adequate numbers of mature males, it may make a good deal of conservation sense only to harvest the oldest males, who have already bred, rather than harvest indiscriminately. This principle is illustrated in lions (Whitman *et al.* 2004). Male lions are attractive as trophies to many tourist hunters, particularly old males with full manes. Lions ordinarily live, feed, and breed in a stable social group called a pride. A typical pride in East Africa comprises a group of six or so breeding females, often sisters, and a coalition of two or three males. Male coalitions come and go, whereas females usually remain within a given pride for their entire life. Given the short pride tenure (2–3 years) that any male can expect to hold, rapid breeding is essential to their reproductive success. Because of this, incoming coalitions often kill all the cubs surviving from their predecessors. This brings all the mothers rapidly into estrous, allowing the new batch of males a chance to sire offspring. However, if males turn over too quickly, infanticide outstrips successful reproduction and the population declines.

Whitman *et al.* (2004) developed a detailed, spatially explicit model of individual lions, each of whom lived, bred, and died within 5–10 computer prides, based on long-term studies conducted in Serengeti National Park and Ngorongoro Crater (Packer *et al.* 2005). They used this model to consider the impact of harvesting of males by sport hunting. At typical quotas for the East African savanna, their model suggests that indiscriminate harvesting of all mature males with full manes (those 4 years and older) is unsustainable. This is because removal of males by hunters before they would ordinarily lose their position at the head of a pride causes new males to come in and there is thus too much infanticide. On the other hand, the model suggests that hunters *could* harvest all the males they might want aged 8 years or older. These males have already bred by and large, so they are expendable. Since they are also the most attractive as trophies it is a win–win situation, so long as hunters can tell how old each male is before shooting him. Whitman *et al.* (2004) showed that the amount of black on a lion's nose provides a reliable indicator of his age. This combination of a simple harvesting strategy and a reliable clue to age might prove vital in conserving African lions in the long term.

Such enlightened harvest policy might prove useful for other trophy species as well. For example, harvesting of horns from male saiga antelopes (*Saiga tatarica*) is an important source of income for people living around the Caspian Sea. Recent data, however, suggest that so many male saiga antelopes are now being removed that breeding by females is compromised (Milner-Gulland *et al.* 2003). This is an example of how harvesting can induce an Allee effect (Petersen and Levitan 2001), whereby population growth rates begin to decline once a population falls below a critical lower threshold. In Chapter 18 we further explore the conservation implications of population harvesting.

16.13 Extinction threat due to habitat loss

The most serious challenge currently facing most threatened bird and mammal species is habitat contraction and loss (see Chapter 21). It is rare that habitat loss would pose a serious risk for a generalist species, capable of living in a wide range of places. It is specialists for whom habitat loss is most crucial – those species whose survival and successful reproduction depend critically on particular, usually rare, habitats. Habitats supply numerous attributes: food, protective cover from predators, denning sites, shelter from inclement weather, and access to mates. This means that habitat needs are probably unique for every species. Nonetheless, there are models that predict the effects of dwindling supply, size, and spatial distribution of habitat patches in a metapopulation framework.

One approach uses incidence functions to characterize probabilities of extinction and recolonization for specific patches (Hanski 1994, 1998). Because extinction is often negatively related to population size and small patches tend to hold small populations at the best of times, extinction is usually modeled as a negative function of patch size. Colonization rates tend to be low when patches are widely spaced, so distance among patches is often a critical variable in incidence functions. Data on the sequence of local extinction and recolonization events allow us to estimate incidence functions across a matrix of possible sites. These functions can then be solved, either using matrix techniques or via simulation, to evaluate the long-term probabilities of persistence (Hanski 1998). Data for the Glanville fritillary (*Melitaea cinxia*) show that this approach has high predictive capability, at least in well-studied species (Hanski 1998; Lindenmayer *et al.* 2003). In Chapter 21 we consider habitat loss and its impact on sustainability in greater detail.

16.14 Summary

A population may, by chance, be forced to extinction by year-to-year variation in weather or other environmental factors. When the population is small it can exhibit a random walk to extinction because its dynamics at low numbers are determined by the unpredictable fortunes of individual members. The probability of population extinction over a specified time span can sometimes be estimated using PVA. There are many different ways to develop PVAs, all of which incorporate stochastic factors into simulations of future population growth. Extinction due to demographic or environmental stochasticity is less common than habitat modification or the introduction, usually by people, of a new element into the environment. This is commonly a new predator, competitor, or pathogen. Sometimes it is simply an unsustainable level of harvesting by humans. The population is then driven to extinction, rather than dropping out by chance. These processes can be incorporated in a variety of modeling frameworks that usefully augment and extend the PVA approach.

17 Conservation in practice

17.1 Introduction

In Chapters 16 and 20 we examine the ways in which demographics and genetics contribute, at least potentially, to the risk that a population will go extinct. The extinction of a species does not differ in kind. The species goes extinct because the last population of that species goes extinct. Here we review actual and near extinctions to show the commonest causes of extinction in practice. We then describe how to detect such problems and how to treat a population in danger.

17.2 How populations go extinct

Extinctions may be divided into two categories, driven extinctions and stochastic extinctions.

1 *Driven extinction*   Whereby a population's environment changes to its detriment and rate of increase falls below zero. The population declines. Perhaps this lowering of density frees up resources to some extent or lowers the rate of predation, but this is not sufficient to counteract the force of the driving variable and the population finally goes extinct. Included in this category are extinctions caused by environmental fluctuation and catastrophes. The latter are viewed here as simply large environmental fluctuations.

2 *Stochastic extinction*   Whereby a population fails to solve *the small population problem*. The effect of chance events, which would be trivial if numbers were high, can have important and sometimes terminal consequences when numbers are low.

   (a) *Extinction by demographic malfunction*   Whereby a population goes extinct by accident (chance) because it is so small that its dynamics are determined critically by the fortunes of individuals rather than by the law of averages. In these circumstances, a population is quite capable, by chance configuration of age distribution or sex ratio, of registering a steep decline to extinction over a couple of years even though its schedules of mortality and fecundity (see Chapter 13) would result in an increase if the age distribution were stable.

   (b) *Extinction by genetic malfunction*   Whereby a population at low numbers for several generations loses heterozygosity to the extent that recessive semilethals are exposed, average fitness therefore drops, and the population declines even further, ultimately to extinction (see Chapter 20). The loss of an allele from the genotype results from the lottery of random mating. Although each individual loss is unpredictable, the average rate of loss, as a function of population size, can be predicted fairly accurately.

Wildlife Ecology, Conservation, and Management, Third Edition.
John M. Fryxell, Anthony R. E. Sinclair and Graeme Caughley.
© 2014 John Wiley & Sons, Ltd. Published 2014 by John Wiley & Sons, Ltd.
Companion Website: www.wiley.com/go/Fryxell/Wildlife

Fig. 17.1 The number of species of birds in isolated patches of eucalypt habitat in Western Australia is related to the distance to roadside strips that act as corridors or stepping stones. (Data from Fortin and Arnold 1997.)

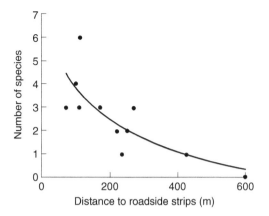

These mechanisms do not exclude each other entirely but they are sufficiently distinct that we treat them separately. Although the relative contribution of these mechanisms is unknown, enough anecdotal information is available to suggest that driven extinction is by far the most prevalent. Extinction by demographic malfunction is probably the second most important but it usually requires that the population is driven to low numbers before demographic stochasticity can operate. Many examples have been documented, particularly for introductions. Extinction by genetic malfunction appears to come a distant third. Genetic problems have a low priority in saving a natural population from extinction. They are more relevant to managing a population in captivity or one whose size is so small and its future so bleak that it should be in captivity.

We will now look at a few examples of species that went extinct, or came close to extinction, to give us a feel for the range of possibilities.

17.2.1 *Effect of habitat change and fragmentation*

Many extinctions appear to have been caused by habitat changes (Griffiths *et al.* 1989; Brooks *et al.* 2002), but the precise mechanism of population decline is usually difficult to determine retrospectively. One form of habitat loss is fragmentation of continuous habitat into patches. Over time, these patches become smaller and the gaps between them become larger, and the ratio of edge to interior habitat increases (Temple 1986). We have seen this clearly in the fragmentation of the eastern hardwood forests of North America since settlement in the 1600s and in the eucalypt woodlands of Australia in the last century (Saunders *et al.* 1993). The hostility of the matrix is important too: a matrix of young regenerating forest or even exotic plantation is less hostile for animals in old-growth forest patches than a matrix of agriculture. Human residential development is yet more hostile (Friesen *et al.* 1995). A further aspect we need to consider is the type of forest involved. In the northern boreal forests of Europe and Canada, which contain widespread migratory bird species, fragmentation has much less effect on species richness and the ability to colonize than it does in tropical forests, with their highly restricted distributions of birds (Haila 1986) Fig. 17.1.

Fragmentation of habitat has a number of consequences:

1 Species that require interior forest habitats (many bird species), away from the edge, experience reduced habitat and hence population reductions (Saunders *et al.*

1993). In a long-term experiment in which forest fragments of different sizes were constructed in the Amazon Forest, the ecosystem showed aspects of degradation within the patches (Laurance *et al.* 2002). Many bird species avoided even small clearings < 100 m across. Edges were avoided and the type of matrix affected movement. In both England and the eastern United States, extinctions of bird species occurred once a critical percentage of the original habitat was destroyed (McLellan *et al.* 1986).

2 Species that need to disperse through intact habitat (many reptiles, amphibians, and ground-dwelling insects) are prevented from doing so and their populations are reduced to isolated pockets, with potential demographic and genetic consequences. In fragmented parts of the northern boreal forests of North America, the foraging movements of the three-toed woodpecker (*Picoides tridactylus*) are highly constrained because this species strongly avoids open areas (Imbeau and Desroches 2002). Dispersal of crested tits (*Parus cristatus*) in Belgium is restricted in pine forest fragments relative to continuous forest, which probably reduces their ability to settle in preferred habitats (Lens and Dhondt 1994). Northern spotted owls (*Strix occidentalis*) also suffer higher mortality when dispersing across unsuitable habitat between patches (Temple 1991).

3 The greater length of habitat edge allows incursions of predators from outside the patch, increasing the predation rate on interior forest species. We discuss such a case for birds in the eastern hardwoods of North America in Section 7.7 (Fig. 7.10b; Wilcove 1985; Temple and Carey 1988). Nesting success of ovenbirds (*Seiurus aurocapillus*), red-eyed vireos (*Vireo olivaceus*), and wood thrushes (*Hylocichla mustelina*) in these deciduous woodlands is reduced by both higher nest predation and increased brood parasitism from the brown-headed cowbird (*Moluthrus ater*) (Donovan *et al.* 1995). In general, fragmentation results in the synergistic interaction of several deleterious factors, particularly habitat decay, reduced dispersal of animal populations, and increased risk of predation (Hobbs 2001; Laurance and Cochrane 2001. However, species respond differently to fragmentation of habitat. Species that do not move far (insects, reptiles, some forest birds) are more restricted than are highly mobile taxa (many birds, mammals, long-lived species, generalist predators) (Debinski and Holt 2000).

There follows some examples of cases in which extinctions or steep declines were associated with a change in habitat and that change probably caused extinction.

The Gull Island vole

The Gull Island vole (*Microtus nesophilus*) was discovered and described in 1889. It was restricted to the 7 ha Gull Island off Long Island, New York. When Fort Michie was built there in 1897, its construction required that most of the island (and thus the voles' habitat) be coated with concrete. The species has not been seen since.

The hispid hare

The hispid hare (*Caprolagus hispidus*) once ranged along the southern Himalayan foothills from Nepal to Assam but is now restricted to a handful of wildlife sanctuaries and forest reserves in Assam, Bengal, and Nepal. This short-limbed rabbit-like hare depends on tall, dense grass formed as a successional stage maintained either by monsoon flooding or periodic burning (Bell *et al.* 1990). Its decline reflects fragmentation of suitable habitat by agricultural encroachment. Most of the surviving populations are now isolated in small pockets of suitable habitats in reserves. Much of the natural grassland has been lost to agriculture, forestry, flood control, and irrigation schemes.

What remains is modified, even within the reserves, by unseasonal burning and grass cutting to provide thatching material.

Like many endangered species, especially those that are small or inconspicuous, neither density nor rate of decline has been measured. The current status was determined from searches of the few remaining pockets of tall grassland. There is some evidence that contraction of the species into pockets of favorable habitat renders individual hares more vulnerable to predation.

Wallabies and kangaroos

Some of the more dramatic examples of driven extinctions involve the ecology of a significant segment of the fauna being disrupted by large-scale and abrupt habitat changes. In Australia there has been a substantial depletion of the Macropodoidea (about 50 species of kangaroos, wallabies, and rat-kangaroos) following European settlement (Calaby and Grigg 1989).

The changes in the macropod fauna include the extinction of six species and the decline of twenty-three, of which two died out completely on the mainland but still occur in Tasmania. The rates of decline are often difficult to estimate because the year in which the decline began is seldom known and indices of population size are seldom available. Land clearing and extensive sheep farming were in full swing in Australia by 1840 and declines in macropods were evident by the late 1800s. Many of the declines followed sweeping changes to habitat as land was cleared for agriculture or as grazing by sheep modified the vegetation. The smaller macropods (< 5 kg) were the most affected, only one species (*Macropus greyi*) of the larger ones going extinct. Calaby and Grigg (1989) emphasized the difficulty of determining the cause of declines retrospectively but considered that the evidence strongly suggested that the declines of nine species could be referred to the effects of land clearing, eight to modification of vegetation by sheep, five to introduced predators (foxes and cats), and seven to unknown causes.

The sheep grazing and woodland clearing that led to the decline or extinction of at least seventeen species concomitantly benefitted five species of the larger macropods, which increased in numbers. A further four large macropods and four of the eleven species of the smaller rock-wallaby (*Petrogale*) changed little in numbers from the time of the European settlement to the present day.

17.2.2   Effect of loss of food

The Mundanthurai Sanctuary in southern India was classified as a tiger reserve in 1988. Tigers (*Panthera tigris*) live in dense vegetation with access to water, but are restricted to core areas of protected reserves and avoid areas frequented by humans. Since 1988, cattle have been removed and fires controlled, allowing a dense vegetation of exotic *Lantana camara* and other species unpalatable to wild ungulates to grow up. Consequently, the large ungulate species that form the tigers' food have declined. Both tigers and leopards (*P. pardus*) in the reserve have declined with their food supply (Ramakrishnan *et al.* 1999).

17.2.3   Effect of introduced predators

The Lord Howe woodhen

The Lord Howe woodhen (*Tricholimnas sylvestris*) is a rail about the size of a chicken. It lives on the 25 km^2 Lord Howe Island in the southwest Pacific, 700 km off the coast of Australia. Lord Howe is one of the few Pacific islands – and the only high one – that was not discovered by Polynesians, Melanesians, or Micronesians before European contact, and it therefore suffered none of the man-induced extinctions common on

Pacific islands over the first millennium AD. Humans set foot on it for the first time in 1788, at which point it hosted thirteen species of land birds, of which nine became extinct over the next 70 years or so.

The story of the Lord Howe woodhen is related by Hutton (1991, 1998). The island was visited regularly for food (which no doubt included woodhen) and water by sailing ships in the late 18th and early 19th centuries, and was finally settled permanently in 1834. Pigs were introduced before 1839, dogs and cats before 1845, domestic goats before 1851, and the black rat (*Rattus rattus*) in 1918, from a shipwreck. By 1853 the woodhen's range was restricted to the mountainous parts of the island and by 1920 it had apparently contracted to the summit plateau (25 ha) of Mt. Gower, a 825 m (2700 ft) mountain almost surrounded by near vertical cliffs rising out of the sea. The summit plateau is a dreary place: dripping moss forest and perpetual cloud, rather different to the coastal flats that used to be the bird's habitat. This was obviously a species on its last legs, but that problem was not recognized until 1969, after which the population was studied intensively. The population was stable at that point at between eight and ten breeding pairs, although in one year it went down to six pairs. Apparently no more than ten territories could be fitted into the 25 ha of space, so we can be confident that the population on Mt. Gower, and probably that of the entire species, did not exceed that size over the 60 years between 1920 and 1980.

The obvious candidate for the contraction of population size and range was the black rat, which has been implicated in the extinction of several species of bird on islands. In this case, however, the rat does not appear to be involved. The woodhen kills rats, and in any event rats are more common on the summit of Mt. Gower than on any other part of the island. The culprit instead seems to have been the feral pig, which will kill and eat incubating birds and destroy their nest and eggs. Pigs cannot accomplish the minor feat of mountaineering needed to reach the summit of Mt. Gower. The pigs were shot out in the 1970s and the cats by 1980, with the consent of the islanders, who now ban domestic cats.

In 1980, a captive breeding center was established on the island at sea level, seeded with three pairs from Mt. Gower. Thirteen chicks were reared in the first season of captivity, nineteen in the second, and thirty-four in the third. The birds were released and the captive breeding was terminated at the end of 1983. The population reached its maximum at about 180 birds, 50–60 breeding pairs, and that number seems to saturate all suitable habitat on the island: mainly endemic palm forest. A byproduct of the pig and cat control is the expansion of breeding colonies of petrels, shearwaters, and terns.

The Stephen Island wren

The Stephen Island wren (*Xenicus lyalli*), the only known completely flightless passerine, was discovered in 1894. It lived on a 150 ha island in Cook Strait, which separates the North and South Islands of New Zealand. Subfossil remains indicate that it was previously widespread on both main islands but became extinct there several centuries before European settlement, part of the extinction event that followed the colonization of New Zealand by the Polynesians in about AD 900. The causal agent of its extinction on the mainland was probably the Polynesian rat (*Rattus exulans*), introduced by the Polynesians.

The wren was extinguished by a single domestic cat, the pet of the lighthouse keeper, Mr. Lyall. He was the only European to see the species alive, and then on but two occasions, both in the evening. He said it ran like a mouse and did not fly, a fact confirmed subsequently from the structure of the primary feathers. The first one he saw was dead, having been brought in by the cat. Subsequently the cat delivered a further 21, 12 of which eventually found their way to museums. Then it brought in no more. The species went extinct in the same year that it was discovered (Galbreath 1989)

17.2.4 *Side effects of pest control*

The effects of pest control often exceed the original intentions of the control exercise.

The black-footed ferret

The sinuously elegant black-footed ferret (*Mustela nigripes*) provides an example of a species paying the price for the control of another. This account of its narrow escape from extinction is taken from Seal *et al.* (1989), Cohn (1991), and Biggins *et al.* (1999).

The black-footed ferret once ranged across most of the central plains of North America from southern Canada to Texas. Its lifestyle is closely linked to that of the prairie dog (*Cynomys leucurus*), a squirrel-like rodent that once lived in huge colonies on the plains. The ferret feeds mainly on prairie dogs but can feed also on mice, ground squirrels, and rabbits. However, 90% of its diet comprises prairie dogs. The ferret lives in the warrens or burrow systems of the prairie dogs and hence that species provides the ferret with both its habitat and a large proportion of its food supply. Around the turn of the century there was a concerted effort to eradicate the prairie dog, which was viewed as vermin by ranchers. It was seen as competing with sheep and cattle for grass and its burrow systems made riding a horse unsafe. The prairie dogs were poisoned, trapped, and shot in their millions by farmers and government pest controllers. As the prairie dogs went, so did the ferrets. By the middle of the century they were judged to be extinct, but in 1964 a small population was discovered in South Dakota. That colony died out in 1973. In 1981 a colony was discovered in Wyoming. Careful censusing produced an estimate of 129 individuals in 1984, but by the middle of 1985 the population had declined to 58 animals, and within a few months it was down to 31. Canine distemper was diagnosed in this population, and it might well have been the cause of that decline (see Chapter 8).

With the population obviously threatened, there was an attempt to capture the remaining animals and install them in a captive breeding colony. Five were caught in 1985, twelve in 1986, and one in 1987. By February 1987 the last known wild black-footed ferret was in captivity. Captive breeding was successful and 49 and 37 ferrets were released in 1991 and 1992, respectively, into their former range in Wyoming (Biggins *et al.* 1999).

17.2.5 *Effect of poorly regulated commercial hunting*

The type example of serious declines caused by hunting is provided by the history of commercial whaling. This demonstrates the effect of the discount rate (see Chapter 18) upon the commercial decision determining whether a sustained yield is taken or the stock is driven to commercial extinction.

Market economics will act to conserve a commercially harvested species only when that species has an intrinsic rate of increase r_m considerably in excess of the commercial discount rate: the interest a bank charges on a loan to a valued customer. Hence, when a species is harvested commercially the yield must be regulated by an organization whose existence and funding are independent of the economics of the industry that it regulates; otherwise it will necessarily endorse the quite rational economic decisions of the industry, which may well be to drive a stock to very low numbers and then switch to another.

Muskoxen in mainland Canada

Unregulated commercial hunting reduced the muskoxen (*Ovibos moschatus*) on the arctic mainland of Canada to about 500 animals by 1917. In that year the species was protected by the Canadian government. The size of the historic populations will never be known but documentation of the purchase of muskox hides from native hunters by trading companies was detailed. Barr (1991) collated the records and estimated that a minimum of 21 000 muskoxen were taken between 1860 and 1916. Their hides were shipped to Europe as sleigh and carriage robes, replacing bison robes after that species had been reduced almost to extinction.

Commercial hunting appears to be the overriding cause of the virtual extinction of the muskox on the Canadian mainland. Legislative protection successfully reversed the trend. Muskoxen now number about 15 000 on the mainland and have reoccupied almost all their historic range (Reynolds 1998). The conservative hunting quotas introduced in the 1970s did not stop that recovery.

17.2.6 *Effect of unregulated recreational hunting*

Recreational hunting is intrinsically safer than commercial hunting because sport hunters operate on an implicit discount rate of zero. Sport hunting hence has an enviable record of conserving hunted stocks. Instances of gross overexploitation are rare but not unknown.

The Arabian oryx

The Arabian oryx (*Oryx leucoryx*) is a spectacular antelope whose demise in the wild and subsequent reestablishment from captive stock are related by Stanley Price (1989) and Gordon (1991). Its original distribution appears to have included most of the Arabian Peninsula, but by the end of the 19th century the remaining individuals were divided into two populations. A northern group lived in and around the sand desert of northern Saudi Arabia known as the Great Nafud and a southern group occupied the Rub' al Khali (the Empty Quarter) of southern Arabia.

The northern population became extinct in about 1950. The range of the southern population declined from about 400 000 km^2 in 1930, through 250 000 km^2 in 1950, to 10 000 km^2 in 1970. Within the next couple of years the population was reduced to six animals in a single herd. They were shot out on 18 October 1972.

Recreational hunting caused this extinction. The countries of the Arabian Peninsula are essentially sea frontages, the inland boundaries being little more than lines on a map. There is scant control over activities in the hinterlands. Oil company employees and their followers used company trucks for hunting trips and seem to have been at least partly responsible for the decline. Then there were the large motorized hunting expeditions originating mainly from Saudi Arabia. These were self-contained convoys that included fuel and water tankers. The vehicles and support facilities allowed large areas to be swept each day, with efficient removal of the wildlife. Their main quarry was bustards and hares, secured by hawking, but gazelles and Arabian oryx were also chased. One such party crossed into the Aden Protectorate (now the People's Democratic Republic of Yemen) in 1961 and killed 48 Arabian oryx, about half the population of that region. In the 1960s, every year large parties from Qatar would capture Arabian oryx with nets in the hinterland of Oman and truck them 900 km back to replenish the captive herd of Shaikh Kasim bin Hamid.

17.2.7 Effect of competition with introduced species

More bird species have been introduced to the Hawaiian Islands than to any other comparable land area. Of 162 species introduced, 45 are fully established and 25 have secured at least a foothold (Scott *et al.* 1986). These exotic species have been suggested as one of the causes of the decline and extinction of native birds.

Mountainspring and Scott (1985) estimated the geographic association within pairs of the more common small- to medium-sized insectivorous forest passerines. After statistically removing the effect of habitat, they showed that a higher proportion of exotic/native pairs of forest birds were negatively associated than were pairs of indigenous birds. They suggested that these results reflected competition, mainly for food.

The Japanese white-eye (*Zosterops japonicus*) became the most abundant land bird in Hawaiian Islands after being introduced to Oahu in 1929 and to the island of Hawaii in 1937. It feeds on a wide variety of foods and is fairly catholic in its choice of habitats. It shares the range of at least three native species with similar food habits. Although causality cannot be demonstrated conclusively, particularly in retrospect, it is possible to draw a strong inference that the Japanese white-eye was involved in the decline of the Hawaii creeper (*Oreomystis mana*) in the 1940s.

17.2.8 Effect of environmental contaminants

Rachel Carson's (1962) classic book *Silent Spring* raised the alarm about the effects on birds of DDT and other organochlorides. In particular, these chemicals caused eggshells to become abnormally thin and fragile (Ratcliffe 1970; Cooke 1973). Because these chemicals accumulated in the food chain, it was the species at the top – the raptors – that suffered the most from the effects. Nesting success declined precipitously and raptor populations collapsed (Hickey and Anderson 1968; Cade *et al.* 1971). The chemical industry initially denied these unwelcome side effects, but by the 1970s the evidence was overwhelming and DDT was banned in most countries. As a result, we have seen a rebound in the populations of several species of raptor, such as the peregrine falcon (*Falco peregrinus*) in North America and the common buzzard (*Buteo buteo*) in Europe.

The issue of contaminants remains with the introduction of new pesticides and herbicides. For example, the so-called second generation of anticoagulants, such as brodifacoum, is both highly toxic to birds and mammals and persistent, so that these chemicals increase the risk of secondary poisoning of non-target species (Eason *et al.* 2002). Monofluoracetate (1080) is commonly used in baits to kill mammal pests in Australia and New Zealand, but it has impacts on non-target birds and mammals (Spurr 1994, 2000).

Indian vultures
In Pakistan and India, cattle and water buffalo (*Bubalus bubalis*) are treated with a nonsteroidal anti-inflammatory drug, diclofenac, to counter the effects of trauma and disease. When vultures feed on the carcasses of these domestic animals they die from the toxic effects of diclofenac, to which they appear particularly sensitive. Three species are affected: the white-backed vulture (*Gyps bengalensis*), the long-billed vulture (*G. indicus*), and the slender-billed vulture (*G. tenuirostris*). The evidence suggests renal failure in the birds.

Between 1990 and 2003, vulture populations declined from tens of thousands to a level where captive breeding is required. At least 95% of the populations died in 10 years. Research is now required to identify alternative drugs that are safe for vultures but remain effective for livestock (Green *et al.* 2004).

The California condor

The story of the California condor (*Gymnogyps californianus*) reveals a little about the realities of conservation: the gaps between theory and practice and the overwhelming need to determine, not assume, the causes of the decline.

The California condor was probably abundant and widely distributed in southern North America during the Pleistocene. Later, it figured in the ceremonies and myths of the prehistoric and historic Indians and caught the eye (and trigger finger) of the early European explorers. California condors ranged from the Columbia River south into New Mexico in the 1800s, but by 1940 their range had contracted to a small area north of Los Angeles. Koford's (1953) estimate, based on sightings, of only 60 individuals surviving by the early 1950s was probably low. Annual surveys by simultaneous observations of known concentrations were begun in 1965 but abandoned in 1981 because they were judged to be subject to unacceptable error. Photographic identifications were then used to generate a total count of 19–21 birds in 1983 (Snyder and Johnson 1985). The decline continued until, in 1985, the last eight wild individuals were caught and added to the captive flock.

The causes of the initial decline were probably shooting and loss of habitat, but the supporting evidence is anecdotal. Low productivity caused by an insufficiency of food was suggested as a cause of the decline during the 1960s. Road-killed deer were cached at feeding stations in 1971–73 to alleviate the perceived shortage of food (Wilbur *et al.* 1974). That program was run for an insufficient time to determine whether supplementary feeding was associated with increased productivity.

The connection between toxic organochlorines and eggshell thinning in birds was established in the late 1960s but the resulting flurry of studies focused on bird-eating and fish-eating birds because avian scavengers were assumed to be less at risk. The possibility of a causal association between environmental toxins and the later decline of the condor was recognized in the mid-1970s but determination of the specific role of organochlorines in that decline was delayed (Kiff 1989). Eggshell samples from California condors had been collected in the late 1960s but for various reasons, including mishaps to the samples, analyses were delayed until the mid-1970s. The negative correlation between eggshell thickness and DDE levels was significant: shells were thinner and had a different structure to those collected before 1944.

It was known that condor eggs often broke, but the cause was open to debate. Even the monitoring activities themselves were suspected as being the cause. The evidence for organochlorines was circumstantial, but it led Kiff (1989) to conclude that "DDE contamination probably had a very serious impact on the breeding success of the remnant population in the 1960s, leading to a subsequent decline in the number of individuals added to the pool of breeding adults in the 1970s." In 1972, DDT was banned in the United States. The few eggs measured after 1975 had thicker eggshells, and this led to guarded optimism. In March 1986, however, an egg laid by the last female to attempt breeding in the wild was found broken. Its thin shell was suspiciously reminiscent of the "DDE thin-eggshell syndrome." In the meantime, analysis of tissue from wild condors found dead in the early 1980s revealed that three of the

five had died from lead poisoning, probably from ingesting bullet fragments in carrion. Other condors had elevated lead levels in their blood (Wiemeyer *et al.* 1988).
Recognition of the deleterious effects of yet another toxin in the condor's food supply led to provision of "clean" carcasses just before the last condors were taken into
captivity.

17.2.9 *Effect*
of introduced diseases

Extinctions caused by disease are particularly difficult to identify in retrospect. Moreover, on theoretical grounds disease is unlikely to be a common agent of extinction. In
their review of pathogens and parasites as invaders, Dobson and May (1986b) noted
the improbability of a parasite or pathogen driving its host to extinction unless it had
access to alternative hosts.

Hawaiian birds

Avian malaria and avian pox have been suggested as contributing to the decline of
Hawaiian birds (Warner 1968). Migratory waterfowl may have provided a reservoir for
avian malaria on the Hawaiian Islands, and the continuous reintroduction by migration
may have maintained a high level of infection in the face of a decline in host numbers
(Dobson and May 1986b). Alternatively, avian malaria may have been carried by introduced birds such as the common myna (*Acridotheres tristis*), which may themselves
have maintained the disease at a high level, as they are not greatly affected by it.

Originally there were no mosquitoes on Hawaii capable of spreading malaria. The
accidental introduction of mosquitoes in 1826 and their rapid spread throughout the
islands coincided with the decline of many species of bird. By 1901, 6 of 11 endemic
passerines on Oahu had died out, before their habitats were modified (Warner 1968).
Experiments showed that the Hawaiian passerines, especially the honeycreepers, were
highly susceptible to malaria: much more so than the introduced species (Warner
1968).

Avian malaria is a factor in restricting the present distribution of native birds on
Hawaii, lending credence to the suggestion that it is implicated in the extinction of
other species. Scott *et al.* (1986) noted that elevations above 1500 m that were free
of mosquitoes hosted the highest densities of native birds, and especially of the rarer
passerines.

17.2.10 *Effect of*
multiple causes

The history of the heath hen is related by Bent (1932). We use here the summary and
interpretation of that history presented by Simberloff (1988).

Probably the best-studied extinction is that of the heath hen (*Tympanuchus cupido
cupido*). This bird was originally common in sandy scrub-oak plains throughout much
of the northeastern United States, but hunting and habitat destruction had eliminated
it everywhere but Martha's Vineyard by 1870. By 1908 there were 50 individuals, for
whom a 1600 acre refuge was established. Habitat was improved and by 1915 the population was estimated to be 2000. However, a gale-driven fire in 1916 killed many birds
and destroyed habitat. The next winter was unusually harsh and was punctuated by a
flight of goshawks; the population fell to 150, mostly males. In addition to the sex ratio
imbalance, there was soon evidence of inbreeding depression: declining sexual vigor.
In 1920 a disease of poultry killed many birds. By 1927 there were 13 heath hens (11
males); the last one died in 1932. It is apparent that, even though hunting and habitat
destruction were minimized, certainly by 1908 and perhaps even earlier the species was

doomed. Catastrophes, inbreeding depression and/or social dysfunction, demographic stochasticity, and environmental stochasticity all played a role in the final demise.

17.3 How to prevent extinction

The previous sections summarized 12 examples of extinction or steep decline. The decline of the heath hen and the Hawaiian birds may be attributable to several factors but research on those species has not adequately revealed the causes of the declines. The extinction of several species of wallaby seems very likely to reflect habitat modification. The extinction of the Stephen Island wren and the decline of the Lord Howe woodhen were unambiguously caused by an introduced predator. The extinction of the Arabian oryx in the wild (despite its subsequent reestablishment from captive stock) and the near extinction of the muskox were also caused by predation, this time by people. The black-footed ferret went extinct in the wild because the source of its habitat and most of its food supply – the prairie dog – was greatly reduced in density by control operations.

These examples implicate only a few potential causes of decline. Probably the most important is modification or destruction of habitat. The local extinction of several mammals from the sheep rangelands of Australia appears to have been caused by habitat changes induced by sheep introduced in the mid-1800s; 12 of the original 38 species of marsupials and 6 of the original 45 species of eutherians (endemic rodents and bats) no longer live in that region (Robertson et al. 1987).

The first step in averting extinction is to recognize the problem. Many species have slid unnoticed to the brink of extinction before their virtual absence has been noticed. The smaller mammals and birds, and frogs and reptiles, are more likely to be overlooked than are large ungulates and carnivores.

The second step is to discover how the population got into its present mess:
- Is the cause of decline a single factor or a combination of factors?
- Are those factors still operating?
- If so, can they be nullified?

The cause of a decline is established by application of the researcher's tools of trade: the listing of possible causes and their sequential elimination individually or in groups, according to whether their predicted effects are observed in fact. This is the standard toolkit of hypothesis production and testing.

It is essential that the logic of the exercise be mapped out before the task is begun. The listing of potential causes is followed by a formulation and then testing of predictions. The efficiency of the exercise is critically dependent on the order in which the hypotheses are tested. Get that wrong and a 3-month job may become a 3-year project. In the meantime, the population may have slid closer to the threshold of extinction, so time is important.

Box 17.1 gives a specimen protocol for determining the cause of a population's decline. The example comes from the decline of caribou on Banks Island in the Canadian Arctic. The first aerial surveys of the island in 1972 revealed an estimated population of 11 000 caribou. Subsequent surveys in the 1980s traced a dwindling population, which numbered barely 900 caribou by 1991. Since then the population has stabilized, being 1195 in 2001 (N. Larter pers. comm.). During the same time, muskoxen increased from 3000 to 46 000, leading to fears that there were too many for the good of the caribou (Gunn et al. 1991). The population continued to increase to 64 600 by 1994 and then slowed to about 69 000 by 2001 (N. Larter pers. comm.).

Box 17.1 Hypotheses to be tested to discover the cause of the decline of caribou on Banks Island, Northwest Territories, Canada.

Hypotheses to account for the decline:

Either:
  A Food shortage
  B Increased predation

If (A) then mechanisms might be:
  A1 Increase in weather events such as freezing rain that affect availability of food
  A2 Competition for food with muskoxen, which are increasing
  A3 Caribou themselves reducing the supply of food

If (B) then mechanisms might be:
  B1 Wolf predation
  B2 Human predation

   The food shortage hypotheses (A) may be tested against the predation hypotheses (B) by checking body condition. Hypotheses A predict poor body condition and low fecundity during a population decline, hypotheses B good condition and high fecundity.
   If this test identifies hypotheses A as the more likely then A1 is separated from A2 and A3 by its predicting a positive rate of increase in some years; A2 and A3 predict negative rates of increase in all years. A2 (competition with another species) is separated from A3 (competition between caribou) by checking for concomitant decline of caribou where musk oxen are not present in the same climatic zone.

Particularly severe winters restricted foraging for the caribou and caused die-offs, at least in 1972–73 and 1976–77. The frequency of severe winters with deep snow and freezing rain increased during the 1970s and 1980s. Caribou and muskoxen differ in lifestyles and responses to winter weather.

   An example of how difficult it can be to get the logic of diagnosis right is provided by research and treatment of the endangered Puerto Rican parrot (*Amazona vittata*). This strikingly attired bird has been the focus of some 50 years of intensive conservation efforts, including some 20 000 hours of observations of ecology and behavior (Snyder *et al.* 1987). The parrot may have numbered more than 1 million historically, but by the early 19th century its distribution had contracted severely with the clearing of much of the forest of Puerto Rica. By the 1930s it was estimated at 2000 and by the mid-1950s, when the first intensive studies started, its numbers had collapsed to 200. Only 24 parrots were left in 1968 when rescue efforts were resumed. Despite a high-profile effort, including a captive breeding program started in 1968, little progress can be reported. The number of parrots in the wild population numbered only 21–23 before the 1992 breeding season (Collar *et al.* 1992) and was still only around 30 birds in 2004 (J. Wunderle pers. comm.). The cause of the decline has not yet been identified unambiguously.

17.4 Rescue and recovery of near-extinctions

Once a decline in a species is recognized and the causes are determined the problem can be treated. The species accounts in the preceding sections give some idea of the range of management actions available to rescue a species from the risk of extinction. Sometimes, all it takes is a legislative change such as a ban on hunting (as with the Canadian muskox). More usually, active management (such as predator control and captive breeding for the Lord Howe woodhen) is necessary. The management actions needed to reverse the fortunes of a declining species are seldom more than conventional management techniques, unless a species is in desperate straits. Then a whole new set of techniques may be called under the heading of *ex situ*. *Ex situ* techniques preserve and amplify a population of an endangered species outside its natural habitat. Thereafter, it can be reintroduced. The Lord Howe woodhen and the Arabian oryx are examples of such reintroductions.

Reintroducing a species to the area from which it died out should not be attempted without some understanding of why the species went extinct there in the first place. Stanley Price (1989) describes the reintroduction of the Arabian oryx with captive stock and details the considerations that should precede a reintroduction. Similar procedures were considered for the release of Przewalski's horse (*Equus przewalskii*) in Mongolia (Ryder 1993) and the addax (*Addax nasomaculatus*) in Niger (Dixon *et al.* 1991). In any event, the liberated nucleus should be large enough to avoid demographic malfunction. Twelve individuals is an absolute minimum for an introduction. Twenty is relatively safe. Cade and Jones (1993) detail the successful captive breeding and reintroduction of the Mauritius kestrel (*Falco punctatus*). In the 1970s it was down to two breeding pairs, but by the 1990s some 235 birds had been reintroduced and established in new habitats.

When the cause of a local extinction is unknown, and when we therefore do not know whether the factor causing the extinction is still operating, a trial liberation should precede any serious attempt to repopulate the area. The 20 or more individuals forming the probe are instrumented where possible (e.g. with radio collars) and monitored carefully to determine whether they survive and multiply or, if not, the cause of their decline. In the latter case, the factor operating against the species can be identified and countermeasures can be formulated. It is worth noting that a closely related species may be used as a probe when it is too risky to use individuals of the endangered species. For example, a successful probe release of Andean condors (*Vultur gryphus*) cleared the way for the release of two Californian condors from a captive breeding population in 1992 (Collar *et al.* 1992).

Short *et al.* (1992) showed the importance of probing for the reintroductions of several wallaby species. Of 10 liberations into areas where the species had once been present but had died out, all failed. Of 16 liberations into areas where the species had not previously occurred, about half were successful. Apparently the factors that had caused the original extinctions of the first category were still operating. The authors suggested that exotic predators were probably the dominant factor.

17.5 Conservation in National Parks and reserves

National Parks and reserves are preeminently important as instruments of conservation. In these areas alone, the conservation of species supposedly takes precedence over all other uses of the land. Debate over whether protected areas such as National Parks or community conservation areas are best for conservation is probably unnecessary because both have their advantages and disadvantages, as outlined in Box 17.2.

17.5.1 *What are National Parks and reserves for?*

On one level, that question is trite, and it leads to the equally trite answer that parks and reserves are for the conservation of nature. When the question is refined to "What are the precise objectives of *this* park?" the answer must be more concrete. However, even the general question is not as trite as it might seem. It is instructive to follow the history of ideas about the function of reserves, of which National Parks can serve as the type example. Here we summarize changing perceptions, as outlined by Shepherd and Caughley (1987).

Box 17.2 Advantages
and disadvantages of
protected areas such as
National Parks compared
to community
conservation areas.

Advantages of protected areas:

1 Will protect fragile habitats (swamps, tundra, islands, endangered species). For example, the only breeding grounds of the whooping crane (*Grus americana*) occur within the Wood Buffalo National Park, Canada, and the only known location of the Madagascan tomato frog (*Dyscophus antongilii*) occurs in a single pond in the north of Madagascar.
2 Will protect large species that cannot coexist with humans, such as large carnivores and herbivores.
3 Can act as ecological baselines or benchmarks to monitor human disturbance outside (Arcese and Sinclair 1997; Sinclair 1998).

Disadvantages of protected areas:

1 Do not represent all ecosystems or communities, often being selected for other reasons.
2 Often too small to maintain viable populations, particularly of species that are adapted to live in large groups or that migrate across international borders (e.g. migrating caribou, bison, saiga antelope (*Saiga tatarica*), shore birds (Charadriidae)).
3 Can alienate locally indigenous peoples excluded by central governments.

Advantages of community conservation areas:

1 Can represent species not included in protected areas, such as noncharismatic species (lower animals, microbes, fungi).
2 Can co-opt support of local peoples if benefits accrue to them.

Disadvantages of community conservation areas:

1 Tend to protect only species of direct benefit to humans and ignores the rest, which is the vast majority.
2 Excludes species that are detrimental to humans
3 Tend to discount the future due to (i) increasing human population demands on the ecosystem and (ii) accelerating economic expectations from the system even with stationary human populations. These result in species loss and ecosystem decline.

The National Park idea has two quite separate philosophical springs, whose streams did not converge until about 1950. The first is American, exemplified by the US Act of 1872 proclaiming Yellowstone as the world's first National Park. The intent was to preserve scenery rather than animals or plants. Public hunting and fishing were at first entirely acceptable.

The second spring is "British colonial," with the Crown asserting ownership over game animals and setting aside large tracts of land for their preservation. The great National Parks of Africa grew out of these game reserves, some physically and the others philosophically. Wildlife was the primary concern and scenery came second, if at all. The first was Kruger National Park, established in 1926 on a game reserve proclaimed in 1898. The Serengeti, in Tanzania, was gazetted in 1947, following from a reserve established in 1927. Kenya's first National Park was established in 1946 on the Nairobi common.

All National Parks established for 40 years or more have had their objectives and their management modified several times. The more influential fashions in park theory, listed here roughly in order of appearance over the last 100 years, are not mutually exclusive. They tend to be added to rather than replaced.

1 *The most important objective is to conserve scenery and "nice" animals*   This aim has translated into restricting roads and railways and attempting to exterminate the carnivores. Banff National Park in Canada has such a history.
2 *The most important objective is the conservation of soil and plants*   This aim was a direct consequence of the rise of the discipline of range management in the United

States during the 1930s. Its axiom was (and still is) that there is a "proper" plant composition and density. Enough herbivores were to be shot each year to hold the pressure of grazing and browsing at the "correct" level. An ecosystem could not manage itself: if left to its own devices it would do the wrong thing (Macnab 1985).

3 *The most important objective is the conservation of the physical and biological state of the park at some arbitrary date*   In the United States, South Africa, and Australia, that date marks the arrival of the first European. This is the source of much of the controversy in Yellowstone National Park.

4 The fashion shifted to the conservation of representative examples of plant and animal associations. The wording is from Bell's (1981) definition of the function of National Parks in Malawi, but the objective underlies the management of National Parks in many other countries.

5 *The most important objective is the conservation of "biological diversity" (or biodiversity)*   This catchphrase had two meanings. It was sometimes used in the sense of "species diversity" (MacArthur 1957, 1960), whereby the information-theory statistic of Shannon and Wiener could be used to estimate the probability that the next animal you saw would differ at the species level from the last. The statistic is maximized for a given number of species when all have the same density. Within park management, the idea translated as "the more species, the better." The second meaning dealt with associations rather than species: the more diverse a set of plant associations, the better the National Park. For example, Porter (1977) defined the objectives of the Hluhluwe Game Reserve in Natal as, "To maintain, modify and/or improve (where necessary) the habitat diversity presently found in the area and thus ensure the perpetuation and natural existence of all species of fauna and flora indigenous to the proclaimed area."

6 *The most important objective is the conservation of "genetic variability"*   This phrase can be defined tightly and usefully (e.g. Frankel and Soulé 1981), but within the theory and practice of park management it lacked focus. It was tossed around with little or no attempt to define or understand what it meant, or whether the variability sought was in heterozygosity, in allelic frequency, or in phenotypic polymorphism. In practice, it again translated into "the more species, the better."

7 The most recent objective differs in kind from the six previous ones. Frankel and Soulé (1981) express it thus: "the purpose of a nature reserve [in which category they include National Parks] is to maintain, hopefully in perpetuity, a highly complex set of ecological, genetic, behavioral, evolutionary and physical processes and the coevolved, compatible populations which participate in these processes." Don Despain (quoted by Schullery 1984) puts it more plainly: "The resource is wildness."

17.5.2 *Processes or states?*

The first six objectives identify biological states as the things to be conserved. The seventh identifies biological processes as the appropriate target of conservation. At first glance, Frankel and Soulé's purpose of a nature reserve appears to require the maintenance of states, because it refers to the conservation of populations. However, populations are not states in the sense that plant associations are states. A plant association has a species composition. Its component populations must have a ratio of densities one to another that remains within defined limits. If those limits are breached, the plant association changes into another kind. A population, however, is not defined by ratios. The ratio of numbers in one age class relative to those in another, or the ratio of males to females, has no bearing on its status as a population.

The management of a National Park will be determined by whether the aim is to conserve biological and physical states by suppressing processes or to preserve processes without worrying too much about the resultant states. There are three options:

1 If the aim is to conserve specified animal and plant associations that may be modified or eliminated by wildfire, grazing, or predation then intervene to reduce the intensity of wildfire, grazing, or predation.

2 If the aim is to give full rein to the processes of the system and to accept the resultant, often transient, states that those processes produce then do not intervene.

3 A combination of both: if the aim is to allow the processes of the system to proceed unhindered unless they produce "unacceptable" states then intervene only when unacceptable outcomes appear likely.

17.5.3 Effects of area

Within any group of islands (e.g. the Antilles, Indonesia, Micronesia), big ones tend to contain more species than do small ones. Size as such is not the only influence on the number of species – distance to the mainland, for example, plays a part – but it does provide a close prediction. The relationship between the number of species and the size of the area within which they were surveyed is known as a species–area curve. Algebraically it takes the form:

$$S = CA^z$$

where S is the number of species of a given taxon (e.g. lizards, forest birds, vascular plants), A is the area, C is the expected number of species on one unit of area (usually $1 \, km^2$), and z indexes the slope of the curve relating the number of species to the number of square kilometers.

Table 17.1 shows the relationship between species number and land area for Tasmania and the islands between it and the Australian mainland (Hope 1972). These were all linked to each other and to the Australian mainland up to about 10 000 years ago; the subsequent fragmentation reflects the rise of sea level at the end of the Pleistocene. The number of marsupial herbivores that they carry therefore reflects differential extinction without reciprocal immigration over the last 10 millennia. The estimated $z = 0.18$ is low for islands, being closer to that expected for areas within continents, and probably reflects the recent continental nature of those islands. Box 17.3 shows how C and z are calculated from these data.

Table 17.1 Relationship between the number of species of herbivorous marsupial and the area of land on Tasmania and the islands between it and the Australian mainland.* The "expected" number is calculated as $S = 1.70A)^{0.18}$. (See Box 17.3 for calculation.)

Island	Area, km^2	Observed species	Expected specis
Tasmania	67 900	10	12.6
Flinders	1330	7	6.3
King	1100	6	6.0
Cape Barren	445	6	5.1
Clarke	115	4	4.0
Deal	20	5	2.9
Badger	10	2	2.6
Prime Seal	9	2	2.5
Erith-Dove	8	3	2.5
Vansittart	8	2	2.5
West Sister	6	2	2.3

*Number of species as at AD 1800. Only islands larger than $5 \, km^2$ are included. (Data from Hope 1972).

Box 17.3 Estimating the constants of a species–area curve.

A species–area curve takes the form $S = CA^z$ where:

S = number of species
A = area, in this case always expressed as km^2
C = expected number of species on an area of 1 km^2
z = slope of the curve relating species number to area

Taking the data of Table 18.1 as our example, we first convert area and species number to logarithms. Any base will do, but we will use logs to the base e. Labeling log area as x and log species number as y:

x	y
11.126	2.3026
7.193	1.9459
7.003	1.7918
6.098	1.7918
4.745	1.3863
2.996	1.6094
2.303	0.6931
2.197	0.6931
2.079	1.0986
2.079	0.6931
1.792	0.6931

We now calculate these:

$$n = 11$$
$$\text{mean } x = 4.510 \qquad\qquad \text{mean } y = 1.336$$
$$\sum x = 49.61 \qquad\qquad \sum y = 14.70$$
$$\sum x^2 = 315.2 \qquad\qquad \sum xy = 82.58$$
$$(\sum x)^2/n = 223.7 \qquad\qquad (\sum x)(\sum y)/n = 66.30$$
$$SS_x = \sum x^2 - (\sum x)^2/n = 91.5 \qquad SS_{xy} = \sum xy - (\sum x)(\sum x)/n = 16.28$$

The constants of the species–area curve are now solved:

$$z = SS_{xy}/SS_x$$
$$= 16.28/91.5$$
$$= 0.18$$
$$C = \text{antilog}\left(\sum y/n - z \sum x/n\right)$$
$$= \exp(1.336 - 0.18 \times 4.51)$$
$$= 1.70$$

Thus, $S = 1.7A^{0.18}$

C, the expected number of species on one unit of area, varies according to latitude, elevation, ecological zone, taxonomic group, and the units in which A is measured. In contrast, z tends to be quite stable. For most taxa and groups of islands, it lies between 0.2 and 0.4. At the mid-point, 0.3, an increase or decrease of area by a factor of 10 results in a doubling or halving, respectively, of the number of species (by virtue of $10^{0.3} = 2$). Thus, when $A = 1$, $S = C$ irrespective of the value of z; and when $A = 10$ and $z = 0.3$, $S = 2C$.

The relationship is the same if we count the number of species on nested areas of progressively larger size on a continent. Here, the value of z tends to be lower, usually around 0.15. This implies that a reduction of area by a factor of 10 reduces the number of species by a factor of only 1.4 ($10^{0.15} = 1.41$). The difference between the exponent of 0.3 for islands and 0.15 for continents probably reflects the easier dispersal between contiguous areas of land as against between islands. These relationships are particularly important for determining optimum sizes of reserve.

17.5.4 Is one big National Park better than two small National Parks?

Suppose we have the money necessary to acquire $100 \, km^2$ of land for conversion into National Parks. If the aim were to conserve the maximum number of species for a long time, should we go for one park of $100 \, km^2$ or two of $50 \, km^2$ each? Obviously, a number of factors would influence our choice, but let us assume that the overriding aim is to maximize the number of species of mammals within the single large reserve or the alternative two smaller reserves. Let us assume that $1 \, km^2$ will on average contain 20 species in this region (i.e. $C = 20$). Further, we know that $z = 0.15$ for mammals in this region. Thus, a National Park of $100 \, km^2$ would contain about 40 species of mammals ($S = CA^z = 20 \times 100^{0.15} = 40$), whereas a park of $50 \, km^2$ would hold about 36 mammals ($S = 20 \times 50^{0.15} = 36$). Whether we favor one park of $100 \, km^2$ or two of $50 \, km^2$ each comes down to how many species are held in common by the two smaller parks. That will depend on the extent to which they differ in habitat and on the distance between them.

The efficacy with which a reserve system conserves species and communities thus depends on the size of the reserves and, more importantly, on where they are: their dispersion relative to the distribution patterns of species. Margules *et al.* (1982) warn against using data-free geometric design strategies (big is better than small, three is better than two, linked is better than unlinked, grouped is better than linear).

17.5.5 Culling in National Parks and reserves

Whether or not the densities of mammals should be controlled artificially in a National Park is a matter of some contention, as illustrated by the papers in the *Wildlife Society Bulletin* No. 3 (1998), which discuss culling in general and the debate on Yellowstone National Park in particular. White *et al.* (1998) present the case for culling, particularly of the elk, while Singer *et al.* (1998), Boyce (1998), Frank (1998), and Detling (1998) present other viewpoints. Our own prejudices are to avoid culling in parks and reserves except in rare, special, and well-defined circumstances.

17.6 Community conservation outside National Parks and reserves

The principles of conservation discussed in the previous section with reference to parks and reserves hold for conservation in other areas too. There are, however, a few important differences. In general, protected areas cover no more than about 10% of the terrestrial global surface, which means from our species–area equation (Section 17.5.3) that only about 50% of the world's species are included. Thus, at least half of our terrestrial biota must be conserved in human-dominated systems. Box 17.2 outlines the pros and cons of community conservation approaches.

Some species and associations of species occur only rarely in reserves, because parks and reserves do not capture a representative sample of the biota. In Australia, for example, few reserves contain forest types that grow on sites of high fertility. Most such sites were incorporated into state forests or alienated from common ownership before the reserve system was established. The koala (*Phascolarctos cinereus*) is dependent on such sites and so almost all attempts to conserve koalas must be made outside the reserve network, where the manager does not have the same control over land-use practices.

Legislation is the main means by which conservation is advanced outside reserves. Various practices, such as the killing of nominated species, are banned. Less commonly, there are controls over land clearing, protecting the habitat of species that dwell in forest and woodland. Activities on land owned by the people as a whole, even though that land is not designated as a conservation reserve, may be subject to environmental impact assessment (EIA). Laws governing conservation outside reserves should take legal precedence over forestry and mining law.

17.7 International conservation

Conservation is the responsibility of sovereign nations unless the issue is subject to international treaty (polar bears, ivory trade, migratory birds) or unless the problem occurs on the high seas (whales and pelagic fish stocks), on essentially unclaimed land (Antarctica), or on land under disputed sovereignty (parts of the high Arctic).

17.7.1 IUCN Red Data Books

The International Union of Nature and Natural Resources (IUCN) issues "Red Data Books" listing threatened species. Five categories are recognized, their exact wording varying according to the taxon. What follows is generalized.

1 *Extinct (Ex)*   Species not definitely located in the wild during the last 50 years.

2 *Endangered (E)*   Taxa in danger of extinction and whose survival is unlikely if the causal factors continue operating. Included are taxa whose numbers have been reduced to a critical level or whose habitats have been so drastically reduced that they are deemed to be in immediate danger of extinction. Also included are taxa that are possibly already extinct but have definitely been seen in the wild in the past 50 years.

3 *Vulnerable (V)*   Taxa believed likely to move into the "endangered" category in the near future if the causal factors continue to operate. Included are taxa of which most or all populations are decreasing because of overexploitation, extensive destruction of habitat, or other environmental disturbance; taxa with populations that have been seriously depleted and whose ultimate security has not yet been assured; and taxa with populations that are still abundant but are under threat from severe adverse factors throughout their range.

4 *Rare (R)*   Taxa with small world populations that are not at present "endangered" or "vulnerable" but are at risk. These taxa are usually localized within restricted geographical areas or habitats, or are thinly scattered over a more extensive range.

5 *Indeterminate (I)*   Taxa known to be "endangered", "vulnerable", or "rare" but where there is not enough information to say which of the three categories is appropriate.

Of these categories, "endangered" and "vulnerable" are the most important, and there is widespread agreement on what these terms mean. "Rare" is not a particularly useful category of extinction risk and probably should not be used as such. If rarity itself is the cause of the risk, in the sense that the population size is at a level low enough to place it in danger of demographic or genetic malfunction, then it should be placed in one of the categories of threat.

The information from which the Red Data Books are produced is extracted largely by the Species Survival Commission (SSC) of IUCN, which is a network of the world's most qualified specialists in species conservation, who serve on a voluntary basis. The various groups and their memberships are listed in the SSC Membership Directory, published by IUCN.

Table 17.2 Number of species covered by Appendices I and II of the Convention on International Trade in Endangered Species of Wild Fauna and Flora (CITES) as of 2004. This can be updated at http://www.cites.org/eng/app/index.php. Roughly 5000 species of animals and 28 000 species of plants are protected by CITES against overexploitation through international trade.

	Appendix I (endangered)	Appendix II (threatened)
Mammals	228 spp. + 21 sspp. + 13 pops	369 spp. + 34 sspp. + 14 pops
Birds	146 spp. + 19 sspp. + 2 pops	1401 spp. + 8 sspp + 1 pop
Reptiles	67 spp. + 3 sspp. + 4 pops	508 spp. + 3 sspp. + 4 pops
Amphibians	16 spp.	90 spp.
Fish	9 spp.	68 spp.
Invertebrates	63 spp. + 5 sspp.	2030 spp. + 1 sspp.
Plants	298 spp. + 4 sspp.	28074 spp. + 3 sspp. + 6 pops
Total	827 spp. + 52 sspp. + 19 pops	32540 spp. + 49 sspp. + 25 pops

spp., species; sspp, subspecies; pops, populations.

17.7.2 *The role of CITES*

The Convention on International Trade in Endangered Species of Wild Fauna and Flora (CITES) regulates trade in species of wildlife that are perceived to be at risk from commercial exploitation. There are 99 countries that are party to the convention.

The teeth of the Convention are contained in its appendices listing the species it covers. Article II of the Convention decrees that:

1 Appendix I shall include all species threatened with extinction which are or may be affected by trade. Trade in specimens of these species must be subject to particularly strict regulation in order not to endanger further their survival and must only be authorized in exceptional circumstances.
2 Appendix II shall include:
(a) all species which although not necessarily now threatened with extinction may become so unless trade in specimens of such species is subject to strict regulation in order to avoid utilization incompatible with their survival; and
(b) other species which must be subject to regulation in order that trade in specimens of certain species referred to in the above sub-paragraph may be brought under effective control.
3 Appendix III shall include all species, which any Party identifies as being subject to regulation within its jurisdiction for the purposes of preventing or restricting exploitation, and as needing the cooperation of other parties in the control of trade.
4 The Parties shall not allow trade in specimens of species included in Appendices I, II, and III, except in accordance with the provisions of the present Convention.

Table 17.2 gives the number of species covered by Appendices I and II of CITES as of 2004.

17.8 **Summary**

Extinctions can be driven by a permanent change to a species' environment (e.g. a new predator, disease, or competitor, or modification of its habitat) or can result from stochastic events. Driven extinctions are the most common. Stochastic extinctions are the chance fate of small populations: factors that would be swamped in a large population can have serious consequences for the individuals of a small one. The critical step in averting extinction is to follow the logical pathway of hypothesis testing to diagnose the cause of the decline. A species can seldom be rescued until the factors driving the decline are identified and removed. Rescue and recovery operations are standard wildlife management practices (e.g. regulation of harvest, predator control) but sometimes more elaborate steps such as captive breeding and translocation are called for. Reserves or National Parks and community conservation play a key role in the nurture and recovery of endangered species.

18 Wildlife harvesting

18.1 Introduction

In this chapter we consider how to estimate an appropriate offtake for a wildlife population. Wildlife is harvested for many different purposes. Sport hunting usually takes a sample of the population during a restricted season, and often with a restriction placed on the sex and age of the harvest. Harvesting for sport is a complex activity whose product is as much a quality of experience as it is meat or trophies. On the other hand, the purpose of commercial hunting or hunting for food is simply to harvest a product, such as meat or skins.

Both recreational and commercial wildlife harvesting are controversial, but it is not the purpose of this chapter to delve into that controversy. Whether or not one considers it appropriate and ethical to harvest a population of a given species depends more on one's view of life than on what may be happening to the population. There is an ethical aspect, however, that is fundamental to wildlife harvesting: the operation, be it for recreation or profit, must result in a sustainable offtake, a yield that can be taken year after year without jeopardizing future yields.

In all but special circumstances, the strategy of sustainable harvesting is simple: to ensure that the number of individuals removed from the population does not exceed the number of new recruits to the population. Hence, a population increasing on average at 20% a year can be harvested at any level up 20% a year, on average, without seriously diminishing the underlying population. Easily said, not so easily done in a world where population size is rarely known with precision, where there is substantial variation in recruitment from year to year due to demographic and environmental stochasticity, and where it is often challenging to tightly control or even monitor the number of animals being harvested. The issue is without a doubt of the utmost conservation importance, however. Overharvesting has been identified as one of the leading factors shared by endangered vertebrate species (Wilcove *et al.* 1998) and we know from historic records of many prominent examples of species that were essentially harvested to extinction (see Chapter 17).

18.2 Fixed-quota harvesting strategy

Most unharvested populations have a rate of increase that, when averaged over several years, is close to zero. The sustained yield for such a population is thus also zero. Before such a population can be harvested for a sustained yield it must be stimulated to increase. A population can most easily be stimulated into a burst of growth by increasing the level of a limiting resource or, much more rarely, by reducing the level of predation to which it is subject. The key resource may be nest sites or cover, but in most cases it is food. For example, red kangaroos increase if the pasture biomass

Wildlife Ecology, Conservation, and Management, Third Edition.
John M. Fryxell, Anthony R. E. Sinclair and Graeme Caughley.
© 2014 John Wiley & Sons, Ltd. Published 2014 by John Wiley & Sons, Ltd.
Companion Website: www.wiley.com/go/Fryxell/Wildlife

exceeds about 200 kg/ha dry weight and decline if it is below that level (Bayliss 1987). The easiest way to increase the average amount of food available to an individual animal is to reduce the number of animals competing with it for that food. The average standing crop of food then rises and the amount available to each individual is thereby increased. As a direct consequence, the fecundity of individuals is enhanced and mortality, particularly juvenile mortality, is reduced. The population enters a regime of increase as it climbs back towards its unharvested density.

18.2.1 *Harvesting*
trades off yield
against density

The tradeoff between yield and density is the most important thing to know about sustained-yield harvesting. In general, the further the density is reduced, the higher the yield as a percentage of population size. The maximum rate of sustainable yield is the population's intrinsic rate of increase, but that rate is obtained only when food (or whatever other resource is limiting) is at maximum, which usually occurs only when the population is at minimum density. Harvesting beyond this limit would inevitably lead to resource collapse and eventual extinction.

Whereas the proportionate offtake (yield divided by population size) tends to increase as density is reduced, the same is not true of the absolute sustained yield. If the population is reduced just a little, the induced rate of increase will be small and the sustained yield will be a small proportion of a relatively large population. The absolute yield will be modest. If the population is drastically reduced, the induced rate of increase will be large and the sustained yield will be a large proportion of what is now a relatively small population. Again, the absolute yield will be modest. The highest yield is taken from a density at which the induced rate of increase multiplied by the density is at a maximum. It tends to be at intermediate density levels. For example, suppose that a population grows in a manner well described by the Ricker logistic model (Chapter 5), with projected population size next year $f(N)$ predicted from current population density in the following way:

$$f(N) = Ne^{r_{max}\left(1-\frac{N}{K}\right)}$$

Net recruitment ($R(N)$) is defined as the difference between population density in year t and population density the previous year, *in the absence of harvesting*:

$$R(N) = f(N) - N = Ne^{r_{max}\left(1-\frac{N}{K}\right)} - N$$

A clear example of this is demonstrated by the elk population in Yellowstone National Park. For many years, elk were regularly culled as a park management strategy, to prevent undesirable effects on vegetation and because of a perceived need to "replace" the ecological impact of natural predators that had been extirpated many years earlier. Growing dissatisfaction with this interventionist policy as a matter of conservation principle emerged in the 1960s and 1970s, leading to the adoption of a much less intrusive approach. As the Yellowstone elk population rebounded from the artificially low level that had been imposed by culling, wildlife biologists annually measured the impact of increased population size on the exponential growth rate and net recruitment of the herd. The Yellowstone elk population exhibited a classic density-dependent response (see Fig. 16.6), leading eventually to population stabilization.

The maximum net recruitment would be expected to occur on average at an elk population of approximately 6000 animals, roughly half of the estimated carrying capacity

Fig. 18.1 Estimated variability in the net recruitment of Yellowstone elk in relation to population density, based on Monte Carlo simulation at observed levels of environmental stochasticity (see Chapter 16). The expected net recruitment is shown by the solid hump-shaped curve.

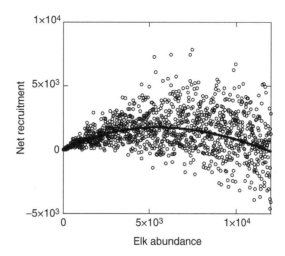

of 11 760 individuals (Fig. 18.1). Using field data to estimate the residual variation around the per capita recruitment (see Chapter 16), one can simulate how widely the net recruitment might be expected to vary over a range of population sizes (Fig. 18.1). This scatterplot makes it clear that even though the net recruitment curve is maximized on average at elk densities of around 6000 animals, the net recruitment actually realized in any single year should be anticipated to vary considerably.

A fixed-quota harvest policy is predicated on sustained use of the net recruitment, treating it as a surplus that can be safely harvested without harming resource sustainability in the long term. For example, imagine that we, as population managers, set the harvest at 1000 elk (Fig. 18.2). For a population conforming to the relationship between sustained yield and population size or density given in Fig. 18.2, a sustained yield of a given size might be taken from either of two densities, at the points at which the horizontal harvest line intersects the hump-shaped net recruitment curve. They form what is known as a sustained-yield pair. The member of the pair taken from the lower density is to be avoided because its harvesting requires more effort than the same yield from the higher density. Even more importantly, any reduction of population

Fig. 18.2 Expected net recruitment of Yellowstone elk in relation to population size (hump-shaped curve) plotted relative to a constant harvest quota of 1000 individuals. At a given harvest quota, there are stable (open circle) and unstable (filled circle) population equilibria. At population densities above the open symbol, the elk population would decline and eventually converge on the stable equilibrium. Perturbation of the population below the filled symbol on the other hand, would lead to further elk decline and eventual extinction. Values in between the equilibria would lead to convergence on the upper equilibrium.

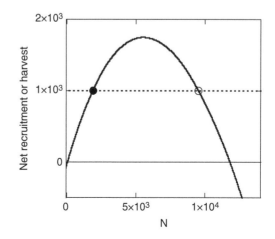

density below the level of the sustained-yield pair will inevitably lead to overharvesting relative to growth capacity of the population. When a constant number of animals is taken each year from a previously unharvested population, provided that there is no stochastic variation in weather or other factors that influence net recruitment, the population will decline and stabilize at the upper density for which that harvest is a sustained yield. Should that number exceed the maximum sustained yield, the population will inevitably decline to extinction.

18.2.2   *Maximum sustained yield*

The harvest that intersects the peak of the hump-shaped net recruitment curve is known as the *maximum sustained yield* (*MSY*). While it is often tempting to try to harvest as many animals as possible, harvesting a population at levels approaching the MSY should never be contemplated. It imparts instability to the population's dynamics. This can be readily understood by using a Ricker logistic model that incorporates both a constant harvest quota (*H*) and environmental stochasticity (ε), drawn from a normal distribution with mean = 0 and standard deviation = 0.19, based on the Yellowstone elk data:

$$N_{t+1} = N_t e^{r_{max}\left(1 - \frac{N_t}{K}\right) + \varepsilon} - H$$

We are now in a suitable position to model the impact of a constant-quota system, with the annual quota set at the MSY (1760 elk per year). The results graphically demonstrate that the Yellowstone elk population could not sustain a constant quota at the MSY for any appreciable length of time (Fig. 18.3). Unless they are much lower than the MSY, fixed quotas are almost always very risky, simply because of the vulnerability of the population to overharvesting over a series of years of below-average net recruitment. When one is removing a large fraction of the population in any given year, which is all too often tempting, it doesn't take many years of overharvesting to cause population collapse.

   The important lesson we obtain from this example is this: given that all harvested wildlife species live in stochastic environments in which weather conditions and food supplies are expected to vary widely, wildlife managers should keep the harvest quota well below MSY. Even then, fixed quotas can run a serious risk of stock collapse if the population should experience a series of years of below-average recruitment. As a result, fixed quotas are almost always unnecessarily risky and should be avoided.

Fig. 18.3 Predicted variation in elk abundance under a constant quota harvest policy, with annual elk harvest set at the maximum sustainable yield (1760 individuals per year). The simulation starts with elk at their ecological carrying capacity (12,000).

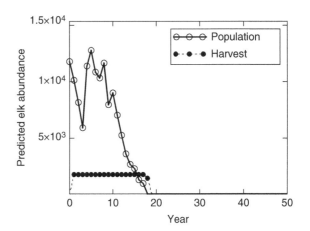

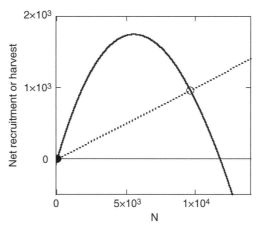

Fig. 18.4 Expected net recruitment of Yellowstone elk (hump-shaped solid curve) in relation to population abundance, plotted relative to a constant proportionate harvest (straight dotted line). The intersection of the net recruitment curve and the harvest function identifies the stable equilibrium, at which off-take equals the growth increment to the population. At any given harvest proportion, there is one stable (open circle) and one unstable (filled circle) equilibrium. At population densities above the stable equilibrium, the population would decline and accordingly converge on the stable equilibrium. Values in between the equilibria would cause increase in elk numbers leading eventually to convergence on the upper stable equilibrium.

18.3 Fixed-proportion harvesting strategy

If a constant proportion of animals is taken each year from a previously unharvested population, the population will decline and stabilize, depending on the rate of harvesting, at any level between unharvested density and the threshold of extinction (Fig. 18.4). So long as the harvest rate (h) does not exceed the maximum intrinsic rate of population growth (r_{max}), harvesting a fixed proportion of the population should settle the population upon a stationary distribution of population densities, generating a sustainable yield, even in the presence of low to moderate stochastic variation in environmental conditions. Provided a wildlife or fisheries manager had perfect information on abundance in any given year prior to the harvest, a proportionate strategy will be guaranteed to allow sustainable harvests forever.

The trouble is, wildlife or fishery managers rarely have accurate and up-to-date information on current abundance and must often set harvest levels (through licensing or allocation of quotas) long before annual recruitment is known. Indeed, the most common situation is that managers know little more about the population than the information gained from harvest success during the previous year. Such information will permit, at best, a forecast of the current recruitment, based on an index of population abundance at the end of the previous harvest. In other words, population managers must guess the current level of abundance in setting an annual quota, whose magnitude is intended to be a constant fraction of the population. The uncertainty thus introduced can considerably increase the risk of unintentional overharvesting.

18.3.1 Fixed proportionate harvesting in a stochastic environment

As an example of applying a fixed-proportion harvest strategy with uncertainty in current population levels, let us consider once again stochastic population growth by Yellowstone elk. As before, we presume that elk population growth tracks changes in elk density as well as being influenced by weather conditions and other stochastic environmental features (see Chapter 5). We presume that the wildlife manager has

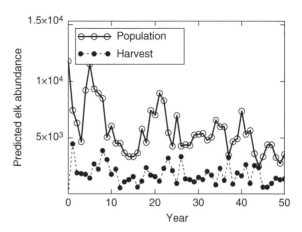

Fig. 18.5 Predicted variation in elk harvests under a fixed mortality harvest policy, with harvest proportion set at the value (0.32) that would produce the maximum sustainable yield (1760 per year) under average conditions. The simulation starts with an elk population at their ecological carrying capacity (12,000).

knowledge of past population abundance, based on the previous year's harvest, which allows a forecast of the current population size with a specified degree of uncertainty, to which we apply an adjustable quota approximating a fixed proportion of the population. To enable direct comparison with the earlier fixed-quota model, we choose a harvest proportion ($h = 32\%$) whose yield under equilibrium conditions is identical to the MSY (1760 individuals per year). The equation of population change is calculated in the following manner:

$$N_{t+1} = N_t e^{r_{max}\left(1-\frac{N_t}{K}\right)+\varepsilon} - hN_t e^{\gamma}$$

where e^{γ} introduces uncertainty to the estimated abundance derived from the previous year. The simulation results (Figs. 18.5) demonstrate that a fixed-proportion harvest strategy is more sustainable than a fixed-quota strategy, even though both would be expected to produce similar yields under average conditions.

Note that simulated harvests vary over time to a greater extent using a fixed-proportion harvest strategy (Fig. 18.5) than they did under the fixed-quota system (Fig. 18.3). This difference in variation arises because harvests are adjusted due to changes in stock size, effectively absorbing the impact of stochastic environmental variation. As a result, harvests drop following years of poor recruitment but improve in years following above-average recruitment. This variation in harvest provides a stabilizing influence to fixed-proportion harvest systems. However, proportionate harvest strategy can still produce overharvesting and therefore collapse when there are several years with unusually low recruitment or when there is pronounced error in a manager's assessment of current population size, because managers do not know with certainty how many individuals have been recruited to a population.

In other words, proportionate harvesting is considerably less risky than fixed-quota harvesting, but not entirely risk-free. In the not-so-distant past, fixed-quota systems tended to be the norm in harvested fish and wildlife populations. Proportionate harvesting policies have become predominant in more recent years, mainly because of recognition of resource conservation benefits (Rosenberg *et al.* 1993).

18.3.2
Constant-effort harvesting strategy

A harvest can be controlled either by placing a quota on offtake or by controlling harvesting effort. Harvest effort can be regulated by restricting the length of the hunting season or by limiting the number of people involved. The essence of controlling effort is that there is no direct attempt to control the number of animals harvested. An important outcome from controlling effort is that a constant *proportion* of the population is harvested. Suppose, for example, that each day harvesters effectively sample 2% of the area inhabited by a harvested species. We can mathematically depict this by saying that the catchability coefficient $q = 0.02$. If hunters remove all animals that are encountered in the area that they sample then the harvest H is related to effort E and population density N in the following manner:

$$H_t = N_t(1 - e^{-qE_t})$$

as illustrated by long-term data for white-tailed deer harvested by Ontario hunters (see Fig. 12.2). As effort climbs to high levels, a larger and larger proportion of the population is harvested, but because of the exponential term this proportion can never exceed 1. Population dynamics are accordingly modeled by the following equation:

$$N_{t+1} = N_t e^{r_{max}\left(1-\frac{N_t}{K}\right)-qE+\varepsilon}$$

Note that fixed-effort policies accordingly have a built-in mechanism to reduce exploitation levels should resource density decline to dangerously low levels, because harvest levels also drop automatically with declines in resource abundance. Note also that a conservative effort level might yield a similar equilibrium harvest to a more extreme effort, which could encourage a more moderate policy (the equilibrium harvest is the point at which the yield function intersects the net recruitment curve). Both of these characteristics tend to have a moderating influence on population dynamics, as we shall demonstrate later in the chapter.

The control of harvest by quotas has an intuitive appeal because there is a direct relationship between the prescription and the result. In contrast, with the harvest regulated by control of harvesting effort an intermediate step has been inserted between prescription and outcome. Administrators tend to favor regulation by quotas because the size of the yield is directly under their control. In fact, the disadvantages of regulating effort are more conceptual than real. Regulation of effort is usually a safer and more efficient means of managing harvested populations than is regulation by quota. Harvesting a constant number of animals each year is inefficient when the population is subjected to large, environmentally induced swings in density. The quota must be set low enough to be safe at the lowest anticipated density, or alternatively the size of the population must be censused each year before the harvesting season, the quota being adjusted according to the estimate. In addition, regulation by quota is unsafe when the quota is near the MSY. As mentioned in Section 18.3.1, the density at equilibrium with that yield is unstable, such that a small environmental perturbation can trigger a population slide towards extinction.

If yield is controlled indirectly by limiting harvesting effort (e.g. by limiting the number of hunters, trappers, or fisherman and/or season length), but there is no further restriction, these dangerous sources of instability are greatly reduced. A fixed-effort system will, within limits, harvest the same proportion of the population at high density as at low. Yield tracks density, the system automatically producing a higher yield when animals are abundant and a lower yield when they are scarce. A regulatory mechanism is built into the harvesting system itself, and it is thus fairly safe so long as the appropriate harvesting effort has been calculated correctly. That is not difficult because fine-tuning of the appropriate effort does not destabilize the system in the way that fine-tuning of a quota can.

18.4 Harvesting in practice: dynamic variation in quotas or effort

Harvesting of wildlife for recreational hunting and fishing is usually managed largely by trial and error. Moreover, there is often a well-established tradition of open access, meaning that virtually any member of the public is free to apply for hunting or fishing licenses. So long as population trends are good (i.e. hunting success and harvests levels haven't changed much), managers tend to stick with the status quo. This approach often works tolerably well when populations and their habitats are good at looking after themselves, when intrinsic rates of increase are high, when rate of increase and density are tightly related by density-dependent negative feedback, and when population size rarely declines to less than half of the ecological carrying capacity.

In practice, however, management by trial and error can also lead to cyclic changes in harvest intensity over time (Fryxell *et al.* 2010). This stems from behavioral responses by both harvesters and wildlife managers to changes in resource abundance. Harvesters share information about their experiences during hunts or fishing trips, just as they do about other important events in their lives. Common sense dictates that when hunting or fishing success is good, more individuals will be encouraged to purchase licenses and participate in recreational harvesting. Hence, hunting effort should tend to increase when resources are abundant but decline when resources are scarce and harvest success declines, as illustrated by long-term data on deer hunters in Ontario (Fig. 18.6a). On the other hand, one would also expect effort to decline as more and more participants join, simply because of the reduced satisfaction that comes from many hunters competing with each other and because the pool of potential new hunters becomes exhausted.

Fig. 18.6 (a) Positive feedback of white-tailed deer density on hunter effort in Ontario, but (b) negative feedback with respect to the current level of effort. (Redrawn from Fryxell *et al.* 2010.)

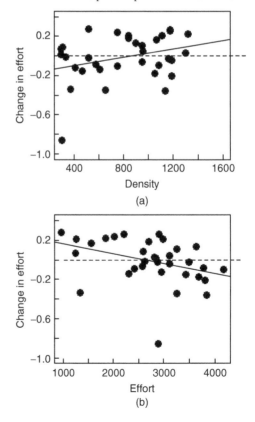

Fig. 18.7 Simulated dynamics of a harvested population with dynamic changes in hunter effort or quotas caused by positive responses to resource levels but negative feedback. (Redrawn from Fryxell *et al.* 2010.)

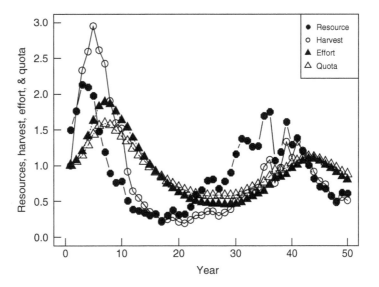

This imposes a negative feedback as hunter effort increases, also well illustrated by the Ontario deer hunter data (Fig. 18.6b).

This dynamic behavioral process can be accommodated by the following model (Fryxell *et al.* 2010), linking the rates of change in resource abundance N and hunter effort E:

$$N_{t+1} = N_t e^{r_{max}\left(1-\frac{N_t}{K}\right)-qE_t+\epsilon}$$
$$E_{t+1} = E_t e^{c+wN_t-uE_t}$$

The most typical outcome of this model is slow cyclic changes in both resources and effort over time (Fig. 18.7), consistent with the long-term cycles documented in white-tailed deer in Ontario (Fig. 18.8a) and some other wildlife and fish species. This is due to much the same process as predator–prey cycles: as resources increase in abundance, hunters or fisherman increase effort levels until they begin to overwhelm the capacity of the resource for renewal. As mortality exceeds recruitment, the resource declines in abundance, leading eventually to decline in hunter effort, and the cycle begins anew.

A similar kind of behavioral response is often seen in management responses to changing wildlife abundance. For example, the rate of change in quotas for moose harvests in Norway is positively related to moose abundance, but with a negative feedback (Fryxell *et al.* 2010). This mix of responses can be depicted by the following model linking the rates of change in both resource abundance N and quota levels Q:

$$N_{t+1} = N_t e^{r_{max}\left(1-\frac{N_t}{K}\right)+\epsilon} - Q_t$$
$$Q_{t+1} = Q_t e^{f+iN_t-jQ_t}$$

Dynamic changes in quotas have much the same effect as dynamic changes in effort, leading to decadal cycles in both harvests and resource abundance (Fryxell *et al.* 2010). Where both effort and quotas are dynamically responsive to changing abundance, they become entrained, once again leading to cycles in resource abundance. The resulting

Fig. 18.8 (a) Cyclic fluctuations of Canadian white-tailed deer (solid circles) caused by variation in harvest levels (open circles) and hunter effort (solid triangles). (b) Cyclic fluctuations of Norwegian moose (filled circles), caused by variation in moose harvest (open circles) and quotas (filled triangles), and moose harvest (open circles).

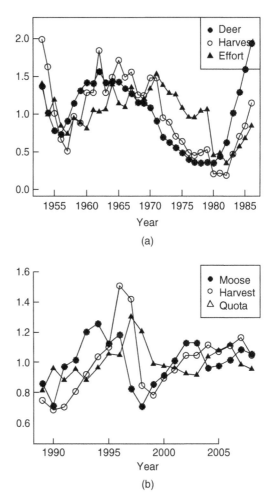

(a)

(b)

cycles can be readily seen in moose data from Norway (Fig. 18.8b), as well as in some freshwater and marine fisheries (Bell *et al.* 1977; Botsford *et al.* 1983; Berryman 1991). Such decadal cycles may not be particularly risky unless they drive the population to low enough levels of abundance that it becomes vulnerable to demographic stochasticity or extreme climatic events. Nonetheless, the propensity for cyclic population variation induced by harvests should be recognized in long-term management plans. It may have other ecological interactions as well, such as with food plants, top predators, or disease agents.

18.5 No-harvest reserves

An alternative means of achieving sustainable harvesting is to set aside some sites as no-harvest zones in close proximity to harvested subpopulations. This policy option has become particularly attractive for marine ecosystems, because of severe challenges in controlling fishing pressure, protecting fish habitat, and monitoring fish stocks in the open ocean (Roberts *et al.* 2005). In the absence of harvesting, populations in no-harvest zones are expected to increase in abundance, provided that individuals don't venture outside the protected area very frequently (Hilborn *et al.* 2004). Obviously, species with small, well-defined home ranges best fit this description, whereas nomadic or migratory species are much harder to protect. Provided that subpopulations in the harvest refuge produce dispersers that spill out into surrounding areas, no-harvest reserves can serve as sources that maintain harvestable sink populations in neighboring areas. Indeed, simple models presuming extremely effective dispersal of larvae demonstrate the potential for comparable or even equivalent long-term yields for systems managed conventionally to achieve MSY and dense networks of no-harvest zones interspersed with harvested subpopulations (Hastings and Botsford 1999; Neubert 2003).

The conservation value of no-harvest reserves is obvious. Where healthy populations are maintained in refugia, long-term sustainability is much easier to ensure than in a species exploited throughout its range. Historical data from a large number of marine reserves suggest that almost all experienced substantial recovery in local fish stock abundance following protection (Halpern 2003; Lester *et al.* 2009), as well as increased species diversity, biomass, and average size of individuals. A growing number of studies have also documented sufficient spillover of dispersing fish to allow sustainable and profitable fisheries outside the reserve boundary (Goni *et al.* 2010; Halpern *et al.* 2010).

There is continuing debate, however, about the effectiveness of no-harvest reserves on long-term yields to the fishery compared to other management strategies (Hilborn *et al.* 2004; Roberts *et al.* 2005). Few systems have been monitored closely enough to determine whether the harvest yields of systems with no-harvest reserves will be comparable to those of systems managed in conventional fashion, although there are encouraging signs of this in some studies (Goni *et al.* 2010; Halpern *et al.* 2010). A major uncertainty is the degree of long- versus short-distance dispersal by larvae and adults and of density-dependent constraints on settlement probability. Another uncertainty is whether clumping by fishermen at the boundary might simply intercept all individuals before they settle. Interestingly, there has been much less discussion to date about the potential role of no-harvest reserves in terrestrial systems, despite similar conservation and management challenges. One reason may be that there is less widespread evidence of severe overharvesting in terrestrial species, in part because of sex- and age-biased harvest controls.

18.6 Age- or sex-biased harvesting

In large mammal species, harvesting is often directed at males rather than females or focused on older rather than younger age groups, often imposed through tag or license restrictions on hunters. The intent of such a policy is to guarantee the perpetuation of the breeding segment of the population. Age- and sex-structured harvest models of Norwegian moose, for example, suggest that the optimal policy would be to concentrate harvests on calves and old males, while rarely removing females from the population (Sæther *et al.* 2001). This makes intuitive sense, if the population is well mixed and all harvested individuals are equally valuable. Male-biased harvesting policy came into common application in North America in the early part of the 1900s, following decades of severe overhunting and the virtual disappearance of many game species. Harvesting of males was correctly seen as a simple means of ensuring the herd would survive in the face of hunting. It continues to be a cornerstone principle of wildlife management, especially in North America.

Before a sex- or age-biased harvest policy is contemplated, however, it is helpful to consider its long-term implications. As an illustration, we will use demographic data for a white-tailed deer population from the George Reserve in Michigan. The George Reserve is a small (4.6 km^2) fenced forest research site maintained and managed by the

University of Michigan. Six white-tailed deer (4 females and 2 males) were introduced into the enclosure in 1928. Annual population counts were subsequently obtained from coordinated drives by a large group of volunteers walking through the reserve, yielding demographic data that had rarely before been gathered in systematic fashion. Within 6 years, the protected population had increased to more than 150 individuals, much to the surprise of wildlife managers, who at the time had limited experience with ungulate populations that were not heavily harvested by humans. This was perhaps one of the earliest demonstrations that ungulates were capable of rapid increase in abundance under appropriate conditions.

Associated with this rise in deer numbers, however, was an obvious change in vegetation characteristics within the reserve, as some of the most palatable browse species either disappeared or were maintained in severely altered growth form. Trees like red cedars or hawthornes were clipped by deer into hedge-like structures, whereas white pines were pruned so severely they resembled bonsai-like dwarves (McCullough 2001). In keeping with conventional wisdom at the time, the wildlife managers of the George Reserve instituted deer culls to safeguard against undesired changes in vegetation.

The long-term data from the George Reserve suggest that per capita growth rates are strongly density-dependent, with an estimated carrying capacity of 176 deer in the initial phase of study and 212 deer during a second phase of increase following a severe reduction in abundance (McCullough 1979, 2001). Hence, the George Reserve deer population had a strong density-dependent recruitment response capable of regulating population abundance around a typical carrying capacity of 200 individuals even in the absence of culling. Survivorship was estimated at roughly 75% for prime-aged individuals and 70% for older adults. Under ideal conditions, white-tailed deer are capable of reproduction at the end of their first year, with twinning normal at low deer densities. One can approximate these demographic rates with the following Lefkovitch transition matrix (Jensen 2000), where demographic rates for males are displayed on odd rows and female vital rates on even ones, with fawns in the top two rows, yearlings in the next two, and adults in the last two:

$$
M = \begin{pmatrix}
0 & 1 & 0 & 1 & 0 & 1 \\
0 & 1 & 0 & 1 & 0 & 1 \\
0.75 & 0 & 0 & 0 & 0 & 0 \\
0 & 0.75 & 0 & 0 & 0 & 0 \\
0 & 0 & 0.75 & 0 & 0.70 & 0 \\
0 & 0 & 0 & 0.75 & 0 & 0.70
\end{pmatrix}
$$

Using this approach, one can consider both male and female segments of the population, as well as different age groups. Note that we lump adults into a single class (as we did with leatherback turtles in Chapter 13) rather than keeping track of 15 or more age groups, which would otherwise be needed for such a long-lived species. The dominant eigenvalue of this transition matrix is 1.74, suggesting that deer in the George Reserve could under ideal conditions increase 74% from one year to the next, an approximation that is consistent with the historical data (Jensen 2000; McCullough 2001). The net recruitment to the deer population can be estimated by subtracting the identity matrix (a 6 × 6 matrix with 0s in every element except down the diagonal, where there are 1s), yielding the following net recruitment matrix (A):

$$A = \begin{pmatrix} -1 & 1 & 0 & 1 & 0 & 1 \\ 0 & 0 & 0 & 1 & 0 & 1 \\ 0.75 & 0 & -1 & 0 & 0 & 0 \\ 0 & 0.75 & 0 & -1 & 0 & 0 \\ 0 & 0 & 0.75 & 0 & -0.3 & 0 \\ 0 & 0 & 0 & 0.75 & 0 & -0.3 \end{pmatrix}$$

If one were only interested in predicting short-term changes, this could be estimated using the Lefkovitch matrix model (see Chapter 13):

$$n_{t+1} = n_t + An_t$$

where n_t is the vector of age-specific abundance, with males in odd rows and females in even ones. As mentioned earlier, there is strong evidence for a density-dependent response by George Reserve deer. Jensen (2000) incorporated this by pre-multiplying transition matrix A by summed densities over all sex and age classes in any given year using the formula $f\left(\sum n_{i,t}\right) = \left(1 - \sum n_{i,t}/K\right)$, where carrying capacity $K = 200$:

$$n_{t+1} = n_t + f\left(\sum n_t\right) An_t$$

This model produces logistic growth over time, such that the population always converges on the carrying capacity of 200 deer and a stable age distribution, as one would expect from any other Leslie matrix or Lefkovitch matrix model (see Chapter 13). With this model in place, it is now possible to apply a given harvest rate h, expressed as a proportion, to each population component. This harvest is incorporated as a matrix with 0s in every element except for the diagonal, which holds the sex- and age-specific harvest rates. For example, the following matrix would apply a 62% harvest rate to males of all ages, but only 12% harvest rate to females:

$$Z = \begin{pmatrix} 0.62 & 0 & 0 & 0 & 0 & 0 \\ 0 & 0.12 & 0 & 0 & 0 & 0 \\ 0 & 0 & 0.62 & 0 & 0 & 0 \\ 0 & 0 & 0 & 0.12 & 0 & 0 \\ 0 & 0 & 0 & 0 & 0.62 & 0 \\ 0 & 0 & 0 & 0 & 0 & 0.12 \end{pmatrix}$$

The harvest matrix Z is multiplied by the vector of sex- and age-specific abundance and then subtracted from the population:

$$n_{t+1} = n_t + f\left(\sum n_{i,t}\right) An_t - Zn_t$$

This sex-biased harvest model still produces logistic growth, but now the difference in harvest rates is reflected in strongly skewed ratios of females/males at any given age (Fig. 18.9). Indeed, adult males become exceedingly rare at such a pronounced bias in exploitation levels. The total population size predicted by the model is well above half of the carrying capacity of 200 individuals, despite the fact that 45 deer are harvested

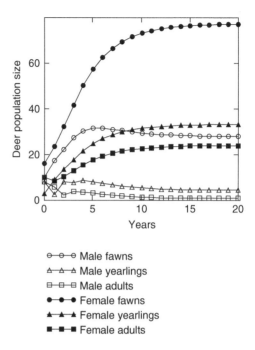

Fig. 18.9 Predicted age- and sex-specific abundance over time for a model population of white-tailed deer exposed to a sex-biased harvest policy with male annual harvest rate = 0.62 and female annual harvest rate = 0.12.

○—○—○ Male fawns
△—△—△ Male yearlings
⊟—⊟—⊟ Male adults
●—●—● Female fawns
▲—▲—▲ Female yearlings
■—■—■ Female adults

every year from the population. This is achievable, of course, because only females produce young. Biasing the harvest towards males ensures a substantial number of female deer are able to reproduce.

Models of male-biased harvest typically produce two outcomes. First, the population is maintained at a higher equilibrium density than would be possible if an identical harvest were applied in balanced fashion across the sexes. Second, the overall harvest yield is usually much lower than that obtained from a balanced harvest, unless females are harvested to some degree. Provided that females are harvested in significant numbers, however, it is possible even to improve long-term harvest yields well above that of the MSY in a balanced harvest system (Fig. 18.10).

Fig. 18.10 Mean harvest expected over time for 2 alternate policies: a constant harvest rate applied to both sexes (dotted line) vs. a sex-biased policy with female harvest rate = 0.74 × male harvest rate (solid line).

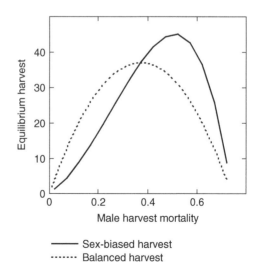

—— Sex-biased harvest
······ Balanced harvest

Male-biased harvesting is not always the best policy. In Scottish red deer, for example, the proportion of males born tends to decline with population density and the mortality of older male animals increases relative to females (Clutton-Brock et al. 2002). As a result, red deer populations have naturally skewed sex ratios, even without sex-biased hunting. However, males bring in far more income to landowners than do females, so highly skewed sex ratios are economically disadvantageous. The remedy is increased culls of females to keep female density < 60% of the ecological carrying capacity (Clutton-Brock et al. 2002).

If taken to extremes, male-biased harvesting could lead to such low levels of male abundance that an appreciable fraction of adult females would be unable to find suitable breeding partners (Ginsburg and Milner-Gulland 1993). For example, years of uncontrolled commercial hunting of Saiga antelope for their valuable horns following collapse of the Soviet Union produced an extreme sex ratio of over 100 females to every 1 male that contributed to a massive collapse in reproduction (Milner-Gulland 1997; Milner-Gulland et al. 2003). Extreme sex ratios have been documented in some other populations with sex-biased harvest policies too (Milner et al. 2007), suggesting that the risk of reproductive collapse is not far-fetched, especially in species that do not normally have extreme variation in male reproductive success, such as species with long-lasting pair bonds. Obviously, reproductive collapse would outweigh any slight advantage accruing from age-biased harvesting.

Restriction of harvesting to older age groups is also beneficial in terms of protecting the breeding "capital" that is necessary to ensure long-term sustainability. The downside of age-biased harvesting, however, is that equilibrium harvest yields are compromised to a considerable degree. For example, the white-tailed deer matrix harvest model suggests that restriction of harvesting to yearling and adult deer (male $h = 0.62$, female $h = 0.12$) would produce a long-term yield of < 2 deer per year, compared to 45 deer when fawns are also harvested. Moreover, the standing age distribution is skewed towards younger individuals. This can create a wide variety of social disturbances if older individuals are socially dominant or more capable of rearing offspring to maturity than are younger ones (Milner et al. 2007). The risk of social strife is particularly pronounced in territorial carnivores, such as bears, wolves, or large cats, because age is often associated with social dominance. Elimination of high-ranking males can lead to increased mortality due to infanticide as lower-ranking males take over newly vacated territories. Such infanticide in a hunted population of European brown bears resulted in a juvenile survival rate of 72%, compared to 98% in unhunted populations (Swenson 2003; Swenson et al. 1997). Infanticide is also common during territory takeovers in lions (Pusey and Packer 1994). Harvesting male lions is thought to increase takeovers, with consequent depression of cub recruitment rates. The simple remedy in this case is to restrict hunting to only the oldest senescent males, which can be readily discriminated in the field by muzzle morphology (Whitman et al. 2004).

In order to develop useful sex- or age-biased models of harvesting, detailed information is required on age- and sex-specific survival and reproductive rates and how these demographic parameters are affected by changes in population density. An additional complication is that harvesting only one segment of a structured population can lead to unanticipated surprises in the density-dependent response compared to a balanced harvest (Benton et al. 2004). The last point is crucial: without reliable information on density-dependent parameters, even when detailed life tables are available misidentification of optimal harvesting levels is a very real possibility. The requisite demographic information is available for few wildlife species, so sex- and male-biased harvests should be viewed with an appropriate degree of caution.

18.7 Commercial harvesting

There is no difference in principle between harvesting for commercial benefit and harvesting for recreational benefit. Both are based on the sustainable yield concept, suitably cushioned by a margin of error. However, in practice there are a number of pitfalls to the management of sustained commercial offtake.

Game harvesting comes in two forms: game ranching (or game farming) and game cropping. The difference is in degree rather than kind, but essentially game ranching seeks to bring the animals under human control, as in the farming of domesticants, whereas game cropping is the harvesting of wild populations. Game ranching is a spectrum of activities that overlap conventional animal husbandry at one end (e.g. the reindeer industry in Finland and Russia) and game cropping at the other. This subject is beyond the scope of this book. The interested reader might try papers in Beasom and Roberson (1985) and Hudson *et al.* (1989) for a general overview of game ranching and the volume edited by Bothma (1989) for a thorough treatment of game ranching in southern Africa. Because game cropping can be an important conservation threat, however, we consider it in greater detail in Sections 18.8 and 18.9.

18.8 Bioeconomics

In addition to the biological complexities inherent in renewable resource systems, there are additional complicating factors when wildlife or fish species are harvested commercially. This is because market dynamics, limitations on harvest controls, and potential conflict between short- and long-term goals also influence the levels of harvest effort. This has important effects on the economic equilibria for effort levels and the risk of resource overexploitation. For example, imagine that a population has a carrying capacity $K = 100$ individuals, maximum intrinsic growth rate $r_{max} = 1$, catchability coefficient $q = 0.02$, and effort $E = 20$ units. Annual net recruitment $R(N)$ as a function of abundance is dictated by the Ricker logistic equation $f(N)$ and proportionate harvest $H(N)$ using a linear-catch-per-unit-effort equation (see Fig. 12.2):

$$R(N) = f(N) - N = Ne^{r_{max}\left(1-\frac{N}{K}\right)} - N$$
$$H = qEN$$

At equilibrium, $H(N) = R(N)$ and $N = H(N)/qE$. By substitution and some algebraic rearrangement (we encourage you to try this yourself), we can obtain the following solution for the harvest at equilibrium as a function of effort $H(E)$:

$$H(E) = qEK\left(1 - \frac{\ln(1+qE)}{r}\right)$$

This implies that there is an equilibrium harvest level for each effort that might be exerted by resource users. We are now going to use this information to calculate the most profitable level of effort to invest. We assume that the revenue scales with the equilibrium harvest (with price per unit catch $p = 0.75$) and that harvesting costs C escalate linearly with effort (with cost per unit effort $c = 0.40$) (Fig. 18.11):

$$costs(E) = cE$$
$$revenue(E) = pH(E)$$

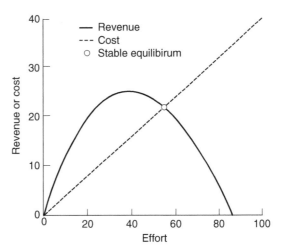

Fig. 18.11 Revenue and costs under constant environmental conditions as a function of effort. The intersection of the revenue curve and the cost line identifies the economic equilibrium, at which cost equals the potential gain.

Presumably commercial resource users should want to maximize profit (which we will symbolize with π), which is the difference between revenue and costs:

$$\pi(E) = pH(E) - cE$$

When economists discuss the cost of a particular activity, they are usually referring to the opportunity cost. This is the difference between a given economic activity and alternative ways of earning income. So, the costs and benefits of resource use are measured relative to other forms of economic investment.

Note that the level of effort that maximizes profits (Fig. 18.12) is below the level of effort that maximizes sustainable harvest (Clark 1976, 1990). This offers yet another argument against maximizing the sustainable yield: it is less profitable than a lower rate of harvesting. That is probably a good thing for conservation as it implies that a monopolistic resource user ought to choose a relatively safe exploitation effort in order to serve its own selfish needs. For this to occur, however, the resource user needs control over decisions regarding resource use. In other words, they need to own the

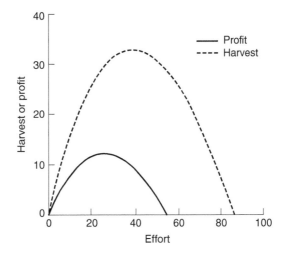

Fig. 18.12 Profit and equilibrium harvest under constant environmental conditions as a function of effort.

property rights to the resource; there would be little point in sustaining the resource unless these property rights were to continue long into the future.

A problem arises when access to the resource industry is unregulated, a situation known as open access. This is common in many game cropping and small-scale fishery operations around the globe. Incoming resource users may not necessarily care about depressing the profit margin, so long as they still find resource exploitation more profitable than other methods of economic investment. The threshold at which exploitation becomes unattractive in this case would be around 50–60 units of effort, rather than the optimal 25 or so. We will term this the "economic equilibrium." Higher effort implies a lower resource abundance than that which optimizes profit. Thus, open access (at least in game cropping and fisheries) can logically lead to depressed harvest levels, depressed profits, and depressed resource abundance (Clark 1976, 1990). Everybody loses.

So far, however, we have not discussed how prices figure into the precarious balance. Before we jump into the special circumstances that influence renewable natural resources, we need a brief reminder about the so-called law of supply and demand. In economic parlance, both supply and demand are assumed to respond to the price of commodities. When the price of a commodity is high, entrepreneurs are encouraged to produce more of it, whereas when they are low there is much less incentive (Fig. 18.13). On the other hand, high prices discourage consumer demand, whereas low prices encourage it (Fig. 18.13). These contrasting responses on the part of consumers and producers lead to an intersection between supply and demand curves. In a free, open, competitive market system, pricing should adjust over time until the economic equilibrium is reached at which supply matches demand. This kind of open market is highly idealized: in practice, there may be collusion between suppliers, artificial price supports imposed by external agencies (such as governments), or other barriers that mitigate against a free market. It may be fairly reasonable, however, as a depiction of economic interactions in a rural economy, such as that supplying game meat in equatorial Africa or South East Asia (Fa and Peres 2001).

Now, we are going to consider how sustainable resource levels are liable to respond to variation in prices (Copes 1970; Clark 1990; Milner-Gulland and Mace 1998). We will use some of the relationships already derived and rearrange terms to calculate a new

Fig. 18.13 Idealized relationship between supply and demand in an open, competitive market.

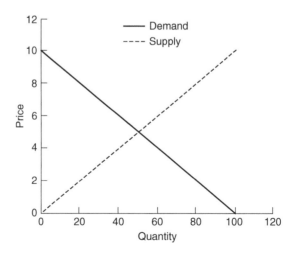

relationship based on price p and equilibrium harvest H, the latter being the relevant measure of production in a renewable-resource system. Recall the following:

$$pqEN - cE = 0$$

$$H(N) = Ne^{r_{max}\left(1-\frac{N}{K}\right)}$$

By dividing both sides of the first equation by effort and then rearranging terms we get $N = c/(pq)$. We can then substitute $c/(pq)$ wherever N appears in the second equation to obtain the following algebraic relationship between equilibrium harvest level and resource price:

$$H(p) = \left(\frac{c}{pq}\right) e^{r_{max}\left(1-\frac{c/pq}{K}\right)}$$

To complete the analysis, we need to add a demand line.

If demand is modest relative to renewable resource supplies then the classic supply–demand relationship occurs. There is a single intersection between the supply and demand curves, defining the economic equilibrium (Fig. 18.14a). Harvests should build over time and resource prices should fall until the supply meets demand. Due to the severe nonlinearities in the supply curve, however, this ideal situation can be rapidly transformed into a more troubling scenario. Imagine, for

Fig. 18.14 Supply (solid) and demand (dotted) curves for a renewable natural resource at low (a) and high (b) levels of demand.

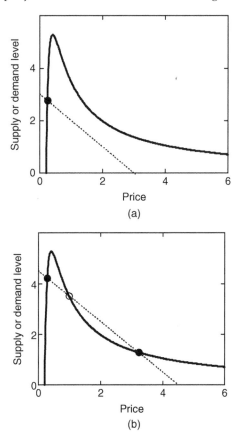

example, that economic development leads to a slight overall increase in consumer demand (Fig. 18.14b). This results in a situation in which there are three points of intersection: at high, moderate, and low prices and low, moderate, and high harvest levels, respectively. Now the system is poised to flip between dramatically different economic equilibria with slight changes in resource levels and economic performance. The high price/low harvest combination is particularly worrisome, as it arises because of a severely overharvested resource, teetering dangerously near extinction. It is nonetheless profitable enough to justify the huge effort required to eke out a meager harvest. Further increase in demand leads to a singular economic equilibrium, this time in the risky zone.

The potential for multiple economic equilibria depends on the degree of elasticity in demand. An economist would characterize a demand curve as being "inelastic" if it required a quite substantial change in price to suitably alter consumption (e.g. a 10% change in price produces a 5% change in demand). For some wildlife products, such as ivory, there is evidence that consumer demand is determined solely by income, regardless of price (Milner-Gulland 1993), so demand is highly elastic. Demand for essential food products like milk and eggs is typically much less inelastic.

Ivory harvesting presents an additional surprising wrinkle. Male elephants have substantially larger tusks than females, so any rational harvesting scheme would be biased towards males (Milner-Gulland and Mace 1998). Rather than culling males at random, however, the optimal economic decision would be to collect ivory from individuals that die naturally or to harvest only senescent males (Basson *et al.* 1991). Elephant tusks increase in size exponentially with age and large tusks are worth more, gram for gram, than small tusks, because large tusks are much more valuable to carvers. As a result, the optimal economic solution should reinforce conservation needs.

18.9 Game cropping and the discount rate

There are two quite distinct phases to a cropping operation: first the population must be reduced below its unharvested density (capital reduction) and then it must be harvested at precisely the rate at which it seeks to bounce back (sustained-yield harvesting). Biologists tend not to think too much about the capital reduction phase because they look forward to the prospect of a yield sustainable into the indefinite future.

If you were offered $1000 now as against $1000 in 10 years' time you would take the money now. However, if you were offered $400 now as against $1000 in 10 years' time the decision is no longer clear-cut. Against money in the hand, you are offered a guarantee of sure future benefit, but the monetary value of that future benefit is unclear. How much is $1000 in 10 years actually worth? A simple answer is that it is worth a present sum which, when prudently invested, yields $1000 10 years hence. If we assume that capital expands at about 10% per year then $1000 in 10 years is worth $385 now, or even less if the currency is inflating. With this knowledge, the answer to $400 now or $1000 in 10 years' time is clear. Take the $400 now; it is worth more. By the same reasoning, a game animal harvested in 10 years' time is worth nothing like an animal harvested now. All future earnings must be discounted by the time it takes to get the money, and the economics of the harvesting operation can be dictated by the ratio of present to future earnings.

Discounting can be represented fairly simply by rearranging the terms of the exponential-growth model. Instead of growing exponentially, however, the value of future profits π declines exponentially at the discount rate θ:

$$V(t) = \pi e^{-\theta t}$$

What this means, of course, is that future profits are not valued as highly as current ones. The higher the discount rate, the less the future is valued. A concrete example may prove illuminating. A mature Bolivian mahogany tree sold in 1998 for about US$396, for the lumber it could supply (Gullison 1998), yet it takes roughly a century for such a tree to mature. At the 17% inflation rate in Bolivia in the early 1990s ($\delta = 0.17$), the present value of a mahogany tree worth US$396 a century later would be $0.01, almost 10 times more than is required for replanting. Small wonder that replanting is a low priority for most logging firms!

The present value of all future harvests can be calculated by integration:

$$V(t) = \int_0^\infty \pi e^{-\theta t} dt$$

If the per unit price $p = 0.75$, cost per unit effort $= 0.5$, $r_{max} = 1$, and $K = 100$ then, for a discount rate of 5%, the total present value of all future harvests would be US$246 at the optimal effort level ($E = 25$). Changing the discount rate to 10% (still very realistic in most economic models) halves the value of all future harvests. Severe discounting can therefore change the attractiveness of sustainable resource utilization for commercial interests.

Biologically, the rational scheme for harvesting is to reduce the population to a density allowing a suitable sustained yield and then take any excess population recruitment that accrues thereafter, via either harvesting a constant proportion of the population or maintaining a constant escapement at this target. However, this biologically sustainable strategy does not necessarily maximize economic gain. Colin Clark's (1976, 1990) superb treatise on the economics of harvesting natural resources shows unambiguously that the best biological strategy and the best economic strategy coincide only when a population's maximum rate of increase greatly exceeds the discount rate. Rabbits and herrings fit into that category. When maximum rate of population increase is lower than the discounting rate, however, the real money is made from capital reduction, not sustained yield. It may even be economically clear-sighted to make a total tradeoff, taking all revenue by capital reduction and sacrificing all future sustained yield. This strategy maximizes net revenue, discounted to present value, when the population's maximum rate of increase is below about 5% per year. This can be the economic justification for the extinction of a slow-growing population (and maybe even a species). Blue whales and redwood trees are obvious examples of such slow-growing organisms, as would be many threatened or endangered species. In such cases, economic incentives clearly cannot be relied upon to serve the greater good.

Discounted valuation of future profits is more likely for a privately owned renewable resource. It should be less likely for a publicly owned resource, because the public's discount rate should be much lower. Indeed, one could argue that environmental stewardship argues against any discounting at all, on purely ethical grounds. Publicly owned resources can sometimes take on the character of privately owned resources, however, when the people managing the resource and the people harvesting the resource imagine that they, and not the people as a whole, are its owners.

Any scheme to harvest a publicly owned renewable resource necessarily involves three parties: the owners of the resource (the people), the harvester of the resource (usually a private company), and the manager of the resource (a government agency), which regulates the harvesting. According to constitutional theory, the people in the agency are supposed to act for the owners of the resource, but often they become locked into a symbiotic relationship with the harvesters, as if those two groups were themselves joint owners. Technical advice on sustainable yield offered by the agency's own research branch is commonly ignored by its policy and planning branch when it conflicts with the short-term requirements of the industry. Thus, the ecological aberrations that necessarily follow from the economic implications of the discount rate, and against which the people in the managing agency are employed to guard, often dominate the harvesting operation. Forestry and fisheries provide numerous examples, and commercial wildlife harvesting is not necessarily immune.

18.10 **Summary**

The way in which a safe sustained yield is estimated depends on the population's growth pattern, which is determined in turn by the relationship between the population and its resources. The yield can be estimated either in terms of a numerical offtake or in terms of an appropriate harvesting proportion. The consequences of any harvesting policy should always be evaluated in light of stochastic variation in population growth. In a stochastic world, harvesting a fixed proportion of a population is much less likely to lead to population collapse than is fixed-quota harvesting. Sex- and age-biased harvesting can considerably improve long-term sustainability but tend to produce much lower yields when selectivity is pronounced. In practice, it is often difficult to apply a constant harvest pressure, because the method of control is typically annual quotas adjusted to crude estimates of current population abundance. Incremental adjustments to effort or quotas can induce cyclical population dynamics in harvested populations. Although from an ecological viewpoint recreational and commercial harvesting do not differ in principle, in practice commercial harvesting sometimes has greater potential to exceed sustainable levels of harvests. This is particularly likely when resources are a common property, for which multiple users compete. Under this circumstance, the economically rational behavior will be for resource users to continue to enter the industry until the resources are reduced to dangerously low levels. The "discount rate" of economic analysis can also encourage overharvesting by imparting a greater value to present yields than to future ones, particularly when the maximum annual rate of resource growth is less than the discount rate. Risk due to discounted valuation once again tends to be most pronounced when resources are common, rather than private property. Particularly stringent conservation measures are called for under these circumstances.

19 Wildlife control

19.1 Introduction

In this chapter we show that a control operation is similar to a sustained-yield exercise but is conceptually more complex. The objective must be defined precisely, not in terms of the number of pest animals removed but according to the benefit derived therefrom. Methods include mortality control, fertility control, and various indirect manipulations. A detailed analysis of this topic can be found in two textbooks by Hone (1994, 2004).

19.2 Definitions

Control has three meanings in wildlife research and management. The first two deal with manipulating animal numbers, the third with experimentation.

It is used first in the sense of a management action designed to restore an errant system to its previously stable state by reducing animal numbers. We speak of controlling an outbreak of mice in a grain store or wheat-growing district. The action is temporary.

The second use of "control" has to do with moving a system away from its stable state to another that is more desirable. The animals are reduced in density and the new density is enforced by continuous control operations. The word is here used in a somewhat different sense than it has in engineering. There, a "control" (e.g. a governor on an engine) stops an intrinsically unstable system from shaking itself apart. It is a regulator. That connotation is inappropriate to wildlife management (although it has been so employed on occasion) because, except in special circumstances, the original state is more stable than that created by the control operation.

"Control" is used in a third sense within the parlance of experimental design. As Chapter 14 explains at length, an experimental control is the absence of an experimental treatment. This meaning of the word is usually obvious from context, except when an experiment tests the efficacy of a control program (i.e. "control" in one of the first two senses). The control operation is then the treatment and the control is the absence of control.

The obvious ambiguity in the previous sentence can easily lead to misunderstandings. For example, in an experiment testing the effect on riverside vegetation of controlling (reducing – the second meaning) hippopotami, they were shot (controlled) periodically in one stretch of river. The vegetation along the bank was compared with that of another stretch of river, where the animals were protected (the control stretch – the third meaning). However, a change of hunting staff led inevitably to the control (protected) stretch being controlled (hunted) one sunny Sunday morning. We have seen similar mistakes (discovered at the last minute) in the testing of rabbit control methods. There is no sure remedy but the chance of

Wildlife Ecology, Conservation, and Management, Third Edition.
John M. Fryxell, Anthony R. E. Sinclair and Graeme Caughley.
© 2014 John Wiley & Sons, Ltd. Published 2014 by John Wiley & Sons, Ltd.
Companion Website: www.wiley.com/go/Fryxell/Wildlife

Table 19.1 Differences between donkey populations on two 225 km² blocks in the Northern Territory of Australia, 3–4 years after one population was reduced by 80%. (Source: Choquenot, 1991. Reproduced with permission of the Ecological Society of America.)

Measurement	High-density block	Low-density block
Initial densities (1982) (donkeys/km²)	>10	>10
Treatment (1983)	None	80% shot
Density (1986)	3.3	1.5
Density (1987)	3.2	1.8
Trend	Nonsignificant decrease	Significant increase (20%)
Sexual maturity (% male, 2.5 year)	43%	100%
Female fecundity (2.5 year)	30%	50%
Juvenile mortality (0.5 year)	62%	21%

a disaster can be reduced somewhat by always linking "experimental" to "control" when discussing experimental design.

19.3 Effects of control

If the density of a population is lowered by control measures, the standing crop of renewable resources (e.g. grass needed by a herbivore) will increase as a result of the lowered use. Nonrenewable resources such as nesting holes will be easier for an individual to find. Hence, control, like harvesting, increases the resources available to the survivors of the operation. Their fecundity, and their survival in the face of other mortality agents, is thereby enhanced. For example, an increase in survival of juvenile rabbits compensated for an experimentally reduced reproduction of females (Williams and Twigg 1996; Twigg and Williams 1999; Twigg et al. 2000). The reduced density, therefore, generates a potential increase that will become manifest if the control or harvesting is terminated. Table 19.1 shows just such an effect generated by control operations against feral donkeys in Australia.

The enhanced demographic vigor following reduction in density is a desirable outcome of a harvesting operation, and in fact the success of the harvesting is determined by such an effect, but it acts against the success of a control operation. The further density is reduced, the more the population seeks to increase. Thus, control, in the sense of enforcing a permanently lowered density, is simply a sustained-yield operation that seldom utilizes the harvest. It is an attempt to drive a negative feedback loop in the opposite direction. In other words, density-dependent effects compensate for the imposed mortality of the control operation.

19.4 Objectives of control

More than the other two areas of wildlife management (conservation and sustained-yield harvesting), control is often flawed by a lack of appropriate and clearly stated objectives.

Control, in contrast to conservation and sustained-yield harvesting, is not itself an objective. It is simply a management action. Its use must be legitimized by a technical objective, such as increasing the density of a food plant of a particular species of bird from one per hectare to three. The control operation in this case would be aimed at a herbivore for which that plant was a preferred food. The success of the operation would be measured by the density of plants, not by the density of the herbivore or by the number of herbivores killed.

Control campaigns in many countries share a common characteristic. Very often the original reason for the management action is forgotten and the control itself (lowering density) becomes the objective. The means become the end.

Table 19.2 Published official justification for government control operations against deer in New Zealand. (Source: Annual Reports of the Department of Internal Affairs and the New Zealand Forest Service.)

Years	Official objectives of deer control
1920–1929	Increase the size of antlers
1930–1931	Reduce competition with sheep
1932–1966	Prevent accelerated erosion generally
1967–1980	Prevent accelerated erosion in the heads of rivers that might flood cities
1981–1992	No verifiable reason offered

A good example is provided by the history of deer control in New Zealand. This is one of the largest and longest-running control operations against vertebrates in any country. Table 19.2 lists the sequence of official justifications for government-funded control of deer from 1920 onwards (Caughley 1983). Whereas the stated justification for the control operations changed with time, those changes had virtually no effect upon the management action. There were certainly changes in control techniques but, with the exception of the change of 1967, these were evolutionary adjustments in the management action itself. They were not driven by changes in policy – the means themselves were the end.

Up until 1980, the reasons given for the control operations were that deer and other species caused erosion of the higher slopes and silting of lower rivers (Table 19.2). However, in 1978 new meteorological, hydrological, geomorphological, and stratigraphic research showed that deer, chamois (*Rupicapra rupicapra*), and tahr (*Hemitragus jemlahicus*) had little or no effect on the rate at which riverbeds silted up or on the frequency and size of floods. Despite these data, deer control continued after 1980 for no verifiable reason. All that changed was the stated objectives, which were variously "for aesthetics," "for proper land use," "to ensure the continuing health of the forest," "to protect intrinsic natural values," and "to maintain the distinctive New Zealand character of our landscapes." These are not open to scientific testing.

Many similar examples could be cited from other countries. Control operations must have clear objectives framed in terms of damage mitigation. *Their success must be measured by how closely those objectives are met, not by the number of animals killed.* The operations must be costed carefully to ensure that their benefit exceeds their cost. Their success or failure must be capable of being independently verified. Table 1.1 gives a matrix of possible objectives and actions. It can be filled in to ensure that the management action is appropriate to the chosen objective.

19.5 Determining whether control is appropriate

There are three circumstances in which control may be an inappropriate management action:

1 Where the cost exceeds the benefit.
2 Where the "pest" is not in fact the cause of the perceived problem.
3 Where the control has an unacceptable effect upon non-target species.

These are best investigated experimentally before a control program is instituted. We give two examples.

In the first example, cats were introduced to the subantarctic Marion Island in 1947 to deal with house mice marooned by shipwrecks. They increased rapidly to 3000 by 1977 and fed mostly on ground-nesting petrels. The breeding success of the petrels, particularly the great-winged petrel (*Pterodroma macroptera*), seemed to be declining and cats were suspected to be the cause. The neighboring island of Prince Edward was conveniently free of cats and became the experimental control. The objective of reducing cats was to increase the breeding success of the petrels. Hence, the success must be defined in terms of the birds' breeding success, not reduced numbers of cats. An introduced disease, shooting, and trapping reduced the cats. The petrel breeding success increased from 0–23% (1979–84) to 100%, while chick mortality decreased from 60% in 1979–84 to 0% in 1990. Comparisons with breeding on Prince Edward Island and within a cat-free enclosure on Marion Island identified the cats as the cause of the initial high mortality and the reduction in cat numbers as the reason for the increase in recruitment (Cooper and Fourie 1991).

The second example deals with non-target species. The insecticide Fenitrothion is a well-known organophosphorus pesticide but its effects on song birds and other non-target animals are little known. The Forestry Commission in Scotland wanted to use Fenitrothion to control the pine beauty moth (*Panolis flammea*) and was required to undertake environmental assessment of the effects of spraying on non-target species. For 3 years Spray *et al.* (1987) monitored the effect on forest birds. Their design comprised two pairs of plots, each measuring about 70 ha. The elements of each pair were matched by soil type, age of planting, and tree composition. One element of each pair was sprayed, all plots being monitored before and after spraying to detect annual variation of the density of breeding birds, short-term changes in abundance within 5 days of spraying, and breeding performance of the coal tit (*Parus ater*). The investigators detected no significant difference in these variables between the insecticide-treated plots and the experimental control plots.

19.6 Methods of control

Animal welfare is an important consideration in any control operation. An animal has the right to be treated in a humane manner, whether it is to be protected or controlled. Unfortunately, the notion of humane treatment is often the first casualty of turning a species into a pest. This is particularly noticeable when the species is an exotic. The wildlife manager's paramount responsibility in any control operation is to ethical conduct rather than to operational efficiency.

Control methods can be divided into those aimed at directly increasing mortality, those aimed at directly reducing fertility, and those that act indirectly to manipulate mortality, fertility, or both. The success of an operation is not gauged by the reduction in the density of the target species but by the reduction in that species' deleterious effects. In all cases, the prime responsibility of the wildlife manager is to determine whether the control adequately reduces deleterious effects and whether its benefit exceeds its cost.

19.6.1 *Control by manipulating mortality*

Control by manipulating mortality may be direct, as in poisoning, trapping, or shooting, or indirect, as in biological control through pathogens.

Direct killing
Four simple principles guide the control of a target population living in an environment that remains reasonably constant from year to year. These are largely independent of the population's pattern of growth and they emphasize the conceptual similarities between control and sustained-yield harvesting.

1 When a constant number of animals is removed from the population each year, the size of the population will be stabilized by the control operations unless the annual offtake exceeds the population's maximum sustained yield (MSY).

2 The level at which the population is stabilized by the removal of a constant number each year is equal to or greater than the density from which the MSY is harvested.

3 Density is stabilized by the removal each year of a constant proportion of the population, providing that proportion is lower than the intrinsic rate of increase r_m.

4 The level at which a population is stabilized by removing a constant proportion of the population each year can be at any density above the threshold of extinction.

5 If animals are removed at an annual rate greater than r_m then the population will decline to extinction.

These simple rules can be sharpened for those populations whose pattern of population growth is approximated by a logistic curve (Caughley 1977b). In general it will serve for populations of large mammals (i.e. low r_m) feeding on vegetation that recovers rapidly from grazing.

1 When the constant number C removed each year is less than the MSY of $r_m K/4$, the population is stabilized at a size of:

$$N = \left(r_m + \sqrt{r_m^2 - \frac{4Cr_m}{K}} \right) / \left(\frac{2r_m}{K} \right)$$

where K is the ecological carrying-capacity density, corresponding to the asymptote of the logistic curve.

2 When a constant proportion of the population is removed each year at a rate less than r_m, the population is stabilized at a size of:

$$N = K - \left(\frac{KH}{r_m} \right)$$

where H is the instantaneous rate of removal.

3 If a constant number of animals greater than $r_m K/4$ is removed each year, the population will eventually become extinct.

Examples of eradication

Even with large mammals, the proportion of a population that must be culled each year in order to eradicate it is substantial: 90% of the feral goats in Egmont National Park, New Zealand, had to be culled annually to achieve eradication in 12 years; if only 50% were culled, eradication would have taken over 50 years, if it had happened at all (Forsyth *et al.* 2003).

By far the most common examples of the eradication of pest species are found in Pacific islands, New Zealand, and Australia, because these places have been subject to the invasion of exotic vertebrates (an important review of these is provided by Veitch and Clout 2002). Many islands were deliberately seeded with pigs, goats, and rabbits by sailors in the 1700s to provide a food source in case of shipwreck. These populations increased rapidly, changed the vegetation, and indirectly caused the extinction of many birds. Possums were introduced to New Zealand for commercial harvesting in the 1800s. Shipwrecks and ordinary landings resulted in the inadvertent introduction of rats and mice to most islands, and of snakes to Guam and Mauritius. Control of rats was the motive for introducing cats (e.g. on Marion Island) and mongooses (e.g. on Mauritius and some Hawaiian and Caribean islands). Control of rabbits in New Zealand was the reason for introducing stoats (ermine; *Mustela erminea*) and ferrets (*M. furo*). All these predators increased rapidly and exterminated much of the native fauna.

In recent decades there has been much effort to eradicate these exotics and repair the ecosystems (see papers in Veitch and Clout 2002). Cats were successfully removed from islands off Mexico, the British West Indies, Marion Island off South Africa (Bloomer and Bester 1992), and Lord Howe Island (Hutton 1998). Foxes have been removed from 39 Alaskan islands (Ebbert and Byrd 2002). Pigs were successfully removed from Lord Howe, as well as Santa Catalina Island in California and some of the Mariana Islands. Goats have also been removed from Mauritius and some Pacific islands. However, despite their large size, both pigs and goats are difficult to eradicate, because they live in difficult terrain; intensive efforts to eradicate goats on Lord Howe left a core of six living on precipitous cliffs (Parkes *et al.* 2002). In general, animals that are hunted will change their behavior, becoming more shy and using refuge habitats, so that a disproportionate effort is required to kill the last few (Choquenot *et al.* 1999; Forsyth *et al.* 2003); it took 1000 hunter-days to kill the last four goats on Raoul Island, north of New Zealand (Parkes 1984). Other species are difficult to remove because of their particular habitat uses and adaptations. Thus, the Indian musk shrew (*Suncus murinus*) is an insectivore expanding rapidly across Asia, Africa, and many islands, where it competes with endemic skinks and geckoes. Difficulties in removing this species include their ability to withstand anticoagulant poisons and the need to use live bait for traps (Varnham *et al.* 2002). Brown tree snakes (*Boiga irregularis*) have been particularly difficult to remove on Guam, as were wolf snakes (*Lycodon aulicus*) on Mauritius, because of their ability to hide in small holes (Rodda *et al.* 2002).

Rats and mice have been removed from several small islands. In general, small species (rodents) with high rates of increase have been removed successfully from islands < 1000 ha, an exception being Campbell Island, south of New Zealand, which is 11 000 ha. Larger species can be removed from larger areas.

Biological control

Biological control, so effective against insects, has a poor record against pest wildlife. One of the few successes is the use of *Myxoma* against rabbits. This holds the density of rabbits in Australia to about 20% of their uncontrolled density, despite declines in the virulence of the virus and in the susceptibility of the rabbits – both a product of massive natural selection.

The chances of finding a biological agent to control vertebrates are always low, largely because the pathogen must be highly host-specific and highly contagious.

19.6.2 *Control by manipulating fertility*

Population control by manipulating fecundity has several advantages over simply killing animals, but it also has problems of its own. It was first suggested as a control method by E.F. Knipling in 1938 (Marsh 1988) but was not applied for another 20 years. Its first use was against the screwworm fly (*Cochliomyia hominivorax*), a serious pest of livestock in the southeast of the United States. Subsequently, it has been used against a number of insect pests in various parts of the world.

The use of contraceptive techniques for population control has been reviewed by Marsh (1988) with respect to rodents and lagomorphs, by Turner and Kirkpatrick (1991) with respect to horses, and by Bomford (1990) with respect to vertebrates in general. Bomford shows that although contraception has often been advocated as a useful control method against vertebrates, and has been tried from time to time, there

Table 19.3 Characteristics of an ideal chemosterilant for rodents. (Source: Marsh, 1988. Reproduced with permission of Taylor & Francis Group LLC.)

1	Orally effective, preferably in a single feeding.
2	Effective in very low doses (not exceeding 10 mg/kg).
3	Permanent or long-lasting sterility (preferably) lasting 6 months or longer, or at least through the major breeding period of the pest species.
4	Effective for both sexes, or preferably for females if only one sex.
5	Rodent-specific or genus-specific.
6	Relatively inexpensive.
7	Wide margin between the chemosterilant effects and lethal doses. (If high in specificity, this may be unimportant, or a narrow margin might be of value.)
8	Well accepted (i.e. highly palatable) in baits at effective concentrations.
9	Biodegradable after a few days in the environment.
10	If not highly specific, rapid elimination from the body of the primary target to avoid secondary effects.
11	No acquired tolerance or genetic or behavioral resistance.
12	Free of behavioral modification (such as altering libido, aggression, or territoriality).
13	Free from production of discomfort or ill feelings that could suppress consumption (i.e. bait shyness) on repeat or subsequent feedings.
14	Humane (i.e. no production of stressful symptoms).
15	Easy to formulate into various kinds of bait.
16	Sufficiently stable when prepared in baits (i.e. adequate shelf life).
17	Not translocated into plants (or only at a very low level), permitting use on crops.

is no clear and well-documented example of unqualified success. "Many tests of fertility control have not been robust enough to allow clear conclusions. Experiments have often failed to include treatment replicates, or have relied on small samples. These results cannot be analyzed statistically to estimate the probability of a treatment effect" (Bomford 1990).

The usual method of use against insects – flooding the population with sterile males – is dependent on the females mating only once. That is common behavior among insects that live for only one year but is rare amongst vertebrates.

Most attempts at control by contraception or sterilization have utilized chemicals such as bromocriptine, quinestrol, mestranol, and cyprosterone. Table 19.3 gives Marsh's (1988) criteria for an ideal rodent chemosterilant.

The effect of a contraceptive or sterilizing agent upon the population's dynamics depends on the breeding system of the species, and particularly upon the form of dominance. In general, a vertebrate population will seldom be controlled adequately by a contraceptive or sterilant specific to males (Bomford 1990), so the target should be either the female segment of the population or both sexes.

Caughley *et al.* (1992) explored the theoretical effect on productivity of three forms of behavioral dominance, two effects of sterilization on dominance, and four modes of transmission. There are 24 possible combinations, of which 17 are feasible, but they lead to only four possible outcomes. Three of these result in lowered productivity. The fourth, where the breeding of a dominant female suppresses breeding in the subordinate females of her social group, leads to a perverse outcome: productivity *increases* with sterilization unless the proportion of females sterilized exceeds $(n - 2)/(n - 1)$, where n is the average number of females in the social group (Fig. 19.1). Hence, knowledge of the social structure and mating system is desirable before population control by suppression of female fertility is attempted. Experimental tests of this dominant-female model using artificially sterilized female red foxes has shown that dominance is not an

Fig. 19.1 Mean number of litters produced during a season of births by a group of females of size *n* subject to varying rates of sterilization. One female is dominant in each group and the other females subordinate to her. Only the dominant female breeds. She relinquishes dominance if sterilized, and the subordinates are then free to breed. (After Caughley *et al.* 1992.)

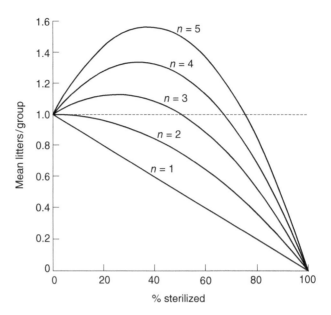

important social suppressor of subordinate female reproduction. Thus, greater female sterility led to lower juvenile recruitment (Saunders and McIlroy 2001).

The theoretically derived examples of reduction of litter production exclude the effect of increased fertility consequent upon lowered density. This cannot be modeled from first principles in the same way as the expected reduction in litter production because its effect is specific to particular species and habitats. It must be examined by way of carefully designed field experiments. Such experiments should follow the effect of the treatment on the population's dynamics rather than simply upon reproduction, because compensatory changes to survival through density-dependent effects may also occur and should always be expected. As mentioned earlier, two rabbit populations in Australia were artificially sterilized to varying degrees. The reduced recruitment of newborn was compensated by a density-dependent survival of the remaining juveniles. Female sterility had to reach 80% before a decline in population was observed (Twigg and Williams 1999; Twigg *et al.* 2000). Similar sterility experiments on brush-tail possums (*Trichosurus vulpecula*) in New Zealand showed population declines with 50 and 80% sterile females, provided immigration was prevented (Ramsey 2000).

Two additional methods of fertility control have been suggested, but these must await further research to demonstrate their general applicability: immunocontraception and genetic engineering.

Immunocontraception
Antibodies can be raised in an individual against some protein or peptide involved in reproduction, hindering the reproductive process. Immunocontraception has been used to reduce reproductive rates and hence mean densities of wild species that have become too numerous, such as elephants, horses, white-tail deer, fallow deer, and seals. One approach uses porcine zona pellucida (PZP) protein antigens, which raise antibodies to block sperm-binding sites on the surface of the ovum in mammals (McShea

et al. 1997; Rudolph *et al.* 2000). This prevents fertilization. The protein must usually be administered by injection or implant because most are broken down by digestion, and the primary inoculation must be boosted once or twice in the following few weeks. Generally, the duration of these vaccines is less than 1 year, so animals must be re-inoculated on an annual basis. This means that the technique is limited to animals that are easy to capture and which come from small populations. This problem has been overcome partly through the use of PZP proteins encapsulated in liposomes, allowing the proteins to be released gradually over a long time span. Ten years of contraception was achieved in grey seals (*Halichoerus grypus*) with a single dose of encapsulated PZP (Brown *et al.* 1997), while Fraker *et al.* (2002) report 3 years of contraception in fallow deer (*Dama dama*). Some of the problems are being overcome, but immunocontraception remains limited to small populations where a significant proportion of the females can be captured for inoculation.

Genetic engineering
Tyndale-Biscoe (1994) suggested using a pathogen of low virulence to vector a foreign gene that would disrupt reproduction. He suggested in particular that the *Myxoma* virus could be used to carry an inserted gene that would reduce the birth rate of the European rabbit (*Oryctolagus cuniculus*) in Australia. Similar approaches are being used for red foxes and house mice (Shellam 1994; Pech 2000).

19.6.3 *Control by indirect methods*

Exclusion
The most obvious way of reducing the deleterious effect of wildlife is to exclude animals from an area. That can be achieved by fencing, by chemical repellents, or by deterrents of one sort or another. Often exclusion is necessary for only part of the time. For example, damage by deer to regenerating pines may be limited to only the first few years after establishment.

Exclosures can be as small as a hectare or of mind-boggling proportions. The first example of the latter category was the Great Wall of China, erected at the instigation of Shih Huang Ti, the first Emperor of the Ch'in Dynasty, between 228 and 210 BC. It protected his northern and western frontiers, the direction from which he was most frequently attacked. The wall traverses about 2400 km of rough country.

Another big one is the Australian Barrier Fence, built to keep dingoes out of sheep country. It runs from the south coast, divides South Australia into two, skirts around the top inland corner of New South Wales, and then loops to enclose all of central Queensland. It was started in 1914 and built in sections, often as an upgrading of previous state border fences and rabbit fences. At its greatest extent it spanned 8614 km – three and a half times longer than the Great Wall of China. In 1980 the loop up through inland Queensland was fenced off about halfway up and the upper fence was abandoned. The present exclosure has a perimeter of 5614 km. In contrast to the numerous rabbit fences that have been built in Australia, the dingo barrier fence has been relatively successful in reducing the spread of a pest species.

The difficulty in reducing populations of introduced predators in New Zealand, Australia, and other islands has promoted the use of fences in those countries, with some success. Although only relatively small areas can be protected, fence maintenance is high, and some predators occasionally get through, this method has worked better than any other so far (Long and Robley 2004).

Sonic deterrents

The modern forms of the scarecrow – sonic devices (bangers, clangers, alarm calls, ultrasonics) – have been reviewed by Bomford and O'Brien (1990), who suggest that, at best, these achieve only short-term reduction in damage. They are particularly critical of the claims made for commercially produced ultrasonic devices and of the standard of experimental testing in this field.

Habitat and food manipulation

This is certainly the most elegant of control techniques, because it does not have to counteract density-dependent compensation within the pest population. The key habitat elements are water and shelter. Red squirrels (*Tamiasciurus hudsonicus*) in British Columbia lodgepole pine forests can be dissuaded from feeding on the stems of very young trees by aerial spreading of sunflower seeds; this alternative food source is preferred over the pines (Sullivan and Klenner 1993). Such diversionary feeding may have reduced predation by raptors on grouse chicks in Britain (Redpath *et al.* 2001) and by small carnivores and corvids on artificial grassland bird nests in Texas (Vander Lee *et al.* 1999). Supplemental food reduced predation by striped skunks (*Mephitis mephitis*) on duck nests, but other carnivores increased their predation, so the results were equivocal (Greenwood *et al.* 1998). Where food is provided in local concentrations at feeders (instead of distributed widely), the high density of carnivores may increase predation of ground-nesting birds (Cooper and Ginnett 2000). Thus, the way food is presented is important to the outcome.

19.7 **Summary**

The control of a pest species, in the sense of holding its density at a reduced level, is essentially a sustained-yield operation in which the yield is not used. Reduction in density is not an end in itself: the success of the operation is measured not by the number of animals removed but by whether the objective was attained, be it an increase in density of an endangered species, an increase in grass biomass, or a reduction in damage to fences. The logic of experimental design must be utilized to determine whether benefits exceed costs, whether the treatment has a deleterious effect on non-target species, and whether the targeted "pest" is really the cause of the perceived problem.

20 Evolution and conservation genetics

20.1 **Introduction**

Evolution and population ecology are naturally linked through their shared dependency on *demographic rates* of a collection of individuals, such as birth and survival rates. Population numbers can only change when the number of new offspring produced does not match the number of deaths that have occurred over the same time-frame. When births exceed deaths, population numbers rise, and when deaths exceed births the opposite occurs. When there is meaningful (nontrivial) variation in the probabilities of birth and death for different individuals and some of that variation traces back to individual traits (genetically determined features of an organism), there is also meaningful potential for evolutionary change in trait frequencies of the population over time. This can arise through gene flow, genetic drift, mutation, cultural selection, or natural selection.

Gene flow refers to movement of individuals from one subpopulation to another that has a different genetic composition. *Genetic drift* is a process closely related to demographic stochasticity (see Chapter 16) in which chance events lead to some individuals leaving behind more offspring than others. Imagine that a pair of albino squirrels manages successfully to discover a previously uncolonized woodlot. The genes for albinism arising from these parents would be passed on and predominate in subsequent generations simply by chance. This would represent genetic drift, which would generally produce loss of some alleles over time. New genetic variation is constantly arising via molecular changes in organismal DNA. So even in the absence of natural or cultural selection, evolutionary change is determined by the relative balance between mutation and genetic drift.

Evolution and natural selection are often used interchangeably in everyday conversation, leading to a great deal of confusion. Fortunately, both concepts can be defined with more care. *Evolution* can be said to occur whenever there is evidence of consistent change over time in the distribution of any trait. *Natural selection* is one agent of such change, which requires several specific conditions (Endler 1986). If (i) there is variation in some trait (termed *phenotypic variation*), (ii) there is a consistent correlation between that trait and survival rate, reproductive capacity, or mating ability (termed *fitness variation*), and (iii) there is a genetically determined tendency for offspring to share the same traits as their parents (termed *heritability*) then (iv) the frequency of occurrence of that trait within a single generation will differ more than would be expected at birth and/or (v) the frequency of occurrence of that trait will change from generation to generation until an equilibrium is reached. In other words, *phenotypic variation + fitness variation + heritability → evolution*.

Wildlife Ecology, Conservation, and Management, Third Edition.
John M. Fryxell, Anthony R. E. Sinclair and Graeme Caughley.
© 2014 John Wiley & Sons, Ltd. Published 2014 by John Wiley & Sons, Ltd.
Companion Website: www.wiley.com/go/Fryxell/Wildlife

Cultural selection differs from natural selection only in the method by which traits are passed on from parents to offspring, with cultural selection involving behavioral traits that are transferred to offspring via learning. *Sexual selection* is the term often used to refer to traits that solely influence breeding success, as opposed to the wide variety of traits that influence survival and reproductive rates. In practice, many studies of natural selection tend to focus on demonstration of the necessary conditions for natural selection (assumptions (i) and (ii)), due to the challenges of measuring heritability and genetic changes over time in the field.

In this chapter we consider the interplay between ecology and evolution via natural selection. We start by developing mathematical models of genetic stasis versus change due to natural selection. This will help us to understand better how natural selection can be measured in the field and how it naturally relates to population ecology. We then consider ways in which changes in environment, ecological relationships, or interaction with humans can lead to evolutionary changes due to natural selection, before flipping things around and considering how rapid evolutionary changes influence first population dynamics and then ecological interactions. Finally, we consider how wildlife populations are affected by loss of genetic variation.

20.2 Maintenance of genetic variation

Suppose the gene pool of a population contains only two alleles at a particular locus. These will be referred to as *A* and *a*. Any individual in that population will thus have one of the three possible combinations of alleles at that locus: *AA* or *Aa* or *aa* . Genotypes *AA* and *aa* show that the individual is *homozygous* at that locus, whereas *Aa* shows that the individual is *heterozygous*. The proportions of the three combinations in the population as a whole are called *genotypic frequencies*, which we will symbolize using p_{AA}, p_{Aa}, and p_{aa}. For example, assume that the frequencies of the three genotypes are as follows:

genotype	*AA*	*Aa*	*aa*
genotypic frequency	$p_{AA} = 0.01$	$p_{Aa} = 0.18$	$p_{aa} = 0.81$

Note that the genotypic frequencies must sum to 1 (because they are proportions). The overall frequency of the *A* allele (symbolized by p_A) will be determined by all the alleles contributed by *AA* individuals and half the alleles carried by *Aa* individuals, so the overall frequency is calculated by $p = p_{AA} + p_{Aa}/2 = 0.01 + 0.18/2 = 0.1$. The corresponding frequency of the alternative allele *a* in the total population $q = p_{aa} + p_{Aa}/2 = 0.81 + 0.18/2 = 0.9$. Note that the sum of a complete set of proportions must equal 1, so $p + q = 1$.

Now we will use a simple model to predict allele frequencies in the next generation. There are six possible genotype combinations of parents that could mate together to produce offspring: (i) *AA* mated with *AA*, (ii) *AA* mated with *Aa*, (iii) *AA* mated with *aa*, (iv) *aa* mated with *Aa*, (v) *aa* mated with *aa*, and (vi) *Aa* mated with *Aa*. If we make the usual assumption that a random assortment of gametes is produced by each parent then the binomial probability distribution (pertaining to chance events involving only two possible outcomes, such as heads versus tails) allows us to readily deduce the fraction of each genotype that could be produced by each mating combination (Table 20.1). The contribution to the next generation can be predicted in turn by simply multiplying the predicted fraction of offspring produced by each possible genotypic combination by the genotypic frequencies for each combination. If we then sum the totals in each column for contribution to the next generation, an interesting

Table 20.1 Predicted probabilities of mating between different genotypes, the fraction of offspring of different genotypes expected as a result of random assortment of gametes, and the contribution of offspring to the next generation, obtained by multiplying the mating frequency by the expected fraction of genotypes among offspring. If there is no selection then the proportion of offspring produced of each genotype is identical to the parental generation (second line from bottom of table). If selection does occur, however, the offspring contribution must be multiplied by the selection coefficient s divided by mean fitness w, leading to a change in genotype frequency from one generation to the next (bottom line of table). (After Hastings 1997.)

Mating	Mating frequency	AA gametes	Aa gametes	aa gametes	aa offspring	Aa offspring	Aa offspring
$AA \times AA$	p^2_{AA}	1	0	0	p^2_{AA}	0	0
$AA \times Aa$	$2p_{AA}p_{Aa}$	1/2	1/2	0	$p_{AA}p_{Aa}$	$p_{AA}p_{Aa}$	0
$AA \times aa$	$2p_{AA}p_{aa}$	0	1	0	0	$2p_{AA}p_{aa}$	0
$Aa \times Aa$	p^2_{Aa}	1/4	1/2	1/4	$p^2_{Aa}/4$	$p^2_{Aa}/2$	$p^2_{Aa}/4$
$Aa \times aa$	$2p_{Aa}p_{aa}$	0	1/2	1.2	0	$p_{Aa}p_{aa}$	$p_{Aa}p_{aa}$
$aa \times aa$	p^2_{aa}	0	0	1	0	0	p^2_{aa}
Offspring produced					$(p_{AA} + p_{Aa})^2 = p^2$	$1 - p^2 - q^2 = 2pq$	$(p_{aa} + p_{Aa})^2 = q^2$
Surviving offspring					$p^2 s_{AA}/w$	$2pq s_{Aa}/w$	$q^2 s_{aa}/w$

pattern emerges: genotype frequencies remain the same from one generation to the next. For example, the frequency of the AA genotype in the next generation $= (p_{AA}^2 + (p_{AA} \times p_{Aa}) + p_{Aa}^2)/4 = ((p_{AA} + p_{Aa})/2)^2 = p^2$: exactly what it was to start with. By similar applications of algebra, the frequency of aa individuals $= q^2$ and the frequency of heterozygous individuals $= 2pq$. This stable recurring relationship between allelic frequencies and genotypic frequencies from generation to generation is known as the Hardy–Weinberg equilibrium law.

Formally, the Hardy–Weinberg equilibrium law holds only when the population is large, its individuals mate at random, and there is no migration, mutation, or selection. In practice, however, it is robust to minor deviations from these assumptions. The Hardy–Weinberg equilibrium holds equally for more than two alleles at a locus (which is common for many alleles) so long as it is calculated in terms of one allele against all the others. Alleles of the two types A and $not\ A$ also take Hardy–Weinberg equilibrium proportions.

The Hardy–Weinberg equilibrium law is important precisely because it explains why there is genetic resistance to change; unless chance events occur frequently (genetic drift), the probability of matings between different genotypes is nonrandom (sexual selection) and the fraction of offspring recruited to the next generation is not determined solely by random assortment of alleles among parents (natural selection).

20.3 Natural selection

Using the same mathematical symbolism we used to derive the Hardy–Weinberg equilibrium, we can determine changes in allele frequency from generation to generation by applying selection coefficients associated with each genotype (Hastings 1997). For example, let's assume that selection coefficients s refer to differential viability of offspring, with homozygous AA offspring having $1.5 \times$ the fitness of heterozygous offspring and $3 \times$ the fitness of homozygous aa individuals, leading to the following selection coefficients: $s_{AA} = 0.9$, $s_{Aa} = 0.6$, and $s_{aa} = 0.3$. The relative contribution of AA individuals in the next generation will be $p_{AA} = p^2 s_{AA}/w$, where mean fitness

$w = p^2 s_{AA} + pq s_{Aa} + q^2 s_{aa}$. In similar fashion, the relative frequency of Aa individuals is predicted by $p_{Aa} = 2pq s_{Aa}/w$ and the relative frequency of aa individuals by $p_{aa} = q^2 s_{aa}/w$. The subsequent contribution of surviving recruits of each genotype to the next generation due to natural selection therefore clearly deviates from the outcomes expected for the Hardy–Weinberg equilibrium in the absence of natural selection (Table 20.1).

Because $s_{AA} > s_{Aa} > s_{aa}$, it is clear that allele A will tend to spread whereas allele a will tend to decline. Changes in the frequency of allele A at time $t + 1$ can be predicted from its frequency at time t by the following formula:

$$p_{t+1} = \frac{p_t(p_t s_{AA} + q_t s_{Aa})}{p_t^2 s_{AA} + 2p_t q_t s_{Aa} + q_t^2 s_{aa}}$$

If there are no changes in selection coefficients, allele A will eventually equilibrate at fixation ($p = 1$) and allele a will disappear (Fig. 20.1a). This process is known as *directional selection*. Thus far, however, we have considered only one possibility – other outcomes can arise depending on the fitness attributes of each of the genotypes. If $s_{aa} > s_{Aa} > s_{AA}$ then allele a will be favored and allele A will disappear over time,

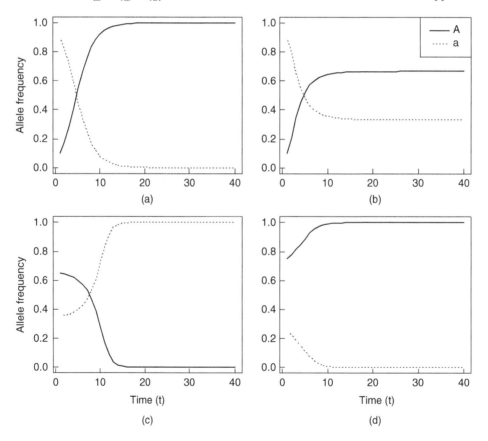

Fig. 20.1 Examples of different outcomes of selection, based on the model in the text. (a) Directional selection for allele A : $s_{AA} = 0.9$, $s_{Aa} = 0.6$, $s_{aa} = 0.3$; (b) Stabilizing selection for both alleles: $s_{Aa} = 0.9$, $s_{AA} = 0.6$, $s_{aa} = 0.3$; (c and d) disruptive selection: $s_{aa} = 0.9$, $s_{AA} = 0.6$, $s_{Aa} = 0.3$, with the outcome dependent on initial allele frequencies.

another example of directional selection. A more intriguing outcome occurs when s_{Aa} is the fittest genotype, in which case both alleles will persist at equilibrium levels determined by the relative fitness of s_{AA} compared to s_{aa} (Fig. 20.1b). This process is known as *stabilizing selection*, which in this case is caused by *heterozygote advantage* maintaining genetic variation in the population. When s_{Aa} is the least fit genotype, either allele A or allele a will become fixed over time, the winner determined by the initial allele frequencies in the population (Fig. 20.1c,d). This process is known as *disruptive selection*.

At any particular time, most selection will be of the stabilizing type. Any tendency for the breeding season to expand, for example, will be attacked continuously by stabilizing selection. This is because an offspring born during a seasonally inappropriate time of the year has little chance of survival and so the mistake, if it has a genetic basis, will be swiftly corrected. The relaxation of that selection in captivity is presumably the reason why the breeding season of captive populations tends to expand after several generations. Stabilizing selection is essential to maintaining the fitness of a wild population. Ironically, it is one of the strongest forces reducing additive genetic variance and heterozygosity.

The model we have been using is rather simplistic, being based essentially on constant demographic rates, which would ordinarily lead to geometric growth or decline over time, but it does give us some idea about the processes that in the short term might lead to changes in trait frequencies. The model is also limited in assuming that the trait shown by the organism (termed the "phenotype") is completely determined by the genotype dictated by a single pair of alleles at a single locus. Genetic effects are almost always much more complex, involving interactions among multiple alleles at multiple loci, leading to continuous trait variation. Population geneticists use the so-called "breeder's equation" to characterize natural selection acting on continuous traits influenced by many loci:

$$R = Sh^2$$

where R is the change in the average value of a trait from one generation to the next, S is the selection differential (analogous to the selection coefficients used earlier), and h^2 is the heritability of the trait, estimated as twice the correlation coefficient for a continuous trait shared by common descent between a parent and its offspring. High levels of both heritability and strength of selection are required to produce rapid evolutionary change in the frequency of a given trait.

20.4 Natural selection and life history tradeoffs

For a long time, biologists assumed that the processes leading to evolutionary change were much, much slower than the processes affecting population dynamics. Indeed, textbooks of the early 2000s often referred to evolution as occurring over tens of thousands of generations. As a result, evolutionary biology and ecology developed as essentially separate disciplines. Yet, the same processes that lead to changes in population numbers are also the raw material on which natural selection can act. A growing number of recent studies conclusively demonstrate that the speed of evolutionary change can indeed be similar to that of changes in population numbers, suggesting that both processes are inextricably bound.

Some of the strongest examples come from *life history* evolution in harvested populations. Life history refers to the lifetime pattern of growth, reproduction, and survival by an organism. The fundamental assumption underlying almost all work on life history evolution is that there are fundamental tradeoffs between these characteristics as a direct consequence of limitations on the energy budget. Hence, increased investment of energy in reproduction at a given age comes at a cost of reduced energetic investment in survival, growth, and future reproduction. This tradeoff of energy is certainly a well-founded assumption. For example, strains of fruit flies that are prevented from breeding in the lab develop a significant increase in longevity; that is, adult survival rates are improved (Müller *et al.* 2001; Novoseltsev *et al.* 2003). Similarly, collared flycatchers that lay eggs at a fast rate early in life show much more rapid decline in egg laying over their lifetime than those that have more modest egg-laying rates from the outset as young adults (Gustafsson and Sutherland 1988).

Concentrating on tradeoffs between current reproduction and adult survival, let us consider some fundamental ways this tradeoff could occur (Gadgil and Bossert 1970). Heavier energetic investment in reproduction should lead to higher recruitment of offspring (obtained by multiplying the birth rate b by juvenile survival s_0). The relationship between realized fecundity ($F(\theta) = bs_0$) and energetic investment could conceivably curve upwards, provided that there is a threshold effect (Fig. 20.2a). For example, a large litter of offspring may be more capable of efficient thermoregulation or more resistant to predators than a small litter, yielding accelerating returns on investment. This shape is termed "concave upwards." On the other hand, the opposite is also possible, such that there are diminishing returns between fecundity and investment (Fig. 20.2b), due to declining proficiency in provisioning the litter or increased disease risk. This shape is known as "convex upwards." At this stage, we aren't concerned with the biological cause, just the shape of the relationship. Survival of the parent (s_a)

Fig. 20.2 Trade-offs between realized recruitment (F, dotted line) and adult survival (s_a, dashed line). Stars indicate optimal fitness combination. In panel a, both F and s_a have concave upward shape, whereas in panels *b-d* both F and s_a have convex upward shape. Juvenile survival in panel *c* is 1/2 that in panel *b* but adult survival is unchanged.

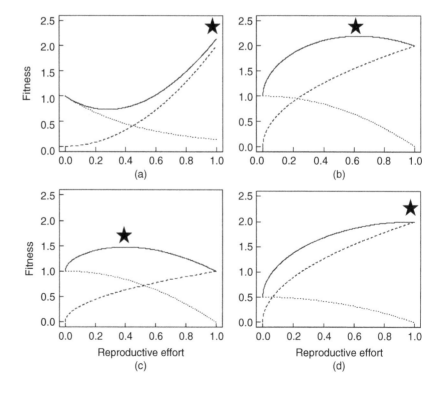

would be expected, in general, to be negatively related to reproductive investment, due to energetic tradeoffs. The form of this tradeoff, however, could be concave (Fig. 20.2a) or convex (Fig. 20.2b) upwards, just as it is for reproduction.

Individual fitness in this simple case can be defined as the sum of adult survival and realized fecundity. This is the same assumption that we have used in deriving the simplest form of the geometric growth model (see Chapters 5), so we will use the annual growth rate λ to symbolize fitness. In the threshold case, where reproductive benefits accelerate with increased energetic investment, echoed by similar threshold effects on adult survival (Fig. 20.2a), the highest fitness will be at either the lowest or the highest value. Which is favored depends entirely on the placement and shape of the curves. In this case, natural selection would favor either a delay in reproduction, with the organism putting all its energy into adult survival, or complete investment in reproduction and no investment in continued survival. Such a life history strategy is known as *semelparity* and is typified by annual plants and migratory salmon, the latter of which spend several years at sea before they migrate back to their natal freshwater stream to reproduce once and then die.

Natural selection should favor a different outcome when reproduction and survival curves are convex upward. In this case, extreme values are the poorest option, with the highest fitness realized at an intermediate level of investment in both reproduction and survival. This is known as an *iteroparous* reproductive strategy, seen in perennial plants and most mammals, fish, and birds, where individuals breed several times in their lifetime. We can go even farther now to consider the impact of changes in juvenile survival versus that of adults. If we imagine a change in ecological circumstances leading to decreased juvenile survival, but little change in that of adults, then this will reduce the realized recruitment curve. As a consequence, the optimal life history will be reduced reproductive investment but increased investment in long-term survival (Fig. 20.2c); that is, if individuals are less secure that their offspring will survive, they should hedge their bets by spreading their reproduction over a longer lifespan. On the other hand, if adult survival is compromised but juvenile survival is little affected then the optimal life history will be to invest more heavily in early reproduction, rather than waiting around for a longer time. The key feature to appreciate here is that the nature of the tradeoffs dictates fundamental changes in selection pressures. A similar logic tends to involve tradeoffs between growth and reproduction, with reduced ratio of juvenile to adult survival favoring greater investment in growth and reduced investment in offspring at any stage in life.

20.5 Natural selection due to hunting

With this background, we can better understand the multiple effects of intense harvesting, say of ungulates (Allendorf and Hard 2009). Recreational hunters, for example, tend to target older, bigger males, which offer more meat and more impressive antlers or horns. Indeed, there is a well-developed sportsman ethic to choosing the biggest and most impressive specimen. For some hunters, the attractiveness of such trophies can be a powerful incentive, with individual hunters investing thousands of dollars in the right to pursue a record trophy. Life history theory suggests that the outcome of this management policy will be to reduce adult survival relative to juvenile survival, with consequent prediction of early onset of reproduction at the expense of further growth and adult survival. One would expect to see smaller, shorter-lived individuals over time.

A meta-analysis of trophy size in 25 different species of hunted ungulates supports this prediction, with significant changes in 11 out of 17 species with antlers and 3 out of 8 species with horns (Monteith *et al.* 2013). In every case, the dimension of registered hunting trophies exhibit slow declines over time, suggesting evolutionary change in phenotype. On the other hand, sustained harvesting may prevent individuals from aging long enough to develop large ornaments, even without any genetic changes. Clearly, more detailed analysis that includes heritability and calculation of selection coefficients on the trait in question is called for.

One of the clearest examples of harvest-induced natural selection comes from a small, isolated population of bighorn sheep in Canada (Coltman *et al.* 2003). Close monitoring of individually recognizable individuals allowed the researchers to evaluate the survival and reproductive success of virtually all animals in the population over a 3-decade study. During this time, trophy hunting was restricted to sexually mature males. Because researchers were able to closely monitor paternity and link it with breeding success, they could assess the loss of future breeding opportunities for those males that were removed early in adult life by hunters versus those that reached their full natural longevity.

Researchers found several pieces of evidence suggesting both strong natural selection and rapid evolutionary change in the population (Coltman *et al.* 2003). Forty percent of the available males were harvested each year, representing a level of mortality several times higher than background natural mortality. Some males were removed as early as 4 years of age. Breeding success was positively linked with horn length and body size, and both of these traits were highly heritable. There was a strong tendency among hunters to remove males that had high potential for breeding, because of their large body size and long horns. Apparently in response to this strong selection pressure, horn length and body mass steadily declined over time (Fig. 20.3). These patterns suggest that intense selection against the biggest rams with the most impressive horns led to faster maturation, smaller body size, and reduced ornamentation in just a matter of years.

Hunters prefer not only male animals with elaborate ornamentation but sometimes also animals with particular coat coloration. For example, the famous evolutionary biologist J.B.S. Haldane (1942) argued that the pronounced increase in the frequency of pelts of normal red-colored morphs of Canadian red foxes relative to those of the more valuable silver morph over the course of the 19th century (Elton 1942) could be explained as the logical outcome of selective preference by trappers for the more valuable commodity. While there are some problems with this explanation, such as how trappers could possibly harvest selectively using the foothold traps in common use at the time, the changes in morph frequency are undeniable. Similar anecdotal evidence applies to the African elephant, whose tusks are highly prized for making jewelry, carvings, and other ornamental objects (Jachmann *et al.* 1995). Over the course of several decades, the proportion of male elephants with tusks in Zambia declined from 90 to 62%. This was interpreted as an evolutionary response to intense poaching pressure.

Recent work suggests that even animal personality might be under selection pressure due to hunting. Analysis of the patterns of movement of Canadian elk suggests that a key variable influencing risk of falling victim to a hunter is simply the boldness of the individual. Wide-ranging animals that are not afraid to explore open areas are significantly more likely to be shot by hunters than are shy animals that stick to more secluded landscapes with dense forest cover (Ciuti *et al.* 2012). Such differences in

Fig. 20.3 Changes in body mass (a), horn length (b), and population size (c) in Alberta bighorn sheep in response to harvest-induced natural selection against big males with large horns. (Source: Coltman *et al.*, 2003. Reproduced with permission of Nature Publishing Group.)

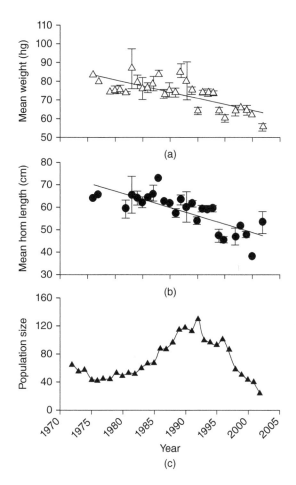

behavior have increasingly come to be seen as part of a behavioral syndrome or animal personality (Sih *et al.* 2004; Reale *et al.* 2007).

In addition to changes in male phenotype, intense harvesting on male ungulates can skew the sex ratio to a considerable degree. It is not uncommon, for example, to have 10 females for every mature male in heavily harvested populations of large mammals (Milner-Gulland *et al.* 2003). This can reduce the effective population size (see Section 20.13) and thereby accelerate loss of genetic variation in the population (Hard *et al.* 2006). Loss of genetic variation can make affected populations more vulnerable to disease, as well as less able to cope with environmental change.

20.6 Natural selection due to fishing

Natural selection through biased harvesting can also apply to many commercial fisheries. This might be induced inadvertently through management restrictions on net size, intended to permit smaller, immature fish to escape harvesters. Early in the 1900s, fishery scientists recognized that this could shift the relative magnitude of adult versus juvenile mortality rates, favoring smaller fish that become reproductively mature at an earlier age and lower level of body weight (Allendorf and Hard 2009). The reason for this is straightforward: if lifetime reproductive success accrues through repeated breeding bouts for long-lived species then reduced survival of adults has a profound

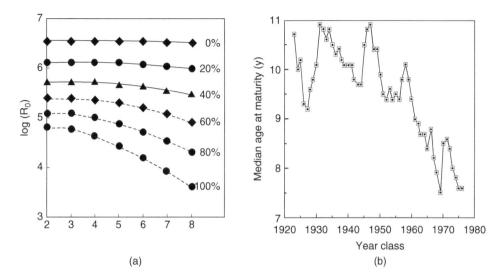

Fig. 20.4 (a) Predictions of the effect on lifetime reproductive output for North Atlantic cod with respect to age in years, based on a range of rates of fishery exploitation levels (0–100%). (Reprinted with permission from Law 2000.) (b) Observed changes in the median age at which Atlantic cod reach sexual maturity over time, combined for both sexes. (Source: Jørgenson, 1990. Reproduced with permission of Oxford University Press.)

effect on fitness. This is a classic outcome from elasticity analysis of life tables (see Chapter 13).

Long-term records for a number of fisheries show substantial variation in life history parameters (Law 2000). The median age at maturation in North Atlantic cod, for example, has declined steadily over the past century (Jørgenson 1990). Cod stocks that matured at 11 years of age in the 1930s were reaching sexual maturity at less than 8 years by the 1980s (Fig. 20.4). Similar patterns have frequently been seen around the world in multiple commercially harvested species (Law 2000). Although this pattern is well known, there is still some debate as to the cause of the life history changes. Many commercial fish stocks are heavily harvested, with some estimates suggesting that a substantial fraction of oceanic fish stocks are currently below half of their pristine carrying capacity. It is well known from general ecological principles that reduction in population density should lead to greater resource availability per capita, with corresponding effects on individual growth, maturation, and productivity (see Chapter 5). Hence, the observed changes in age and weight at maturation might simply stem from reduced intraspecific competition.

Most estimates taken from commercial fisheries suggest that heritability for the relevant life history traits (growth and age at maturation) is rather low (Law 2007; Allendorf and Hard 2009). Moreover, until recently there were few published estimates of selection coefficients, and those that were available were often rather low. It has been argued, accordingly, that selection for earlier maturation caused by fishing could be rather weak (Anderson and Brander 2009). A recent meta-analysis of 29 harvested species, including 21 commercial fish stocks, showed that rates of change in life history parameters were substantially greater than those in unharvested species, with average changes of 20–25% in relevant life history traits. Clearly, this is a contentious issue, but the evidence in favor of human-induced evolutionary change on fish life history parameters is hard to dismiss.

Recreational fishing might also be a potent agent of natural selection. Many fishing regulations have a minimum size restriction on fish that requires release of any small fish that are captured. There are abundant records suggesting that the size of fish caught by hook and line has dramatically declined in many lakes (Post *et al.* 2002). The process of hook-and-line fishing is in some ways reminiscent of predation. We might accordingly expect rapid evolution of behavioral phenotypes to avoid fishermen given sufficiently high levels of angling. This has been documented in large-mouth bass, where a significant level of heritability in behavior leading to capture was detected in a bass population suddenly exposed to anglers (Philipp *et al.* 2009).

The management implications of these kinds of evolutionary response to harvesting could be profound. Controlled trials have shown that selective changes in growth rates quickly reduce yields in experimentally harvested populations of fish (Conover and Munch 2002). This could cause overly affected populations to remain in an unproductive and unprofitable state for a considerable period even after harvest quotas are reduced. Such an argument has been made to explain why collapsed Atlantic cod stocks off the eastern coast of Canada have failed to recover following the imposition of much reduced harvest levels (Swain *et al.* 2007).

Several management options are potentially available, although few management agencies have seriously embraced the issue of undesirable evolutionary change imposed through selective harvesting. Imposition of upper size limits in addition to lower ones seems like a sensible means to counter the negative impact of age- and size-biased mortality (Law 2007). How this might be achieved is less obvious, but extruder devices to shunt large fish out of the nets, like those used to reduce by-catch of sea turtles (Chapter 13), might be one option. In other cases, such as lobster fisheries, modification of trap design could readily ensure that large fish escape capture. Finally, no-harvest zones that allow more natural patterns of age-specific survival might be a simpler means of maintaining genetic variation than trying to prevent capture of big fish in an open-access fishery.

Evolutionary issues might well prove easier to manage in recreational hunting and fishing, simply by altering management policy to de-emphasize male-only licenses, eliminating typical restrictions on harvest of young animals, or imposing a ban on trophy hunting itself. The latter might well be economically unpalatable, however, given the enormous revenues that are often associated with the trophy-hunting market.

20.7 Selection due to environmental change

One of the major challenges for field studies of natural selection is linking strong causal factors in the environment with genetically influenced traits. Ideally, one would need to measure a wide suite of attributes (coloration, body size, foraging ability, etc.) for a large fraction of the population for a long time, such that the population experienced substantial variation in selection pressure. This is a difficult task on many fronts: few species can be monitored so closely, unless the study area is small and highly isolated, the latter to restrict movement in and out of the population, while following the fates of all individuals in even a small area can be daunting, unless each individual is recognizable on the basis of physical features or has been permanently marked early in life by the researcher. Such longitudinal studies of individually known populations have been enormously useful in helping explain the role of natural selection in ecological time. One of the best-known examples is the long-term studies of Soay sheep on the island of St. Kilda, off the west coast of Scotland (Clutton-Brock and Pemberton 2004).

An early form of domesticated sheep was introduced on to St. Kilda by Celtic settlers from Scotland in the early medieval period. The remote location combined with the harshness of the environment made it difficult for pastoralists to thrive and the island was finally abandoned by them in the middle of the 20th century, leaving the remaining members of the sheep flock to evolve without human influence. It is possible to monitor closely the fate of each individual on this windswept island. Continuous monitoring dating back to the 1950s shows that this population exhibits extreme fluctuations in abundance from year to year, triggered by drastic changes in survival (see Chapter 13). Combinations of high population density, limited food availability, and wet, cold weather conditions during the winter lead to massive individual die-offs. In the harsh environment of St. Kilda, few sheep live more than 6 years, and many die in their first year of life.

Detailed demographic studies have been conducted, in which most of the lambs born on the island are caught by hand, weighed, and have their body measurements taken; the lambs are then fitted with numbered ear tags before being released back into the population. At regular intervals, a large fraction of the population is caught and remeasured, allowing detailed records to be made of the lifetime trajectory of changes in body size and weight across several years. Since the reproductive success and survival of most individuals is well known, it is possible to tease out the relative contribution of each individual to the mean fitness of the entire population (sometimes termed *Malthusian fitness*). This can be estimated by the annual growth rate:

$$w_t = \frac{N_{t+1}}{N_t}$$

where N_t is population abundance in any given year.

The contribution of a given individual i to population fitness can be similarly estimated by imagining that individual i and its offspring had simply disappeared (Coulson *et al.* 2006), in which case mean fitness would have been reduced to:

$$w_{t(-i)} = \frac{N_{t+1} - \varphi_i}{N_t - 1}$$

where φ is the net contribution of offspring from the individual plus 1 if the individual itself survived and 0 if it died over the course of the year. Relative fitness is therefore $p_{t(i)} = w_t - w_{t(-i)}$. This formula can be expressed in terms of adult survival and reproductive contribution:

$$p_{t(i)} = \frac{s_{t(i)} - \text{mean}(s_t)}{N_t - 1} + \frac{f_{t(i)} - \text{mean}(f_t)}{N_t - 1}$$

where $s_{t(i)}$ is the survival of individual i in year t and $f_{t(i)}$ is the number of surviving offspring produced by individual i in year t.

Using this so-called de-lifing approach, Pelletier *et al.* (2007) examined a number of different traits that might influence fitness in Soay sheep. They found that body size, measured via either weight or length of the femur, was a strong determinant of survival of sheep from year to year. Such phenotypic traits are highly heritable in Soay sheep. The combination of directional selection and demonstrable variation in a trait that is controlled to an appreciable degree by genotype implies strong potential for rapid evolution in body size.

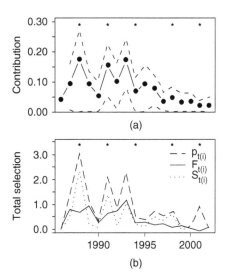

Fig. 20.5 (a) Variation over time in the contribution of body mass to fitness in Soay sheep and (b) variation over time in the strength of selection on body mass that stems from effects on survival (S) and fecundity (F). (Source: Pelletier *et al.*, 2007. Reproduced with permission of the American Association for the Advancement of Science.)

The strength of selection on Soay sheep varies considerably, however, from year to year (Fig. 20.5). Not surprisingly, large body size is most beneficial in years of harsh winter weather, with little detectable benefit in benign years. This suggests that evolution might be most rapid in poor years, or perhaps in years where the population is particularly challenged in other ways by poor feeding conditions, high costs of thermoregulation, or high population density. The key point is that natural selection in this population is strong enough to make a meaningful impact in a matter of years, rather than over millennia.

In addition to competition for limited resources, survival in wildlife species is often influenced by predation, creating a different avenue on which natural selection might act. One of the clearest examples of the impact of predation on life history strategies comes from wild guppy populations in Trinidad. In the headwaters of some streams there are waterfalls which act as a natural barrier to dispersal by guppies and their predators. Reznick *et al.* (1997) took advantage of this to create an elegant experimental design to test the magnitude of natural selection. They collected guppies from sites in two different streams that had been exposed to high predation rates for quite some time. These individuals were transplanted above the waterfall into new sites on the same stream without predators and changes in their phenotype were monitored over 7 and 11 years. After several generations, samples of both experimental and control populations of fish were then brought back into the lab and reared under common conditions for two generations, long enough to eliminate any maternal conditional effects that might stem from the natal environment. They found that relaxation in predation risk led to substantial changes in life history, such that guppies matured at a later age and larger body size than in the original stream reach. Predation-free guppies also produced a smaller number of larger eggs than were produced by the population only a few generations earlier (Reznick *et al.* 1997). The rate of change in these life history parameters was several times faster than expected from phenotypic rates of change from the fossil record. Because the trials were conducted under the same lab conditions (a so-called "common garden experimental design"), the differences in life history trace back to genetic differences that had clearly evolved in only a few years.

These changes in the relative allocation of effort to growth and maturation versus reproduction were exactly as one might predict from life history theory. Interestingly, however, later experiments showed that the decrease in reproductive allocation induced by relaxation in predation risk had no detectable effect on longevity of guppies (Reznick *et al.* 2004). That is, guppies from the riskier environment matured earlier and at a smaller size, but this did not necessarily bring on more rapid senescence. Nonetheless, experiments did show that the guppies from the riskier environment did not perform as well in tests of swimming endurance, which could well be indicative of accelerated physiological costs.

Since both competition and predation are common to most species, it would be informative to understand better which is the more potent agent of natural selection. Interestingly, few such studies have been done, especially under normal field conditions. A superb example, however, comes from Anolis lizards in the Caribbean. These lizards naturally occur on a wide variety of islands, and a number of different studies have been conducted using lizard populations on different islands as a convenient tool to manipulate ecological conditions.

Calsbeek and Cox (2010) transplanted over 1300 Anolis lizards from a single-source population on to a series of six small, isolated islands from which lizards had been completely removed. Half of the islands were seeded with high population densities of lizards, while the other half received low-density populations. One-third of the islands were kept predator free by placing netting around the island perimeter and canopy, one-third were exposed to bird predators, and one-third were exposed to predation by both birds and snakes transplanted there. Before introduction, each lizard was marked and weighed and a series of anatomical measurements were taken. During the experiment, researchers monitored habitat-use behavior, particularly with respect to time spent on the ground versus high on vegetation, where lizards were free of snake predators. Four months after transplantation, surviving lizards were recaptured and the survival rate was calculated for lizards of different sizes and behaviors.

Results showed that lizards recognized the threat imposed by predators, those predated by snakes reacting by perching higher in the vegetation than those introduced on to islands devoid of them (Calsbeek and Cox 2010). Even though they tended to avoid risky ground habitat where they were more exposed to snake predators, lizards in the snake treatment had significantly lower rates of survival than those in predator-free islands or those with just bird predators. Despite this increased risk factor, there was no detectable effect of predation as a source of natural selection on body weight, hind-limb length, or running endurance. All lizards were seemingly of equal vulnerability to predators. On the other hand, variation in lizard density affected all of these variables, so was a demonstrable source of natural selection in these experimental populations, such that bigger lizards with longer hind legs, especially those with greater running endurance, were at a selective advantage. In sum, competition seemed to be the stronger agent of natural selection in these trials (Fig. 20.6). Similar transplant experiments elsewhere in the Caribbean do suggest that predation is still a demonstrable source of selection, particularly on hind-limb length (Losos *et al.* 2004, 2006), but these changes took longer to be realized, suggesting that while the intensity of natural selection might have been highest for competitive interactions, predation risk also plays a role, albeit at a slower pace.

These natural and human-caused experiments suggest that natural selection is detectable even on short timescales. They also suggest that realistic variation in

Fig. 20.6 Variation in the strength of natural selection imposed by predation (left panels) and intra-specific competition (right panels). Three anatomical measures were compared: body length (top panels), hind leg length (middle panels), and running endurance (bottom panels). Only competition had significant effects. (Source: Calsbeek and Cox, 2010. Reproduced with permission of Nature Publishing Group.)

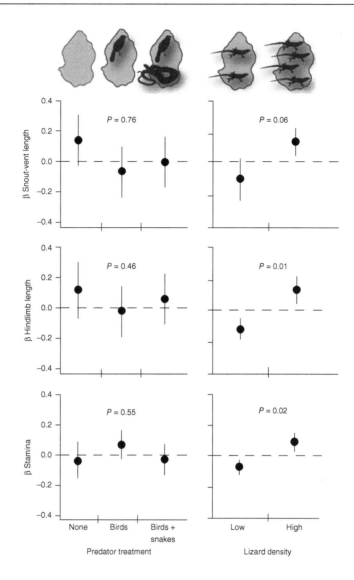

weather conditions, population density, and predation risk over time could cause rapid enough changes in the phenotype of wildlife species to be detectable. This suggests that in addition to routinely measuring the normal demographic parameters (population numbers, age distribution, and spatial patterns), wildlife studies might well benefit from monitoring of phenotypic and perhaps even genotypic constitution. Although this research approach is still in its infancy, there is clearly potential to weave these processes more routinely into wildlife management.

Interestingly, there is some evidence that rates of trait change are fastest in studies conducted at short, rather than long, timescales. Hence, the rates of life history evolution detected in the Anolis lizard or Soay sheep studies tend to be faster than those routinely picked up from body-size changes recorded from the fossil record. It is not entirely clear whether this short-term effect is a real phenomenon or not, because short-term studies that do not detect any trait change may be less likely to be published.

A more plausible explanation is that it is rare for conditions to remain constant for a long time, leading to rapid directional selection, but with frequent reversals, leaving little evidence of net trait change over the long term.

20.8 Ecological dynamics due to evolutionary changes

Given the growing body of evidence that natural selection in response to changing ecological conditions can occur rapidly, there is a clear need for better understanding of the impact of rapid evolutionary change on population and community dynamics (often termed *eco-evolutionary dynamics*). This topic is an active area of research interest, but there are enormous challenges. A complete demonstration of eco-evolutionary dynamics requires long-term data, ideally from multiple populations, for genetic, phenotypic, and demographic responses to well-understood ecological conditions. One would also hope for a clear demonstration of the actual mechanism by which evolutionary processes influence demographic responses, as well as some means of assessing the relative importance of evolutionary versus ecological sources of variation. A conclusive test would also require a control treatment in which the eco-evolutionary mechanism has been disengaged. Currently, few studies meet such stringent demands (Fussmann *et al.* 2007; Schoener 2011), but several come tantalizingly close.

One of the best-known examinations of phenotypic responses to environmental fluctuations comes from the long-term studies of ground finches in the Galapagos Islands (Grant and Grant 2006). As Darwin himself noted, the morphology of these finches is strongly shaped by resource availability on the islands. Finches with robust, thick beaks can process big, extremely hard seeds, but they are not very efficient at handling smaller seeds that require dexterity (Schluter *et al.* 1985; Grant and Grant 2006). Finches with narrow beaks are more efficient at handling small seeds but have insufficient leverage to crack heavy ones.

Populations of the finch *Geospiza fortis* crashed dramatically in the mid-1970s in response to a drought that eliminated most of the smaller seeds on the island where the study was conducted (Fig. 20.7a). In the dry years that followed, big individuals with thick beaks began to predominate in the finch population, likely in response to the augmented supply of big seeds (Fig. 20.7b,c). As weather conditions returned to normal, the fraction of small seeds became more numerous and finch size and beak dimensions declined to their original levels.

In order to evaluate these historical changes, we need a means of estimating the impact of evolutionary versus ecological sources of variation on fitness. Hairston *et al.* (2005) provided a neat solution to this issue that is quite applicable to the kinds of population census data routinely gathered by wildlife biologists. Working with the exponential growth rate *r* as a measure of Malthusian fitness, they pointed out that the contribution to variation in fitness from year to year caused by variation in a selected trait (*z*) compared to that from an ecological factor like weather (*x*) can be expressed by the following formula:

$$\Delta r = \frac{\partial r}{\partial x}\Delta x + \frac{\partial r}{\partial z}\Delta z$$

where $\frac{\partial r}{\partial x}$ is a coefficient measuring the effect of a slight change in weather on population growth rate, $\frac{\partial r}{\partial z}$ is a coefficient measuring the effect of a slight change in trait value on population growth rate, and Δr, Δx, and Δz represent changes in exponential growth rate (see Chapter 5), weather conditions, and trait value from year to year,

Fig. 20.7 Variation over time in (a) the abundance of the ground finch Geospiza fortis, (b) body and beak size, (c) seed abundance, and (d) the proportion of large seeds. These data were used to estimate (e) the impact of variation in evolutionary (solid line) and ecological variables (dotted line) on population growth by the finch. (Source: Hairston *et al.*, 2005. Reproduced with permission of John Wiley & Sons Ltd.)

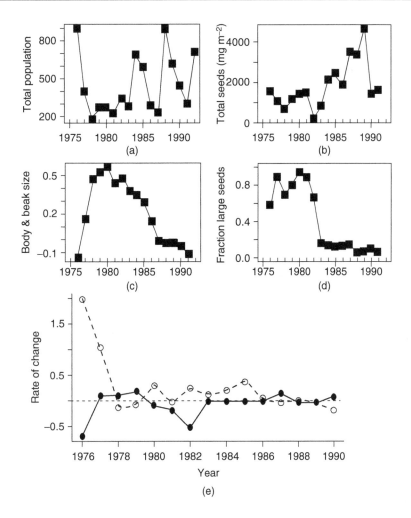

respectively. The magnitude of these coefficients can be readily estimated using standard regression techniques.

Using this measure, Hairston *et al.* (2005) evaluated the relative magnitude of evolutionary versus ecological impacts on the demography of the Galapagos ground finch *Geospiza fortis*. They estimated that evolutionary variables such as body mass and beak depth explained more of the variation in population growth rate in 10 out of 15 years of monitoring than did ecological variables such as rainfall. In other words, evolutionary changes from year to year were at least as important in understanding finch population dynamics as ecological ones. This is a remarkable finding, supporting not only Darwin's intuition that survival of the fittest is a potent agent of evolutionary change but also the equally logical corollary that evolutionary change can be a potent agent determining finch survival.

Long-term data of a similar nature are available for a number of wildlife species. Taking advantage of these and using Hairston *et al.*'s (2005) method, Ezard *et al.* (2009) evaluated the magnitude of evolutionary versus ecological influences on growth rate in five different populations of large herbivores (Soay sheep, roe deer, mountain goats, and two populations of bighorn sheep) for which detailed data were available on phenotype and annual growth rate for a large number of years. Their results suggested that phenotypic variation was at least as important as ecological variation in explaining population growth rates in four out of five of these populations. Clearly, more such analyses need to be performed, but the available evidence certainly suggests that there is much to be gained by looking more closely at phenotypic change in order to understand wildlife population dynamics. We will now consider how loss of genetic variation can affect wildlife populations.

20.9 Heterozygosity

The number and frequency of different alleles at a locus can be determined fairly easily by electrophoresis or DNA sequencing. If p_{ij} is the frequency of allele i at locus j in the population as a whole, the expected proportion of individuals heterozygous at that locus h_j, assuming random mating, may be estimated as:

$$h_j = 1 - \sum p_{ij}^2$$

providing that the number of individuals n_j examined for locus j is greater than 30. If fewer, h_j is underestimated, but it can be corrected by multiplying by $2n_j/(2n_j - 1)$. Thus, the more alleles at a locus, the higher the value of h_j, and the less diverse the frequencies of the alleles, the higher h_j.

The expected level of heterozygosity, assuming random mating, is estimated as:

$$H = \left(\frac{1}{L}\right) \sum_j h_j$$

where L is the number of loci examined and H is a measure of gene diversity that varies considerably between species for reasons that are not understood. As estimated by one-dimensional electrophoresis of loci controlling production of proteins, H ranges between 0.00 and 0.26 for mammals, with an average at about 0.04. Heterozygosity estimated in the same way yields $H = 0.036$ for both white-tailed and mule deer (Gavin and May 1988) and $H = 0.029$ for leopards. Fig. 20.8 shows the frequency

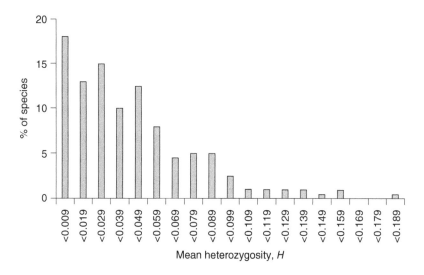

Fig. 20.8 Frequency distribution of mean heterozygosity H for 169 mammalian species. (Data from Nevo *et al.* 1984.)

distribution of H for 169 species of mammals. Note first that the distribution is shaped like a reverse J—most species have a low H but a few break out of that pattern to return a high one – and second that a substantial proportion (10.6%) of mammalian species are homozygous at all loci that were examined by electrophoresis.

Genetic variability can also be reported as the proportion of polymorphic loci in a population (i.e. the proportion of loci for which there is more than one allele within the population as a whole). This is not the same as H. If all but one individual in the population is homozygous at locus A, that locus is nonetheless scored as polymorphic. The proportion of polymorphic loci within the population is therefore higher than the average heterozygosity (the proportion of heterozygous loci in an average individual) – usually about three times higher. Further, the proportion of polymorphic loci is an unstable statistic tending to increase as sample size increases; we recommend against its use.

At this stage we make an important point on a subject that is widely misunderstood: the nature of *genetic diversity*. This is usually conceptualized by conservationists as the number of distinct alleles in a population. The loss of one of those alleles is seen as a reduction in genetic diversity and therefore a bad thing. This notion of genetic diversity is trivial and entirely inappropriate to conservation management. Rather, the important measure is *genetic variance*, which can be conceptualized as a parameter closely akin to heterozygosity H, the proportion of loci that are heterozygous in an average individual. Consequently, the amount of heterozygosity carried within a sample comprising a couple of dozen individuals will closely approximate the amount of heterozygosity within the population from which that sample was drawn. By extension, the amount of genetic variance carried by a relatively small sample of the population will closely approximate the magnitude of the genetic variance of that population as a whole. It is the genetic variance of the population that we seek to conserve, not the "genetic diversity" represented by the total number of distinct alleles within the population.

This misunderstanding carries over to populations established by translocation. It is often argued that since these were usually started by a nucleus of only a few individuals, and since those individuals must carry only a small fraction of the genetic variability of the population from which they were drawn, the subsequent robust health of the population that developed from the liberation indicates that a population needs little genetic variability. These populations tend either to increase rapidly or go extinct (often apparently by simple demographic stochasticity) within the first few years following translocation. If the former, they are soon free of any further loss of genetic diversity. The condition that gets a population into the greatest trouble is a history of small population size that persists for many generations.

20.10 Genetic drift and mutation

In the absence of immigration and mutation, the number of different alleles at a locus in the population as a whole can either remain constant or decrease. It cannot increase. In practice it will always decrease, because alleles will be lost under the influence of nonrandom mating and unequal reproductive success between individuals. Heterozygosity thus decreases also. Its rate of decline is a function of population size N, the proportion of heterozygous loci in the population as a whole being reduced by the fraction $1/(2N)$ per generation. Over one generation, H changes according to:

$$H_1 = H_0(1 - 1/(2N))$$

and over t generations:

$$H_t = H_0(1 - 1/(2N))^t$$

which may alternatively be written:

$$H_t = H_0 e^{-t/2N}$$

After $t = 2N$ generations, the population's heterozygosity will have dropped to 0.37 (i.e. e^{-1}) of its initial value at time $t = 0$. This holds as well for a single locus as it does across all loci. The loss of additive genetic variance is exactly analogous and conforms to the same equations. This process is called *random genetic drift*.

The rate of mutation at a single locus is about 10^{-6} per gamete per generation. However, most evolution is in terms of phenotypic characters that change in very small steps. These quantitative characters are controlled by many genes. Mutation within such a gene complex is much more frequent than at a single locus, closer to 10^{-3} or even 10^{-2} per gamete per generation.

Heterozygosity (and hence additive genetic variance) will change over one generation by an amount ΔH according to:

$$\Delta H = -\frac{H}{2N} + m$$

where m is the input of heterozygosity by mutation. Its equilibrium is solved by setting ΔH to zero:

$$H = 2Nm$$

which informs us that for any population size N there will be an equilibrium between mutational input of additive genetic variance and loss of it by drift. What varies, however, is the value of H at equilibrium; this will be higher when N is large and lower when N is small. The population size N must remain constant for many generations for such an equilibrium to establish.

20.11 Inbreeding depression

Suppose a particular allele is recessive, its effect being masked by a dominant allele when the two occur together at the locus. If additionally the recessive allele is slightly deleterious when its effect is expressed, it can have damaging effects on fecundity and survival of the population if its frequency increases by the statistical luck of random pairing of alleles in each generation. The gene pools of most populations contain many of these sublethal recessives, approximately enough to kill an individual three times over if by chance they all occurred on its chromosomes in homozygous form and were therefore all expressed in its phenotype. Thus, a decline in heterozygosity tends to lead to a decline in fitness.

Genetic malfunction may follow as a consequence of small population size. The following sequence can be triggered if a population becomes too small:

1 the frequency of mating between close relatives rises and random genetic drift increases;
2 which leads to reduced heterozygosity in the offspring;
3 which exposes the effect of semi-lethal recessive alleles;
4 which reduces fecundity and increases mortality;
5 which causes the population to become smaller still.

This trend may continue until extinction. The population must be held at low numbers for several generations before the process is initiated; a short bout of low population size has little effect on heterozygosity.

Loss of fitness during inbreeding can be traced largely to the process of fixation (i.e. reduction of alleles at a locus to one type) of deleterious recessive alleles. In mammals, mortality is 33% higher for the offspring of parent–offspring or full-sibling matings than for the offspring of unrelated parents (Ralls *et al.* 1988). Hybrid vigor is largely the reverse – the masking of the effect of those recessives – but it might also contain a component of *heterosis*, where the heterozygote is fitter than either homozygote.

Inbreeding does not automatically lead to inbreeding depression. It is seldom reported for populations larger than a couple of dozen individuals. Nor does low heterozygosity necessarily lead to inbreeding depression. Note that the average individual of most wild populations is heterozygous at less than 10% of loci. A population that has survived a bout of inbreeding may come out of it with enhanced fitness, because inbreeding exposes deleterious recessives and allows them to be selected out of the population. That is precisely the method used by animal breeders to remove deleterious alleles. Homozygosity causes an immediate problem only when the allele is deleterious. Nonetheless, inbreeding often does produce inbreeding depression. The possibility must always be kept in mind if a population is small. The next section addresses the question, "How small is too small?".

20.12   How much genetic variation is needed?

The cheetah (*Acinonyx jubatus*) has a low level of heterozygosity. Stephen O'Brien and his colleagues (O'Brien *et al.* 1985, 1986; Cohn 1986) studied the genetic variance of captive populations of this species originating from southern Africa. A standard electrophoretic analysis of 52 loci ($n = 55$ individuals) discovered a heterozygosity value of $H = 0.00$, as compared with $H = 0.063$ for people and 0.037 for lions. A more refined "two-dimensional" electrophoretic analysis, separating the proteins first by electrical charge and then by molecular weight, uncovered rather more variability and yielded $H = 0.013$ for cheetahs, as against 0.024 for people analyzed by the same method. A further sample from East Africa returned $H = 0.014$ (O'Brien *et al.* 1987).

Is the cheetah in peril? It is possible that as a direct consequence of the low heterozygosity the cheetah produces sperm of low viability, has an abnormally high rate of juvenile mortality, and is particularly susceptible to disease. All these claims have been made, but no causal relationship has been established between these putative defects and the peculiarities of the genotype. Alternatively, the cheetah may be in no danger of demographic collapse despite its low genetic variability. In support of this is its widespread distribution, which was even wider in the recent past, particularly in Asia. Contraction of range over the last 1000 years has been no greater than that of the lion, another widespread species but one with a standard level of heterozygosity. For both species, the contraction of range seems to be a result of excessive human predation rather than of a diminished genetic fitness. As far as we know there is no evidence from the wild suggesting that the cheetah is faced by a level of risk beyond those hazards imparted by a rising human population (Caughley 1994).

The cheetah clearly has low genetic variance, but it is well within the range exhibited by mammalian species (see Fig. 20.8). The suggestion that it is in demographic peril as a consequence of this earns no support from what is known of its ecology. This example provides two messages. The first is special: we need more disciplined information on the cheetah in the field to determine whether its diminished genetic variance is associated with demographic malfunction. The second is general: by the genetic theory currently followed in conservation biology, the cheetah should be in demographic trouble, but there is no convincing evidence for that and considerable but circumstantial evidence to the contrary. A plausible alternative hypothesis is that present genetic theory overestimates the amount of genetic variation needed to sustain an adequate level of individual fitness. One should not jump to the conclusion that a species is in danger simply because it has a low H. There is too much evidence to the contrary.

We cannot yet lay down general rules as to the minimum genetic variance required for adequate demographic fitness. Nor can we define a minimum population size needed to maintain genetic variation. We need much more research on the incidence of inbreeding depression in the field, the population size and the period over which it must be maintained before inbreeding depression becomes a problem, and the relationship between heterozygosity and fitness.

20.13 Effective population size

It can happen that the size of a population appears large enough to avoid genetic malfunction but that it acts genetically as if it were much smaller. The proportion of genetic variability lost by random genetic drift may be higher than the computed theoretical $1/(2N)$ per generation because that formulation is correct only for an "ideal population." In this sense, "ideal" means that family size is distributed as a Poisson, sex ratio is $50:50$, generations do not overlap, mating is strictly at random, and rate of increase is zero. This introduces the notion of *effective population size* in the genetic sense: the size of an ideal population that loses genetic variance at the same rate as the real population. The population's effective size (genetic) will be less than its census size except in special and unusual circumstances.

Perhaps the greatest source of disparity between census size and effective size is the difference between individuals in the number of offspring they contribute to the next generation. In the ideal population, their contribution has a Poisson distribution, the fundamental property of which is that the variance equals the mean. Should the variance of offspring production between individuals exceed the mean number of offspring produced per individual, the effective population size will be smaller than the census size. In the unlikely event of variance being less than the mean, the effective population size will be greater than the census size and the population will be coping better genetically than one might naively have expected. The effective population size N_e corrected for this demographic character can be calculated as:

$$N_e = (NF - 1)/(F + (s^2/F) - 1)$$

where F is the mean lifetime production of offspring per individual and s^2 is the variance of production between individuals. This indicates that when mean and variance are equal, N_e approximates to N. Since males and females sometimes differ in mean and variance of offspring production, the equation is often solved for each sex separately and the sex-specific N_e values summed.

Genetic drift is minimized when the sex ratio is $50:50$. Effective population size (genetic) in terms of sex ratio is given by:

$$N_e = 4/(1/N_{em}) + (1/N_{ef}))$$

where N_{em} and N_{ef} are the effective numbers of males and females as corrected for variation in production of offspring. The following series shows the relationship numerically:

Sex ratio 50 : 50	60 : 40	70 : 30	80 : 20	90 : 10
$N_e N$	0.96N	0.84N	0.64N	0.36N

Further corrections can be made for other sources of disparity between real and ideal populations. These considerations are often important in *Drosophila* research and in managing the very small populations in zoos but they have little utility in conservation. Rather than attempting to estimate N_e for a threatened population one should simply assume as a rule of thumb that $N_e = 0.4N$ and that the censused population loses genetic variability at a rate appropriate to an ideal population less than half its size.

20.14 Effect of sex ratio

If a species is polygamous – and most species of mammalian wildlife are (though birds are largely monogamous) – a disparate sex ratio may have a large effect upon net change in numbers over a year, and hence on effective population size (demographic) at the beginning of that year. Net change in numbers over a year can be calculated as:

$$\Delta N = NpbP_f - N(1 - p)$$

where P_f is the proportion of females in the population, p is the probability of surviving the year averaged over individuals of all ages within a stable age distribution, and b is the number of live births produced per female at the birth pulse terminating the year. This indicates that net change in numbers in a population of any given size is a linear function of the proportion of females in the population. The regression of ΔN on P_f has a slope of Npb and an intercept (i.e. the value of ΔN when $P_f = 0$) of $-N(1 - p)$.

For example, let us assume a hypothetical population of $N = 1000$ individuals, a probability of survival per individual per year of $p = 0.9$, and a fecundity rate of $b = 0.95$ live births per female per year. Feeding these values into the equation yields a ΔN of 541 when $P_f = 0.75$, of 328 when $P_f = 0.5$, and of 114 when $P_f = 0.25$.

The relationship can be rearranged to estimate N_e, the effective population size (demographic), as:

$$N_e = N(pbP_f + p - 1)/(0.5pb + p - 1)$$

For the preceding example, N_e is solved as 1653 when $P_f = 0.75$, as 1000 when $P_f = 0.5$, and as 347 when $P_f = 0.25$. Thus, a disparate sex ratio may have a significant effect on a population's ability to increase from low numbers, enhancing that ability when females predominate and depressing it when males dominate. Ungulate populations that have crashed because they have eaten out their food, or because a drought has cut it from under them, often end the population slide with a preponderance of females. They are thus in better shape demographically to recover from the decline than if parity of sex ratio had been retained. Note here an important point: the sex ratio minimizing genetic drift (50 : 50) is not that maximizing rate of increase (disparity of females). Hence, the appropriate "effective population size" depends on context. The genetic version is appropriate for a small, capped population in a zoo. The demographic version is often more appropriate in the wild, where the aim is usually to stimulate the growth of an endangered species.

20.15 How small is too small?

Two values for minimum viable population size (genetic) are commonly quoted: 500 and 50. The difference between them reflects the differing assumptions upon which they are based. The 500 figure (Franklin 1980) is the effective population size at which the heritability of a quantitative character stabilizes at 0.5 on average (i.e. 50% of quantitative phenotypic variation is inherited and 50% is environmentally induced). The figure of 0.5 is the heritability coefficient for bristle number in *Drosophila*, and quantitative characters in farm animals often have a heritability coefficient of that general magnitude. The assumption is that such a level of heritability reflects a genetically healthy population. The genetic variance needed to enforce such a heritability coefficient has an equilibrium (where loss by genetic drift is balanced by gain from mutation) of $N_e = 500$ in the absence of selection, and that is taken as a safe lower limit for population size.

The estimate of 50 comes from the observation by animal breeders that a loss of genetic variance of 1% per generation causes no genetic problems. Since that rate is $1/(2N_e)$, we can write $0.01 = 1/(2N_e)$, rearrange it as $N_e = 1/(2 \times 0.01)$, and solve it as $N_e = 50$. The arbitrary nature of both estimates of minimum viable population size will be readily apparent. Neither should be accepted as anything more than general speculation.

20.16 Summary

Natural selection and ecology have traditionally been viewed as fundamentally different disciplines. Recent studies show that this assumption is simply incorrect, because natural selection can be pronounced from year to year, leading to rapid change in phenotype. Some of the fastest processes of natural selection known have been caused by hunting and fishing of wildlife populations, leading to pronounced evolutionary changes in life history parameters such as age of maturity, reproductive rate, and longevity. Environmental variation can also trigger rapid evolutionary changes in morphology or life history traits, through fitness effects arising from competitive ability and predation risk. The speed of evolutionary response may sometimes be great enough to influence demographic parameters and thereby ecological processes. Loss of genetic variation due to genetic drift and inbreeding depression can occur when populations are maintained at low numbers. Such loss of genetic variation can reduce overall fitness and thereby lower numbers even further, contributing towards an extinction vortex.

21 Habitat loss and metapopulation dynamics

21.1 Introduction

The world is a very heterogeneous place. Some *habitat heterogeneity* stems naturally from physical processes such as volcanism, geological formation, erosion, and fire. Humans also have a massive impact on the spatial structure of many ecosystems, through resource harvesting, pollution, and land clearing for agriculture, industry, or residential development. Such processes contribute to both *habitat loss* and *habitat fragmentation*. In this chapter we consider some of the ways in which habitat disturbance influences population dynamics within a species and community interactions among species. *Corridors* can be one management tool for countering the negative effects of habitat fragmentation.

There are many different ways spatial environmental variation can play out in an ecological setting. One might have true islands of useable habitat, surrounded by a matrix of hostile, unlivable habitat. If a collection of similar islands of suitable habitat is large enough to support meaningful local populations and there is some shared dispersal among the islands, we refer to this as a *metapopulation*. A metapopulation can occur across a mixture of highly useable versus poorly useable islands of habitat, often termed a *source–sink* system. *Island–mainland* systems represent a special case of source–sink dynamics, comprising a single, usually large, mainland population that feeds emigrants to a number of outlying, usually small, islands. Sometimes a single individual, a mated pair, or a small family group lives in isolated patches of suitable habitat. In such cases, we can think of this as a form of *territoriality*, with the useable patch comprising a defensible resource. If a group of different species interacts ecologically on such islands, comprising a set of metapopulations of different species, we refer to this as a *metacommunity*.

21.2 Habitat loss and fragmentation

The human population explosion has made enormous inroads into the amount of undisturbed habitat available virtually anywhere in the world. This can be graphically demonstrated by changes in forest cover over time, as recorded for example by satellite imagery (Fig. 21.1). As disturbance increases, the total amount of useful habitat typically declines, whereas inadequate habitat (sometimes termed *matrix* in the landscape ecology literature) increases. As habitat is lost from a formerly contiguous landscape, a wide range of patterns can develop. If one imagines disturbance as cutting holes in the fabric of a large landscape, initial stages of habitat disturbance don't change habitat configuration very dramatically (Fig. 21.1). As disturbance proceeds further, however,

Wildlife Ecology, Conservation, and Management, Third Edition.
John M. Fryxell, Anthony R. E. Sinclair and Graeme Caughley.
© 2014 John Wiley & Sons, Ltd. Published 2014 by John Wiley & Sons, Ltd.
Companion Website: www.wiley.com/go/Fryxell/Wildlife

Fig. 21.1 Google earth image showing fragmentation due to deforestation in Altamira region of Brazil. Mature forest and newly cleared areas of deforestation differ in depth of shading.

more and more of the removed patches of habitat are inevitably side by side, a process that eventually leads to isolation of a large number of small patches of useable habitat. Computer simulations of this sort of random habitat destruction suggest that there is often a critical threshold of 70–80% disturbance at which the number of isolated fragments rapidly increases (Andren 1994; Fahrig 2003). Further levels of habitat loss result in shrinking patches of useful habitat that become ever more widely spaced in a growing matrix of inadequate habitat.

The island of Madagascar is well known as a hotspot of biodiversity, with a spectacular number of endemic species that do not occur elsewhere, including the well-known lemurs. Pollen samples recorded over the island suggest that substantial amounts of the forest present in prehistory had already disappeared by the 1950s, when aerial photography and satellite imagery first became available for the monitoring of landscape changes. Deforestation has led to a further loss of 41–43% of humid and dry forest habitat and a slightly lower 28% of spiny forest habitat since the 1950s (Harper *et al.* 2007). Accompanying the overall loss of forest habita has been substantial change in the structure of the remaining habitat. Fewer large forest patches remain, and the fraction of forested area that is within 500 m of nonforest habitat has dramatically increased. Many of these forest stands have become increasingly isolated (Fig. 21.2). In other words, continuous forest tracts have been gradually replaced by smaller relict forest patches surrounded by nonforest habitat. Much of the nonforest habitat is now agricultural fields or is densely inhabited by humans.

Several important predictions arise from this idealized process of disturbance. First, one would expect a threshold effect, with loss of even 50% of useful habitat having

Fig. 21.2 Possible ways in which the number of patches, size of patches, and patch spacing could vary in a system undergoing habitat fragmentation. (Source: Fahrig, 2003. Reproduced with permission of Annual Reviews Inc.)

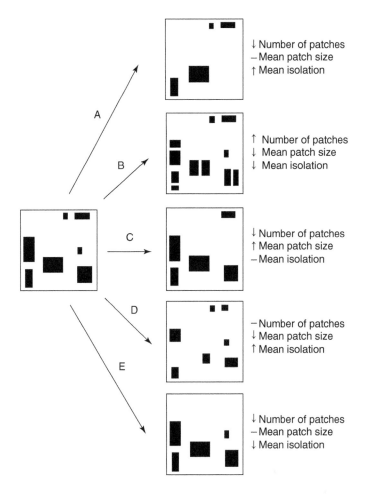

A

↓ Number of patches
− Mean patch size
↑ Mean isolation

B

↑ Number of patches
↓ Mean patch size
↓ Mean isolation

C

↓ Number of patches
↑ Mean patch size
− Mean isolation

D

− Number of patches
↓ Mean patch size
↑ Mean isolation

E

↓ Number of patches
− Mean patch size
↓ Mean isolation

little major impact on habitat configuration. At high levels of loss, however, useful habitat fragments become essentially islands in a hostile sea of matrix habitat. Each of these small islands will have small local populations even if their local density is little affected. This can lead to local extinctions due to demographic stochasticity (see Chapter 16), particularly if individuals are reluctant to cross hostile matrix habitat. Repeated often enough, such local extinctions reduce overall mean density in remaining habitat fragments below that recorded in a pristine environment. In addition, fragmentation can impose additional demographic challenges due to reduced dispersal success, edge effects on crucial resources, and increased risk of predation.

It is far from likely, however, that habitat fragmentation should ever occur in the sort of random fashion presumed by most theoretical studies. Habitat loss might occur through subdivision of existing patches rather than random allocation across the system. Hence, one can imagine all possible combinations of changes in patch size, patch isolation, and patch number depending on the precise manner with which disturbance occurs (Fig. 21.2). It is scarcely surprising, therefore, that the strongest consistent effect on demography and biodiversity of wildlife is typically due to habitat loss itself, with much weaker effects typically arising from variation in patch size, patch isolation, and patch number (Andren 1994; Fahrig 2003). The devil is clearly in the details when it comes to habitat loss and fragmentation.

21.3 Ecological effects of habitat loss

While the pattern of habitat loss is often graphically obvious, the demographic implications might well be more subtle. Consider a lemur species whose feeding has evolved to specialize on a particular food plant. A change in the total forested area would represent a loss of a critical, essential resource for the specialized lemur. It is likely that such a change would lead to decline in the overall abundance of Madagascar lemurs. Destruction of some forest stands might well have little impact on the carrying capacity of the remaining stands, in which case local population density of lemurs would be little affected. Density-dependent interactions at a local level (home range spacing, population regulation, or consumer–resource interactions) might be little affected by habitat loss, provided the fragments are large enough to be inhabitable by sustainable local groups. On the other hand, theory suggests that local extinctions due to demographic stochasticity (see Chapter 16) should be dramatically increased if total population size within each patch declines at pronounced stages of habitat loss.

Reviews of a large number of field studies strongly suggest that both total population size and local densities are often (but not always) reduced in ecosystems subjected to major habitat disturbance (Andren 1994; Fahrig 2003). Moreover, habitat loss is the most common contributory factor for endangered species (Wilcove *et al.* 1998). But habitat loss can also change the spatial configuration of habitat, which introduces its own set of problems.

For example, many pool-breeding frogs and salamanders spend most of their adult life in forested environments. Pool-breeding species are sometimes reluctant to cross open, clear-cut patches compared to mature forest stands with abundant foliage at ground level (deMaynadier *et al.* 1999). If matrix habitat offers poorer probability of survival then increased loss of juveniles during dispersal can itself reduce population viability. Replicated experiments in logged versus unlogged landscapes in the United States indicate that survival and growth rate of juvenile and adult amphibians are consistently depressed by habitat disturbance (Semlitsch *et al.* 2009). One of the biggest challenges to amphibian survival is simply staying alive while crossing roads. Sjögren Gulve (1994) showed that reduced survival during dispersal and reduced local abundance at breeding sites contributed to local extinction rates of 2% per year in Swedish ponds. As a result of such interrelated factors, many amphibian populations have suffered demonstrable impact from forest habitat loss and forest fragmentation (Cushman 2006).

The increased ratio of patch edge to patch area in highly fragmented ecosystems may similarly reduce average density recorded over a collection of fragments (Fahrig 2003). Close proximity to habitat edges can lead to increased risk of predation or parasitism, a process that is of particular concern for nesting birds (Robinson *et al.* 1995). For example, forests in the US Midwest have become highly fragmented after decades of intense agriculture. Nine species of passerine bird show a consistent negative relationship between forest cover and rate of nest parasitism by cowbirds and nest predation by a suite of species, including blue jays, raccoons, and crows. Increase in the amount of edge relative to core habitat is thought to be a key factor, as is increased risk of exposure to generalist nest parasites and predators whose densities are high in landscapes with a broad variety of habitat types.

Spectacular cascading effects have been documented in habitat islands created through large-scale disturbance. Perhaps the best-known example is a series of islands created when a hydroelectric dam was constructed in the Caroni Valley of Venezuela. Many of these hilltop islands were too small to sustain large carnivores like pumas and jaguars, or even large raptors. As a result of relaxed predation, many species of herbivore exploded in abundance, a process termed a *trophic cascade* (Terborgh and Estes 2010). This was particularly evident in red howler monkeys, common iguanas, and leaf-cutter ants, each of which increased 10- to 100-fold over their earlier densities (Terborgh *et al.* 2001). Increased herbivore pressure in turn led to lower standing crops of ground-level vegetation, leaving a park-like denuded understory in the previously luxuriant tropical forest. The ripple effect of this trophic cascade was felt at multiple trophic levels (Terborgh and Estes 2010).

At the community level, there is also growing evidence that biodiversity is often compromised by habitat loss and/or fragmentation (Fahrig 2003). Adequate testing of this idea, however, requires detailed sampling of species diversity over multiple landscapes. This is so time-consuming and difficult that it requires participation by large numbers of nonscientists with a strong personal interest in natural history. Good examples of such "citizen science" are the Christmas Bird Counts and the Breeding Bird Atlas in North America. Boulinier *et al.* (2001) used historic data from the Breeding Bird Atlas to assess sources of variation in passerine bird diversity with respect to landscape variables. Breeding bird species richness was positively related to average forest patch size. This quite likely stemmed from increased local extinction rates estimated within the same forest patches, leading to high rates of species turnover in the smaller patches, as predicted by *island biogeography theory* (see Chapter 22).

A number of more recent studies have used manipulative experimental comparisons before and after deforestation to assess the impact on biodiversity. Such an ambitious project was initiated in 1979 in the Amazonian rainforest near Manaus, Brazil. For 3 decades, ecologists have recorded changes in the diversity and abundance of a wide range of organisms, ranging from plants to termites to birds (Laurance *et al.* 2002) in forest fragments of 1, 10, 100, and 1000 hectares. For many groups, the results are not encouraging, showing substantial loss of diversity in the smaller patches. Army ants, euglossine bees, and dung beetles decline sharply in isolated forest stands (Laurance *et al.* 2002). Perhaps not surprisingly, many species of insectivorous birds also decline precipitously, especially those that forage on the forest floor, leading to substantial loss of bird diversity. For example, ant birds feed on the swarms of insects disturbed by army ant incursions. Because each army ant colony alternates between active and quiescent phases, ant bird home ranges tend to encompass multiple army ant colonies. With diminished army ant abundance, ant birds rapidly disappeared from forest fragments (Stouffer and Bierregaard 1995). Fragmentation also changes dispersal rates for many species, with a few birds simply capable of crossing even minor expanses of cutover matrix (Van Houtan *et al.* 2007). Larger vertebrates, such as peccaries and some monkeys, typically become much rarer in forest fragments, although other species, such as howler monkeys, can paradoxically increase in abundance (Laurance *et al.* 2002).

Many fragmentation effects are caused by alteration of forest structure and physiological conditions. The increased amount of edge habitat leads to substantial penetration of wind and light into edges of forest fragments, causing pronounced temperature variation, reduced humidity, increased leaf litter, and elevated tree mortality. Pioneer species, such as *Cecropia*, thrive in the resulting edge habitat, but many late successional species simply cannot cope with the altered physical conditions. As a result, fragmentation essentially leads to altered successional pathways in the remaining patches of suitable habitat. The result is a highly unstable assemblage, making long-term speculation about sustainability and biodiversity hazardous. Given the rapid rate of deforestation occurring through Amazonia (Skole and Tucker 1993), such impacts may be considerable (Laurance *et al.* 2002).

21.4 Metapopulation dynamics

Dispersal plays a key role in explaining the dynamics of species that are subdivided because of habitat fragmentation or other natural causes into discrete subpopulations. Provided that there is some degree of dispersal among subpopulations, ecologists refer to the larger aggregate as a *metapopulation*, or population of populations (Hanski and Gilpin 1997). Metapopulations can occur in a variety of contexts. Bird species on continental islands are an obvious example (Sæther *et al.* 1999). However, it is just as valid to think of butterflies inhabiting grassy glades in a matrix of boreal forest as a metapopulation (Hanski *et al.* 1994). Since 1980 there has been a surge in interest in metapopulation dynamics, fuelled in part by the recognition that human environmental impacts often lead to fragmentation of natural areas, creating effective metapopulations from populations that were continuously distributed in the not-so-distant past. Here we outline some of the basic principles of metapopulation dynamics, particularly with relation to the impact of further habitat loss.

There are many ways in which one can represent metapopulations, but the Levins model (Levins 1969) and its subsequent modifications (reviewed by Gyllenberg *et al.* 1997) have perhaps been the most influential. Let p = proportion of occupied sites, c = probability of successful colonization, and e = probability of extinction of an occupied site. The rate of change in the number of occupied sites is calculated in the following manner:

$$\frac{dp}{dt} = cp(1 - p) - ep$$

The first term represents colonization of new sites, the second extinction. Provided that $c > e$, this model predicts that the proportion of occupied sites will converge to the following equilibrium over time:

$$p_{eq} = 1 - \frac{e}{c}$$

This can be seen clearly in the simulation shown in Fig. 21.3. The stability of the metapopulation at equilibrium belies a constant turnover of subpopulations. A substantial fraction of sites (45%, in fact) go extinct per unit time. This extinction rate does not get translated into a dangerous collapse of population, however, because of the steady stream of colonists from the remaining occupied sites. As we discuss in Chapter 16, local extinction is expected to become common when subpopulations have been reduced to low numbers, simply due to chance demographic events or rapid genetic loss. Hence, fragmentation of the environment into numerous small patches creates a situation in which local extinction risk is a very real possibility.

Empirical data consistent with the metapopulation scenario are accumulating. One of the best-documented examples is Hanski and coworkers' studies of the Glanville fritillary (*Melitaea cinxia*), an endangered butterfly inhabiting the Åland Islands in the eastern Baltic Sea, off the coast of Finland (Hanski *et al.* 1994). The spatial distribution

Fig. 21.3 Dynamics over time of a metapopulation with colonization rate $c = 0.90$ and extinction rate $e = 0.45$.

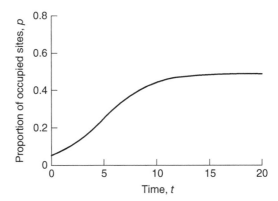

of fritillary butterflies is quite patchy, in keeping with the patchy distribution of the larval food plants in dry meadows interspersed by a woodland matrix. Most butterflies spend their entire lives in a single meadow, but some individuals disperse, typically 600 m from the natal site. This level of dispersal is sufficient to locate neighboring meadows, which are typically < 300 m from the nearest neighbor. The Finnish team repeatedly censused the number of butterfly larvae at each of several hundred locales. As the larvae are colonial and quite conspicuous, it is relatively straightforward to ascertain whether local extinction has taken place in the small grassy meadows, typically much less than 1 ha in size. Results of the repeated counts demonstrated that extinction was common amongst these local subpopulations, in accordance with metapopulation theory, but that extinction events were not synchronized simultaneously across all subpopulations. Of equal importance, locally extinct populations were often rapidly recolonized by dispersers from occupied sites nearby.

As we might expect, many factors influenced the risk of extinction, including the size of the local subpopulation, the degree of genetic variability, and the degree of isolation from neighboring sites (Hanski *et al.* 1994). A high degree of turnover of local populations was normal, with the overall prevalence determined by the probabilities of colonization versus extinction. Hanski (1994) showed how such additional functional relationships can be used to develop more spatially realistic models of metapopulation dynamics, and in a series of later papers variations of this model have been applied to the consideration of various aspects of Glanville fritillary conservation biology. An important feature of these more realistic models is that it is possible to estimate the probability of chance extinctions under conditions in which the simpler Levins' model would predict a steady-state equilibrium. In this way, chance dispersal events at the local population level have a similar effect to demographic stochasticity (see Chapter 16). Unfortunately, both the spatially realistic models and recent population trends suggest that the Glanville fritillary may be fighting a losing battle against extinction.

Well-studied examples of metapopulation dynamics in vertebrates are less common. Long-term studies of pool frogs along the coast of Sweden demonstrate a steady pattern of subpopulation turnover (Sjögren Gulve 1994). Similar patterns of extinction and recolonization have been shown in a number of other systems: cougars inhabiting chapparal shrub patches in urban southern California (Beier 1996), pikas living on mine tailings in the Sierra Nevada of California (Smith and Gilpin 1997), and sparrows on windswept islands off the coast of Norway (Saether *et al.* 1999). There is no doubt that the preconditions for metapopulation dynamics exist for many wildlife

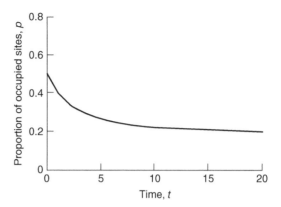

Fig. 21.4 Metapopulation dynamics over time for the same parameters as in Fig. 7.9 ($c = 0.90$, $e = 0.45$), now with substantial habitat loss ($H = 0.3$) among the previously inhabitable sites. This leads to a reduction in the metapopulation equilibrium from 50 to 20%.

species. The unresolved questions are how common metapopulation dynamics might be in nature and what biological factors determine their importance to long-term persistence.

Levins' (1969) simple metapopulation model can be readily modified to predict the effect of habitat loss. Let h reflect the proportion of sites that are useable and $1 - h$ represent the proportion of disturbed sites, so that any propagule that lands in a disturbed site cannot persist. The dynamics of this degraded environment are as follows:

$$\frac{dp}{dt} = cp(h - p) - ep$$

As a result of habitat loss, the equilibrium level of occupancy is reduced to $p = h - e/c$. If h falls below a critical threshold ($h \leq e/c$) then the metapopulation will not be able to persist at all (Fig. 21.4). This is a simple but nonetheless useful way of looking at the potential costs of habitat degradation. Empirical examples of habitat degradation leading to extinction are largely anecdotal, but nonetheless abound in the natural history literature. The importance of habitat loss may be magnified in the future, if people do not learn to limit habitat fragmentation and prevent further alienation of subdivided patches of habitat.

The Levins metapopulation model can be extended to consider the ecological dynamics of multiple competing species with respect to spatial occupancy (Nee and May 1992). For example, species are likely to vary in both their dispersal capacity and competitive ability. Assume that a superior interspecific competitor will take over a patch occupied by an inferior competitor but that otherwise the dynamics of the two species (termed A and B) are dictated by their respective colonization rates (c_A and c_B) balanced against their extinction rates (e_A and e_B). We seek to predict changes in the proportions of patches occupied by the competitive dominant species 1 versus those occupied by the competitively poorer species 2, which can be modeled by the following set of equations:

$$\frac{dp_1}{dt} = c_1 p_1(h - p_1) - e_1 p_1$$
$$\frac{dp_2}{dt} = c_2 p_2(h - p_2) - c_1 p_1 p_2 - e_2 p_2$$

While this looks slightly more complicated than the simple model we worked with before, the underlying logic is identical. The only significant difference is that competitive dominants can expand either by landing in empty patches or by landing in patches already occupied by species 2, an inferior competitor. Species 2 can only expand through discovery of unpopulated patches. Species 1 dispersers immediately eliminate species 2 local populations in co-occupied patches. The dynamics of this model inevitably lead to competitive exclusion of the poorer competitor, unless the ratio of colonization to extinction rates of the poor competitor (c_2/e_2) exceeds that of the good competitor (c_1/e_1) and both of these ratios exceed 1. In other words, the *metacommunity* can sustain poor competitors that are nonetheless efficient as dispersers as well as species that are strong competitors but which are inefficient at dispersing, in the sense that the probability of patch discovery is considerably greater than that of local extinction. In nature, it is often the case that early-succession species are good dispersers but poor competitors, and this allows them to coexist with later-succession competitive dominants.

If habitat loss occurs across a landscape, such that some of the colonists land in disturbed patches that do not allow reproduction, then the better competitor (and weakest disperser) will be most heavily affected by habitat loss. This same process can be applied to a larger community with any number of competing species, leading to the alarming prediction that habitat loss threatens the best competitors – the very species whose ecological adaptations would ordinarily allow them best to thrive (Tilman 1994). The full impact of species extinction due to habitat loss might take many generations to unfold completely; this process has therefore been called the *extinction debt* (Tilman 1994).

Many of the habitat loss and fragmentation studies in the literature might be viewed as examples of metapopulations, played out at either the single-species or the community level. To test this assertion conclusively, however, requires unusually detailed information on long-term probabilities of colonization, extinction, and turnover. Even under these circumstances, it is far from obvious that all patches are likely candidates for extinction, if only because some patches are much larger or have markedly better resources than others, as seen in the Glanville frittilary example. Moreover, in some species patches are solely occupied by a mating pair and their offspring or a group of related individuals. We go on to consider these special varieties of metapopulation in the following sections.

21.5 Territorial metapopulations

Through natural or human-influenced patterns of habitat disturbance, suitable patches of habitat for breeding can become widely separated. Individuals setting up breeding territories must locate these suitable patches in an unfavorable matrix before they can set up a territory to attract mates (Lande 1987). Under these circumstances, the dynamics of territory occupancy by breeding pairs and their offspring are logically similar to those of a classic metapopulation (Lande 1987; Lamberson *et al.* 1992, 1994; Noon and McKelvey 1996). Extinction is amplified when the probability of territory (patch) colonization is low relative to the probability of territory vacancy (local extinction) arising due to mortality. A key difference, however, is that successful breeding requires that both a male and a female independently discover a suitable territory site. At low probabilities of discovery, some individuals may never find mates. This *Allee effect* (named after W.C. Allee (1938), who defined the process) can cause extinction if the overall level of territory occupancy falls below a critical level (Lande 1987).

A good example of this kind of situation involves the northern spotted owl (*Strix occidentalis*) of the western United States. Spotted owls require substantial tracts of old-growth forest for their breeding territories, but 80% or more of the mature forest has been logged in the northwestern United States over the past half century. As a consequence, local populations of owls are increasingly isolated from each other by large areas of clear-cutting. Variation in forest structure at the local level can also influence territory occupancy. Concern about the long-term viability of northern spotted owl populations led to intense debates about appropriate management schemes for the public forest lands in the Pacific Northwest (Doak 1989; Lamberson *et al.* 1992, 1994). In regions where forest extraction is an important source of jobs and local business revenue it is no simple matter to find the best compromise between human interests and those of endangered wildlife. In the case of the spotted owl, the best compromise solution is to both control the future loss of mature forest and manage the spatial pattern of forest utilization in such a way as to maintain an effective metapopulation structure (Murphy and Noon 1992; Noon and Mckelvey 1996). This kind of management controversy will only become more common with further fragmentation of existing wild lands.

21.6 Mainland–island metapopulations

Special kinds of metapopulation dynamics occur when some patches are large enough or productive enough to sustain permanent subpopulations and others are small enough or unproductive enough that local extinction is common. If both the permanent patches and the transient patches can support positive population growth, this is termed a *mainland–island system*. Such systems differ from conventional metapopulation systems in having wide variation in probabilities of both extinction and colonization, mediated by population size.

A well-studied example is the checkerspot butterfly (*Euphydryas editha bayensis*), which inhabits widely scattered (typically several kilometers apart) patches of serpentine soil in coastal California (Harrison *et al.* 1988). This system contains one large (> 400 000 individuals) and many small populations (< 500 individuals) of butterflies. Historical records suggest that catastrophic weather events can eliminate virtually all small populations simultaneously, and chance extinctions of small populations can occur presumably as a result of demographic stochasticity (see Chapter 16). The large mainland population is unlikely to disappear through either source of variation, so it represents a dependable source of colonists in a given year. As in many such studies, the probability of successful dispersal was negatively related to distance between suitable habitat patches but positively related to local patch characteristics such as temperature and the presence of suitable food plants (Harrison *et al.* 1988). This system demonstrates that spacing between patches has strong bearing on the probability of colonization.

A second example relates to a long-term study of orb-weaving spiders on a set of small islands in the Bahamas. Monitoring of 108 islands showed that complete disappearance of spiders was a frequent event on some islands but exceedingly rare on others (Schoener and Spiller 1987). This important result casts a different light on a second basic assumption of simple metapopulation models: that all local populations are prone to extinction. Long-term records from the Bahamian spider populations suggest that population size is an important predictor of extinction risk. Some local populations of orb-weaving spiders in the Bahamas are simply large enough to be at little risk of chance extinction events. Islands that have substantial numbers of predators, such as lizards, are also more vulnerable to chance extinction because of reduced density or altered patterns of habitat use (Schoener *et al.* 2001). Islands with substantial populations probably resupply other islands that are at greater risk of chance extinction. As a result of this disparity in population size, of course, the entire spider metapopulation is substantially less vulnerable to extinction than if it were configured as a simple Levins-style metapopulation.

21.7 Source–sink metapopulations

In some ecosystems, some patches can sustain positive population growth even when local density is well below the ecological carrying capacity (termed *source* patches), while others always experience higher rates of mortality than birth, so local populations would be unsustainable in isolation (*sink* patches). A collection of demographically variable subpopulations in patches linked together via dispersal is known as a *source–sink system*, with source sites supplying a steady stream of dispersers, which fan out to fill surrounding sinks (Pulliam 1988; Pulliam and Danielson 1991). Immigration from sources can be frequent enough to maintain substantial numbers of individuals in sinks, even though they cannot replace themselves at these inferior sites.

Mainland–island systems such as the California checkerspot butterflies or Bahamian orb-weaving spiders are logical candidates for source–sink dynamics. Large mainland populations are presumably capable of sustaining themselves even if they were to be completely isolated. The size of small satellite populations suggests that they might not be sustainable without a fresh supply of immigrants, but this assertion can only be tested by measuring expectations of reproductive success versus mortality. In other systems, however, the key consideration is spatial variation in habitat quality, rather than variation in the carrying capacity of the local populations.

Because of the difficulty of measuring mortality and reproductive success in many different habitat patches, well-documented examples of source–sink dynamics in natural systems are rare. Perhaps one of the most thoroughly studied examples is the Florida scrub jay. This large bird lives in communal social groups in thinly forested scrub oak stands on sandy infertile soils in southern Florida. Each family group defends a 10 ha territory against other families. Scrub jays are cooperative breeders, whose young remain and help in rearing siblings at the natal nest for a year or more before dispersing to seek a new territory for themselves. Scrub oak habitats are rapidly dwindling in abundance with the burgeoning human population in southern Florida, so it is perhaps not surprising that the Florida scrub jay has been designated as being threatened, with less than 4000 pairs remaining in the early 1990s. There are reasons to believe that the species may be declining by as much as 25–50% per decade due to habitat loss (Stith *et al.* 1996).

Not all scrub oak habitat is equally suitable for scrub jays. Records of fledgling recruitment and adult mortality measured over 13 years at a cross-section of sites indicate that recruitment of young scrub jays only exceeds mortality in scrub oak stands of intermediate height, with female jays unable to replace themselves in shorter or taller stands (Breininger and Carter 2003). Even source territories have a great deal of variation in reproductive success from year to year, perhaps because of variation in acorns, their main food resource.

Florida scrub jays have limited dispersal, with few individuals travelling beyond the nearest five neighboring territories before settling (Woolfenden and Fitzpatrick 1984; Breininger and Carter 2003). Because of the limited availability of good habitat, many birds originating from source colonies will land in less-than-optimal habitat. However, a few lucky birds reared at sink sites will land in optimal habitat (Breininger and Carter

2003), maintaining dispersal as an evolutionarily stable strategy. The combination of limited habitat availability, short dispersal distances, and pronounced spatial variation in reproductive success and survival characterize a classic source–sink system.

Beavers (*Castor canadensis*) inhabiting shallow lakes in the mixed deciduous and boreal forest of southern Ontario provide another example of a wildlife species with source–sink dynamics (Fryxell 1999). This species lives in territorial family groups in lakes and streams in temperate North America and Europe. It feeds on a variety of terrestrial and aquatic plants, particularly the bark, leaves, and twigs of deciduous shrubs and trees. Because plants vary considerably in nutritional value, much of the vegetation available to a beaver colony may be incapable of meeting their energetic needs. This is particularly true of colonies that have been occupied for many generations, because tree cutting by previous cohorts has removed most of the highly nutritious deciduous trees from the surrounding forest and forest succession has led to replacement with poorer-quality conifers. Records of offspring production over a decade at a number of different colonies in Ontario demonstrate that only a small fraction of colonies had sufficient food supplies to support enough offspring production to balance adult mortality (Fryxell 1999). These source colonies apparently populate the surrounding area when they disperse, because most of the other colonies rarely produced young.

Clearly, the conservation needs of mainland–island and source–sink systems differ from those of classic metapopulations. Mainland and source sites take on disproportionate importance in sustaining viable populations over the larger landscape. Loss of even small amounts of these critical source or mainland habitats could be unsustainable.

21.8 Metacommunity dynamics of competitors

The recognition that long-term viability for a single species can depend on spatial interactions leads naturally to the possibility that community structure and dynamics are similarly dependent on spatial interactions. This is an area of much recent work by community ecologists (Leibold *et al.* 2004). For example, if processes of colonization and extinction determine how widespread a given species is in a given system, it is a short step to include competitive interactions across a range of species drawn from the regional species pool. Theoretical models suggest that there are several fundamental ways in which spatial structure can influence the structures of competitor guilds. First, spatial or temporal variation in the suitability of different patches can dramatically increase the number of species that can coexist in perpetuity in a larger metapopulation, with different assemblages associated with different habitat features. Second, if there are tradeoffs between dispersal ability and competitive ability then there is essentially no limit to the number of species that can coexist in a metacommunity (Tilman 1994; Amarasekare and Nisbet 2001). Strong colonists perpetuate themselves by more readily locating vacant patches, whereas strong competitors persist by virtue of their ability to take over an already-occupied patch. Unsynchronized subpopulation dynamics can be key in allowing the coexistence of multiple competitors for extended periods. Alternatively, it takes a long time for the combination of competing species to reach an equilibrium, and this slow rate of equilibration of metacommunities might permit many competitors to coexist for a very long time before being replaced by competitive dominants. In this context, even a modest rate of speciation combined with slow non-equilibrium metacommunity dynamics might be sufficient to maintain high levels of biodiversity, the so-called "neutral theory of biodiversity" (Bell 2000; Hubbell 2001).

Metacommunity research is still in its infancy, so theory greatly outstrips empirical studies, which by their very nature are highly challenging. Several empirical examples are outlined in Holyoak *et al.* (2005). An excellent example is the study by Gonzalez *et al.* (1998) of arthropod communities living on moss mats. As in many ecosystems, abundance of soil arthropods tends to be correlated with broad habitat occupancy. When moss mats are cut into small fragments, however, both the percentage of patches occupied and overall arthropod abundance drop, as predicted by metacommunity models. Intriguingly, arthropod diversity within patches also drops, due to local extinction events (Gonzalez *et al.* 1998). It is not hard to see how a similar approach might be taken in analyzing the effects of habitat fragmentation on wildlife species in forest environments, although it would certainly be challenging to estimate rates of dispersal and demographic parameters across a broad community of wildlife species.

21.9 Metacommunity dynamics of predators and prey

Extinction of local populations could be caused by a number of factors: demographic stochasticity, local environmental stochasticity, competition, disease, or predation. The latter has received particular attention in both the theoretical and experimental literature. A well-established maxim in population ecology is that predator and prey populations, particularly those within a restricted area, can be spectacularly unstable (Gause 1934). Test-tube populations of protist prey, for example, often increase faster than their predators, and so thrive at early stages of trials. As protist predators increase in abundance, however, they often drive prey to extinction, before disappearing themselves. Several features arising from predator–prey theory (see Chapter 9) are consistent with such trials. First, there is often a mismatch in maximum growth rates, causing prey to increase faster than their predators. Second, consumption and mortality patterns of predators seem to be unresponsive until prey decline to extremely low numbers. Finally, the combination of these two features produces cyclical changes in abundance of both predators and prey, with predators lagging somewhat behind prey (see Chapter 9). In extreme cases, extinction then occurs at low points in the cycle, no doubt enhanced by the increased risk of chance mortality events when populations are small.

Several elegant laboratory studies have tested whether metapopulations of such predators and prey, living in small cells connected by dispersal tunnels, are better able to persist than single local populations. For example, Holyoak and Lawlor (1996) used plastic tubing to link a network of 25 small plastic bottles containing populations of protist predators and prey. One set of trials showed that predator and prey populations raised in isolated bottles rarely survived more than 18 days without predators extinguishing prey. In bottles that were 25 × larger, extinction took longer (an average of 70 days) but was still inevitable. In contrast, metapopulations of small bottles with the same total volume never went extinct over 130 days. Variance in predator abundance within vials and variance of the metapopulation itself were markedly less than those in single containers. Recurrent rescue of dwindling or even locally extinct predator populations was responsible for the more persistent dynamics of the metapopulation. Clearly, movement between vials was a key factor influencing persistence.

Similar metapopulation experiments have been conducted with populations of herbivorous and predatory insects. In a landmark experiment, Huffaker (1958) showed that persistence of populations of predatory and herbivorous mites could be markedly improved by simply making it harder for predators to travel among

Fig. 21.5 Dynamics of laboratory populations of herbivorous mites (open circles) and predatory mites (filled triangles) in a single patch (a) or in a metapopulation of 8 patches connected by bridges (b). In a single patch, predators completely eliminated their prey by 150 days, leading to local extinction of both populations, whereas the metapopulation persisted for over a year. (Source: Ellner *et al.*, 2001. Reproduced with permission of Nature Publishing Group.)

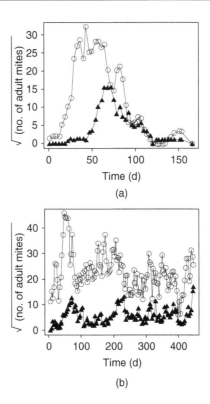

vegetation patches. This classic experiment was replicated recently, in an experiment comparing the dynamics of a single island of potted plants with those of eight floating islands that supported populations of a herbivorous mite and a predatory mite (Ellner *et al.* 2001). Their results demonstrated that whereas a single population of plants was vulnerable to a single outbreak cycle of mites before collapsing, a metapopulation of plants less readily discovered by the predatory mites was much less variable in abundance of mites and less prone to extinction (Fig. 21.5). Population models designed to mimic the experimental system suggest that a key factor in extending metapopulation persistence in this experimental system is simply the degree of asynchrony in local population fluctuations in different patches (Ellner *et al.* 2001). This naturally arises because of the probabilistic nature of discovery by predators of different patches in the landscape, suggesting that a clear understanding of factors influencing patterns of movement among patches may be essential in understanding the long-term ecological impact.

21.10   Corridors

Concerns about the speed with which fragmentation of natural habitats has occurred have naturally led to intense discussion about management interventions. One of the most hotly contested management options is corridors. If restriction of organisms to small patches is inevitable but multiple patches remain in a given landscape then surely demographic rates can only improve if these patches are linked together via corridors? Such linkage would certainly be expected to improve probabilities of successful dispersal, reestablishing locally extinct patches or "rescuing" dwindling local populations before they went extinct (Brown and Kodric-Brown 1977). These issues are particularly relevant to many wildlife species, because large animals almost always have large home ranges, which may not neatly fit within a single area of suitable habitat. Even if relict habitat fragments are large enough to support a mating pair of large terrestrial animals, natal dispersal might well put juveniles at great risk if they have to traverse a hostile matrix habitat, not to mention the public if the matrix is heavily used by humans (Chetkiewicz *et al.* 2006). Such is the logic of the pro-corridor school, and it has had great influence in public debates regarding networks of natural reserves, travel corridors to circumvent roads, and spatial configuration of development plans through zoning regulations.

On the other hand, there are good ecological reasons to doubt the wisdom of corridors, at least as a ubiquitous solution. Linking together patches might simply accelerate the spread of predators, invasive exotic competitors, or disease organisms (Simberloff *et al.* 1992), leading to synchronized dynamics and extinction en masse, rather than rescue. Clearly, there is evidence from both laboratory and field studies (Holyoak and Lawlor 1996; Ellner *et al.* 2001) that asynchrony in local population dynamics can be key in stabilizing metapopulations. Predators could lay in wait along corridors to ambush unsuspecting prey, undoing any possible benefit of improved dispersal efficiency. Finally, the extra expenditure on lands suitable as corridors might simply cut into already-strained budgets for acquisition of prime habitat itself (Simberloff *et al.* 1992). These are important issues, hotly debated because of the urgency of the conservation challenges involved and the time, money, and energy required for such major conservation interventions (Mann and Plummer 1995).

A key question is whether humans can ever know what would comprise an effective corridor. Chetkiewicz *et al.* (2006) recount the sad story of a young woman needlessly killed by a grizzly bear unable to negotiate the array of roads, residential zones, and tourist developments in the Canadian Rockies. While corridors for grizzlies had been put in place, ostensibly to prevent just such a catastrophe, radio-telemetry records show that they were little used. This case history clearly emphasizes the importance of obtaining basic biological understanding before establishing habitat corridors. There are several ways this can be achieved. Perhaps the most straightforward is to perform detailed studies of animal movement and habitat preference (see Chapter 3), before proceeding with any development project (Chetkiewicz *et al.* 2006). Follow-up studies to examine movement, habitat use, and demography in sites receiving corridors versus control sites can then be used as an appropriately controlled experimental design to test the corridors' efficacy (Beier and Noss 1998). While such designs lack replication at a landscape level, repetition over time would provide a feasible means of obtaining reliable knowledge with which to inform future developments.

So what is the track record for corridors? Perhaps the best lessons come from large-scale studies in which controlled habitat removals have taken place at ecologically relevant scales. One of the most thoroughly documented is in the pine forests of South Carolina (Tewksbury *et al.* 2002). Through controlled timber harvesting in the forest, the research team created patches of open grassy habitat used by birds, butterflies, and plants. Some of these patches were connected to each other with sizeable corridors, others had dead ends that added the same amount of area but couldn't improve dispersal success because they led nowhere. Results showed that corridors did have a detectable impact on the movement rates of some butterflies and birds (Haddad and Baum 1999; Tewksbury *et al.* 2002; Levy *et al.* 2005). Improved dispersal success in birds even had a demonstrable effect on plant biodiversity, encouraging improved seed dispersal via bird feces transported from patch to patch (Levy *et al.* 2005). Somewhat surprisingly, direct observations showed that birds did not use the corridors themselves, but rather flew through the forest along their edges (Levy *et al.* 2005). This is a nice example of the potential danger of preconceived notions about corridor use. Increased abundance of plants and butterflies in patches with corridors suggests that improved dispersal efficiency paid off in terms of improved demographic performance (Haddad and Baum 1999).

The overall record for the effectiveness of corridors is somewhat mixed, however. There are many studies in the literature demonstrating use of corridors, some of which also yield practical insight into factors that inhibit use by wildlife. For example, comparisons of wildlife use of highway underpasses in Banff National Park in the Canadian Rocky Mountains suggest that a key factor is simply whether such corridors have to be shared with humans. Where human use of corridors is heavy, wildlife use tends to fall off (Clevenger 2000). While underpasses are popular with some species, such as elk and wolves, others, such as voles, rarely use them. There are far fewer studies showing demonstrated improvement in demographic performance as a result of corridors. Direct studies of movement itself are also rare in corridor studies, however, no doubt in part because of the cost and effort required to gather movement data. Hence, it is often hard to say whether it is common for most wildlife species to use corridors any more than they would use the matrix. Clearly there is great need for such studies, given the costs and potential benefits (Chetkiewicz *et al.* 2006). Lindenmayer (1994) lists the factors that might influence the use of corridors:

1 The biology, ecology, and life history of the species.
2 Habitat suitability, including the degree of original vegetation integrity, length, and width.
3 Location of the corridors in the landscape.
4 The type of disturbance in the matrix surrounding the fragments and corridors.
5 Suitability of the matrix habitat.

There is often an additional conceptual problem with corridors. By definition, these are strips of habitat that are too small for the species of interest to live in permanently (e.g. too close to the edge of forest for interior forest birds, or too narrow to support a territory). Such strips may be suitable for wide-ranging species, such as rodents, which will benefit from the cover provided by forest or shrubs, allowing safer movement than that over fields. However, such species would probably traverse these open habitats if corridors did not exist. In contrast, sedentary species such as interior forest birds (the New Zealand kokako (*Callaeas cinerea*) is a good example of a highly territorial bird that flies poorly and moves little through dense forest) are unlikely to venture into corridors because they are unsuitable habitat. Thus, those species that would benefit most from corridors – the reluctant travelers – are the least likely to use them, and vice versa. Corridors might also act as sinks, trapping animals in them but preventing successful breeding (Saunders and Hobbs 1991).

Narrow strips of eucalypt woodland act as corridors, or more accurately as stepping stones, connecting isolated fragments of original eucalypt woodland in Western Australia. These corridors result in higher species richness of birds, so that the closer the corridors are to a patch, the higher isthe species number (Fortin and Arnold 1997). Kangaroos (*Macropus robustus*) also use corridors to move between remnant patches of eucalypt woodland in Western Australia (Arnold *et al.* 1993).

Box 21.1 Potential advantages and disadvantages of conservation corridors. (From Noss 1987).

Potential advantages

1 Increased immigration rate to a reserve, which could:
(a) increase or maintain species richness (as predicted by island biogeography theory);
(b) increase population sizes of particular species and decrease probability of extinction (provide a "rescue effect") or permit reestablishment of extinct local populations;
(c) prevent inbreeding depression and maintain genetic variation within populations.

2 Increased foraging area for wide-ranging species.

3 Provision of predator-escape cover for movements between patches.

4 A mix of habitats and successional stages accessible to species that require a variety of habitats for different activities or stages of their life cycles.

5 Alternative refugia from large disturbances (a "fire escape").

6 "Greenbelts" to limit urban sprawl, abate pollution, provide recreational opportunities, and enhance scenery and land values.

Potential disadvantages

1 Increased immigration rate to a reserve, which could:
(a) facilitate the spread of endemic diseases, insect pests, exotic species, weeds, and other undesirable species into reserves and across the landscape;
(b) decrease the level of genetic variation among populations or subpopulations, or disrupt local adaptations and co-adapted gene complexes ("outbreeding depression").

2 Facilitation of spread of fire and other abiotic disturbances ("contagious catastrophes").

3 Increased exposure of wildlife to hunters, poachers, and other predators.

4 Possible lack of enhancement of dispersal or survival of upland species by riparian strips, which are often recommended as corridor sites.

5 Cost, and conflict with conventional land preservation strategy to preserve endangered species habitat (when the inherent quality of corridor habitat is low).

Golden lion tamarins (*Leontopithecus r. rosalia*) need corridors between patches of tropical forest habitat – all that remains within a sea of sugarcane fields in Brazil. Rodents, such as chipmunks and voles, can use fencerows and hedgerows as corridors (La Polla and Barrett 1993; Bennett *et al.* 1994). Hill (1995) showed how some poorly dispersing insects (butterflies, dung beetles) use corridors but others do not.

All of this suggests that the efficacy of corridors needs to be assessed on a case-by-case basis. Thus, Simberloff and Cox (1987) used the Seychelles Islands of the Indian Ocean to make the point that corridors are not always beneficial. The Seychelles contained 14 endemic land birds when Europeans arrived in 1770. Land clearing, fires, and the introduction of rats and cats devastated the archipelago over the subsequent 200 years but resulted in the extinction of only two of those species. Losses were limited partly because no corridors (isthmuses) linked the islands: Introduced predators and fires were unable to reach all of them.

The potential advantages and disadvantages of conservation corridors as summarized by Noss (1987) and Saunders and Hobbs (1991) are presented in Box 21.1.

Sometimes the best option might be simply to improve the matrix so that it is more readily traversed. A good example comes from eastern collared lizards in Missouri (Templeton *et al.* 2011). These lizards occupy open glades in forests of the Ozark Mountains. Although there is evidence that lizards were once common at the glades, by the early 1980s they were rare; at that point they were reintroduced into a small number of sites. Since then, collared lizards have been regularly captured and marked, and information from these animals has been used to estimate dispersal rates, demographic rates, and local population abundance. Despite glades being close to one another, very little dispersal was recorded during initial stages of the project: only 1 out of 66 individuals successfully found another glade. In the early 1990s, however, conservation biologists tried a different tack. Recognizing that fire had once been a common feature in these forests, the biologists instituted a prescribed burning program. This had the effect of opening up the understory, creating a more favorable thermal environment for poikilotherms like the collared lizard, and elevating densities of grasshoppers, their favored food item. In response to these changes to the forest matrix, annual dispersal rates increased 10-fold and lizards rapidly occupied many of the patches. Initially, new immigrants had no trouble finding unoccupied glade patches. Over time, however, vacant patches declined in abundance and so colonization rates dropped, just as one might expect based on metapopulation theory. Eventually colonization rates reached the level of local extinction rates, at which point patch occupancy stabilized, again in keeping with metapopulation theory. A key point in this successful program was the recognition that the system could only function as a viable metapopulation when patches were made more accessible through prescribed burning as a management tool.

21.11 Summary

The expansion of human populations across the globe has led to both loss of wildlife habitat and fragmentation of the remaining habitat into smaller parcels. These small patches often have more edge than core and are sometimes widely spaced. The effects of habitat loss and fragmentation can be complex because of the wide variety of patterns that can arise through habitat destruction and the difficulty in measuring local abundance and community richness in multiple patches. A growing number of studies suggest that both the abundance and the diversity of wildlife can be greatly affected by habitat loss, and to a lesser extent habitat fragmentation.

Dispersal is integral to the dynamics of organisms occupying spatially subdivided habitats forming metapopulations. Simple models demonstrate that the long-term persistence of metapopulations depends on the relative probabilities of extinction versus dispersive colonization. There is some empirical evidence for regular turnover of colonies and equilibrial balancing of rates of extinction and colonization. Variations on the metapopulation theme include source–sink systems, island–mainland systems, and metapopulations with internal territory structure.

Both competition and predator–prey interactions are influenced by spatial structure. When multiple species are involved, the resulting interactions are referred to as "metacommunity dynamics." Key features influencing metacommunity dynamics are the relative mobility of different species in the community, the degree of asynchrony in local population dynamics, heterogeneity in patch characteristics, and tradeoffs between dispersal ability and demographic parameters.

Corridors and improvements to the matrix are often discussed as remedies for habitat loss and fragmentation. While corridors can improve movement rates and population densities in remaining habitat patches, they may not be attractive to dispersing animals and they can have a downside if they serve to stimulate, rather than mute, synchronized population fluctuations due to competition, predation, parasitism, or disease transmission. Given their expense, the costs versus benefits of corridors require close examination on a case-by-case basis

22 Ecosystem management and conservation

22.1 Introduction

Most of the previous chapters have focused on management and conservation of individual species and their direct interactions with the next trophic level. Management and conservation threats are usually stated in terms of overharvesting (Chapter 18), overpredation (Chapters 9, 16), disease (Chapter 8), pests (Chapter 19), or habitat loss (Chapter 21, Griffith *et al.* 1989). Although these issues are correct as far as they go, they hide a common feature: population declines occur through a combination of factors, derived from complex interactions between environment and biota, which together overwhelm the ability of a species to withstand them. These are the effects of ecosystem dynamics and they often dictate the fates of individual species. Natural ecosystem complexity arises from factors such as nonlinear biotic interactions, evolutionary and assembly history of species, and often-unpredictable environmental disturbance.

Human interference, both from exploitation and from stewardship, can alter ecosystem dynamics in ways that are the opposite of what is intended. For example, fences put up in central Botswana and in Kruger Park in South Africa with the intention of protecting the cattle industry in the former and wildlife in the latter both altered migrations and caused a collapse of wildlife populations; placement of artificial waterholes in Hwangi Park in Zimbabwe, Kruger Park (du Toit *et al.* 2003), and the Sahel of northern Africa resulted in local overuse of resources and unwanted ecosystem effects (Sinclair and Fryxell 1985); and crude attempts at biological control through the introduction of exotic predators to control exotic prey in Hawaii, Australia, New Zealand, and many other areas resulted in the extinction, or near extinction, of endemic species and the complete failure of the control of pests (Serena 1994). Thus, what at first seems an obvious conservation solution may not prove to be so upon closer examination.

The fundamental problem is that ignoring some aspects of ecosystem dynamics – historical legacy, community interactions, disturbance at different scales – can lead to inappropriate conservation efforts, curing the symptoms rather than the cause. Ecosystem dynamics are now well described (e.g. Boyce and Haney 1997). In this chapter we put the various aspects that we have discussed in the rest of the book into the context of the ecosystem to show how it is pertinent to management and conservation. We start by providing some definitions.

Wildlife Ecology, Conservation, and Management, Third Edition.
John M. Fryxell, Anthony R. E. Sinclair and Graeme Caughley.
© 2014 John Wiley & Sons, Ltd. Published 2014 by John Wiley & Sons, Ltd.
Companion Website: www.wiley.com/go/Fryxell/Wildlife

22.2 Definitions

Communities are complexes of interacting populations. They involve both direct and indirect effects of competition, predation, and parasitism. They are made up of major players, called *dominant* and *keystone* species, and combinations of many other minor species. Although species combinations are fluid, there are usually identifiable characteristic groups that define a community. Communities exist within the abiotic physical and chemical environment, and the combination of these forms the *ecosystem*. All ecosystems are maintained by external inputs of energy and certain levels of nutrients. Some receive most of their nutrients from outside (*allochthanous* supply), such as leaf debris from river banks for stream communities or the rain of detritus to the abyssal zone of the ocean. Others are more self-contained, with good internal nutrient supplies (*autochthanous* supply). Ecosystems can be small or large, ranging from a stream complex or watershed to the $25\,000\,km^2$ Serengeti or segments of the vast boreal forest of Canada and Russia. Although ecosystems are open, they are identified by some form of boundary across which the combination of abiotic and biotic factors changes. The study of the large-scale ecology of ecosystems is called *landscape ecology*.

22.3 Gradients of communities

The early classic studies of plant communities suggested that communities of plants existed as discrete units with sharp boundaries, something like a superorganism. It was generally agreed that sharp divisions in the environment – different geologies, soils, or other environmental factors – caused discrete boundaries between communities. However, it was also suggested that such boundaries could be caused by groups of co-evolved species occurring together. Another school of thought suggested that plants generally existed independently of each other, so that a gradual change of species took place along a gradient in the abiotic environment, such as a gradient of moisture, of altitude and temperature on mountains, or of exposure and salinity on the seashore (Whittaker 1967). Present understanding (Austin 1985) suggests that species do form a continuum along gradients, but not uniformly. Groups of species appear and disappear together for two reasons. First, some species depend upon each other. Where plants go, so too do their associated animals, and therefore these groups are found in similar places. Second, the abiotic gradient is not usually uniform: there is often a break or rapid change in geology, soil substrate, or exposure, and at these points there are rapid changes in species complexes.

22.4 Niches

Where a species is found is determined by its tolerance and adaptation to the abiotic environment. This is the *fundamental niche*, which is constrained by the biotic processes of competition and predation to form the *realized niche*. See Sections 6.4–6.6 for a more detailed explanation of niches and how species may divide up the niche space along gradients.

22.5 Food webs and intertrophic interactions

A community can be divided into *trophic levels*, defined as the *location of energy or nutrients within a food chain or food web* (Fig. 22.1). In terrestrial systems we describe these as plants that capture energy from the sun (*autotrophs*), herbivores that feed on plants, carnivores that feed on herbivores or other carnivores (the last two levels being called *heterotrophs*), and detritivores that receive dead products from all of the other trophic levels. Energy and nutrients enter a system from outside and flow through it. Energy is progressively lost through the levels but much of the nutrient bulk cycles back via the detritivores into the plants. A fraction is lost by water leaching of soils,

Fig. 22.1 Hypothetical flow diagram showing the passage of nutrients through the trophic levels (boxes). The numbers are in units/ha/year. Inputs and outputs for each box and for the whole system must balance.

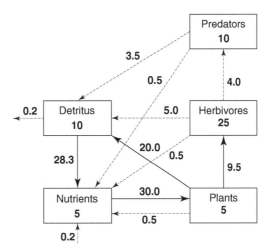

runoff into rivers, burning, wind transport, and so on. In a stable system inputs and outputs at all levels must balance.

Indirect effects are those where one trophic level affects components above or below it. This can take two forms: *trophic cascades*, where linear predation influences the next-but-one level, and *nonlinear effects*, where competitors within the predator or prey levels respond to changes in another such level (Wootton 1994a, b). Two examples are illustrated here.

22.5.1
Hyperpredation and apparent competition

Top predators can affect the diversity of their prey through changes in the abundance of primary prey. This is particularly important where both exotic predators and their primary prey cause increased predation of endemic species. The introduction of feral cats and European rabbits to islands has resulted in catastrophic declines of many seabird populations (many examples are given in Courchamp *et al.* 2000b). In New Zealand, introduced ferrets (*Mustela furo*), cats, and rabbits have caused predation of two endemic skink species. Predation on skinks is inversely density-dependent, as one expects for secondary prey (see Section 7.7.2) (Norbury 2001). We saw in Chapter 16 what happened when feral pigs (*Sus scrofa*) were released into the California Channel in the early 20th century (Roemer *et al.* 2002). As they became more abundant, mainland golden eagles (*Aquila chrysaetos*) were able to colonize the islands in the 1970s, building to high numbers by the 1990s (see Fig. 16.12). These raptors caused a rapid decline in the endemic island fox (*Urocyon littoralis*), which in turn caused a change in the competitive balance of the fox with the endemic spotted skunk (*Spilogale gracilis amphiala*), which was able to increase. Thus, hyperpredation – where top predators consume lower predators – resulted in a change in competition between predators.

22.5.2
Mesopredator release

The opposite process occurs when top predators are removed and a linear trophic cascade results in the increase of lower predators and a decrease in some of their prey, leading to a change in species composition. For example, on islands where both cats (the top predator) and rats (their prey) are present, removal of the former can increase the latter, which then prey on seabird nests (Courchamp *et al.* 1999). In southwest Spain, the Iberian lynx (*Felis pardalis*) depredates both Egyptian mongooses (*Herpestes ichneumon*) and rabbits. In areas where lynx are absent, mongoose predation results in rabbit densities two to four times lower than in areas where they occur (Palomares *et al.* 1995). In southern California, the disappearance of coyotes (*Canis latrans*) has allowed an increase in several smaller predators (e.g. striped skunk, grey fox, feral cats), which in turn have reduced the diversity of scrub breeding birds (Crooks and Soule 1999).

A more complex situation is found on Marion Island off South Africa (Huyser *et al.* 2000). There, feral cats fed on exotic house mice (*Mus musculus*) and burrowing petrels, reducing petrel numbers considerably. Mice consume terrestrial macroinvertebrates, which the endemic lesser sheathbill (*Chionis minor*), an aberrant shorebird of the Chionididae related to plovers, depends on in winter. Macroinvertebrates are maintained by the feces of burrowing petrels, and the loss of petrels also results in the loss of invertebrates. Removal of cats caused an increase in house mice (but not petrels, because their habitat for burrows had changed) and thus declines in macroinvertebrates. The sheathbill population collapsed on that island relative to the neighboring Prince Edward Island, where mice were at much lower density. Thus, nonlinear indirect effects involved an increase in competition between mice and sheathbills for invertebrates through mesopredator release.

In all of these examples, ecosystem disturbances through the introduction of a new species results in unpredicted outcomes due to indirect food web interactions, with unexpected consequences for conservation.

22.6 Community features and management consequences

22.6.1 *Dominant and keystone species*

Within groups of co-occurring species, some play a larger role in group definition. In terrestrial environments, it is the plants that dictate the structure and function of the community, and they in turn are determined by the abiotic environment. Thus, in arctic and alpine areas we get tundra, with herbs, grasses, and shrubs determining a low-lying structure. In contrast, in very wet temperate and tropical areas we find forests, with tall, woody plants creating a complex three-dimensional structure. We say that the most common plants (most numerous, highest biomass), such as white spruce in boreal forest, are *dominant* species that determine the structure and function of the ecosystem.

Some species can have a considerable influence on the community even though they are relatively rare. Often these are predators, and they can determine not just the species composition of prey but, indirectly, many other components of the ecosystem. Such species have been called *keystone* (Paine 1969; Power *et al.* 1996) and are defined as *those that have a greater role in maintaining ecosystem structure and function than one would predict based on their abundance or biomass.* Top predators are often presented as keystone species: for example, the presence or absence of sea otters (*Enhyra lutris*) as top predators of inshore marine communities determines the abundance and species composition of other members (Estes and Duggins 1995). Herbivores, however, can also act as keystones; for example, rodents can structure desert plant communities (Brown and Heske 1990), snowshoe hares (*Lepus americanus*) structure boreal forest vertebrate communities (Krebs *et al.* 2001b), and wildebeest (*Connochaetes taurinus*) structure the Serengeti ecosystem (Sinclair 2003).

There are thorny problems with the keystone concept. First, there are operational problems with identifying keystone species. We need to define which parameter we will measure: abundance, biomass, species composition, or something else. We need to specify the degree of change in the community expected from the loss of the species. Communities are open-ended and we must state how far into the food web we should trace impacts; the impacts of top predators might be traced only as far as the herbivores and plants, or through other indirect links to more distant herbivores, detritivores, protozoans, or even microbes (Mills *et al.* 1993; Power *et al.* 1996).

Despite these problems, we recognize that some species define community composition and that the removal of these can produce changes in state, whereas the loss of other species has little effect on the rest of the community. The conservation consequences of the loss of wildebeest as a keystone species in the Serengeti would be disproportionately greater than those due to the loss of black rhino (*Diceros bicornis*) or wild dog (*Lycaon pictus*), both of which have occurred with little impact on the system.

Keystone species can also have counterintuitive effects. Bell miners (*Manorina melanphrys*) are dominant territorial insectivorous birds that feed on psyllids (plant sucking Homoptera) in Australian eucalypt forests. Where bell miners occur, trees appear unhealthy, the foliage infested with these insects. When bell miners were removed, 11 other insectivore bird species moved in, fed on the psyllids, and within 4 months eradicated the infestation (Loyn *et al.* 1983). Interspecific territoriality by the miners maintained their food supply but reduced the diversity of competing predators. The management and conservation implications for keystones such as wildebeest or bell miners are very different from those for other species in their ecosystems. The task of managers is to identify such species.

22.6.2
Overpredation

Top predators can increase the diversity of prey species through intermediate disturbance effects (Connell 1978). However, predators can also have the opposite effect and reduce diversity of prey. Such effects arise because rare species are secondary prey, essentially *by-catch* for predators that depend on more common prey (Chapters 7 and 16). It is, of course, the rare species that attract the attention of managers.

Exotic predators are of special management concern because they can threaten rare species. We see this where exotic predators such as red fox (*Vulpes vulpes*) in Australia, stoats (*Mustela erminea*) in New Zealand, and many other species on islands in the Pacific and Indian Oceans, supported by exotic prey or carrion, have caused the extinction of numerous marsupial mammals, birds, and invertebrates (Serena 1994; Atkinson 2001).

The concept of by-catch stems from harvesting, particularly fisheries harvests. Many fish species are caught as by-catch in fisheries that focus on more abundant species. The latter can be maintained while the by-catch species decline, either because they are easier to catch or because they have a lower intrinsic rate of increase (Hilborn *et al.* 2003). In the Atlantic, both common skates (*Raja batis*) and barndoor skates (*Raja laevis*) almost went extinct because of by-catch (Casey and Myers 1998), and shark numbers have declined by over 50% since 1990 due to by-catch in swordfish and tuna fisheries (Baum *et al.* 2003).

A prey species' role can change with the presence of other species. In British Columbia, moose (*Alces alces*) are primary prey for wolves, which are driving mountain caribou (*Rangifer tarandus*) to extinction (Seip 1992; Wittmer *et al.* 2005); but in the nearby Banff ecosystem, moose are secondary prey that are being exterminated by wolves which are dependent primarily on elk (*Cervus elaphus*).

In general, the dynamics of predation, whether by natural predators or by humans, affect prey species differently depending on their role in the system, their abundance, and their intrinsic adaptations (Courchamp *et al.* 1999, 2000b). Therefore, the conservation threat to a species, particularly due to predation, depends on the ecosystem in which it is found.

22.6.3 *Co-evolved associations*

A species may not be a keystone yet be essential to the survival of other species if it has closely evolved associations with those species. Many tropical flowers have special adaptations to attract specific insect pollinators. In Hawaii, a whole group of plants, *Hibiscadelphus*, has become extinct, or nearly extinct, following the extinction of its honeycreeper pollinators. On Mauritius, the tree *Calvaria major* almost went extinct following the extinction of the dodo (*Raphus cucullatus*), which ate its seeds and promoted their germination, probably by cleaning the fruit (Temple 1977; Traveset 1998).

Such co-evolved associations have important flow-on effects on ecosystem dynamics. New Zealand is relatively depauperate in animal pollinators, allowing the evolution of some unusual co-evolved associations (Webb and Kelly 1993). One such involves the rare dioecious obligate root parasite of forest trees, *Dactylanthus taylorii*, which produces strongly scented brownish flowers on the forest floor. This species has adaptations for pollination by short-tailed bats (*Mystacina* spp.), which in the total absence of terrestrial mammals forage on the forest floor like rodents. Multiple threats, such as limited pollination of *Dactylanthus* due to declining bat populations (Ecroyd 1996) and severe herbivory by the exotic brush-tailed possum (*Trichosurus vulpecula*), feral pig (*Sus scrofa*), and rats (*Rattus* spp.), have put the plant in serious decline.

Even common species can be threatened if there is a co-evolved association with a vulnerable species. For example, pollination limitation is becoming increasingly evident in mainland New Zealand flora compared to that on offshore islands (Montgomery *et al.* 2001; Anderson 2003) and the decline in avian pollinators appears to be the main cause. Thus, the mistletoe (*Peraxilla tetrapetala*) is showing declining pollination rates because its main pollinator, the bellbird (*Anthornis melanura*), an endemic honeyeater, is declining in number (Murphy and Kelly 2001). In general, these plant species were not themselves under threat until the species that they depended upon declined. Thus, the conservation needs of one species must take into account the requirements of others. Prior knowledge of such obligatory associations would allow conservationists to predict threats to survival of species in a wider ecosystem context.

22.7 **Multiple states**

Operationally, *multiple states can be identified when an external perturbation changes a system from one state to another and the system does not return to the original state once the perturbation has ceased.* This definition excludes situations where different states occur under different environmental conditions (Holling 1973; Sinclair 1989; Knowlton 1992; Beisner *et al.* 2003). Changes in state are characterized by nonlinear dynamics between trophic levels, exhibiting initial slow change followed by fast, catastrophic change (May 1977; Scheffer *et al.* 2001). Predation is one process that can produce such multiple states. Under special circumstances of Type III functional responses, predators can theoretically hold prey populations at two levels under the same environmental conditions (see Chapter 7 and Fig. 7.7). Multiple states can also arise through switches in competitive ability between species, while environmental disturbances such as storms or fire that change soil conditions can result in permanent changes in state.

Evidence for multiple states in nature is extremely sparse. Forest insects may be held at low density by warblers but can erupt to high density where warblers do not regulate them (Ludwig *et al.* 1978; Crawford and Jennings 1989). There are a few examples where mammals act as the predator. White-tailed deer maintain different plant communities by feeding on young trees. Two tree densities can be found depending on whether young trees can escape this herbivory or not (Augustine *et al.* 1998). Similarly, elephants in the Serengeti can maintain two different densities of *Acacia* tree. When fire prevents regeneration and mature trees die of old age, a grassland is produced; elephants maintain this by feeding on it and regulating juvenile trees. When overhunting by humans removed elephants (in both the 1880s and 1980s), trees escaped herbivory and formed a mature savanna. After both periods of removal, elephant numbers increased and they fed on the mature trees, but they did not return the woodland to grassland (Sinclair and Krebs 2002). Examples of multiple states where mammals are prey are also rare (Section 7.7.1). Outbreaks of house mice and European rabbits in Australia may be interpreted as changes from a predator-regulated to a food-regulated state (Pech *et al.* 1992). The collapse of the "forty mile" caribou herd of the Yukon may be evidence of multiple states. Similarly, managers culled wildebeest in the Kruger National Park, South Africa to reduce their numbers. When culling ceased, wildebeest numbers continued to decline through lion predation (Smuts 1978). In general, these examples illustrate that more than one state can occur under a given set of climatic conditions.

In some cases, the change in state has undesirable consequences for management and conservation. For example, in the semi-arid regions of the Negev-Sinai in Israel and Egypt and in the Sahel of Africa, a shrub and herb layer acts as a blanket on the soil, retaining moisture and heat overnight. During the day, thermal upcurrents carry moisture from both the soil and transpiring plants to upper levels, where it condenses as rain. This supplies the plants and soil, completing a positive-feedback, self-sustaining system when in an undisturbed state. In contrast, overgrazing by livestock leaves a denuded soil surface, higher surface albedo, and cooling at night. There are fewer thermals, and these carry less moisture. Thus, overgrazed areas have much lower precipitation; this is also a positive feedback (Otterman 1974; Sinclair and Fryxell 1985). The vegetated state switches to the denuded one through the disturbance of overgrazing; that is, there is a threshold level of disturbance (grazing) at which one state switches to another. A similar positive-feedback switch in vegetation state occurs in Niger, where overgrazing has altered vegetation structure, leading to reduced water retention, increased soil loss, and further vegetation loss. The system is now locked into this reduced state (Wu *et al.* 2000).

However, good examples of multiple states are rare, with a few known from lakes, rivers, coral reefs, grasslands, and forests (Knowlton 1992; Augustine *et al.* 1998; Dent *et al.* 2002). The relevance to management is that multiple states are an emerging property of ecosystems that will rarely be predicted from the study of single species. Some states arise from excessive disturbance. Thus, conservation needs to plan for more than one natural state, while avoiding unnatural states caused by excessive human disturbance.

22.8 Regulation of top-down and bottom-up processes

Fig. 22.1 illustrates the pathways for energy and nutrients. However, it does not indicate where the regulation occurs in the system. Regulation can come via the food supply (bottom-up), from predation (top-down), or both. In the absence of predators (or parasites), bottom-up processes must regulate all populations. Regulation through resources must be the basic rule, and it clearly applies to all top predators. There are several factors that affect resource production and biomass in the world (Polis 1999). However, four main conditions provide a refuge from predation and so allow bottom-up regulation.

22.8.1 *Body size*

Small prey species are vulnerable to predation, whereas very large species, especially in mammals, have outgrown all present-day predators and so are regulated by food supply. Thus, a suite of predators accounts for virtually all mortality of adult snowshoe hares in northern Canada (Hodges *et al.* 2001), while the wood bison (*B. bison athabascae*) population in the Mackenzie Bison Sanctuary of Canada appears to be regulated by food supply, despite wolf predation of juveniles (Larter *et al.* 2000).

In Africa, we see a similar effect of body size on causes of regulation. Elephants, rhinos, and hippos are too large for predators. Although predators kill a few newborn animals, they have no effect on the population (Sinclair 1977). Even animals the size of African buffalo and giraffe are large enough that predators have difficulty killing them: predation accounts for a small proportion of adult mortality, with undernutrition the predominant cause (Sinclair 1977, 1979b).

22.8.2 *Migration*

Migration is an adaptation that overcomes the constraints imposed by body size (see Section 7.8.1). Predators cannot follow migrating herbivores because they are confined to territories where they raise and protect their young. This general rule is evident in all mammal migration systems, such as wildebeest and gazelles in the Serengeti and Botswana, white-eared kob (*Kobus kob*) in Sudan, caribou (*Rangifer tarandus*) in northern Canada, and most probably the original plains bison of the North American prairies (Fryxell and Sinclair 1988a). It might also apply to the migration of marine mammals. Migrating species, therefore, escape from predator regulation even when they are relatively small in size, as with gazelles. In addition, migration allows access to ephemeral, high-quality food resources not available to nonmigrants. These two features allow migratory populations to become an order of magnitude greater in number than resident ones.

22.8.3 *Low-diversity ecosystems*

In higher latitudes there are often predator–prey systems with only one major predator and one or a few mammal prey species. We see such systems in temperate woodlands and tundra, and even in mammals of tropical forest (although not in other groups). In these ecosystems we normally see bottom-up regulation of the prey. Nevertheless, there are a few cases of top-down regulation. Wolves might regulate moose in some parts of Canada and Alaska (Gasaway *et al.* 1992; Messier and Joly 2000). In contrast, on Isle Royale in Lake Superior, wolf numbers appear to track moose numbers and do not regulate that population (Peterson and Vucetich 2003). Thus, we have evidence of regulation of herbivores by both predators and food supply. Particular features of the ecosystem and the species involved determine the direction of regulation. In addition, multiple states (see Section 22.6) may occur, where regulation can switch in a single system from resource limitation to predation or vice versa. Alternatively, regulation may be determined by the presence or absence of alternative prey for the predator (see Section 7.7.1).

22.8.4 *High-diversity communities*

In some systems there is a high diversity of herbivores and carnivores. Nearly all are associated with tropical ecosystems. Whether a herbivore species is limited by predators is determined by its place in the hierarchy of herbivores. In African savanna there are as many as 10 coexisting canid or felid carnivores feeding on ungulates, lagomorphs, and rodents. They vary in size from the 200 kg lion (*Panthera leo*) to the 10 kg wild cat (*Felis sylvestris*). The larger the carnivore, the greater its range of prey sizes. Thus, the lion's diet ranges from buffalo (450 kg) to dikdik (*Madoqua kirkii*), a small antelope (10 kg), whereas that of the 16 kg caracal (*Felis caracal*) ranges from duiker (15 kg) to 100 g rodents. The consequence of this is that smaller ungulates have many more predators than do larger ungulates; smaller ungulates experience more predation and top-down regulation (Sinclair *et al.* 2003).

22.9 Ecosystem consequences of bottom-up processes

Regulation of herbivore populations through their food supply has profound consequences for the ecosystems where they occur. Mammals may not be numerous compared to other animal groups but their impact is considerable. Perhaps more than any other group, they can determine the physical structure of the habitats, alter the rates of ecosystem processes such as nutrient flow, growth rate, and decomposition, and dictate species diversity. These large-scale effects – at the level of ecosystems, watersheds, and biomes – can be thought of as *ecological landscaping* (Sinclair 2003).

22.9.1 *Vegetation structuring*

Plants determine the physical structure of habitats, the particular type being a function of the abiotic conditions. Some periodic environmental effects, such as fire, hurricanes, and floods, can interrupt the normal succession of plant species towards a climax. In savanna, fire typically impedes the succession of trees to produce a "fire disclimax" of grassland and fire-tolerant herbs, shrubs, and trees. Herbivorous mammals can have analogous effects to fire in savanna systems (Hobbs 1996) and so produce a "mammal disclimax." Plant succession is held in a different state as a result of the restructuring imposed by mammals. Such impacts are evident in most terrestrial biomes where mammals are abundant. However, mammals have their greatest impacts in the tropical savannahs, particularly through feeding by megaherbivores (Owen-Smith 1988), and in grasslands throughout temperate and tropical regions, due to grazing and browsing by ungulates (Sinclair 2003).

In recent times, mammal herbivores have had little structuring effect in high-latitude tundra biomes. However, the Pleistocene tundra supported a substantial biomass of mammoths, woolly rhinos, and bison that fed upon the shrubs and sedges. Herbivorous mammals also do not substantially alter tropical forest, although mammals do influence the dispersal of tree seedlings (Janzen 1970). In both arctic tundra and tropical forest the low impact of herbivores may be due to the top-down effects of mammal carnivores, which limit herbivore densities in these systems (Terborgh 1988; Oksanen 1990).

22.9.2 *Ecosystem rates*

Mammals and birds influence rates of nutrient cycling, in addition to altering physical structure. High densities of mammals and birds can influence soil processes through their deposition of feces and urine. Before the arrival of Pacific rats (*Rattus exulans*) in New Zealand some 700 years ago, billions of shearwaters lived on the forest floor and provided a considerable nutrient input. The extinction of shearwaters from mainland New Zealand due to the rats has altered the nutrient dynamics of that country (Worthy and Holdaway 2002). The volcanic Serengeti plains have very high nutrient turnover rates due to the large herds of ungulates grazing them in the wet season and the plethora of dungbeetle species that act to bury their dung. This process leads to high protein and mineral content in the grasses eaten by the grazing herds. In essence, ungulates fertilize their own food, and so create a positive feedback, increasing their own density (Botkin *et al.* 1981; McNaughton *et al.* 1997). In the low-nutrient granitic soils of Southern Africa, the vegetation is also low in nutrients, and only very large mammals such as elephants can feed in those woodlands. Nutrient recycling is slow (Bell 1982).

On the Canadian arctic shorelines, high densities of snow geese (*Chen hyperborea*) influence the growth of their food plants. Moderate grazing promotes growth through fecal nutrient cycling. Recent population increases of geese have resulted in overgrazing, which overwhelms the positive effects (Jefferies *et al.* 2004). In boreal forests, moose decrease nitrogen mineralization of the soil by decreasing the return of high-quality litter: their browsing on deciduous trees reduces their leaf fall while promoting low-quality white spruce inputs (Pastor *et al.* 1993). In contrast, soil nitrogen cycling in Yellowstone and other prairie areas of the United States is increased by large mammal grazers (Hobbs 1996; Frank and Evans 1997).

22.9.3 *Plant species composition*

Herbivory alters not only structure but also the type of plant that can withstand such impacts. On the North American prairies, rodents such as black-tailed prairie dogs (*Cynomys ludovicianus*) live in large colonies. These species graze grasses to a low level (a few cm) around their colonies. Grazing changes the composition of grass species to low-growing forms, and many dicot species survive due to reduced competition from grass. American plains bison preferentially graze these short grasses and pronghorn antelope (*Antilocapra americana*) feed on the dicots (Huntly and Inouye 1988; Miller *et al.* 1994).

Rabbits maintain short grasslands with many dicots on the South Downs of Sussex, UK. When the epizootic myxomatosis removed rabbits in 1953, plant species composition changed to one of tall tussock grasses with few dicots, and there were subsequent changes in the ants and lizards dependent on these plant forms (Ross 1982). A whole range of plant species evolved in New Zealand with special structural defenses against moa browsing not seen elsewhere (Bond *et al.* 2004).

22.10 Ecosystem disturbance and heterogeneity

22.10.1 *Degrees of disturbance*

Ecosystems, left undisturbed, will change through succession to a plant community dominated by good competitors and their associated fauna. A few wet tropical forests may exhibit this situation. However, it is rare that ecosystems experience such constancy of environment. Disturbances disrupt this succession, and the community reflects this history of disturbance (Pickett and White 1985). If disturbances are severe and frequent then a few hardy plants that can tolerate these stressful environments will characterize the community. The boundary between the alpine tussock grasslands and nival (snow) zone of the Southern Alps of New Zealand is characterized by a few plants adapted to steep loose scree (tallus) that moves frequently through heavy rain, earthquakes, and trampling or feeding by exotic mountain ungulates: the tahr (*Hemitragus jemlahicus*) and chamois (*Rupicapra rupicapra*).

More commonly, disturbances are both less frequent and less extreme, allowing a combination of good dispersing plants and good competitors (see Section 22.12). Forests experience treefalls that create canopy gaps, letting in light and opportunities for light-seeking species. This disturbance is particularly important in tropical forests where fruiting plants are fed upon by birds (Levey 1990). Temperate conifer forests experience fire at frequencies of 50–200 years, maintaining a mosaic of stands of different age and species composition and a diversity of habitats for birds and mammals (Bunnell 1995).

In tropical savannas, frequent fire dominates the system, impeding plant succession and maintaining an open tree canopy (< 30% cover) and a grass understory. This fire regime provides the optimum habitat for the high diversity of ungulates in East Africa (Frost 1985). Fire is also required to maintain specific habitats: in North America, the endangered Kirtland's warbler (*Dendroica kirtlandii*) requires fire to create its jack pine (*Pinus banksiana*) habitat (Probst 1986).

Other forms of disturbance include hurricanes and floods. Hurricanes are important at the 10–20° latitudes along coastlines. Their periodicity is usually measured in decades and they have physical restructuring effects through the opening up of forests, altering of shores, estuaries, and riverbanks, and sedimentation rates.

Flooding of rivers and estuaries is more frequent and widespread. Some river flooding is necessary to maintain nesting habitat on sandbars for least terns (*Sterna albifrons*) and piping plovers (*Charadrius melodius*) in the United States and for endangered black stilts (*Himantopus novaezealandiae*) on braided rivers in New Zealand (Boyce and Payne 1997).

One of the important consequences of disturbance is that it creates heterogeneity, or patchiness, in habitats, particularly because it reverses succession. A mosaic of patches of different ages from the time of disturbance leads to different combinations of habitats and species, an aspect that is important for ecosystem management and conservation. Heterogeneity creates mosaics of sources and sinks (see Chapter 21; Pulliam 1988). *Sources* are good habitats where species are self-supporting. Surplus animals emigrate from these to *sinks*, which are poorer habitats that are not self-supporting. However, sinks provide a vital role in allowing nonbreeding individuals to survive while they wait for opportunities to obtain territories in source habitats. Sinks provide compensation in a population for unpredictable disturbances that reduce breeding populations. Protected areas, therefore, should contain both source and sink habitats.

22.10.2 *Disturbance and ecosystem management*

Disturbances that are too frequent or too extreme can radically alter an ecosystem, changing it to a different state. Persistent overgrazing can result in denudation, as mentioned earlier for semi-arid areas (Wu *et al.* 2000), or in a change from grassland to woodland, as in savanna areas (Walker *et al.* 1981). Other forms of human overdisturbance are often the underlying cause of invasions of exotic species, which can take over and maintain new states (Vitousek *et al.* 1996).

More rarely, single extreme natural disturbances can change ecosystem states: the "Wahine" storm of 1968 destroyed the beds of aquatic macrophytes in Lake Ellesmere on the South Island of New Zealand, and the physical change in the lake sediments has prevented their return. The resident population of black swans (*Cygnus atratus*) numbered 40000–80000 in the 1950s and 1960s. They used the weed beds for food and to raise young. Mortality from the storm itself, from starvation, and from reduced breeding due to a lack of suitable habitat rapidly reduced the population to < 10000 in subsequent years and they have never returned to their original numbers (Williams 1979; A. Byrom pers. comm.).

Earthquakes and their resulting tsunamis are another form of disturbance that can cause sudden effects in an ecosystem with long-term consequences. Botswana in southern Africa is so flat that even minor tectonic shifts can change the direction of river flows (Cooke 1980; Shaw 1985). The Savuti River is a side channel of the Kwando/Linyanti river system that flows into the Zambezi. The Savuti was dry from the 1860s to the 1950s. In 1956 the channel started flowing due to a tectonic shift, and it continued to flow, with floods in 1979, until 1983, when just as suddenly it stopped. The channel has been dry since, and now both the Kwando and Linyanti are drying, exacerbated by climate change and human use. The Savuti was used extensively by large numbers of ungulate species in the dry season. When the river dried, it altered the ecology of a large region, with ungulate migration patterns changing to other areas on the Linyanti (M. Vanderwalle pers. comm.). In New Zealand, earthquake disturbances are relatively frequent. Their ecosystem effects are of a lower intensity but can be widespread. Landslides resulting from earthquakes cause sudden catastrophic mortality of forest trees that is scale-dependent and has important effects on forest dynamics: multiple small patches have high mortality, resulting in a mosaic of different–aged stands in the forest (Allen *et al.* 1999).

Climatic and temperature changes in physical oceanic conditions, as occurred suddenly in the North Pacific in the late 1970s, appear to have long-term ecosystem consequences. In the North Pacific the complex of fish species changed after the regime shift, and it appears the new fish community cannot provide sufficient quality food for the Steller's sea lion (*Eurometopias jubatus*). In the mid-1970s this sea lion population had been increasing and was around 250 000. It dropped rapidly to 100 000 by 1990 and to 50 000 by 2000 (Trites and Donnelly 2003). It is possible, though not yet established, that killer whales (*Orcinus orca*) are exacerbating the decline through predation on sea lions at these low numbers (Springer *et al.* 2003).

Conservation of ecosystems, therefore, has to be sufficiently flexible to accommodate major natural disturbances such as earthquakes, fires, floods, and storms, and to allow recovery from human disturbances such as overgrazing and overharvesting. Such approaches will require a better understanding not only of the impacts of disturbances but also of the temporal and spatial scales at which those disturbances operate.

22.11 Ecosystem management at multiple scales

22.11.1 *Management at large spatial scales*

Ecosystems function at multiple scales, with small scales affecting large scales and vice versa. The migration of Serengeti wildebeest covers the entire ecosystem of 25 000 km^2. Wildebeest move several hundred kilometers to the short grass plains in the wet season because these plains support the most nutritious grasses in the system (Holdo *et al.* 2009). Dung beetles, of which there are some 80 species, rapidly bury feces (within a few minutes) and hence expedite nutrient cycling on these plains. They promote the high-quality nutrition of the grasses, producing a positive feedback. Dung beetles can only function when the soil is damp, so they have a negligible effect on returning nutrients to the soil in the dry season when wildebeest are in woodland areas. Thus, the very local-scale functions of the beetles influence the large-scale movements of the ungulates.

The recent collapse of the Canadian arctic grazing ecosystem has occurred through subsidies to snow geese (*Chen caerulescens*) on winter feeding grounds as far away as the southern United States, resulting in overpopulation, overgrazing, and a new ecosystem state in the Arctic (Jefferies *et al.* 2004).

Although population declines can often be attributed to immediate proximate causes such as predation and habitat loss, ultimately large-scale remote causes may underlie these events. Such fundamental causes only become apparent when the large-scale ecosystem is considered. Thus, in North America the brown-headed cowbird (*Molothrus ater*) is a nest parasite of many small passerines. Its population is increasing and it has caused the decline of at least two species, the least Bell's vireo (*Vireo bellii*) and the black-capped vireo (*V. atricapillus*) (Smith *et al.* 2000). The ultimate cause is events that date back to the 1800s (Rothstein 1994). This cowbird's original range lies in forests of the American northeast, where it prefers open areas for foraging. Expansion of open land through agriculture across North America has allowed this species to spread into new areas and parasitize species that have few adaptive traits to counter it. Both large-scale and long-term events underlie the spread of this nest parasite.

22.11.2
Management at long timescales

Large-scale temporal patterns are also important in ecosystem dynamics. The classic snowshoe hare (*Lepus americanus*) cycle of North America is synchronized spatially by decadal weather events (Stenseth *et al.* 2002). This cycle of numbers then influences the rest of the ecosystem (Krebs *et al.* 2001b). Synchrony is enhanced by environmental correlation across sites and reduced by dispersal between sites (Kendall *et al.* 2000). There is increasing evidence that complex ecosystem processes are influenced by multi-year fluctuations in climate such as the North Atlantic Oscillation and the Southern and Northern Pacific Oscillations (Post and Stenseth 1998; Coulson *et al.* 2001a). In the context of long-term conservation, spatial and temporal synchronies are of particular concern when a species is rare, because of its increased probability of extinction.

Human-induced changes in climate can be considered a very long-term, persistent disturbance. Global climate change is now having measurable effects on ecosystems, altering community composition by shifting species ranges differentially towards the poles, upwards in altitude (especially in the tropics; Pounds *et al.* 1999), and away from the tropics (Schneider and Root 2002; Parmesan and Yohe 2003). In Britain, birds and other groups are breeding earlier. Changes in community structure mean that species will experience changes in food supply and predation rates (Crick *et al.* 1997).

In general, ecosystem processes at different spatial and temporal scales, including disturbance and long-term trends, must be considered as part of any conservation strategy for the ecosystem in which a threatened species exists. Such processes operate at larger scales than we have traditionally planned for. Moreover, conservation is no longer a short-term exercise: we have to consider timescales of a hundred years or more.

22.12 Biodiversity

22.12.1
Determinants of diversity

Biodiversity is defined as the complement of living organisms in an ecosystem. In terrestrial systems there is a general tendency for animal groups to be more species-diverse in tropical latitudes (MacArthur 1972). Fig. 22.2 shows such a distribution for birds. There are local anomalies in this pattern; for example, for reptiles in Australia 1982 (Schall and Pianka 1978).

Many hypotheses have been offered to explain this latitudinal gradient in biodiversity from tropics to poles. These include:

1 *Structural heterogeneity* Tropical vegetation is more complex structurally, providing more niches for animals compared to arctic tundra or even boreal forest.
2 *Community age* Ice caps once covered large areas of northern North America and Eurasia, and these melted only 10 000 years ago. There has been insufficient time to reinvade these higher latitudes and evolve new species. This process of eradication and replacement has been repeated many times in the Pliocene and Pleistocene, and probably earlier, while the tropics have been extant throughout.
3 *Productivity* The warmer temperatures and higher rainfall of tropical regions allows species to fit into narrow niches over small areas while still maintaining a large enough population to avoid extinction. In contrast, low productivity at high latitudes

Fig. 22.2 The number of terrestrial bird species in North America declines from the tropics to the Arctic. (Data from MacArthur 1972.)

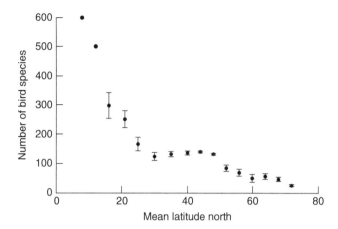

means that fewer species must maintain broad niches over wide areas to maintain the same population size.

4 *Predation*   The higher biodiversity of the tropics includes a higher number of predators. These impose a top-down regulation on prey species (see Section 22.8.4), which allows a greater number of prey species to coexist because they are not competing with one another. This process is called *predator-mediated coexistence*.

5 *Environmental stability*   High latitudes fluctuate considerably between summer and winter in environmental parameters such as temperature. Large populations are required to withstand such fluctuations and this means few species can live there. Stable environments in the tropics allow smaller populations to survive. These small populations can have smaller niches and so more species can fit into the system.

6 *Intermediate disturbance*   This is a variation of **5**. While agreeing that higher latitudes experience frequent major disturbances and so support fewer species, it differs in considering that the tropics have some smaller disturbances (hence intermediate between few and many) that both allow early succession and enable highly competitive species to coexist (Connell 1978).

In summary, no single hypothesis explains all of the observed distributions. It is likely that different hypotheses apply in different locations, and also that more than one process occurs at any one location.

22.12.2 *Local and regional diversity*

The total diversity within a large area – a region – is called the *gamma* diversity. This is determined by two components: (i) *alpha* diversity, which is the number of species in a local area or habitat; and (ii) *beta* diversity, which is the reciprocal of the mean number of habitats or localities occupied by a species. Thus:

gamma diversity = average diversity per habitat(alpha)

    × 1/mean number of habitats occupied by a species(beta)

    × total number of habitats

This allows us to differentiate between the separate aspects of species (habitat breadth through beta diversity) and the regional heterogeneity through the number of available habitats. These values allow a mechanism for comparing different communities (Schluter and Ricklefs 1993).

If a species lives in a local area or habitat independently of any other species that is present then one would expect that the more species there are in the surrounding region, the more there will be in the local area. Thus, there should be a linear relationship between species richness locally and that on the larger regional scale. At face value, such an assumption may seem unlikely, since we know there are many interactions between species. However, a linear relationship could occur whereby a young community that is still evolving does not yet have its full complement of species, so that local-area richness reflects that in the wider region; biotic interactions may not have come fully into play. Alternatively, a local patch that receives a large number of immigrants (such as one with many sink populations) relative to the competitive abilities of the residents will also reflect the richness of the region that produces the dispersers: dispersing species overwhelm the residents. This linear relationship has been termed an *unsaturated* pattern.

In contrast, if there are strong biotic interactions between species locally, such that many species are excluded from the community, one would expect that after an initial colonization period a limit to the number of species will occur locally, irrespective of the number available in the region. The curve of local versus regional species richness will have an upper limit. This has been called the *saturated* pattern, because no more species can be added to the local area (Srivastava 1999; Hillebrand and Blenckner 2002).

Processes such as dispersal and interspecific competition for space underlie these patterns, which were initially used to infer the mechanism. However, several different processes can produce either pattern (Chave *et al.* 2002; Shurin and Srivastava 2005). For example, facilitation through a keystone predator can increase local richness, the amount dependent on the regional pool of species. This process could override the saturating pattern of strong interspecific competition.

Another problem arises when one tries to identify local and regional scales. Shurin and Srivastava (2005) show that the pattern changes from saturated to unsaturated as the ratio of local to regional increases. Thus, the pattern depends on the scale of the study. Many saturated communities have been overlooked because the local scale was too large. In general, such patterns of biodiversity provide only weak evidence for the underlying mechanisms structuring that diversity.

22.13 Island biogeography and dynamic processes of diversity

MacArthur and Wilson (1967) and MacArthur (1972) proposed that the diversity of species in discreet ecosystems can be considered as a dynamic equilibrium, a balance between the rate of immigrating species and the rate of local extinction of species. They chose islands to illustrate this principle (Fig.22.3). Starting with a newly formed island, the rate of immigration of new species declines as the full complement from the source (the mainland) is achieved. However, as species accumulate on the island, some go extinct, and the probability of extinction increases as more species arrive. Thus, where the two rates are equal we achieve the equilibrium of species number S for that island.

The immigration rate must be determined by the *distance* the island lies from the mainland, so a distant island will have a lower immigration rate than a near island. Thus, for islands of equal size the expected number of species for distant islands S_{far} should be lower than that for near islands S_{near}.

The extinction rate is determined by several factors, such as the size of the population and the number of competing and predatory species. All of these are related to

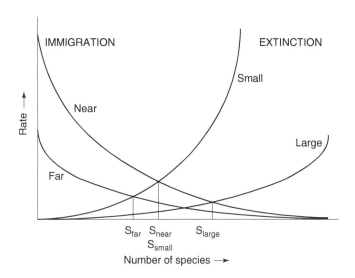

Fig. 22.3 Hypothetical relationship between the rates of immigration or extinction of species and the number of species in a habitat patch or island. Where the immigration and extinction curves cross we obtain an equilibrium number of species, S. For a given extinction rate, islands closer to the source of immigrants have a higher number (S_{near}) than islands further away (S_{far}). For a given immigration rate, larger islands have a higher number of species (S_{large}) than smaller islands (S_{small}).

the area of the ecosystem, such that islands of larger *size* should have lower extinction rates. Thus, the expected number of species for a large island S_{large} should be greater than that for a small island S_{small}. This leads to the classic species–area equation that we explored in Section 17.5.3.

Some classic studies provide circumstantial evidence for this *theory of island biogeography*. Diamond (1969) showed that bird species on the California Channel Islands maintained similar numbers of species over a 50-year period. There was a turnover of species because some went extinct in the interval and others arrived to replace them. Simberloff and Wilson (1970) showed that small mangrove islands, artificially denuded of their invertebrates, regained a similar number of species as occurred before the removal, but the types differed to some extent. These results concur with the theory.

We should recognize that this is an idealized concept and there are several factors that could distort it. The complexity of the habitat is one (Lack 1971), but several of the other factors determining biodiversity that we discussed in Section 22.12.1 also apply. The important point is to understand that communities are dynamic, with species coming and going, and not to place too much emphasis on looking for an equilibrium.

One important prediction from this theory concerns what happens when a piece of a large mainland ecosystem is suddenly isolated. Isolation means that the immigration rate should be reduced and hence the species number should decline to a lower value. This process, called *faunal relaxation* (Diamond 1972), was documented for islands that were isolated from the mainland by the 100 m rise in seawater level after the last ice age, 10 000 years ago. Examples of such islands, called *landbridge islands*, are found in the Malay Archipelago (the Sunda Islands), among others (MacArthur 1972).

The process of faunal relaxation has important sequences for conservation. Protected areas (National Parks and other reserves) are often small islands carved out of much larger ecosystems. These are formed when habitat surrounding them is altered in such a way that many species find it difficult or impossible to cross, particularly large mammals and sedentary birds, reptiles, and amphibians. This creates a natural barrier. A park should lose species over time, with the rate of loss higher and the eventual total number of losses lower for smaller parks. There is some evidence that this is already occurring (Newmark 1987).

22.14 **Ecosystem function**

Ecosystem function is a general term that covers three different components: processes, products or "goods," and services. Box 22.1 lists some examples of each. Processes involve rates of flow of nutrients, growth of biomass, and so on. Services are processes that are of benefit to human society, such as pollution control. Goods are the end result of these, such as clean water (Schulze and Mooney 1993; Mooney and Ehrlich 1997; Kinzig *et al.* 2001; Loreau *et al.* 2002; Srivastava and Vellend 2005).

Ecosystem function is often linked to biodiversity (Schulze and Mooney 1993). The argument is that the more species in a system, the greater its ability to withstand shocks (Walker 1992, 1995; Naeem 1998; Tilman 1999; Loreau 2000). This can be illustrated in two ways, as in Fig. 22.4.

First, ecosystem function can be related to the number of species linearly if all species make equal contribution to the function (e.g. productivity) and act independently (Fig. 22.4a). However, we know such equality of function is unlikely to occur, due to dominance and keystone effects. If we lose species from a system, we are unlikely initially to see much reduction in function, for two reasons. One is that the species lost might be minor players in the system, and a drop in function will not be observed until a keystone or dominant is lost. Thus, a drop in function can occur anywhere in the sequence (even at the beginning) depending on when the important species are lost. A second reason is that in a diverse community the first few species to be lost can be replaced by others, which take over their function. This involves the idea that there are redundant species in a system (Walker 1992, 1995). An indication of this effect is seen in the progressive loss of fish species due to human harvesting of the oceans. Initially, productivity of the fisheries was maintained through the replacement of lost species by others, which increased. However, after several decades so many species were lost that the remaining ones were unable to replace them and fish harvests collapsed (Jackson *et al.* 2001).

Box 22.1 Ecosystem function can be divided into three categories.

Ecosystem processes

1 Hydrology
2 Biological productivity
3 Biogeographical cycling and storage
4 Decomposition
5 Resilience
6 Robustness–fragility

Ecosystem services

1 Maintaining hydrological cycles
2 Regulating climate
3 Purifying water and air
4 Pollinating crops and natural systems
5 Generating and maintaining soils
6 Cycling nutrients
7 Detoxifying pollutants
8 Providing aesthetic quality
9 Providing baseline research

Ecosystem goods

1 Food and clean water
2 Construction materials
3 Medicinal plants
4 Wild genes
5 Replacement species
6 Biological control agents
7 Tourism

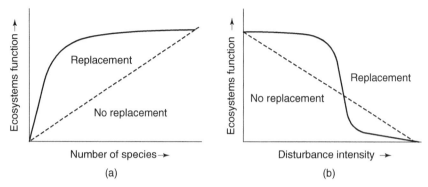

Fig. 22.4 (a) Hypothetical relationship between ecosystem function and species number. If lost species can be replaced by others then system function follows the solid line. If there is no replacement then function follows the broken line. (b) Ecosystem function relative to the degree of disturbance that can deplete species diversity. Low disturbance (left on x-axis) is tolerated because species can replace each other. Higher disturbance can cause a rapid decline in function when redundanc of species is used up.

The second way of looking at ecosystem function is to consider how it changes with degree of disturbance (Fig. 22.4b). With minor disturbances, some species can compensate for lost ones and maintain the function. However, at some point the disturbance is so large that there are insufficient species to replace any lost ones.

So far, most of the evidence for these concepts comes from plants and the stability of productivity and nutrient cycles (Vitousek and Hooper 1993; Naeem 1998, 2002; Naeem and Li 1997; Naeem *et al.* 1995).

The distortion of ecosystem processes can lead to unwanted ecosystem effects, conservation threats to individual species, and expensive ecosystem management. For example, before European settlement Australia was largely covered in eucalypt woodland. In the past century agriculture has removed nearly all of this woodland, especially in Western Australia. Originally, eucalypts kept groundwater levels down through transpiration processes. Once the trees were removed, groundwater levels rose, water evaporated at the soil surface, and saline deposits made the soil unsuitable not only for crops but also for native biota. Now large areas of Australia have a major problem with salinization of soil and groundwater upwelling, with a resultant decline in agricultural productivity. In response to this ecological (and economic) problem, Australia has had to adopt the expensive policy of revegetation (Nulsen 1993; McFarlane *et al.* 1993).

Disturbed systems provide evidence for the function of biota in creating resilience to disturbance. Thus, intact eucalypt forest in Australia supports many species of coexisting endemic honeyeaters. In contrast, the noisy miner (*Manorina melanocephala*), the geographical replacement for the bell miner mentioned in Section 22.6.1, dominates the bird community in fragmented forest in which an open canopy has developed through a combination of logging, persistent livestock grazing, and agriculture. The noisy miner aggressively excludes most of the smaller honeyeaters. Consequently, exposed trees suffer chronic infestations of psyllids, die-back of their main limbs, and death. In essence, exposed trees are no longer viable due to the dearth of insectivorous birds (Landsberg 1988; Grey *et al.* 1997, 1998).

We still need to evaluate whether it is the biodiversity itself (richness) or the composition of the community that is important; that is, whether who is present is as important as how many species (Srivastava and Vellend 2005). Much still needs to be done to determine whether biodiversity alters the processes within ecosystems, and if so what the mechanisms are. Nevertheless, evidence is accumulating to suggest that where ecosystem processes have been ignored or distorted they can lead to perverse effects for both ecosystems and individual species and to expensive remedial action for conservation (Schwartz *et al.* 2000, Hector *et al.* 2001).

22.15 Summary

Management of populations has to take into account that they are embedded within a matrix of other competitors, predators, and prey. These form the community and their environments. Thus, we need to consider the management of individuals, populations, and species in the context of the ecosystem. The main points are:

1 Ecosystems involve long-term events related to the environment. Management needs to account for infrequent and unpredictable events, such as earthquakes, hurricanes, floods, fire, and droughts.

2 These events provide insight into the mechanisms of ecosystem regulation and stability. The term "long-term" is a function of the slowest variable in the system and is not related to the life history of the organisms concerned. Management should maintain these regimes either naturally or by mimicking their effects. Planning should consider natural periodicities from the very long term (200 years for earthquake and fire cycles) to the short term of a few years for the effects of El Niño Southern Oscillation and the North Atlantic Oscillation (e.g. Post and Stenseth 1998; Coulson *et al.* 2001a). Thus, the periods of both unpredictable, sudden events and slow change dictate the timescale for conservation planning. In most cases, planning should be for 30–50-year periods, or sometimes even longer.

3 Ecosystem management needs to consider that slow trends occur through environmental change, plant succession, and animal population fluctuation. These slow trends show an interaction between abiotic and biotic processes, each affecting the other.

4 Ecosystems should be managed at an appropriate spatial scale. Small patches of forest are insufficient to support viable populations of predators such as northern spotted owls (*Strix occidentalis*). Large areas are required for migrating ungulates moving between summer and winter ranges. Sufficient area is required to produce the mosaic of burns of different ages. This mosaic creates habitat heterogeneity, which provides animal sources and sinks. Sinks are required as holding areas for nonbreeding animals waiting to obtain territories.

5 Management of target populations, such as pest species, can result in indirect interactions through hyperpredation, apparent competition, and mesopredator release. These can produce unexpected consequences.

6 Slow change can become an irreversible rapid shift into a new state. Thus, management needs to consider that an ecosystem can occur in multiple states. Some of these can be natural but others can be artifacts of human disturbance.

7 Ecosystems are not static, so management cannot aim to maintain the status quo, but rather should allow natural change to take place. It is likely that such change is oscillatory, in that previous conditions will be reverted to after a time.

8 Within protected areas, management should distinguish between natural change and direct human-induced change. Protected areas can act as ecological baselines where human-induced change is kept to a minimum, and the system can then be compared to areas outside that are influenced by human activity.

9 Long-term baseline data are fundamental to conservation management, because they provide the background against which to interpret causes of change and hence determine the course of management.

Appendices

Wildlife Ecology, Conservation, and Management, Third Edition.
John M. Fryxell, Anthony R. E. Sinclair and Graeme Caughley.
© 2014 John Wiley & Sons, Ltd. Published 2014 by John Wiley & Sons, Ltd.
Companion Website: www.wiley.com/go/Fryxell/Wildlife

Appendix 1 Cumulative F distribution. The body of the table contains critical values of F for both "1-sided" and "2-sided" significance probabilities.

d.f. in denominator	"1-side" tests	Degrees of freedom in numerator																"2-sided" tests
1	.10	39.9	49.5	53.6	55.8	57.2	58.2	58.9	59.4	59.9	60.2	60.7	61.2	61.7	62.3	62.8	63.3	.20
	.05	161	200	216	225	230	234	237	239	241	242	244	246	248	250	252	254	.10
	.01	4050	5000	5400	5620	5760	5860	5930	5980	6020	6060	6110	6160	6210	6260	6310	6370	.02
2	.10	8.53	9.00	9.16	9.24	9.29	9.33	9.35	9.37	9.38	9.39	9.41	9.42	9.44	9.46	9.47	9.49	.20
	.05	18.5	19.0	19.2	19.2	19.3	19.3	19.4	19.4	19.4	19.4	19.4	19.4	19.5	19.5	19.5	19.5	.10
	.01	98.5	99.0	99.2	99.2	99.3	99.3	99.4	99.4	99.4	99.4	99.4	99.4	99.4	99.5	99.5	99.5	.02
3	.10	5.54	5.46	5.39	5.34	5.31	5.28	5.27	5.25	5.24	5.23	5.22	5.20	5.18	5.17	5.15	5.13	.20
	.05	10.1	9.55	9.28	9.12	9.01	8.94	8.89	8.85	8.81	8.79	8.74	8.70	8.66	8.62	8.57	8.53	.10
	.01	34.1	30.8	29.5	28.7	28.2	27.9	27.7	27.5	27.3	27.2	27.1	26.9	26.7	26.5	26.3	26.1	.02
4	.10	4.54	4.32	4.19	4.11	4.05	4.01	3.98	3.95	3.94	3.92	3.90	3.87	3.84	3.82	3.79	3.76	.20
	.05	7.71	6.94	6.59	6.39	6.26	6.16	6.09	6.04	6.00	5.96	5.91	5.86	5.80	5.75	5.69	5.63	.10
	.01	21.2	18.0	16.7	16.0	15.5	15.2	15.0	14.8	14.7	14.5	14.4	14.2	14.0	13.8	13.7	13.5	.02
5	.10	4.06	3.78	3.62	3.52	3.45	3.40	3.37	3.34	3.32	3.30	3.27	3.24	3.21	3.17	3.14	3.10	.20
	.05	6.61	5.79	5.41	5.19	5.05	4.95	4.88	4.82	4.77	4.74	4.68	4.62	4.56	4.50	4.43	4.37	.10
	.01	16.3	13.3	12.1	11.4	11.0	10.7	10.5	10.3	10.2	10.1	9.89	9.72	9.55	9.38	9.20	9.02	.02
6	.10	3.78	3.46	3.29	3.18	3.11	3.05	3.01	2.98	2.96	2.94	2.90	2.87	2.84	2.80	2.76	2.72	.20
	.05	5.99	5.14	4.76	4.53	4.39	4.28	4.21	4.15	4.10	4.06	4.00	3.94	3.87	3.81	3.74	3.67	.10
	.01	13.7	10.9	9.78	9.15	8.75	8.47	8.26	8.10	7.98	7.87	7.72	7.56	7.40	7.23	7.06	6.88	.02
7	.10	3.59	3.26	3.07	2.96	2.88	2.83	2.78	2.75	2.72	2.70	2.67	2.63	2.59	2.56	2.51	2.47	.20
	.05	5.59	4.74	4.35	4.12	3.97	3.87	3.79	3.73	3.68	3.64	3.57	3.51	3.44	3.38	3.30	3.23	.10
	.01	12.2	9.55	8.45	7.85	7.46	7.19	6.99	6.84	6.72	6.62	6.47	6.31	6.16	5.99	5.82	5.65	.02
8	.10	3.46	3.11	2.92	2.81	2.73	2.67	2.62	2.59	2.56	2.54	2.50	2.46	2.42	2.38	2.34	2.29	.20
	.05	5.32	4.46	4.07	3.84	3.69	3.58	3.50	3.44	3.39	3.35	3.28	3.22	3.15	3.08	3.01	2.93	.10
	.01	11.3	8.65	7.59	7.01	6.63	6.37	6.18	6.03	5.91	5.81	5.67	5.52	5.36	5.20	5.03	4.86	.02

df	α																		α
9	.10	3.36	3.01	2.81	2.69	2.61	2.55	2.51	2.47	2.44	2.42	2.38	2.34	2.30	2.25	2.21	2.16	.20
	.05	5.12	4.26	3.86	3.63	3.48	3.37	3.29	3.23	3.18	3.14	3.07	3.01	2.94	2.86	2.79	2.71	.10
	.01	10.6	8.02	6.99	6.42	6.06	5.80	5.61	5.47	5.35	5.26	5.11	4.96	4.81	4.65	4.48	4.31	.02
10	.10	3.28	2.92	2.73	2.61	2.52	2.46	2.41	2.38	2.35	2.32	2.28	2.24	2.20	2.16	2.11	2.06	.20
	.05	4.96	4.10	3.71	3.48	3.33	3.22	3.14	3.07	3.02	2.98	2.91	2.84	2.77	2.70	2.62	2.54	.10
	.01	10.0	7.56	6.55	5.99	5.64	5.39	5.20	5.06	4.94	4.85	4.71	4.56	4.41	4.25	4.08	3.91	.02
12	.10	3.18	2.81	2.61	2.48	2.39	2.33	2.28	2.24	2.21	2.19	2.15	2.10	2.06	2.01	1.96	1.90	.20
	.05	4.75	3.89	3.49	3.26	3.11	3.00	2.91	2.85	2.80	2.75	2.69	2.62	2.54	2.47	2.38	2.30	.10
	.01	9.33	6.93	5.95	5.41	5.06	4.82	4.64	4.50	4.39	4.30	4.16	4.01	3.86	3.70	3.54	3.36	.02
15	.10	3.07	2.70	2.49	2.36	2.27	2.21	2.16	2.12	2.09	2.06	2.02	1.97	1.92	1.87	1.82	1.76	.20
	.05	4.54	3.68	3.29	3.06	2.90	2.79	2.71	2.64	2.59	2.54	2.48	2.40	2.33	2.25	2.16	2.07	.10
	.01	8.68	6.36	5.42	4.89	4.56	4.32	4.14	4.00	3.89	3.80	3.67	3.52	3.37	3.21	3.05	2.87	.02
20	.10	2.97	2.59	2.38	2.25	2.16	2.09	2.04	2.00	1.96	1.94	1.89	1.84	1.79	1.74	1.68	1.61	.20
	.05	4.35	3.49	3.10	2.87	2.71	2.60	2.51	2.45	2.39	2.35	2.28	2.20	2.12	2.04	1.95	1.84	.10
	.01	8.10	5.85	4.94	4.43	4.10	3.87	3.70	3.56	3.46	3.37	3.23	3.09	2.94	2.78	2.61	2.42	.02
30	.10	2.88	2.49	2.28	2.14	2.05	1.98	1.93	1.88	1.85	1.82	1.77	1.72	1.67	1.61	1.54	1.46	.20
	.05	4.17	3.32	2.92	2.69	2.53	2.42	2.33	2.27	2.21	2.16	2.09	2.01	1.93	1.84	1.74	1.62	.10
	.01	7.56	5.39	4.51	4.02	3.70	3.47	3.30	3.17	3.07	2.98	2.84	2.70	2.55	2.39	2.21	2.01	.02
60	.10	2.79	2.39	2.18	2.04	1.95	1.87	1.82	1.77	1.74	1.71	1.66	1.60	1.54	1.48	1.40	1.29	.20
	.05	4.00	3.15	2.76	2.53	2.37	2.25	2.17	2.10	2.04	1.99	1.92	1.84	1.75	1.65	1.53	1.39	.10
	.01	7.08	4.98	4.13	3.65	3.34	3.12	2.95	2.82	2.72	2.63	2.50	2.35	2.20	2.03	1.84	1.60	.02
120	.10	2.75	2.35	2.13	1.99	1.90	1.82	1.77	1.72	1.68	1.65	1.60	1.54	1.48	1.41	1.32	1.19	.20
	.05	3.92	3.07	2.68	2.45	2.29	2.18	2.09	2.02	1.96	1.91	1.83	1.75	1.66	1.55	1.43	1.25	.10
	.01	6.85	4.79	3.95	3.48	3.17	2.96	2.79	2.66	2.56	2.47	2.34	2.19	2.03	1.86	1.66	1.38	.02
—	.10	2.71	2.30	2.08	1.94	1.85	1.77	1.72	1.67	1.63	1.60	1.55	1.49	1.42	1.34	1.24	1.00	.20
	.05	3.84	3.00	2.60	2.37	2.21	2.10	2.01	1.94	1.88	1.83	1.75	1.67	1.57	1.46	1.32	1.00	.10
	.01	6.63	4.61	3.78	3.32	3.02	2.80	2.64	2.51	2.41	2.32	2.18	2.04	1.88	1.70	1.47	1.00	.02

Appendix 2 Critical 5% values for Cochran's test for homogeneity of variance.

df for s_j^2	2	3	4	5	6	7	8	9	10	15	20
1	.9985	.9669	.9065	.8412	.7808	.7271	.6798	.6385	.6020	.4709	.3894
2	.9750	.8709	.7679	.6838	.6161	.5612	.5157	.4775	.4450	.3346	.2705
3	.9392	.7977	.6841	.5981	.5321	.4800	.4377	.4027	.3733	.2758	.2205
4	.9057	.7457	.6287	.5441	.4803	.4307	.3910	.3584	.3311	.2419	.1921
5	.8772	.7071	.5895	.5065	.4447	.3974	.3595	.3286	.3029	.2195	.1735
6	.8534	.6771	.5598	.4783	.4184	.3726	.3362	.3067	.2823	.2034	.1602
7	.8332	.6530	.5365	.4564	.3980	.3535	.3185	.2901	.2666	.1911	.1501
8	.8159	.6333	.5175	.4387	.3817	.3384	.3043	.2768	.2541	.1815	.1422
9	.8010	.6167	.5017	.4241	.3682	.3259	.2926	.2659	.2439	.1736	.1357
10	.7880	.6025	.4884	.4118	.3568	.3154	.2829	.2568	.2353	.1671	.1303

$C = (\text{largest } s^2)/(\Sigma s_j^2)$, k = number of variances.

Glossary

Adapted from Watt *et al.* (1995), Ricklefs and Miller (2000) and Krebs (2001)

Abiotic factors Characterized by the absence of life; include temperature, humidity, pH, and other physical and chemical influences.

Adaptation A genetically determined characteristic that enhances the ability of an individual to cope with its environment; an evolutionary process by which organisms become better suited to their environments.

Adaptive radiation The evolution of ecological and phenotypic diversity within a rapidly multiplying lineage. It is the differentiation of a single ancestor into an array of species that inhabit a variety of environments and which differ in the morphological, physiological, and behavioral traits used to exploit those environments.

Age class The individuals in a population of a particular age.

Age structure The relative proportions of a population in different age classes.

Aggregation Organisms show an aggregated spatial distribution when they co-occur significantly more than would be expected from a (completely random) *Poisson distribution*. This clumping is reflected in a variance mean ratio significantly greater than unity. *Macroparasites* are usually aggregated in their host population, the majority of hosts harboring a few or no parasites and a few harboring large parasite burdens. Aggregated distributions are often well described empirically by the *negative binomial distribution*; the degree of aggregation is inversely proportional to the negative binomial parameter, k.

Allele One of a pair of characters that are alternative to each other in inheritance, being governed by genes situated at the same locus in homologous chromosomes.

Allochthonous Originating outside of a system, such as minerals and organic matter transported from marine to terrestrial habitats or from land into streams and lakes. (Compare with *Autochthonous*.)

Allopatric Occurring in different places; usually referring to geographic separation of populations. (See *Sympatric*.)

Alpha diversity The mean variety of organisms occurring in a particular place or habitat; often called local diversity.

Ambient Conditions of the *abiotic* environment surrounding an organism.

Antibody A protein produced in the blood of vertebrates in response to an *antigen*. The antibody produced is able to bind specifically to that antigen, and plays a role in its inactivation or removal by the *immune system*.

Antigen A substance, generally foreign, capable of inducing *antibody* formation.

Wildlife Ecology, Conservation, and Management, Third Edition.
John M. Fryxell, Anthony R. E. Sinclair and Graeme Caughley.
© 2014 John Wiley & Sons, Ltd. Published 2014 by John Wiley & Sons, Ltd.
Companion Website: www.wiley.com/go/Fryxell/Wildlife

Apparent competition   A situation in which two or more species negatively affect one another indirectly through their interaction with a common predator.

Assimilation efficiency   A percentage expressing the proportion of ingested energy that is absorbed into the bloodstream.

Autecology   Study of the individual in relation to environmental conditions.

Autochthonous   Originating within a system, such as organic matter produced or minerals cycled within streams and lakes. (Compare with *Allochthonous*.)

Autotroph   An organism that obtains energy from the sun and materials from inorganic sources. Most plants are autotrophs. (Contrast with *Heterotroph*.)

Basal metabolic rate (BMR)   The energy expenditure of an organism that is at rest, fasting, and in a thermally neutral environment.

Beta diversity   The reciprocal of the mean number of habitats or localities occupied by a species. Also defined as the proportional difference in species composition between habitats; that is, the turnover of species among habitats.

Biodiversity   The variety of types of organisms, habitats, and ecosystems on earth or in a particular place.

Biomass   Weight of living material, usually expressed as a dry weight, in all or part of an organism, population, or community. Commonly presented as weight per unit of area, or biomass density.

Biome   A major category of ecological communities (e.g. tundra biome.).

Biosphere   The whole-earth ecosystem, also called the *ecosphere*. Divided into *biomes*.

Biota   All species living in a defined area.

Biotic factors   Environmental influences caused by plants or animals; opposite of *abiotic factors*.

Birth rate (b_x)   The average number of offspring produced per individual per unit of time, often expressed as a function of age (x).

Browsers   Organisms that consume parts of woody plants. (See *Grazers*.)

By-catch   The incidental capture of prey by predators or humans whose efforts are dependent on, or focus on, other more abundant prey.

Carnivore   Flesh eater; an organism that eats other animals. (Contrast with *Herbivore*.)

Carrying capacity   The number of individuals in a population that the resources of a habitat can support; the asymptote, or plateau, of the logistic and other sigmoid equations for population growth.

Chromosomes   Rodlike structures in eukaryotic cells on which genes reside.

Climax   The end point of a successional sequence, or sere; a community that has reached a steady state under a particular set of environmental conditions.

Cline   A gradual change in population characteristics or adaptations over a geographic area.

Coexistence   Occurrence of two or more species in the same habitat; usually applied to potentially competing species.

Cohort   A group of individuals of the same age recruited into a population at the same time.

Community   An association of interacting populations, usually defined by the nature of their interaction or by the place in which they live.

Competition   Occurs when a number of organisms of the same or of different species utilize common resources that are in short supply (*exploitation*); if the resources are not in short supply, competition occurs when the organisms seeking that resource harm one another in the process (*interference*).

Competitive exclusion principle The hypothesis that two or more species cannot coexist on a single resource that is scarce relative to the demand for it. Also called *Gause's principle*.

Consumer–resource interactions Interactions in which individuals of one species consume individuals of another. Consumer–resource interactions affect the consumer positively and the resource negatively.

Contact rate The average frequency per unit time with which infected individuals contact, or otherwise put themselves in a position to transmit an *infection* to, *susceptible* individuals.

Contagious disease See *Infectious disease*.

Death rate (d_x) The percentage of newborns dying during a specified interval, often expressed as a function of age (x). (Compare with *Mortality*.)

Deme A local population within which mating occurs among individuals more or less at random.

Demographic stochasticity Random variation in birth and death rates in a population.

Demography The study of population structure and growth.

Density The number of individuals relative to the space, volume, or other resources that they need.

Density dependent Having an influence on individuals in a population that varies with the density of that population. Often applied to birth and death rates.

Density independent Having an influence on individuals in a population that does not vary with the density of that population.

Depensatory See *Inverse density dependent*.

Deterministic Having an outcome that is not subject to stochastic (random) variation.

Deterministic model Mathematical model in which all relationships are fixed and the concept of probability does not play a part: a given input produces an exact prediction as output. (Opposite of *Stochastic model*.)

Detritus Freshly dead or partially decomposed organic matter.

Direct interaction An interaction between organisms that involves direct physical contact between the interactors (e.g. predation and herbivory). (Compare with *Indirect interaction*.)

Disease The debilitating effects of *infection* by a *parasite*; sometimes incorrectly used to refer to the disease-causing parasite. It is possible for a host to be infected by a parasite but to show no *symptoms* of disease.

Dispersal Movement of organisms away from their place of birth or from centers of population density.

Dispersion The spatial pattern of distribution of individuals within populations.

Distribution The geographic extent of a population or other ecological unit.

Diversity The number of species in a local area (*alpha diversity*) or region (*gamma diversity*). Also, a measure of the variety of species in a community that takes into account the relative abundance of each.

Dominance condition The situation where one or a few species, by means of number, coverage, or size, have considerable influence upon or control of the conditions of existence of associated species.

Dominants   The few species that attain high abundances in a community.

Dynamic life table   The age-specific survival and fecundity of a cohort followed from birth to the death of the last individual. Also called a *cohort life table*.

Ecological efficiency   The percentage of energy in the biomass produced by one trophic level that is incorporated into the biomass produced by the next higher trophic level.

Ecological longevity   Average length of life of individuals of a population under stated conditions.

Ecosphere   See *Biosphere*.

Ecosystem   A biotic community and its abiotic environment. Can occur at various scales, but at larger scales this is synonymous with a *Landscape*.

Ecosystem diversity   The variety of different ecosystems.

Ecosystems ecology   The study of ecosystems, particularly the interactions of the whole biota and their environments.

Ecotone   Transition zone between two diverse communities (e.g. the tundra–boreal forest ecotone).

Effective population size (N_e)   The average size of a population expressed in terms of the number of individuals assumed to contribute genes equally to the next generation; generally smaller than the actual size of the population, depending on the variation in reproductive success among individuals.

Emergent property   A feature of a system not deducible from lower-order processes.

Endemic   (i) Biodiversity: a species whose range is confined to a defined area. (See *Indigenous, Exotic*.) (ii) Epidemiology: a *parasite* whose *prevalence* does not exhibit wide fluctuations through time in a defined host, host species, or host population.

Environment   All the biotic and abiotic factors that actually affect an individual organism at any point in its life cycle.

Environmental stochasticity   Random variation in the abiotic environment.

Epidemic   A sudden, rapid spread or increase of a disease-causing *parasite* through a human population. An epidemic is often the result of a change in circumstances that favors pathogen transmission, such as a rapid increase of host population density or the introduction of a new parasite (or genetic strain of a parasite) to a previously unexposed host population.

Epizootic   The sudden spread of a disease-causing *parasite* through a nonhuman population; equivalent to an *epidemic* in human populations.

Equilibrium isocline   A line on a population graph designating combinations of competing populations, or predator and prey populations, for which the growth rate of one of the populations is zero.

Eruption   The sudden increase in a species population in a defined area. (See *Irruption*.)

Eutrophic   Rich in the mineral nutrients required by green plants; usually applied to an aquatic habitat with high productivity.

Evapotranspiration   Sum total of water lost from the land by evaporation and plant transpiration.

Exotic   A species found outside of its normal habitats. (See *Endemic, Indigenous*.)

Experiment   A test of a hypothesis, either observational or manipulative. The experimental method is the scientific method.

Exponential rate of increase (r) The natural log of the *finite rate of increase*. Also called the *instantaneous rate of increase.*

Extinction The disappearance of a species or other taxon from a region or biota.

Facilitation The enhancement of conditions for a population of one species by the activities of another, particularly during early succession.

Fecundity The potential reproductive capacity of an organism, measured by the number of gametes produced.

Fertility An ecological concept of the actual number of viable offspring produced by an organism, equivalent to realized fecundity.

Finite rate of increase (λ) The ratio of the population density in one year to that in the previous year (N_t/N_{t-1}). (See *Intrinsic* and *Exponential growth rate.*)

Fitness The genetic contribution by an individual's descendants to future generations of a population relative to those of other individuals.

Food chain The sequence of energy or nutrient transfer through the trophic levels, beginning with plants and ending with the largest carnivores.

Food web A representation of the various paths of energy flow through populations in a community; the complex of *food chains.*

Fragility Narrow tolerance to disturbance. (See *Robustness, Resilience.*)

Functional response A change in the rate of exploitation of prey by an individual predator as a result of a change in prey density. (See *Numerical response.*)

Fundamental niche See *Niche.*

Gamma diversity The average diversity per habitat (*alpha*) × 1/mean number of habitats occupied by a species (*beta*) × the total number of habitats.

Gause's principle See *Competitive exclusion principle.*

Gene A unit of genetic inheritance. (i) Biochemistry: the part of the DNA molecule that encodes a single enzyme or structural protein. (ii) Evolutionary ecology: that which segregates independently.

Generation time The average age at which a female gives birth to her first offspring, or the average time it takes for a population to increase by a factor equal to the net reproductive rate.

Genotype The entire genetic constitution of an organism. (Contrast with *Phenotype.*)

Global stability Ability to withstand perturbations of a large magnitude and not be affected. (See *Local stability.*)

Grazers (i) Organisms that eat grasses or non-woody herbs. (ii) Organisms that feed on many other individuals but usually do not kill them.

Gross production Production before respiration losses are subtracted; photosynthetic production for plants and metabolizable production for animals.

Habitat The place where an animal or plant normally lives, often characterized by a dominant plant form or physical characteristic (e.g. soil habitat, forest habitat).

Harvesting Removal of animals or plants from a population, usually by humans.

Helminths Members of the five classes of parasitic worm: Monogenea, Digenea, Cestodes, Nematodes, and Acanthocephalans.

Herbivore An organism that eats plants. (Contrast with *Carnivore.*)

Herbivory A consumer–resource interaction involving the consumption of plants or plant parts.

Heterosis A situation in which the heterozygous genotype is more fit than either homozygote.

Heterotroph An organism that obtains energy and materials by eating other organisms. (Contrast with *Autotroph.*)

Heterozygous Having two different alleles of a gene, one derived from each parent. (Compare with *Homozygous.*)

Homeothermic Pertaining to warm-blooded animals that regulate their body temperature. (Contrast with *Poikilothermic.*)

Homozygous Having two identical alleles of a gene. (See *Heterozygous*).

Hypothesis A universal proposition that suggests an explanation for some observed ecological situation.

Hysteresis In ecosystems, the time-lag in the effect of a disturbance between the increase in disturbance and the decrease in disturbance.

Immunity The ability to combat *infection* or *disease* due to the presence of *antibodies* or activated cells. Essentially, it can be divided into three types:(a) *acquired immunity* is conferred on an individual after recovery from a disease; (b) *natural* or *innate immunity* is inherited from parents (in some cases antibodies may be passed across the placenta and therefore are present in the blood at birth); (c) *artificial immunity* may be induced by the injection of either a *vaccine*, denatured *antigens* of a *parasite* (which induces production of antibodies and thus gives active artificial immunity), or antiserum, which contains antibodies and thus may be used when the host is already infected. As well as strengthening the host's resistance, this also confers passive artificial immunity against any subsequent infection. (Sometimes confusingly termed *resistance*, or more correctly, *specific resistance*).

Inbreeding Mating among related individuals.

Inbreeding depression Reduction in fitness caused by inbreeding.

Inclusive fitness The sum of fitness of an individual and the fitnesses of its relatives, the latter weighted according to the coefficient of relationship.

Indigenous A species that occurs in its normal habitats but is not necessarily confined to those areas. (See *Exotic, Endemic.*)

Indirect interaction An interaction between two individuals that involves one or more intermediary species. (See *Direct interaction.*)

Infection The presence of a *parasite* within a *host*, where it may not cause *disease*.

Infectious disease A *disease* caused by *infection* with a *parasite* that can be *transmitted* from one individual to another, either directly (e.g. measles) or indirectly by a *vector* (e.g. malaria). *Contagious disease* is a specific subset of infectious disease and pertains exclusively to those diseases transmitted directly between hosts through close bodily contact (this includes aerosols).

Infectious period Period during which the infected individual is able to *transmit* an *infection* to a *susceptible* host or *vector*. The infectious period may or may not coincide with the disease.

Instantaneous rate of increase (r) The rate of increase of a population undergoing exponential growth under a given set of ecological conditions; can be positive, negative, or zero, and if birth and death rates are constant for sufficient time will produce a *stable age distribution*. In the *logistic equation*, $r = r_{max} (1 - N/K)$. (See *Intrinsic* and *Exponential rate of increase*).

Intensity   Either (i) the mean number of *parasites* within infected members of a host population or (ii) the mean parasite burden of the entire population. It is important to distinguish between these two usages, because unless *prevalence* is 100% the latter will be smaller than the former.

Interaction   See *Direct* and *Indirect interaction*.

Intermediate host   A host organism that acquires an infectious agent and in which an obligatory period of development or multiplication/replication occurs before the agent becomes infectious to another host. For example, tabanids flies are vectors (only) of Surra (*Trypanosoma evansi*). Their habit of interrupted feeding allows the direct mechanical transmission of the trypanosome from one host to another on the flies' mouthparts. Tabanids are both vectors and intermediate hosts of the nematode (*Pelecitus roemeri*) and act in a similar fashion to mosquitoes playing the role of intermediate hosts and vectors of malaria. Snails are the intermediate hosts of schistosomes but they are not vectors; the infectious stage of the parasite escapes from the snail and finds and penetrates the definitive host.

Interspecific competition   Competition between members of different species.

Intrinsic rate of increase (r_{max})   The rate of increase of a population undergoing exponential growth under optimum ecological conditions; the maximum instantaneous rate that a species is capable of. It is a characteristic of a species. (See *Instantaneous* and *Exponential rate of increase*.)

Inverse density dependent   A rate (births or deaths) that decreases proportionately as population size or density increases. Also called *Depensatory*.

Irruption   A sudden expansion of the range of a species, which may or may not be accompanied by an increase in population. (See *Eruption*.)

Isocline   In a population graph, a line designating combinations of competing populations, or predator and prey populations, for which the growth rate of one of the populations is zero.

Keystone species   A species whose functional role in a community is disproportionately greater than that predicted by its abundance; usually one whose removal has strong effects on community diversity and composition. Keystone species are often top predators.

Landscape   A large-scale *ecosystem*.

Leslie matrix   A matrix of values of age-specific fecundity and survivorship used to project the size and age structure of a population through time; a population matrix.

Life history   The set of adaptations of an organism that influence the life table values of age-specific survival and fecundity, such as reproductive rate or age at first reproduction.

Life table   A summary by age of the survivorship and fecundity of individuals in a population.

Limitation   A process that determines the size of the equilibrium population.

Limiting factor   Any factor that causes population *limitation*.

Limiting resource   A resource that is scarce relative to the demand for it.

Local stability   Ability to withstand perturbations of a small magnitude and remain unaffected. (See *Global stability, Fragility, Robustness*.)

Logistic equation   Model of *population growth rate (dN/dt)* described by the two constants r_{max} and K, the *carrying capacity*. Thus, $dN/dt = r_{max} \cdot N(1 - N/K)$. It produces a symmetrical S-shaped curve, with K the upper asymptote.

Log-normal distribution   Frequency distribution of species abundances in which the *x*-axis is expressed as a logarithmic scale. The *x*-axis is the (log) number of individuals represented in a sample and the *y*-axis is the number of species.

Macroparasites   *Parasites* that do not in general multiply within their *definitive hosts* but instead produce transmission stages (eggs and larvae) that pass into the external environment or to *vectors* (i.e. the parasitic *helminths* and arthropods). The immune response elicited by these metazoans generally depends on the number of parasites present in a given host and tends to be of a relatively transient nature.

Maximum sustained yield (MSY)   The greatest sustainable rate at which individuals may be harvested from a population without causing a decline to extinction; that is, the harvest at which recruitment equals harvesting.

Mesic   A habitat with plentiful rainfall and well-drained soils. (See *Xeric*.)

Metapopulation   A set of local populations linked together through dispersal.

Microparasites   *Parasites* that undergo direct multiplication within their *definitive hosts* (e.g. viruses, rickettsia, bacteria, fungi, and protozoa). Microparasites are characterized by small size, short generation time, and a tendency to induce *immunity* to reinfection in those hosts that survive the primary *infection*. Duration of infection is usually short in relation to the *expected life span* of the host (there are, however, important exceptions, such as the slow viruses).

Mineral nutrients   Elements that are necessary for the growth and development of plants, such as nitrogen, phosphorus, sulfur, potassium, and calcium.

Minimum viable population (MVP)   The smallest population that can persist for some arbitrarily long time, usually 1000 years.

Mortality (m_x)   Ratio of number of deaths to individuals at risk, often described as a function of age (x). (Compare with *Death rate*.)

Multiple states   The situation when communities exist in different combinations of species abundance under the same environmental conditions. These multiple states within an ecosystem are detected when a perturbation radically alters the abundance of many species in the community, which then do not return to their original abundance when the perturbation is removed

Mutation   Any change in the genotype of an organism occurring at the gene, chromosome, or genome level; usually applied to changes in genes to new allelic forms.

Mutualism   A relationship between two species that benefits both.

Natal dispersal   Dispersal of young animals from their place of birth.

Natural selection   Change in the frequency of genetic traits in a population through differential survival and reproduction of individuals bearing those traits.

Net production   Production after respiration losses are subtracted.

Net reproductive rate (R_e)   Expected number of offspring produced by a female during her lifetime.

Niche   The set of resources and environmental conditions that allows a single species to persist in a particular region, often conceived as a multidimensional space. Also called *Fundamental niche*. (See *Realized niche*.)

Niche breadth The variety of resources utilized and the range of conditions tolerated by an individual, population, or species. (Compare with *Niche width*.)

Niche complementarity A situation in which species that overlap extensively in their use of one resource differ substantially in their use of another.

Niche overlap The sharing of niche space by two or more species; similarity of resource requirements and tolerance of ecological conditions.

Niche width The standard deviation of the distribution of resource use. (Compare with *Niche breadth*.)

Numerical response A change in the population size of a predator species as a result of a change in the density of its prey. (Compare with *Functional response*.)

Nutrient Any substance required by organisms for normal growth and maintenance.

Nutrient cycle The path of an element through the ecosystem, including its assimilation by organisms and its regeneration in a reusable inorganic form.

Oligotrophic Being poor in the mineral nutrients required by green plants; applied to an aquatic habitat with low productivity. (See *Eutrophic*.)

Omnivore An organism whose diet is broad, including both plant and animal foods; specifically, an organism that feeds at more than one trophic level.

Pandemic A widely distributed *epidemic*.

Panmixia Random mating within a population; a panmictic population.

Panzootic A widely distributed epizootic, often affecting more than one host species.

Parameter In statistics, an unknown true characteristic of a statistical population. It is usually impossible to know the value of a parameter. A statistic estimates a parameter.

Parasite An organism exhibiting a varying but obligatory dependence on another organism, its *host*, and which is detrimental to the survival and/or *fecundity* of that host. (See *Microparasite* and *Macroparasite*.)

Parasitoid Any of a number of insect species whose larvae live within and consume their host, usually another insect.

Perennial An organism that lives for more than one year; lasting throughout the year.

Phenology Study of the periodic (seasonal) phenomena of animal and plant life and their relations to the weather and climate (e.g. the time of flowering in plants).

Phenotype Expression of the characteristics of an organism as determined by the interaction of its genetic constitution with the environment. (Contrast with *Genotype*.)

Phylogeny The pattern of evolutionary relationships among species or other taxonomic groups.

Pleiotropy The influence of one gene on the expression of more than one trait in the phenotype.

Poikilothermic Cold-blooded animals; organisms that have no rapidly operating heat-regulatory mechanism. (See *Homeothermic*.)

Polygenic Determined by the expression of more than one gene.

Polymorphism The occurrence together in the same habitat of two or more discontinuous forms of a species in such proportions that the rarest of them cannot merely be maintained by recurrent mutation or immigration.

Population A group of individuals of a single species in a defined area.

Population viability analysis (PVA) The strategic analysis of the ecological, economic, and political issues and challenges related to the conservation of an endangered species, community, or ecosystem.

Population growth rate (dN/dt) The rate of growth of a population over a short period of time, defined by the product of population size, N, and the *instantaneous rate of increase, r*. (See *Logistic equation*.)

Prevalence The proportion of the host population with *infection* or *disease*, often expressed as a percentage. A measure of how widespread is the infection or disease.

Primary production Production by green plants.

Production Accumulation of energy or biomass.

Productivity The rate at which energy is accumulated.

Realized niche The set of resources and environmental conditions constrained by competition or predation that allows a single species to persist in a particular region. A subset of the *fundamental niche*.

Recruitment Increment to a natural population, usually from young animals or plants entering the adult population.

Regulation Occurs when a population experiences density-dependent mortality or birth rates.

Rescue effect Prevention of the extinction of a local population by immigration of individuals from elsewhere, often from a more productive habitat.

Resilience (i) The rate at which a population returns to equilibrium after a disturbance. (ii) The ability to withstand disturbance.

Resource A substance or object required by an organism for normal maintenance, growth, and reproduction. (See *Limiting resource*.)

Respiration The complex series of chemical reactions by which energy is made available for use, found in all organisms; carbon dioxide, water, and energy are the end products.

Robustness A relatively wide tolerance to disturbance. (See *Fragility*, *Resilience*.)

Secondary plant compounds Chemical products of plant metabolism specifically for the purpose of defense against herbivores and disease organisms.

Secondary production Production by herbivores, carnivores, or detritus feeders. (Contrast with *primary production*.)

Senescence Process of aging.

Sere A series of stages of community change in a particular area leading towards a stable state, or climax.

Sex ratio Ratio of the number of individuals of one sex to that of the other within a population.

Shared predation A type of apparent competition in which two species fall victim to a single predator and perhaps compete for enemy-free space in which to avoid the predator.

Sink An ecosystem, habitat, population, or community that receives input of materials or individual organisms. The population is not self-sustaining. (See *Source*.)

Source An ecosystem, habitat, population, or community from which materials or organisms move. The population is self-sustaining. (See *Sink*.)

Source–sink metapopulation   A *metapopulation* in which some local populations (*sources*) have a positive growth rate at low densities and others (*sink*) have a negative growth rate in the absence of immigration.

Stability   Absence of fluctuations in populations; ability to withstand perturbations without large changes in composition. (See *Resilience.*)

Stable age distribution   The proportions of individuals in various age classes within a population that has a constant instantaneous rate of growth r.

Stable equilibrium   The state to which a system returns if displaced by an outside force.

Stage-classified population   A population containing individuals of different developmental states (e.g. adults and larvae) in the same or different habitats.

Static life table   The age-specific survival and fecundity of individuals of different ages within a population at a given time; also called a *time-specific life table*. A cross-section of all cohorts in the population at a given time.

Stochastic model   Mathematical model based on probabilities. The prediction of the model is not a single fixed number but rather a range of possible numbers. (Contrast with *Deterministic model.*)

Succession   Replacement of one kind of community by another; the progressive changes in vegetation and animal life that may culminate in the climax community.

Survival (l_x)   Proportion of newborn individuals alive at age x; also called *survivorship*.

Survivorship curve   Curve showing the number of individuals surviving to age x (log scale) plotted against age.

Susceptible   Accessible or liable to infection by a particular *parasite*.

Susceptible individual   Either naive (previously uninfected) or having lost *immunity*.

Switching   A change in diet to favor food items of increasing suitability or abundance.

Sympatric   Occurring in the same place; usually refers to areas of overlap in species distributions. (See *Allopatric.*)

Synecology   The study of a groups of organisms in relation to their environment; includes population, community, and ecosystem ecology.

Time-specific life table   See *Static life table*.

Total response of predator   The product of functional and numerical responses plotted as per capita mortality of prey against prey density.

Transmission   The process by which a *parasite* passes from a source of *infection* to a new *host*. There are two major types: *horizontal* and *vertical* transmission. The majority of transmission processes operate horizontally, for example by direct contact between infected and *susceptible* individuals or between disease *vectors* and susceptibles. There are six main methods of horizontal transmission: (i) ingestion of contaminated food or drink; (ii) inhalation of contaminated air droplets; (iii) direct contact; (iv) injection into a tissue via an animal's saliva or bite; (v) invasion via open wounds; and (vi) penetration of the host by active parasite transmission stages (e.g. schistosome miracidia or cercariae). Vertical transmission occurs when a parent conveys an infection to its unborn offspring, as occurs in HIV in humans and in many *arboviruses*.

Transmission threshold   Level of transmission below which an *infection* is unable to maintain itself within the host population (or populations, in the case of *indirectly transmitted infections*).

Trophic level Position in the food chain, determined by the number of energy transfer steps to that level. The first trophic level includes green plants, the second herbivores, and so on.

Type I error The rejection of a true null hypothesis.

Type II error The acceptance of a false null hypothesis.

Unstable equilibrium The state of a system at which forces are precisely balanced, but away from which the system moves when displaced.

Variable A characteristic or measure of the natural world that may take on any of a number of different values.

Vector An organism that acquires, transports, and delivers an infectious agent to a host. Sometimes there is development or multiplication/replication of the infectious agent in a vector, but this is not obligatory. (See *Intermediate host*.)

Watershed (i) North America: the drainage basin of a stream or river. (2) Europe and elsewhere: the line demarcating different drainage basins.

Wildlife Wild animals, usually terrestrial vertebrates whose populations are monitored and managed for exploitation or conservation.

Xeric A habitat in which water is in short supply to plants. (See *Mesic*.)

Zoonosis Natural occurrence of a parasite in animals and natural transmission between animals and humans.

References

Aarts, G., MacKenzie, M., McConnell, B., Fedak, M., and Matthiopoulos, J. 2008. Estimating space use and habitat preference from wildlife telemetry data. *Ecography* 31:140–160.

Abaturov, B.D. and Magomedov, M.-R.D. 1988. Food value and dynamics of food resources as a factor characterizing the state of populations of herbivorous mammals. *Zoologicheskii Zhurnal* 76:223–234.

Abramsky, Z. 1981. Habitat relationships and competition in two Mediterranean *Apodemus* spp. *Oikos* 36:219–225.

Abramsky, Z. and Sellah, C. 1982. Competition and the role of habitat selection in *Gerbillus allenbyi* and *Meriones tristrami*: a removal experiment. *Ecology* 63:1242–1247.

Abramsky, Z., Dyer, M.I., and Harrison, P.D. 1979. Competition among small mammals in experimentally perturbed areas of the shortgrass prairie. *Ecology* 60:530–536.

Abramsky, Z., Bowers, M.A., and Rosenzweig M.L. 1986. Detecting interspecific competition in the field: testing the regression method. *Oikos* 47:199–204.

Acevedo, P., Juimenez-Valverde, A., Meolo-Ferreira, J., Real, R., and Alves, P.C. 2012. Parapatric species and the implications for climate change studies: A case study on hares in Europe. *Global Change Biology* 18:1509–1519.

Aebischer, N.J., Robertson, P.A., and Kenward, R.E. 1993. Compositional analysis of animal habitat-use from radio-tracking data. *Ecology* 74:1313–1325.

Aiello-Lammens, M.E., Chu-Agor, M.L., Convertino, M., Fischer, R.A., Linkov, I., and Akcakaya, H.R. 2011. The impact of sea-level rise on Snowy Plovers in Florida: Integrating geomorphological, habitat, and metapopulation models. *Global Change Biology* 17:3644–3654.

Akaike, H. 1973. Information theory as an extension of the maximum likelihood principle. In: Petrov, B.N. and Csaki, F. (eds) *Second International Symposium on Information Theory*, pp. 267–281. Akademiai Kiado, Budapest.

Akçakaya, H.R. and Atwood, J.L. 1997. A habitat-based metapopulation model of the California gnatcatcher. *Conservation Biology* 11:422–434.

Albright, T.P., Pigeon, A.M., Rittenhouse, C.D., Clayton, M.K., Flather, C.H., Culbert, P.D., Wardlow, B.D., and Radeloff, V.C. 2010. Effects of drought on avian community structure. *Global Change Biology* 16:2158–2170.

Alexander, R.D. 1974. The evolution of social behavior. *Annual Review of Ecology and Systematics* 5:325–383.

Allee, W.C. 1938. *The Social Life of Animals*. W.W. Norton, New York, NY.

Allen, R.B., Bellingham, P.J., and Wiser, S.K. 1999. Immediate damage by an earthquake to a temperate montane forest. *Ecology* 80:708–714.

Allendorf, F.W. and Hard, J.J. 2009. Human-induced evolution caused by unnatural selection through harvest of wild animals. *Proceedings of the National Academy of Sciences USA* 106:9987–9994.

Wildlife Ecology, Conservation, and Management, Third Edition.
John M. Fryxell, Anthony R. E. Sinclair and Graeme Caughley.
© 2014 John Wiley & Sons, Ltd. Published 2014 by John Wiley & Sons, Ltd.
Companion Website: www.wiley.com/go/Fryxell/Wildlife

Amarasekare, P. and Nisbet, R.M. 2001. Spatial heterogeneity, source-sink dynamics, and the local coexistence of competing species. *American Naturalist* 158:572–584.

Anderson, A.E., Medin, D.E., and Ochs, D.P. 1969. Relationships of carcass fat indices in 18 wintering mule deer. *Proceedings of the Western Association of State Game and Fish Commissioners* 49:329–340.

Anderson, A.E., Medin, D.E., and Bowden, D.C. 1972. Indices of carcass fat in a Colorado mule deer population. *Journal of Wildlife Management* 36:579–594.

Anderson, A.E., Bowden, D.C., and Medin, D.E. 1990. Indexing the annual fat cycle in a mule deer population. *Journal of Wildlife Management* 54:550–556.

Anderson, D.R. 1975. Optimal exploitation strategies for an animal population in a Markovian environment: a theory and an example. *Ecology* 56:1281–1297.

Anderson, D.R., Burnham K.P., White, G.C., and Otis, D.L. 1983. Density estimation of small-mammal populations using a trapping web and distance sampling methods. *Ecology* 64:674–680.

Anderson, D.R., Burnham, K.P., and Thompson, W.L. 2000. Null hypothesis testing: problems, prevalence, and an alternative. *Journal of Wildlife Management* 64:912–923.

Anderson, E.W. and Sherzinger, R.J. 1975. Improving quality of winter forage for elk by cattle grazing. *Journal of Range Management* 28:120–125.

Anderson, R.C. 1972. The ecological relationships of meningeal worm and native cervids in North America. *Journal of Wildlife Diseases* 8:304–310.

Anderson, R.M. 1991. Populations and infectious diseases: ecology or epidemiology? *Journal of Animal Ecology* 60:1–50.

Anderson, R.M. and May, R.M. 1979. Population biology of infectious diseases: Part I. *Nature* 280:361–367.

Anderson, R.M. and May, R.M. 1986. The invasion, persistence and spread of infectious diseases within animal and plant communities. *Philosophical Transactions of the Royal Society of London B, Biological Sciences* 314:533–570.

Anderson, S.H. 2003. The relative importance of birds and insects as pollinators of the New Zealand flora. *New Zealand Journal of Ecology* 27:83–94.

Andow, D.A., Karieva, P.M., Levin, S.A., and Okubo, A. 1990. Spread of invading organisms. *Landscape Ecology* 4:177–188.

Andren, H. 1992. Corvid density and nest predation in relation to forest fragmentation: a landscape perspective. *Ecology* 73:794–804.

Andren, H. 1994. Effects of habitat fragmentation on birds and mammals in landscapes with different proportions of suitable hbaitat: a review. *Oikos* 71:355–366.

Arcese, P. and Sinclair, A.R.E. 1997. The role of protected areas as ecological baselines. *Journal of Wildlife Management* 61:587–602.

Arcese, P., Smith, J.N.M., Hochachka, W.M., Rogers, C.M., and Ludwig, D. 1992. Stability, regulation, and the determination of abundance in an insular song sparrow population. *Ecology* 73:805–822.

Arditi, R. and Dacorogna, B. 1988. Optimal foraging on arbitrary food distributions and the definition of habitat patches. *American Naturalist* 131:837–846.

Arditi, R. and Ginzburg, L. 2012. *How Species Interact*. Oxford University Press, Oxford, UK.

Arlettaz, R., Perrin, N., and Hausser, J. 1997. Trophic resource partitioning and competition between the two sibling bat species *Myotis myotis* and *Myotis blythii*. *Journal of Animal Ecology* 66:897–911.

Arneberg, P., Folstad, I., and Karter, A.J. 1996. Gastrointestinal nematodes depress food intake in naturally infected reindeer. *Parasitology* 112:213–219.

Arnold, G.W., Steven, D.E., and Weeldenburg, J.R. 1993. Influences of remnant size, spacing pattern and connectivity on population boundaries and demography in Euros *Macropus robustus* living in a fragmented landscape. *Biological Conservation* 64:219–230.

Arnold, T.W. 1994. A roadside transect for censusing breeding Coots and Grebes. *Wildlife Society Bulletin* 22:437–443.

Arthur, S.M., Paragi, T.F., and Krohn, W.B. 1993. Dispersal of juvenile fishers in Maine. *Journal of Wildlife Management* 57:868–874.

Atkinson, I.A.E. 2001. Introduced mammals and models for restoration. *Biological Conservation* 99:81–86.

Augustine, D.J., Frelich, L.E., and Jordan, P.A. 1998. Evidence for two alternate stable states in an ungulate grazing system. *Ecological Applications* 8:1260–1269.

Austin, M.P. 1985. Contiuum concept, ordination methods, and niche theory. *Annual Review of Ecology and Systematics* 16:39–61.

Bailey, J.A. 1968. A weight-length relationship for evaluating physical condition of cottontails. *Journal of Wildlife Management* 32:835–841.

Bailey, N.T.J. 1951. On estimating the size of mobile populations from recapture data. *Biometrika* 38:293–306.

Bailey, N.T.J. 1952. Improvements in the interpretation of recapture data. *Journal of Animal Ecology* 21:120–127.

Baker, J.R. 1938. The evolution of breeding seasons. In: de Beer, G.R. (ed) *Evolution: Essays on Aspects of Evolutionary Biology Presented to Professor E.S. Goodrich on his Seventieth Birthday*, pp. 161–177. Clarendon Press, Oxford.

Ballard, W.B., Whitman, J.S. and Gardner, C.L. 1987. Ecology of an exploited wolf population in south-central Alaska. *Wildlife Monographs* 98:1–54.

Bamford, J. 1970. Estimating fat reserves in the brush-tailed possum, *Trichosurus vulpecula* Kerr (Marsupialia : Phalangeridae). *Australian Journal of Zoology* 18:415–425.

Barbet-Massin, M., Thuiller, W., and Jiguet, F. 2012. The fate of European breeding birds under climate, land use and dispersal scenarios. *Global Change Biology* 18:881–890.

Barker, S.C., Singleton, G.R., and Spratt, D.M. 1991. Can the nematode *Capillaria hepatica* regulate abundance of wild house mice? Results of enclosure experiments in southeastern Australia. *Parasitology* 103:439–449.

Barr, W. 1991. *Back from the Brink: The Road to Muskox Conservation in the Northwest Territories.* Komatik Series, No. 3. The Arctic Institute of North America, University of Calgary, AB.

Basson, M., Beddington, J.R., and May, R.M. 1991. An assessment of the maximum sustainable yield of ivory from African elephant populations. *Mathematical Biosciences* 104:73–95.

Baum, J.K., Myers, R.A., Kehler, D.G., Worm, B., Harley, S.J., and Doherty, P.A. 2003. Collapse and conservation of shark populations in the northwest Atlantic. *Science* 299:389–392.

Bayliss, M., Mellor, P.S., and Meiswinkel, R. 1999. Horse sickness and ENSO in South Africa. *Nature* 397:574.

Bayliss, P. 1987. Kangaroo dynamics. In: Caughley, G., Shepherd, N., and Short, J. (eds) *Kangaroos: Their Ecology and Management in the Sheep Rangelands of Australia*, pp. 119–134. Cambridge University Press, Cambridge.

Bazely, D.R. and Jefferies, R.L. 1989. Lesser snow geese and the nitrogen economy of a grazed salt marsh. *Journal of Ecology* 77:24–34.

Beale, C.M., Lennon, J.J., and Gimona, A. 2008. Opening the climate envelope reveals no macroscale associations with climate in European birds. *Proceedings of the National Academy of Sciences USA.* 105:14908–14912.

Beasom S.L. and Roberson, S.F. (eds) 1985. *Game Harvest Management.* Caesar Kleberg Wildlife Research Institute, Kingsville, TX.

Beaumont, L.J., McAllan, I.A.W., and Hughes, L. 2006. A matter of timing: Changes in the first date of arrival and last date of departure of Australian migratory birds. *Global Change Biology* 12:1339–1354.

Beddington, J.R. 1975. Mutual interference between parasites or predators and its effect on searching efficiency. *Journal of Animal Ecology* 44:331–340.

Beier, P. 1996. Metapopulation models, tenacious tracking, cougar conservation. In: McCullough, D. (ed) *Metapopulations and Wildlife Conservation*, pp. 293–323. Island Press, Washington, DC.

Beier, P. and Noss, R.F. 1998. Do habitat corridors provide connectivity? *Conservation Biology* 12:1241–1252.

Beisner, B.E., Haydon, D.T., and Cuddington, K. 2003. Alternative stable states in ecology. *Frontiers in Ecology and Environment* 1:376–382.

Beissinger, S.R. and McCullough, D.R. (eds) 2002. *Population Viability Analysis*. University of Chicago Press, Chicago, IL.

Bell, D.J., Oliver, W.L.R., and Goose, R.K. 1990. The hispid hare *Caprolagus hispidus*. In: Chapman, J.A. and Flux, J.E.C. (eds) *Rabbits, Hares and Pikas: Status Survey and Conservation Action Plan*, pp. 128–136. IUCN, Gland, Switzerland.

Bell, G., Handford, P., and Dietz, C. 1977. Dynamics of an exploited population of Lake Whitefish. *Journal of the Fisheries Research Board of Canada* 34:942–953.

Bell, R.H.V. 1970. The use of the herb layer by grazing ungulates in the Serengeti. In: Watson, A., (ed) *Animal Populations in Relation to their Food Resources*, pp. 111–124. The 10th Symposium of the British Ecological Society, Blackwell, Oxford.

Bell, R.H.V. 1971. A grazing ecosystem in the Serengeti. *Scientific American* 225(1):86–93.

Bell, R.H.V. 1981. An outline of a management plan for Kasungu National Park, Malawi. In: Jewell, P.A., Holt, S., and Hart, D. (eds) *Problems in Management of Locally Abundant Wild Mammals*, pp. 69–89. Academic Press, New York, NY.

Bell, R.H.V. 1982. The effect of soil nutrient availability on community structure in African ecosystems. In: Huntly B.J. and Walker, B.H. (eds) *Ecology of Tropical Savannas*, pp. 193–216. Springer-Verlag, New York, NY.

Belovsky, G.E. 1978. Diet optimization in a generalist herbivore: the moose. *Theoretical Population Biology* 14:105–134.

Belovsky, G.E. 1986. Generalist herbivore foraging and its role in competitive interactons. *American Zoologist* 26:51–69.

Belovsky, G.E. 1988. Food plant selection by a generalist herbivore: the moose. *Ecology* 62:1020–1030.

Belovsky, G.E., Ritchie, M.E.,and Moorehead, J. 1989. Foraging in complex environments: when prey availability varies over time and space. *Theoretical Population Biology* 36:144–160.

Belsky, A.J. 1987. The effects of grazing: confounding of ecosystem, community and organism scales. *American Naturalist* 129:777–783.

Bender, E.A., Case, T.J. and Gilpin, M.E. 1984. Perturbation experiments in community ecology: theory and practice. *Ecology* 65:1–13.

Benhamou, S. 1992. Efficiency of area-concentrated searching behaviour in a continuous patchy environment. *Journal of Theoretical Biology* 159:67–81.

Bennett, A.F., Henein, K., and Merriam, G. 1994. Corridor use and the elements of corridor quality: Chipmunks and fencecrows in a farmland mosaic. *Biological Conservation* 68:155–165.

Bent, A.C. 1932. *Life Histories of North American Gallinaceous Birds*. Smithsonian Institution, US National Museum Bulletin 162.

Benton, T.G., Cameron, T.C., and Grant, A. 2004. Population responses to perturbations: predictions and responses from laboratory mite populations. *Journal of Animal Ecology* 73:983–995.

Berger, J., Cain, S.L., and Berger, K.M. 2006. Connecting the dots: an invariant migration corridor links the Holocene to the present. *Biology Letters* 2:528–531.

Berger, L., Speare, R., Daszak, P., Earl Green, D., Cunningham, A.A., Goggin, C.L., Slocombe, R., Ragan, M.A., Hyatt, A.D., McDonald, K.R., Hines, H.B., Lips, K.R., Marantelli, G., and Parkes, H. 1998. Chytridiomycosis causes amphibian mortality associated with population declines in the rain forests of Australia and Central America. *Proceedings National Academy of Science* 95:9031–9036.

Bergerud, A.T. and Page, R.E. 1987. Displacement and dispersion of parturient caribou at calving as antipredator tactics. *Canadian Journal of Zoology* 65:1597–1606.

Bergman, C.M., Fryxell, J.M., Gates, C.C., and Fortin, D. 2001 Ungulate foraging strategies: energy maximizing or time minimizing? *Journal of Animal Ecology* 70:289–300.

Berryman, A.A. 1991. Can economic forces cause ecological chaos? The case of the Northern California Dungeness crab fishery. *Oikos* 62:106–109.

Beyer, H.L., Haydon, D.T., Morales, J.M., Frair, J.L., Hebblewhite, M., Mitchell, M., and Matthiopoulos, J. 2010. The interpretation of habitat preference metrics under use-availability designs. *Philosophical Transactions of the Royal Society (B)* 365:2245–2254.

Biggins, D.E., Vargas, A., Godbey, J.L., and Anderson, S.H. 1999. Influence of prerelease experience on reintroduced black-footed ferrets (*Mustela nigripes*). *Biological Conservation* 89:121–129.

Birch, L.C. 1957. The meanings of competition. *American Naturalist* 91:5–18.

Bischof, R., Loe, L.E., Meisingset, E.L., Zimmerman, B., Van Moorter, B., and Mysterud, A. 2012. A migratory northern ungulate in the pursuit of spring: jumping or surfing the green wave? *American Naturalist* 180:407–424.

Björnhag, G. 1987. Comparative aspects of digestion in the hindgut of mammals. The colonic separation mechanism (CSM) (a review). *Deutsche Tieraerztliche Wochenschrift* 94:33–36.

Blank, T.H., Southwood, T.R.E., and Cross, D.J. 1967. The ecology of the partridge: I. Outline of the population processes with particular reference to chick mortality and nest density. *Journal of Animal Ecology* 36:549–556.

Bloomer, J.P. and Bester, M.N. 1992. Control of feral cats on sub-Antarctic Marion Island, Indian Ocean. *Biological Conservation* 60:211–219.

Blower, J.G., Cook, L.M., and Bishop, J.A. 1981. *Estimating the Size of Animal Populations*. George Allen and Unwin, London.

Bobek, B. 1977. Summer food as the factor limiting roe deer population size. *Nature* 268:47–49.

Bock, C.E., Smith, H.M., and Bock, J.H. 1990. The effect of livestock grazing upon abundance of the lizard, *Sceloporus scalaris*, in southeastern Arizona. *Journal of Herpetology* 24:445–446.

Boertje, R.D. Valkenburg, P., and MacNay, M.E. 1996. Increases in moose, caribou, and wolves following wolf control in Alaska. *Journal of Wildlife Management* 60:474–489.

Bomford, M. 1988. Effect of wild ducks on rice production. In: Norton, G.A. Pech, R.P. (eds) *Vertebrate Pest Management in Australia: A Decision Analysis/Systems Analysis Approach*, pp. 53–57. Project Report No. 5, CSIRO, Australia.

Bomford, M. 1990. *A Role for Fertility Control in Wildlife Management*. Bureau of Rural Resources, Bulletin No. 7, Canberra.

Bomford, M. and O'Brien, P.H. 1990. Sonic deterrents in animal damage control: a review of device tests and effectiveness. *Wildlife Society Bulletin* 18:411–422.

Bond W.J., Lee, W.G., and Craine, J.M. 2004. Plant structural defences against browsing birds: a legacy of New Zealand's extinct moas. *Oikos* 104:500–508.

Boonstra, R., Krebs, C.J., and Beacham, T.D. 1980. Impact of botfly parasitism on *Microtus townsendii* populations. *Canadian Journal of Zoology* 58:1683–1692.

Both, C. and Visser, M. 2001. Adjustment to climate change is constrained by arrival date in a long-distance migrant bird. *Nature* 411:296–298.

Both, C., Bouwhuis, S., Lessells, C.M., and Visser, M.E. 2006. Climate change and population declines in a long-distance migratory bird. *Nature* 441:81–83.

Bothma, J. du P. (ed.) 1989. *Game Ranch Management*. J.L. van Schaik, Pretoria.

Botkin, D.B., Mellilo, J.M., and Wu, L. S.-Y. 1981. How ecosystem processes are linked to large mammal population dynamics. In: Fowler, C.W. and Smith T.D. (eds) *Dynamics of Large Mammal Populations*, pp. 373–387. John Wiley & Sons Ltd., New York, NY.

Botsford, L.W., Methot, R.D., and Johnson, W.E. 1983. Effort dynamics of the Northern California Dungeness crab (*Cancer magister*) fishery. *Canadian Journal of Fisheries and Aquatic Sciences* 40:337–346.

Boulinier, T., Nichols, J.D., Hines, J.E., Sauer, J.R., Flather, C.H., and Pollock, K.H. 2001. Forest fragmentation and bird community dynamics: inference at regional scales. *Ecology* 82:1159–1169.

Boutin, S. 1992. Predation and moose population dynamics: a critique. *Journal of Wildlife Management* 56:116–127.

Boutin, S., Krebs, C.J, Sinclair, A.R.E. and Smith, J.N.M. 1986. Proximate causes of losses in a snowshoe hare population. *Canadian Journal of Zoology* 64:606–610.

Bowden, D.C. and Kufeld, R.C. 1995. Generalized mark-sight population size estimation applied to Colorado moose. *Journal of Wildlife Management* 59:840–851.

Bowen, Z. and Read, J. 1998. Population and demographic patterns of rabbits (*Oryctolagus cuniculus*) at Roxby Downs in arid South Australia and the influence of rabbit haemorrhagic disease. *Wildlife Research* 25:655–662.

Boyce, M.S. 1989. *The Jackson Elk Herd: Intensive Wildlife Management in North America*. Cambridge University Press, Cambridge.

Boyce, M. 1992. Population viability analysis. *Annual Reviews of Ecology and Systematics* 23:481–506.

Boyce, M. 1998. Ecological-process management and ungulates: Yellowstone's conservation paradigm. *Wildlife Society Bulletin* 26:381–398.

Boyce, M.S. and Haney, A. 1997. *Ecosystem Management*. Yale University Press, New Haven.

Boyce, M.S. and McDonald, L.L. 1999. Relating populations to habitats using resource selection functions. *Trends in Ecology and Evolution* 14:268–272.

Boyce , M.S. and Payne, N.F. 1997. Applied disequilibriums: Riparian habitat management for wildlife. In: Boyce, M.S. and Haney A. (eds) *Ecosystem Management*, pp. 133–146. Yale University Press, New Haven, CT and London.

Boyce, M.S., Haridas, C.V., and Lee, C.T. 2006. Demography in an increasingly variable world. *Trends in Ecology and Evolution* 21:141–148.

Boyd, D.K. and Pletscher, D.H. 1999. Characteristics of dispersal in a colonizing wolf population in the central Rocky Mountains. *Journal of Wildlife Management* 63:1094–1108.

Bradley, N.L., Leopold, A.C., Ross, J., and Huffaker, W. 1999. Phenological changes reflect climate change in Wisconsin. *Proceedings of the National Academy of Sciences USA*. 96:9701–9704.

Bredon, R.M., Harker, K.W., and Marshall, B. 1963. The nutritive value of grasses grown in Uganda when fed to Zebu cattle: I. The relation between the percentage of crude protein and other nutrients. *The Journal of Agricultural Science (Cambridge)* 61:101–104.

Breininger, D.R. and Carter, G.M. 2003. Territory quality transitions and source-sink dynamics in a Florida scrub-jay population. *Ecological Applications* 13:516–529.

Briggs, S.V. 1991. Effects of breeding and environment on body condition of maned ducks *Chenonetta jubata*. *Wildlife Research* 18:577–588.

Brightwell, L.R. 1951. Some experiments with the common hermit crab (*Eupagurus bernhardus*) Linn, and transparent univalve shells. *Proceedings of the Zoological Society of London* 121:279–283.

Britannica Guide to Climate Change. 2008. Constable and Robinson, London.

Bromley, R.G., Heard, D.C., and Croft B. 1995. Visibility bias in aerial surveys relating to nest success of Arctic geese. *Journal of Wildlife Management* 59:364–371.

Brook, B.W., O'Grady, J.J., Chapman, A.P., Burgman, M.A., Akçakaya, H.R., and Frankham, R. 2000. Predictive accuracy of population viability analysis in conservation biology. *Nature* 404:385–387.

Brooks, T.M., Mittermeier, R.A., Mittermeier, C.G., da Fonseca, G.A.B., Rylands, A.B., Konstant, W.R., Flick, P., Pilgrim, J., Oldfield, S. Magin, G., and Hilton-Taylor, C. 2002. Habitat loss and extinction in the hotspots of biodiversity. *Conservation Biology* 16:900–923.

Brower, L.P. 1984. Chemical defense in butterflies. In: Vane-Wright, R.I. and Ackery, P.R. (eds) *The Biology of Butterflies*, pp. 109–134. Academic Press, London.

Brown, J.H. 1995. *Macroecology*, pp. 269. University of Chicago Press, Chicago, IL.

Brown, J.H. and Heske, E.J. 1990. Control of a desert-grassland transition by a keystone rodent guild. *Science* 250:1705–1707.

Brown, J.H. and Kodric-Brown, A. 1977. Turnover rates in insular biogeography: effect of immigration on extinction. *Ecology* 58:445–449.

Brown, J.L. 1964. The evolution of diversity in avian territorial systems. *Wilson Bulletin* 76:160–169.

Brown, J.L. 1969. The buffer effect and productivity in tit populations. *American Naturalist* 103:347–354.

Brown, J.S. 1988. Patch use as an indicator of habitat preference, predation risk, and competition. *Behavioral Ecology and Sociobiology* 22:37–47.

Brown, J.S., Kotler, B.P., and Mitchell, W.A. 1994. Foraging theory, patch use, and the structure of a Negev desert granivore community. *Ecology* 75:2286–2300.

Brown, R.D., Hellgren, E.C., Abbott, M. Ruthven, D.C. and Bingham, R.L. 1995. Effects of dietary energy and protein restriction on nutritional indices of female white-tailed deer. *Journal of Wildlife Management* 59:595–609.

Brown, R.G., Bowen, W.D., Eddington, J.D., Kimmins, W.C., Mezei, M., Parsons, J.L., and Pohajdak, B. 1997. Evidence for a long-lasting single administration contraceptive vaccine in wild grey seals. *Journal of Reproductive Immunology* 35:43–51.

Bryant, D.M. 1989. Determination of respiration rates of free-living animals by the double-labelling technique. In: Grubb, P.J. and Whittaker, J.B. (eds) *Toward a More Exact Ecology*, pp. 95–109. 30th Symposium of the British Ecological Society. Blackwell Scientific Publication, Cambridge.

Bryant, J.P. 1981. Phytochemical deterrence of snowshoe hare browsing by adventitious shoots of four Alaskan trees. *Science (Washington)* 213:889–890.

Bryant, J.P. and Kuropat, P.J. 1980. Selection of winter forage by subarctic browsing vertebrates: the role of plant chemistry. *Annual Review of Ecology and Systematics* 11:261–285.

Buckland, S.T. Anderson, D.R. Burnham, K.P., and Laake, J.L. 1993. *Distance Sampling: Estimating Abundance of Biological Populations*. Chapman and Hall, London.

Buckland, S.T., Anderson, D.R., Burnham, K.P., Laake, J.L., Borchers, D.L., and Thomas, L. 2001. *Introduction to Distance Sampling: Estimating Abundance of Biological Populations*. Oxford University Press, New York, NY.

Buckner, C.H. and Turnock, W.J. 1965. Avian predation on the larch sawfly, *Pristiphora erichsonii* (Htg.), (Hymenoptera : Tenthredinidae). *Ecology* 46:223–236.

Bunnell, F.L. 1995. Forest-dwelling vertebrate faunas and natural fire regimes in British Columbia: patterns and implications for conservation. *Conservation Biology* 9:636–644.

Burger, C., Belskii, E., Eeva, T., Laaksonen, T., Magi, M., Mand, R., Qvarnstrom, A., Slagsvold, T., Veen, T., Visser, M.E., Wiebe, K.L., Wiley, C., Wright, J., and Both, C. 2012. Climate change, breeding date and nestling diet: how temperature differentially affects seasonal changes in pied flycatcher diet depending on habitat variation. *Journal of Animal Ecology* 81:926–936.

Burnham, K.P. and Anderson, D.R. 1998. *Model Selection and Inference*. Springer-Verlag, New York, NY.

Burnham, K.P. and Overton, W.S. 1978. Estimation of the size of a closed population when capture probabilities vary among animals. *Biometrika* 65:625–633.

Burnham, K.P., Anderson, D.R., and Laake, J.L. 1980. Estimation of density from line transect sampling of biological populations. *Wildlife Monographs* 72:1–202.

Burt, W.H. 1943. Territoriality and home range concepts as applied to mammals. *Journal of Mammology* 24:346–352.

Byrom, A. E. 2002. Dispersal and survival of juvenile feral ferrets *Mustela furo* in New Zealand. *Journal of Applied Ecology* 39:67–78.

Byrom, A.E. and Krebs, C.J. 1999. Natal dispersal of juvenile arctic ground squirrels in the boreal forest. *Canadian Journal of Zoology* 77:1048–1059.

Cade, T.J. and Jones C.G. 1993. Progress in restoration of the Mauritius kestrel. *Conservation Biology* 7:169–175.

Cade, T.J. Lincer, J.L., and White, C.M. 1971. DDE residues and eggshell changes in Alaskan Falcons and Hawks. *Science* 172:955–957.

Calaby, J.H. and Grigg, C.C. 1989. Changes in macropodoid communities and populations in the past 200 years, and the future. In: Grigg, G., Jarman, P. and Hume, I. (eds) *Kangaroos, Wallabies and Rat-Kangaroos*, pp. 813–820. Surrey Beatty & Sons, New South Wales, Australia.

Calsbeek, R. and Cox, R.M. 2010. Experimentally assessing the relative importance of predation and competition as agents of selection. *Nature* 465:613–616.

Cann, R.L. and Douglas, L.J. 1999. Parasites and conservation of Hawaiin birds. In:Landweber L.F. and Dobson A.P. (eds) *Genetics and the Extinction of Species*, pp. 121–161. Princeton University Press, Princeton, NJ.

Carson, R. 1962. *Silent Spring* Hamish Hamilton, London.

Campbell, R.D., Nouvellet, P., Newman, C., MacDonald, D.W., and Rosell, F. 2012. Influence of mean climate trends and climate variance on beaver survival and recruitment dynamics. *Global Change Biology* 18:2730–2742.

Cosner, C., DeAngelis, D.L., Ault, J.S., and Olson, D.B. 1999. Effects of spatial grouping on the functional response of predators. *Theoretical Population Biology* 56:65–75.

Carpenter, J.W., Clark, G.G., and Watts, D.M. 1989. The impact of eastern equine encephalitis virus on efforts to recover the endangered whooping crane. In: Cooper, J.E. (ed.) *Disease and Threatened Birds*, pp. 115–120. International Coujncil for Bird Preservation, Technical Publication no 10. ICBP, Cambridge.

Case, T.J. 2000. *An Illustrated Guide to Theoretical Ecology*. Oxford University Press, New York, NY and Oxford.

Casey, J.M. and Myers, RA. 1998. Near extinction of a large widely distributed fish. *Science* 281:690–692.

Caswell, H. 1978. A general formula for the sensitivity of population growth rate to changes in life history parameters. *Theoretical Population Biology* 14:215–230.

Caswell, H. 2001. *Matrix population models*. Sinauer Associates, Inc., Sunderland, MA.

Catchpole, E.A., Morgan, B.J.T., Coulson, T.N., Freeman, S.N., and Albon, S.D. 2000. Factors influencing Saoy sheep survival. *Applied Statistics* 49:453–472.

Caughley, G. 1970. Population statistics of chamois. *Mammalia* 34:194–199.

Caughley, G. 1974. Bias in aerial survey. *Journal of Wildlife Management* 38:921–933.

Caughley, G. 1976. Wildlife management and the dynamics of ungulate populations. In: Coaker, T.H. (ed.) *Applied Biology*, Vol 1. pp. 183–246. Academic Press, New York, NY.

Caughley, G. 1977a. Sampling in aerial survey. *Journal of Wildlife Management* 41:605–615.

Caughley, G. 1977b. *Analysis of Vertebrate Populations*. John Wiley & Sons Ltd., London.

Caughley, G. 1981. Overpopulation. In: Jewell, P.A., Holt, S. and Hart, D. (eds) *Problems in Management of Locally Abundant Wild Mammals*, pp. 7–19. Academic Press, New York, NY.

Caughley, G. 1983. *The Deer Wars: The Story of Deer in New Zealand*. Heinemann, Auckland, New Zealand.

Caughley, G. 1985. Harvesting of wildlife: past, present and future. In: Beasom, S.L. and Roberson, S.F. (eds) *Game Harvest Management*, pp. 3–14. Caesar Kleberg Wildlife Research Institute, Kingsville, TX.

Caughley, G. 1987. Ecological relationships. In: Caughley, G., Shepherd, N., and Short, J. (eds) *Kangaroos: Their Ecology and Management in the Sheep Rangelands of Australia*, pp. 159–187. Cambridge University Press, Cambridge.

Caughley, G. 1994. Directions in conservation biology. *Journal of Animal Ecology* 63:215–244.

Caughley, G. and Goddard, J. 1972. Improving the estimates from inaccurate censuses. *Journal of Wildlife Management* 36:135–140.

Caughley, G. and Grice, D. 1982. A correction factor for counting emus from the air, and its application to counts in Western Australia. *Australian Wildlife Research* 9:253–259.

Caughley, G. and Krebs, C.J. 1983. Are big mammals simply little mammals writ large? *Oecologia* 59:7–17.

Caughley, G., Grice, D., Barker, R., and Brown, R. 1988. The edge of the range. *Journal of Animal Ecology* 57:771–785.

Caughley, G., Pech, R., and Grice, D. 1992. Effect of fertility control on a population's productivity. *Wildlife Research*19:623–627.

Chapman, D.G. 1951. Some properties of the hypergeometric distribution with application to zoological sample censuses. *University of California Publications in Statistics* 1:131–160.

Channell, R. and Lomolino, M.V. 2000. Trajectories to extinction: spatial dynamics of the contraction of geographical ranges. *Journal of Biogeography* 27:169–179.

Charnov, E.L. 1976a. Optimal foraging: attack strategy of a mantid. *American Naturalist* 110:141–151.

Charnov, E.L. 1976b. Optimal foraging: the marginal value theorem. *Theoretical Population Biology* 9:129–136.

Chave, J., Muller-Landau, H.C., and Levin, S.A. 2002. Comparing classical community models: theoretical consequences for patterns of diversity. *American Naturalist* 159:1–23.

Cheatum, E.L. 1949. Bone marrow as an index of malnutrition in deer. *New York State Conservationist* 3(5):19–22.

Chepko-Sade, B.D., Shields, W.M., Berger, J., Halpin, Z.T., Jones, W.T., Rogers, L.L., Rood, J.P., and Smith, A.T. 1987. The effects of dispersal and social structure on effective population size. In: Chepko-Sade, B.D. and Halpin, Z.T. (eds) *Mammalian Dispersal Patterns: The Effects of Social Structure on Population Genetics*, pp. 287–321. University of Chicago Press, Chicago, IL.

Cherfas, J. 1988. *The Hunting of the Whale*. The Bodley Head, London.

Chetkiewicz, C.-L.B., St. Clair, C.C., and Boyce, M.S. 2006. Corridors for conservation: integrating pattern and process. *Annual Review of Ecology and Systematics* 37:317–342.

Chilcott, M.J. and Hume, I.D. 1985. Coprophagy and selective retention of fluid digesta: their role in the nutrition of the common ringtail possum, *Pseudocheirus peregrinus*. *Australian Journal of Zoology* 33:1–15.

Child, G. and von Richter, W. 1969. Observations on ecology and behaviour of lechwe, puku, and waterbuck along the Chobe River, Botswana. *Zeitschrift für Säugetierkunde* 34:275–295.

Chitty, D. 1967. The natural selection of self-regulatory behavior in animal populations. *Proceedings of the Ecological Society of Australia* 2:51–78.

Chivers, D.J. and Langer, P. 1994. *The Digestive System in Mammals: Food, Form and Function*, Cambridge University Press, Cambridge.

Choquenot, D. 1991. Density-dependent growth, body condition, and demography in feral donkeys: testing the food hypothesis. *Ecology* 72:805–813.

Choquenot, D., Hone, J., and Saunders, G. 1999. Using apsects of predator–prey theory to evaluate helicopter shooting for feral pig control. *Wildlife Research* 26:251–261.

Ciuti, S., Muhly, T.B., Paton, D.G., McDevitt, D., Musiani, M., and Boyce, M.S. 2012. Human selection of elk behavioural triats in a landscape of fear. *Proceedings of the Royal Society of of London (B)* 279:4407–4416.

Clark, C.W. 1976. *Mathematical Bioeconomics: The Optimal Management of Renewable Resources*. John Wiley & Sons Ltd., New York, NY.

Clark, C.W. 1990. *Mathematical Bioeconomics: The Optimal Management of Renewable Resources*. John Wiley & Sons Ltd., New York, NY.

Clark, C.W. and Mangel, M. 2000. *Dynamic State Variable Models in Ecology*. Oxford University Press, Oxford.

Clark, W.C., Jones, D.D., and Holling, C.S. 1979. Lessons for ecological policy design: a case study of ecosystem management. *Ecological Modelling* 7:1–53.

Clevenger, A.P. and Waltho, N. 2000. Factors influencing the effectiveness of wildlife underpasses in Banff national park, Alberta, Canada. *Conservation Biology* 14:47–56.

Clifton-Hadley, R.S., Sauter-Louis, C.M., Lugton, I.W., Jackson, R., Durr, P.A., and Wilesmith J.W. 2001. Mycobacterial Diseases. In: Williams, E.S. and Barker, I.K. *Infectious Diseases of Wild Mammals*, pp. 340–371. Manson Publishing/The Veterinary Press.

Clutton-Brock, T.H. and Harvey, P.H. 1983. The functional significance of variation in body size among mammals. In: Eisenberg J.F. and Kleiman, D.G. (eds) *Advances in the Study of Mammalian Behaviour*, pp. 632–663. Special Publication no. 7, The American Society of Mammalogists, Shippensburg State College, Shippensburg, PA.

Clutton-Brock, T.H. and Pemberton, J.M. 2004. *Soay Sheep: Dynamics and Selection in an Island Population*. Cambridge University Press, Cambridge.

Clutton-Brock, T.H., Guinness, F.E., and Albon, S.D. 1982. *Red Deer: Behaviour and Ecology of Two Sexes*. University of Chicago Press, Chicago, IL.

Clutton-Brock, T.H., Major, M., and Guiness, F.E. 1985. Population regulation in male and female red deer. *Journal of Animal Ecology* 54:831–846.

Clutton-Brock, T.H., Price, O.F, Albon, S.D., and Jewell, P.A. 1991. Persistent instability and population regulation in Soay sheep. *Journal of Animal Ecology* 60:593–608.

Clutton-Brock, T.H., Illius, A.W., Wilson, K., Grenfell, B.T., MacColl, A.D.C., and Albon, S.D. 1997. Stability and instability in ungulate populations: an empirical analysis. *American Naturalist* 149:195–219.

Clutton-Brock, T.H., Coulson, T.N. Milner-Gulland, E.J., Thomson, D., and Armstrong, H.M. 2002. Sex differences in emigration and mortality affect optimal management of deer populations. *Nature* 415:633–637.

Cochran, W.G. 1954. The combination of estimates from different experiments. *Biometrics* 10:101–129.

Cochran, W.G. 1977. *Sampling Techniques*, 3 edn. John Wiley & Sons Ltd., New York, NY.

Cockburn, A., Scott, M.P., and Scotts, D.J. 1985. Inbreeding avoidance and male-biased natal disperal in *Antechinus* sp. (*Marsupilia: Dasyuridae*). *Animal Behaviour* 33:908–915.

Cohn, J.P. 1986. Surprising cheetah genetics. *Bioscience* 36:358–362.

Cohn, J.P. 1991. Ferrets return from near extinction. *Bioscience* 41:132–135.

Collar, N.J., Gonzaga, L.P., Krabbe, N., Madroño Nieto, A., Naranjo, L.G., Parker, T.A., and Wege, D.C. 1992. *Threatened Birds of the North Americas*. International Council for Bird Preservation, Cambridge.

Coltman, D.W., O'Donoghue, P., Jorgenson, J.T., Hogg, J.T., Strobeck, C., and Festa-Binachet, M. 2003. Undesirable evolutionary consequences of trophy hunting. *Nature* 426:655–658.

Connell, J.H. 1978. Diversity in tropical rainforests and coral reefs. *Science* 199:1302–1310.

Conner, M.L. and Leopold, B.D. 2001. A Euclidean distance metric to index dispersion from radiotelemetry data. *Wildlife Society Bulletin* 29:783–786.

Cooke, A.S. 1973. Shell thinning in avian eggs by environmental pollutants. *Environmental Pollution* 4:85–152.

Cooke, H.J. 1980. Landform evolution in the context of climatic change and neo-tectonism in the Middle Kalahari of north-central Botswana. *Transactions of the Institute of British Geography, New Series* 5:80–99.

Cooper, A.B., Hilborn, R., and Unsworth, J.W. 2003. An approach for population assessment in the absence of abundance indices. *Ecological Applications* 13:814–828.

Cooper, J. and Fourie, A. 1991. Improved breeding success of great-winged petrels *Pterodroma macroptera* following control of feral cats *Felis catus* at subantarctic Marion Island. *Bird Conservation International* 1:171–175.

Cooper, S.M. and Ginnett, T.F. 2000. Potential effects of supplemental feeding of deer on nest predation. *Wildlife Society Bulletin* 28:660–666.

Cooper, S.M., Owen-Smith, N., and Bryant, J.P. 1988. Foliage acceptability to browsing ruminants in relation to seasonal changes in the leaf chemistry of woody plants in a South African savanna. *Oecologia* 75:336–342.

Copes, P. 1970. The backward-bending supply curve of the fishing industry. *Scottish Journal of Political Economics* 17:69–77.

Coppock, D.L., Ellis, J.E., Detling, J.K., and Dyer, M.I. 1983. Plant-herbivore interactions in a North American mixed-grass prairie: II Responses of bison to modification of vegetation by prairie dogs. *Oecologia* 56:10–15.

Corfield, T.F. 1973. Elephant mortality in Tsavo National Park, Kenya. *East African Wildlife Journal* 11:339–368.

Cork, S.J. and Warner, A.C.I. 1983. The passage of digesta markers through the gut of a folivorous marsupial, the koala *Phascolarctos cinereus*. *Journal of Comparative Physiology B* 152:43–51.

Coughenour, M.B. and Singer, F.J. 1996. Elk population processes in Yellowstone National Park under the policy of natural regulation. *Ethological Applications* 6:573–593.

Coulson, T., Catchpole, E.A., Albon, S.D., Morgan, B.J.T., Pemberton, J.M., Clutton-Brock, T.H., Crawley, M.J., and Grenfell, B.T. 2001a. Age, sex, density, winter weather, and population crashes in Soay sheep. *Science* 292:1528–1531.

Coulson, T., Mace, G.M., Hudson, E., and Possingham, H. 2001b. The use and abuse of population viability analysis. *Trends in Ecology and Evolution* 16:219–221.

Coulson, T., Guiness, F., Pemberton, J., and Clutton-Brock, T. 2003. The demographic consequences of releasing a population of red deer from culling. *Ecology* 85:411–422.

Coulson, T., Benton, T.G., Lundberg, P., Dall, S.R.X., Kendall, B.E., and Gaillard, J.-M. 2006. Estimating individual contributions to population growth: evolutionary fitness in ecological time. *Proceedings of the Royal Society of London (B)* 273:547–555.

Courchamp, F., Langlais, M., and Sugihara, G. 1999. Cats protecting birds: modelling the mesopredator release effect. *Journal of Animal Ecology* 68:282–292.

Courchamp, F., Grenfell, B.T., and Clutton-Brock, T.H. 2000a. Inverse density dependence and the Allee effect. *Trends in Ecology and Evolution* 14:405–410.

Courchamp, F., Langlais, M. and Sugihara, G. 2000b. Rabbits killing birds: modelling the hyperpredation process. *Journal of Animal Ecology* 69:154–164.

Courchamp, F., Woodruffe, R. and Roemer, G. 2003. Removing protected populations to save endangered species. *Science* 302:1532.

Cowie, R.J. 1977. Optimal foraging in great tits (*Parus major*). *Nature* 268:137–139.

Cox, D.R. 1969. Some sampling problems in technology. In: Johnson, N.L. and Smith, H. (eds) *New Developments in Survey Sampling*, pp. 506–527. John Wiley & Sons Ltd., New York, NY.

Cramp, S. 1972. One hundred and fifty years of Mute Swans on the Thames. *Wildfowl* 23:119–124.

Crawford, H.S. and Jennings, D.T. 1989. Predation by birds on spruce budworm *Choristoneura fumiferana*: functional, numerical and total responses. *American Naturalist* 70:152–163.

Crawley, M.J. 1987. Benevolent herbivores? *Trends in Ecology and Evolution* 2:167–168.

Crête, M. and Bédard, J. 1975. Daily browse consumption by moose in the Gaspé Peninsula, Quebec. *Journal of Wildlife Management* 39:368–373.

Crick, Q.P., Dudley, C., Glue, D.E., and Thomson, D.L. 1997. UK birds are laying eggs earlier. *Nature* 388:526.

Crooks, K.R. and Soule, M.E. 1999. Mesopredator release and avifaunal extinctions in a fragmented system. *Nature* 400:563–566.

Crouse, D., Crowder, L., and Caswell, H. 1987. A stage-based population model for loggerhead sea turtles and implications for conservation. *Ecology* 68:1412–1423.

Crowell, K.L., and Pimm, S.L. 1976. Competition and niche shifts of mice introduced onto small islands. *Oikos* 27:251–258.

Cushman, S.A. 2006. Effects of habitat loss and fragmentation on amphibians: a review and prospectus. *Biological Conservation* 128:231–240.

Dailey, T.V., Thompson Hobbs, N., and Woodard, T.N. 1984. Experimental comparisons of diet selection by mountain goats and mountain sheep in Colorado. *Journal of Wildlife Management* 48:799–806.

Dale, B.W., Adams, L.G., and Bowyer, R.T. 1994. Functional response of wolves preying on barren-ground caribou in a multiple-prey ecosystem. *Journal of Animal Ecology* 63:644–652.

Darimont, C.T., Carlson, S.M., Kinnison, M.T., Paquet, P.C., Reimchen, T.E., and Wilmers, C.C. 2009. Human predator outpace other agents of trait change in the wild. *Proceedings of the National Academy of Sciences USA* 106:952–954.

Dark, J., Forger, N.G. and Zucker, I. 1986. Regulation and function of lipid mass during the annual cycle of the golden-mantled ground squirrel. In: Heller, H.C., Musacchia, X.J., and Wang, L.C.H. (eds) *Living in the Cold: Physiological and Biochemical Adaptations*, pp. 445–451. Elsevier Scientific, New York, NY.

Dasmann, R.F. 1956. Fluctuations in a deer population in California chaparral. *Transactions of the North American Wildlife Conference* 21:487–499.

Daszak, P. Cunningham, A.A., and Hyatt, A. 2000. Emerging infectious diseases of wildlife threats to biodiversity and human health. *Science* 287:443–449.

Davidson, N.C. and Evans, P.R. 1988. Prebreeding accumulation of fat and muscle protein by arctic-breeding shorebirds. *Proceedings of the International Ornithological Congress* 19:342–363.

Davies, S.J.J.F. 1976. Environmental variables and the biology of Australian arid zone birds. In: Frith, H.J. and Calaby, J.H. (eds) *Proceedings of the 16th International Ornithological Congress, Canberra*, pp 481–488. Australian Academy of Science, Canberra.

Davis, M.B. 1986. Climatic instability, time lags, and community disequilibrium. In: Diamond, J. and Case, T.J. (eds) *Community Ecology*, pp. 269–284. Harper and Row, New York, NY.

Davis, M. and Shaw, R.G. 2001. Range shifts and adaptive responses to Quaternary climate change. *Science* 292:673–679.

Dawson, T.J. and Ellis, B.A. 1979. Comparison of the diets of yellow-footed rock-wallabies and sympatric herbivores in western New South Wales. *Australian Wildlife Research* 6:245–254.

Dawson, T.J. and Hulbert, A.J. 1970. Standard metabolism, body temperature, and surface areas of Australian marsupials. *American Journal of Physiology* 218:1233–1238.

DeAngelis, D.L. and Waterhouse, J.C. 1987. Equilibrium and nonequilibrium concepts in ecological models. *Ecological Monographs* 57:1–21.

DeAngelis, D.L., Goldstein, R.A., and O'Neill, R.V. 1975. A model for trophic interactions. *Ecology* 56:881–892.

Debinski, D.M. and Holt, R.D. 2000. A survey and overview of habitat fragmentation experiments. *Conservation Biology* 14:342–355.

DeCesare, N.J., Hebblewhite, M., Schmiegelow, F., Hervieux, D., McDermid, G.J., Neufeld, L., Bradley, M., Whittington, J., Smith, K.G., Morgantini, L.E., Wheatley, M., and Musiani, M. 2012. Transcending scale dependence in identifying habitat with resource selection functions. *Ecological Applications* 22:1068–1083.

De Kroon, H., Plaiser, A., van Groenendael, J., and Caswell., H. 1986. Elasticity: the relative contribution of demographic parameters to population growth. *Ecology* 67:1427–1431.

DelGiudice, G.D. and Seal, U.S. 1988. Classifying winter undernutrition in deer via serum and urinary urea nitrogen. *Wildlife Society Bulletin* 16:27–32.

DelGiudice, G.D., Mech, L.D., and Seal, U.S. 1990. Effects of winter undernutrition on body composition and physiological profiles of white-tailed deer. *Journal of Wildlife Management* 54:539–550.

DelGiudice, G.D., Asleson, M.A., Varner, L., Hellgren, E.C., and Riggs, M.R. 1996. Creatinine ratios in random sampled and 24-hour urines of white-tailed deer. *Journal of Wildlife Management* 60:381–387.

deMayanier, P.G. and Hunter, M.L. 1999. Forest canopy closure and juvenile emigration by pool-breeding amphibians in Maine. *Journal of Wildlife Management* 63:441–450.

Dent, C.L., Cumming, G.S., and Carpenter, S.R. 2002. Multiple states in river and lake ecosystems. *Philosophical Transactions of the Royal Society B* 357:635–645.

Depperschmidt, J.D., Torbit, S.C., Alldredge, A.W. and Deblinger, R.D. 1987. Body condition indices for starved pronghorns. *Journal of Wildlife Management* 51:675–678.

Detling, J.K. 1998. Mammalian herbivores: ecosystem-level effects in two grassland national parks. *Wildlife Society Bulletin* 26:438–448.

De Roos, A., McCauley, E., and Wilson, W. 1991. Mobility versus density limited predator–prey dynamics on different spatial scales. *Proceedings of the Royal Society of London.* 246:117–122.

De Vries, M.F.W., Laca, E.A., and Demment, M.W. 1999. The importance of scale of patchiness for selectivity in grazing herbivores. *Oecologia (Berlin)* 121:355–363.

Diamond, J.M. 1969. Avifaunal equilibria and species turnover rates on the Channel Islands of California. *National Academy of Science, Proceedings.* 64:57–63.

Diamond, J.M. 1972. Biogeographic kinetics: estimation of relaxation times for avifaunas of southwest Pacific islands. *Proceedings of the National Academy of Sciences of the USA* 69:3199–3202.

Diamond, J.M. 1975. Assembly of species communities. In: Cody, M.L. and Diamond, J.M. (eds) *Ecology and Evolution of Communities*, pp. 342–444. Belknap, Cambridge, Massachusetts.

Dice, L.R. 1938. Some census methods for mammals. *Journal of Wildlife Management* 2:119–130.

Dinius, D.A. and Baumgardt, B.R. 1970. Regulation of food intake in ruminants: 6. Influence of caloric density of pelleted rations. *Journal of Dairy Science* 53:311–316.

Dill, L.M. 1978. An energy-based model of optimal feeding-territory size. *Theoretical Population Biology* 14:396–429.

Distel, R.A., Laca, E.A., Griggs, T.C., and Demment, M.W. 1995. Patch selection by cattle: maximization of intake rate in horizontally heterogeneous pastures. *Applied Behaviour Science* 45:11–21.

Dixon, A.M., Mace, G.M., Newby, J.E., and Olney, P.J.S. 1991. Planning for the re-introduction of scimitar-horned oryx (*Oryx dammah*) and addax (*Addax nasomaculatus*) into Niger. *Symposium Zoological Society London* 62:201–216.

Doak, D.F. 1989. Spotted owls and old growth logging in the Pacific Northwest. *Conservation Biology* 3:389–396.

Dobson, A.P. and Hudson, P.J. 1986. Parsites, disease and the structure of ecological communities. *Trends in Ecology and Evolution* 1:11–15.

Dobson, A.P. and Hudson, P.J. 1992. Regulation and stability of a free-living host–parasite system: *Trichostrongylus tenuis* in red grouse. II. Population models. *Journal of Animal Ecology* 61:487–498.

Dobson A. and Hudson, P. 1994. The interaction between the parasites and predators of Red Grouse *Lagopus lagopus scoticus*. *Ibis* 137:S87–S96.

Dobson, A.P. and May, R.M. 1986a. Disease and conservation. In: Soulé, M.E. (ed.) *Conservation Biology: The Science of Scarcity and Diversity*, pp. 345–365. Sinauer Associates, Sunderland, MA.

Dobson, A.P. and May, R.M. 1986b. Patterns of invasions by pathogens and parasites. In: Mooney H.A. and Drake, J.A. (eds) *Ecology of Biological Invasions of North America and Hawaii*, pp. 58–76. Ecological Studies 58, Springer-Verlag, New York, NY.

Dobson, A. and Meagher, M. 1996. The population dynamics of Brucellosis in the Yellowstone National Park. *Ecology* 77:1026–1036.

Dobson, F.S. 1982. Competition for mates and predominant juvenile male dispersal in mammals. *Animal Behaviour* 30:1183–1192.

Dobson, F.S. and Jones, W.T. 1985. Multiple causes of dispersal. *American Naturalist* 126:855–858.

Dodd, M.G. and Murphy, T.M. 1995. Accuracy and precision of techniques for counting great blue heron nests. *Journal of Wildlife Management* 59:667–673.

Donovan, T.M., Thompson, F.R. III,, Faaborg, J., and Probst, J.R. 1995. Reproductive Success of Migratory Birds in Habitat Sources and Sinks. *Conservation Biology* 9:1380–1395.

Droege, S. and Sauer, J.R. 1989. North American Breeding Bird Survey annual summary 1988. *United States Fish and Wildlife Service Biological Report* 89.

Drever, M.C., Clark, R.G., Derksen, C., Slatterly, S.M., Toose, P., and Nudds, T.D. 2012. Population vulnerability to climate change linked to timing of breeding of boreal ducks. *Global Change Biology* 18:480–492.

du Toit, J.T. and Cumming, D.H.M. 1999. Functional significance of ungulate diversity in African savanna and the ecological implications of the spread of pastoralism. *Biodiversity and Conservation* 8:1643–1661.

du Toit, J.T., Rogers, K.H., and Biggs, H.C. (eds) 2003. *The Kruger Experience.* Island Press, Washington, DC.

Duarte, H., Tejedo, M., Katzenberger, M., Marangoni, F., Baldo, D., Beltran, J.F., Marti, D.A., Richter-Boix, A., and Gonzalez-Voyer, A. 2012. Can amphibians take the heat? Vulnerability to climate warming in subtropical and temperate larval amphibian communities. *Global Change Biology* 18:412–421.

Dublin, H.T., Sinclair, A.R.E., and MCGlade, J. 1990. Elephants and fire as causes of multiple stable states in the Serengeti-Mara woodlands. *Journal Animal Ecology* 59:1147–1164.

DuBowy, P.J. 1988. Waterfowl communities and seasonal environments: temporal variability in interspecific competition. *Ecology* 69:1439–1453.

Durant, S.M., Kelly, M., and Caro, T.M. 2004. Factors affecting life and death in Serengeti cheetahs: environment, age, and sociality. *Behavioral Ecology* 15:11–22.

Eason C.T., Murphy, E.C., Wright, G.R.G., and Spurr E.B. 2002. Assessment of risks of brodifacoum to non-target birds and mammals in New Zealand. *Ecotoxicology* 11:35–48.

Ebbert, S.E. and Byrd, G.V. 2002. Eradications of invasive species to restore natural biological diversity on Alaska Maritime National Wildlife Refuge. In: Veitch, C.R. and Clout, M.N. (eds) *Turning the Tide: The Eradication of Invasive Species*, pp. 102–109. Occasional Paper of the IUCN Species Survival Commission No. 27 IUCN – The World Conservation Union 2002, Gland, Switzerland and Cambridge, UK.

Eberhardt, L.L. 1969. Population estimates from recapture frequencies. *Journal of Wildlife Management* 33:28–39.

Eberhardt, L.L. 1978. Transect methods for population studies. *Journal of Wildlife Management* 42:1–31.

Eberhardt, L.L. 1982. Calibrating an index by using removal data. *Journal of Wildlife Management* 46:734–740.

Eberhardt, L.L. and Simmons, M.A. 1987. Calibrating population indices by double sampling. *Journal of Wildife Management* 51:665–675.

Eberhardt, L.L. and Thomas, J.M. 1991. Designing environmental field studies. *Ecological Monographs* 6:53–73.

Eberhardt, L.L., Knight, R.R., and Blanchard, B. 1986. Monitoring grizzly bear population trends. *Journal of Wildlife Managament* 50:613–618.

Ecroyd, C.E. 1996. The ecology of *Dactylanthus taylorii* and threats to its survival. *New Zealand Journal of Ecology* 20:81–100.

Edwards, W.R. and Eberhardt, L. 1967. Estimating cottontail abundance from livetrapping data. *Journal of Wildlife Management* 31:87–96.

Edwards, M. and Richardson, A.J. 2004. Impact of climate change on marine pelagic phenology and trophic mismatch. *Nature* 430:881–884.

Elliott, G.P. 1996. Mohua and stoats: a population viability analysis. *New Zealand Journal of Zoology* 23:239–247.

Elton, C.S. 1924. Periodic fluctuations in the numbers of animals: their causes and effects. *British Journal of Experimental Biology* 2:119–163.

Elton, C. 1927. *Animal Ecology.* Macmillan, New York, NY.

Elton, C.S. 1942. *Voles, Mice, and Lemmings: Problems in Population Dynamics.* Clarendon Press, Oxford.

Elton, C.S. 1958. *The Ecology of Invasions by Animals and Plants*. Methuen, London.

Emlen, J.T., Dejong, M.J. Jaeger, M.J., Moermond, T.C., Rusterholz, K.A., and White, R.P. 1986. Density trends and range boundary constraints of forest birds along a latitudinal gradient. *Auk* 103:791–803.

Endler, J.A. 1986. *Natural Selection in the Wild*. Princeton University Press, Princeton, NJ.

Erlinge, S. 1975. Feeding habits of the weasel *Mustela nivalis* in relation to prey abundance. *Oikos* 26:378–384.

Erlinge, S., Goransson, G., Hansson, L., Hogstedt, G., Liberg, O., Nilsson, I.N., von Schantz, T., and Sylven, M. 1983. Predation as a regulating factor on small rodent populations in southern Sweden. *Oikos* 40:36–52.

Estes, J.A. and Duggins, D.O. 1995. Sea otters and kelp forests in Alaska: generality and variation in a community ecology paradigm. *Ecological Monographs* 65:75–100.

Ewins, P.J. and Miller, M.J.R. 1995. Measurement error in aerial surveys of osprey productivity. *Journal of Wildlife Management* 59:333–338.

Ezard, T.H.G., Cote, S.D., and Pelletier, F. 2009. Eco-evolutionary dynamics: disentangling phenotypic, environmental and population fluctuations. *Philosophical Transactions of the Royal Society of London (B)* 364:1491–1498.

Fa, J.E. and Peres, C.A. 2001. Game vertebrate extraction in African and Neotropical forests: an intercontinental comparison. In: Reynolds, J.D., Mace, G.M., Redford, K.H., and Robinson, J.G. (eds) *Conservation of Exploited Species*, pp. 203–241. University of Cambridge Press, Cambridge.

Fahrig, L. 2003. Effects of habitat fragmentation on biodiversity. *Annual Review of Ecology and Systematics* 34:487–515.

Fancy, S.G. and White, R.G. 1985. Energy expenditures by caribou while cratering in snow. *Journal of Wildlife Management* 49:987–993.

Farentinos, R.C., Capretta, P.J., Kepner, R.E., and Littlefield, R.E. 1981. Selective herbivory in tassle-eared squirrels: role of monoterpenes in ponderosa pines chosen as feeding trees. *Science (Washington)* 213:1273–1275.

Farley, S.D. and Robbins, C.T. 1994. Development of two methods to predict body composition of bears. *Canadian Journal of Zoology* 72:220–226.

Fedak, M.A. and Seeherman, H.J. 1979. Reappraisal of energetics of locomotion shows identical cost in bipeds and quadrupeds including ostrich and horse. *Nature (London)* 282:713–716.

Feeny, P.P. and Bostock, H. 1968. Seasonal changes in the tannin content of oak leaves. *Phytochemistry (Oxford)* 7:871–880.

Fenner, F. 1983. The Florey lecture, 1983: biological control, as exemplified by smallpox eradication and myxomatosis. *Proceedings of the Royal Society of London B, Biological Sciences* 218:259–286.

Fenner, F. and Fantini, B. 1999. *Biological Control of Vertebrate Pests. The History of Myxomatosis, an Experiment in Evolution*. CABI Pubs, Oxon, UK.

Fenton, M.B. 2012. Bats and white-nose syndrome. *Proceedings of the National Academy of Sciences USA* 109:6794–6795.

Ferrar, A.A. and Walker, B.H. 1974. An analysis of herbivore/habitat relationships in Kyle National Park, Rhodesia. *Journal of the Southern African Wildlife Management Association* 4:137–147.

Fieberg, J. and Ellner, S.P. 2000. When is it meaningful to estimate an extinction probability? *Ecology* 81:2040–2047.

Fielder, P.C. 1986. Implications of selenium levels in Washington Mountain goats, mule deer, and Rocky Mountain goats. *Northwest Science* 60:15–20.

Finch, V.A. 1972. Energy exchanges with the environment of two East African antelopes, the eland and the hartebeest. In: Maloiy, G.M.O. (ed.) *Comparative Physiology of Desert Animals. Symposia of the Zoological Society of London*, No. 31, pp. 315–326. Academic Press, London.

Findlay, C.S. and Cooke, F. 1982. Synchrony in the lesser snow goose (*Anser caerulescens*): II. The adaptive value of reproductive synchrony. *Evolution* 36:786–799.

Finger, S.E., Brisbin I.L. Jr,, Smith, M.H., and Urbston, D.F. 1981. Kidney fat as a predictor of body condition in white-tailed deer. *Journal of Wildlife Management* 45:964–968.

Fisher, R.A. 1930. *The Genetical Theory of Natural Selection*. Clarendon Press, Oxford.

Flockhart, D.T.T., Wassenaar, L.I., Martin, T.G., Hobson, K.A., Wunder, M.B., and Norris, D.R. 2013. Tracking multi-generational colonization of the breeding grounds by monarch butterflies in eastern North America. *Proceedings of the Royal Society of London (B)* 280:1087.

Flueck, W.T. 1994. Effect of trace elements on population dynamics: selenium deficiency in free-ranging black-tailed deer. *Ecology* 75:807–812.

Focardi, S. and Marcellini, P. 1996. A mathematical framework for optimal foraging of herbivores. *Journal of Mathematical Biology* 33:365–387.

Fogden, M.P.L. and Fogden, P.M. 1979. The role of fat and protein reserves in the annual cycle of the grey-backed camaroptera in Uganda (Aves: Sylvidae). *Journal of Zoology (London)* 189:233–258.

Foley, W.J. and Cork, S.J. 1992. Use of fibrous diets by small herbivores: how far can the rules be "bent"? *Trends in Ecology and Evolution* 7:159–172.

Foley, W.J. and Hume, I.D. 1987. Passage of digesta markers in two species of arboreal folivorous marsupials – the great glider (*Petauroides volans*) and the brushtail possum (*Trichosurus vulpecula*). *Physiological Zoology* 60:103–113.

Forsyth, D.M. and Hickling, G.J. 1998. Increasing Himalayan tahr and decreasing chamois densities in the eastern Southern Alps, New Zealand: evidence for interspecific competition. *Oecologica* 113:377–382.

Forsyth, D.M., Hone, J., Parkes, J.P. Reid, G.H., and Stronge, D. 2003. Feral goat control in Egmont National Park, New Zealand. *Wildlife Research* 30:437–450.

Fortin, D. and Arnold, G.W. 1997. The influence of road verges on the use of nearby small shrubland remnants by birds in the central wheatbelt of Western Australia. *Wildlife Research* 24:679–689.

Fortin, D., Beyer, H.L., Boyce, M.S., Smith, D.W., Duschesne, T., and Mao, J.S. 2005. Wolves influence elk movements: behavior shapes a trophic cascade in Yellowstone National Park. *Ecology* 86:1320–1330.

Forys, E.A. and Humphrey, S.R. 1997. Comparison of 2 methods to estimate density of an endangered lagomorph. *Journal of Wildlife Management* 61:86–92.

Fowler, C.W. 1981. Density dependence as related to life history strategy. *Ecology* 62:602–610.

Fowler, C.W. 1987. A review of density dependence in populations of large mammals. In: Genoways, H.H. (ed.) *Current Mammalogy*, pp. 401–441. Plenum Press, New York, NY.

Fox, G.A. and Gurevitch, J. 2000. Population numbers count: tools for near-term demographic analysis. *American Naturalist* 156:242–256.

Fraker, M.A., Brown, R.G., Gaunt, G.E., Kerr, J.A., and Pohajdak, B. 2002. Long-lasting, single-dose immunocontraception of feral fallow deer in British Columbia. *Journal of Wildlife Management* 66:1141–1147.

Frank, D.A. 1998. Ungulate regulation of ecosystem processes in Yellowstone National Park: direct and feedback effects. *Wildlife Society Bulletin* 26:410–418.

Frank, D.A. and Evans, R.D. 1997. Effects of native grazers on grassland N cycling in Yellowstone National Park. *Ecology* 78:2238–2248.

Frankel, O.H. and Soulé, M.E. 1981. *Conservation and Evolution*. Cambridge University Press, Cambridge.

Franklin, I.R. 1980. Evolutionary change in small populations. In: Soulé, M.E. and Wilcox, B.A. (eds) *Conservation Biology: an Evolutionary-Ecological Perspective*, pp. 135–149. Sinauer Associates, Sunderland, MA.

Franzmann, A.W and LeResche, R.E. 1978. Alaskan moose blood studies with emphasis on condition evaluation. *Journal of Wildlife Management* 42:334–351.

Fraser, D. and Reardon, E. 1980. Attraction of wild ungulates to mineral-rich springs in central Canada. *Holarctic Ecology* 3:36–39.

Fretwell, S.D. 1972. *Populations in a Seasonal Environment*. Princeton University Press, Princeton, NJ.

Fretwell, S.D. and Lucas, H.L. 1970. On territorial behavior and other factors influencing habitat distribution in birds. I. Theoretical development. *Acta Biotheoretica* 19:16–36.

Frick, W.F., Pollock, J.F., Hicks, A.C., Langwig, K.E., Reynolds, D.S., Turner, G.G., Butchkoski, C.M., and Kunz, T.H. 2010. An emerging disease causes regional population collapse of a common North American bat species. *Science* 329:679–682.

Friesen, L.E., Eagles, P.F., and Mackay, R.J. 1995. Effects of residential development on forest-dwelling neotropical migrant songbirds. *Conservation Biology* 9:1408–1414.

Frost, P.G.H. 1985. The responses of savanna organisms to fire. In: Tothill, J.C. and Mott, J.J. (eds) *Ecology and Management of the World's Savannas*, pp. 232–237. Australian Academy of Science, Canberra.

Fryxell, J.M. 1999. Functional responses to resource complexity: an experimental analysis of foraging by beavers. In: Olff, H., Brown, V.K., and Drent, R.H. (eds) *Herbivores: Between Plants and Predators*, pp. 371–396. Blackwell, Oxford.

Fryxell, J.M. and Doucet, C.M. 1993. Diet choice and the functional response of beavers. *Ecology* 74:1297–1306.

Fryxell, J.M. and Lundberg, P. 1993. Optimal patch use and metapopulation dynamics. *Evolutionary Ecology* 7:379–393.

Fryxell, J.M. and Lundberg, P. 1994. Optimal diet choice and predator–prey dynamics. *Evolutionary Ecology* 8:407–421.

Fryxell, J.M. and Lundberg, P. 1997. *Individual Behavior and Community Dynamics*. Chapman and Hall, New York, NY.

Fryxell, J.M. and Sinclair, A.R.E. 1988a. Causes and consequences of migration by large herbivores. *Trends in Ecology and Evolution* 3:237–241.

Fryxell, J.M. and Sinclair, A.R.E. 1988b. Seasonal migration by white-eared kob in relation to resources. *African Journal of Ecology* 26:17–31.

Fryxell, J.M., Mercer, W.E., and Gellately, R.B. 1988a. Population dynamics of Newfoundland moose using cohort analysis. *Journal of Wildlife Management* 52:14–21.

Fryxell, J.M., Greever, J. and Sinclair, A.R.E. 1988b. Why are migratory ungulates so abundant? *American Naturalist* 131:781–798.

Fryxell, J.M., Hussell, D.J.T., Lambert, A.B., and Smith, P.C. 1991. Time lags and population fluctuations in white-tailed deer. *Journal of Wildlife Management* 55:377–385.

Fryxell, J.M., Falls, J.B., Falls, E.A., Brooks, R.J., Dix, L. and Strickland, M.A. 1999. Density-dependence, prey-dependence and population dynamics of martens in Ontario. *Ecology* 80:1311–1321.

Fryxell, J.M., Wilmshurst, J.F., and Sinclair, A.R.E. 2004. Predictive models of movement by Serengeti grazers. *Ecology* 85:2429–2435.

Fryxell, J.M., Mosser, A., Sinclair, A.R.E., and Packer, C. 2007. Group formation stabilizes predator–prey dynamics. *Nature* 449:1041–1043.

Fryxell, J.M., Hazell, M., Börger, L., Dalziel, B.D., Haydon, D.T., Morales, J.M., McIntosh, T., and Rosatte, R.C. 2008. Multiple movement modes by large herbivores at multiple spatiotemporal scales. *Proceedings of the National Academy of Sciences USA* 105:19 114–19 119.

Fryxell, J.M., Packer, C., McCann, K., Solberg, E.J., and Sæther, B.-E. 2010. Resource management cycles and the sustainability of harvested wildlife populations. *Science* 328:903–906.

Fuller, T.K. 1989. Population dynamics of wolves in north-central Minnesota. *Wildlife Monographs* 105:1–41.

Fuller, T.K. and Keith, L.B. 1980. Wolf population dynamics and prey relationships in northeastern Alberta. *Journal of Wildlife Management* 44:583–602.

Fussmann, G.F., Loreau, M., and Abrams, P.A. 2007. Eco-evolutionary dynamics of communities and ecosystems. *Functional Ecology* 21:465–477.

Gadgil, M. and Bossert, W.H. 1970. Life historical consequences of natural selection. *American Naturalist* 104:1–24.

Gaillard, J.M., Festa-Bianchet, M., Yoccoz, N.G., and Toigo, C. 2000. Temporal variation in fitness components and population dynamics of large herbivores. *Annual Review of Ecology and Systematics* 31:367–393.

Galbreath, R. 1989. The lighthouse keepers cat. In: *Walter Buller The Reluctant Conservationist*, pp. 207–215. G.P. Books, Wellington.

Gales, R., Renouf, D., and Worthy, G.A.J. 1994. Use of biolectrical impedance analysis to assess body composition of seals. *Marine Mammal Science* 10:1–12.

Gasaway, W.C., Boertje, R.D., Grangaard, D.V., Kelleyhouse, D.G., Stephenson, R.O., and Larsen, D.G. 1992. The role of predation in limiting moose at low densities in Alaska and Yukon and implications for conservation. *Wildlife Monographs* 120:1–59.

Gascoigne, J.C. and Lipcius, R.N. 2004. Allee effects driven by predation. *Journal of Applied Ecology* 41:801–810.

Gass, C.L., Angehr, G., and Centa, J. 1976. Regulation of food supply by feeding territoriality in the rufous hummingbird. *Canadaian Journal of Zoology* 54:2046–2054.

Gaston, K.J. and Blackburn, T.M. 2000. *Pattern and Process in Macroecology*. Blackwell Science, Oxford.

Gates, C.C. and Hudson, R.J. 1981. Weight dynamics of wapiti in the boreal forest. *Acta Theriologica* 26:407–418.

Gause, G.F. 1934. *The Struggle for Existence*. Williams and Wilkins, Baltimore, MD. (Reprinted 1964 by Hafner, New York, NY.)

Gavin, T.A. and May, B. 1988. Taxonomic status and genetic purity of Columbian white-tailed deer. *Journal of Wildlife Management* 52:1–10.

Geist, V. and McTaggert-Cowan, I. (eds) 1995. *Wildlife Conservation Policy*. Deselig Press, Calgary, AB.

Gentilli, J. 1992. Numerical clines and escarpments in the geographical occurrence of avian species; and a search for relevant environmental factors. *Emu* 92:129–140.

Gerhart, K.L., White, R.G., Cameron, R.D.s and Russell D.E. 1996. Estimating fat content of caribou from body condition scores. *Journal of Wildlife Management* 60:713–718.

Getz. W.M. and Wilmers, C.C. 2004. A local nearest-neighbor convex-hull construction of home ranges and habitat utilization. *Ecography* 27:489–505.

Gibbs, J.P. and Karraker, N.E. 2006. Effects of warming conditions in eastern North American forests on red-backed salamander morphology. *Conservation Biology* 20:913–917.

Gibson, L.A., Wilson, B.A., Cahill, D.M., and Hill, J. 2004. Spatial prediction of rufous bristlebird habitat in a coastal heathland: a GIS-based approach. *Journal of Applied Ecology* 41:213–223.

Gill, R.B., Carpenter, L.H., and Bowden, D.C. 1983. Monitoring large animal populations: the Colorado experience. In: *Transactions of the 48th North American Wildlife and Natural Resources Conference*, pp. 330–441. Wildlife Management Institute, Washington, DC.

Gillingham, M.P., Parker, K.L., and Hanley T.A. 1997. Forage intake by black-tailed deer in a natural environment: bout dynamics. *Canadian Journal of Zoology.* 75:1118–1128.

Ginsburg, J.R. and Milner-Gulland, E.J. 1993. Sex-biased harvesting and population dynamics: implications for conservation and sustainable use. *Conservation Biology* 8:157–166.

Ginsberg, J.R., Mace, G.M., and Albon, S. 1995. Local extinction in a small and declining population: wild dogs in the Serengeti. *Proceedings of the Royal Society, London B* 262:221–228.

Gleeson, S.K. and Wilson, D.S. 1986. Equilibrium diet: optimal foraging and prey coexistence. *Oikos* 46:139–144.

Gochfeld, M. 1980. Mechanisms and adaptive value of reproductive synchrony in colonial seabirds. In: Burger, J., Olla, B.L., and Winn, H.E. (eds) *Behavior of Marine Animals: Current Perspectives in Research, Vol. 4: Marine Birds*, pp. 207–270. Plenum, New York, NY.

Godin, J.-G. and Keenleyside, M.H.A. 1984. Foraging on patchily distributed prey by a cichlid fish (Teleostei: Cichlidae): a test of the ideal free distribution theory. *Animal Behaviour* 32:120–131.

Golley, F.B. 1961. Energy values of ecological materials. *Ecology* 42:581–584.

Goni, R., Hilborn, R., Diaz, D., Mallol, S., and Adlerstein, S. 2010. Net contribution of spillover from a marine reserve to fishery catches. *Marine Ecology Progress Series* 400:233–243.

Gonzalez, A., Lawton, J.H., Gilbert, F.S., Blackburn, T.M., and Evans-Freke, I. 1998. Metapopulation dynamics, abundance, and distribution in a microecosystem. *Science* 281:2045–2047.

Goodman, R.E., Lebuhn, G., Seavy, N.E., Gardali, T., and Bluso-Demers, J.D. 2012. Avian body size changes and climate change: warming or increasing variability? *Global Change Biology* 18:63–73.

Gordon, I.J. 1988. Facilitation of red deer grazing by cattle and its impact on red deer performance. *Journal of Applied Ecology* 25:1–10.

Gordon I.J. 1991. Ungulate re-introductions: the case of the scimitar-horned oryx. *Symposium Zoological London* 62:217–240.

Goss-Custard, J.D. and Durrell, S.E.A. le V. dit. 1987. Age-related effects in oystercatchers, *Haematopus ostralegus*, feeding on mussels, *Mytilus edulis*. I. Foraging efficiency and interference. *Journal of Animal Ecology* 56:521–536.

Grant, P.R. and Grant, B.R. 2006. Evolution of character displacement in Darwin's finches. *Science* 313:224–226.

Green, A.J. 1998. Comparative feeding behaviour and niche organization in a Mediterranean duck community. *Canadian Journal of Zoology* 76:500–507.

Green, B. 1978. Estimation of food consumption in the dingo, *Canis familiaris dingo*, by means of ^{22}Na turnover. *Ecology* 59:207–210.

Green, G. and Brothers, N. 1989. Water and sodium turnover and estimated food consumption rates in free-living fairy prions (*Pachyptila turtur*) and common diving petrels (*Pelecanoides urinatrix*). *Physiological Zoology* 62:702–715.

Green, B., Anderson, J. and Whateley, T. 1984. Water and sodium turnover and estimated food consumption in free-living lions (*Panthera leo*) and spotted hyaenas (*Crocuta crocuta*). *Journal of Mammalogy* 65:593–599.

Green, R.H., Newton, I., Shutlz, S., Cunningham, A.A., Gilbert, M., Pain, D.J., and Prakash, V. 2004. Diclofenac poisoning as a cause of vulture population declines across the Indian subcontinent. *Journal of Applied Ecology* 41:793–800.

Greenwood, P.J. 1980. Mating systems, philopatry and dispersal in birds and mammals. *Animal Behaviour* 28:1140–1162.

Greenwood, P.J. 1983. Mating systems and the evolutionary consequences of dispersal. In: Swingland, I.R. and Greenwood, P.J. (eds) *The Ecology of Animals Movement*. Clarenden Press, Oxford.

Greenwood, P.J. and Harvey P.H. 1982. The natal and breeding dispersal of birds. *Annual Review of Ecology and Systematics* 13:1–21.

Greenwood, R.J., Pietruszewski, D.G., and Crawford, R.D. 1998. Effects of food supplementation on depredation of duck nests in upland habitat. *Wildlife Society Bulletin* 26:219–226.

Grenfell, B.T., Price, O.F., Albon, S.D., and Clutton-Brock, T.H. 1992. Overcompensation and population cycles in ungulates. *Nature* 355:823–826.

Grenfell, B.T., Wilson, K., Finkenstädt, B.F., Coulson, T.N., Murray, S., Albon, S.D., Pemberton, J.M., Clutton-Brock, T.H., and Crawley, M.J. 1998. Noise and determinism in synchronized sheep dynamics. *Nature* 394:674–677.

Grey, M.J., Clarke, M.F., and Loyn, R.H. 1997. Initial changes in the avian community of remnant eucalypt woodlands following a reduction in the abundance of Noisy Miners, *Manorina melanocephala*. *Wildlife Research* 24:631–648.

Grey, M.J., Clarke, M.F., and Loyn, R.H. 1998. Influence of the Noisy Miner, *Manorina melanocephala* on avian diversity and abundance in remnant Grey Box woodland. *Pacific Conservation Biology* 4:55–69.

Griffith, B., Scott, J.M., Carpenter, J.W., and Reed C. 1989. Translocation as a species conservation tool: status and strategy. *Science* 245:477–480.

Grigg, G.C., Taplin, L.E., Green, B., and Harlow, P. 1986. Sodium and water fluxes in free-living *Crocodylus porosus* in marine and brackish conditions. *Physiological Zoology* 59:240–253.

Gulland F.M.D. 1992. The role of nematode parasites in Soay sheep (*Ovis aries L.*) mortality during a population crash. *Parasitology* 105:493–503.

Gullion, G.W. 1965. A critique concerning foreign game bird introductions. *Wilson Bulletin* 77:409–414.

Gullison, R.E. 1998. Bigleaf mahogany. In: Milner-Gulland, E.J. and Mace, R. (eds) *Conservation of Biological Resources*, pp. 193–205. Blackwell Science, Oxford.

Gunn, A., Shank, C., and McLean, B. 1991. The history, status and management of muskoxen on Banks Island. *Arctic* 44:188–195.

Gustafsson, L. and Sutherland, W.J. 1988. The costs of reproduction in the collared flycatcher *Ficedula albicollis*. *Nature* 335:813–815.

Guthery, F.S., Lusk, J.L., and Peterson, M.L. 2001. The fall of the null hypothesis: liabilities and opportunities. *Journal of Wildlife Management* 65:379–384.

Gyllenberg, M., Hastings, A., and Hanski, I.A. 1997. Structured metapopulation models. In: Hanski, I. and Gilpin, M.E. (eds) *Metapopulation Biology*, pp. 93–122. Academic Press, San Diego, CA.

Haddad, N.M. and Baum, K.A. 1999. An experimental test of corridor effects on butterfly densities. *Ecological Applications* 9:623–633.

Haila, Y. 1986. North European land birds in forest fragments: evidence for area effects? In: Verner, J., Morrison, M.L., and Ralph, J.C. (eds) *Wildlife 2000: Modeling Habitat Relationships of Terrestiral Vertebrates*, pp. 315–319. University of Wisconsin Press, Madison, WI.

Hairston, N.G., Ellner, S.P., Geber, M.A., Yoshida, T., and Fox, J.A. 2005. Rapid evolution and the convergence of ecological and evolutionary time. *Ecology Letters* 8:1114–1127.

Haldane, J.B.S. 1942. The selective elimination of silver foxes in Canada. *Journal of Genetics* 44:296–304.

Hallett, J.G. 1982. Habitat selection and the community matrix of a desert small-mammal fauna. *Ecology* 63:1400–1410.

Halpern, B.S. 2003. The impact of marine reserves: do reserves work and does reserve size matter? *Ecological Applications* 13:S117–S137.

Halpern, B.S., Lester, S.E., and Kellner, J.B. 2010. Spillover from marine reserves and the replenishment of fished stocks. *Environmental Conservation* 36:268–276.

Hamilton, W.D. 1971. Geometry of the selfish herd. *Journal of Theoretical Biology* 31:295–311.

Hanks, J. 1981. Characterization of population condition. In: Fowler, C.W. and Smith, T.D. (eds) *Dynamics of Large Mammal Populations*, pp. 47–73. John Wiley & Sons Ltd., New York, NY.

Hansen, R.M., Mugambi, M.M., and Bauni, S.M. 1985. Diets and trophic ranking of ungulates of the northern Serengeti. *Journal of Wildlife Management* 49:823–829.

Hanski, I. 1994. A practical model of metapopulation dynamics. *Journal of Animal Ecology* 63:151–162.

Hanski, I. 1998. Metapopulation dynamics. *Nature* 396:41–49.

Hanski, I.A. and Gilpin, M.E. 1997. *Metapopulation Biology*. Academic Press, San Diego, CA and London.

Hanski, I., Kuusaari, M., and Nieminen . 1994. Metapopulation structure and migration in the butterfly *Melitaea cinxia*. *Ecology* 75:747–762.

Hanson, W.R. 1967. Estimating the density of an animal population. *Journal of Research on the Lepidoptera* 6:203–247.

Hansson, L. 1987. An interpretation of rodent cycles as due to trophic interactions. *Oikos* 50:308–318.

Hansson, L. and Henttonen, H. 1985. Gradients in density variations of rodents: the importance of latitude and snow cover. *Oecologia* 67:394–402.

Harcourt, A.H. 1995. Population viability estimates: theory and practice for a wild gorilla population. *Conservation Biology* 9:134–142.

Hard, J.J., Mills, L.S., and Peek, J.M. 2006. Genetic implications of reduced survival of male red deer Cervus elaphus under harvest. *Wildlife Biology* 12:427–441.

Hardin, G. 1960. The competitive exclusion principle. *Science* 131:1292–1297.

Harper, D.G.C. 1982. Competitive foraging in mallards: "ideal free" ducks. *Animal Behavior* 30:575–584.

Harper, G.J., Steininger, M.K., Tucker, C.J., Juhn, D., and Hawkins, F. 2007. Fifty years of deforestation and forest fragmentation in Madagascar. *Environmental Conservation* 34:325–333.

Harrington, R.N., Owen-Smith, N., Viljoen, P., Biggs, H. and Mason, D. 1999. Establishing the causes of the roan antelope decline in the Kruger National Park, South Africa. *Biological Conservation* 90:69–78.

Harris, J.B.C., Fordham, D.A., Mooney, P.A., Pedler, L.P., Araujo, M.B., Paton, D.C., Stead, M.G., Watts, M.J., Akcakaya, H.R., and Brooks, B.W. 2012. Managing the long-term persistence of a rare cockatoo under climate change. *Journal of Applied Ecology* 49:785–794.

Harrison, S., Murphy, D.D., and Ehrlich, P.R. 1988. Distribution of the bay checkerspot butterfly, *Euphydryas editha bayensis*: evidence for a metapopulation model. *American Naturalist* 132:360–382.

Hastings, A. 1997. *Population Biology*. Springer, New York, NY.

Hastings, A. and Botsford, L.W. 1999. Equivalence of yield from marine reserves and traditional fisheries. *Science* 284:1537–1538.

Hastings, A. and Powell, T. 1991. Chaos in a three-species food chain. *Ecology* 72:896–903.

Hayes, R.D. and Harestad, A.S. 2000. Wolf functional response and regulation of moose in the Yukon. *Canadian Journal of Zoology* 78:60–66.

Hayes, R.D. Baer, A.M. Wotschikowsky, U., and Harestad, A.S. 2000. Kill rate by wolves on moose in the Yukon. *Canadian Journal of Zoology* 78:49–59.

Hayne, D.W. 1949. An examination of the strip census method for estimating animal populations. *Journal of Wildlife Management* 13:145–157.

Hayward, B.J., Heske, E.J., and Painter, C.W. 1997. Effects of livestock grazing on small mammals at a desert cienaga. *Journal of Wildlife Management* 61:123–129.

Heard, D.C. 1992. The effect of wolf predation and snow cover on musk-ox group size. *American Naturalist* 139:190–204.

Heard, D.C. and Williams, W.D. 1991. Wolf den distribution on migratory barren ground caribou ranges in the NWT. In: Butler, C.E. and Mahoney, S.P. (eds) *Proceedings of the 4th North American Caribou Workshop, St. John's, Newfoundland*, pp. 249–250.

Hebblewhite, M., Pletscher, D.H., and Paquet, P.C. 2002. Elk population dynamics in Banff National Park, Alberta. *Canadian Journal of Zoology* 80:789–799.

Hebblewhite, M., Merrill, E., and McDermid, G. 2008. A multi-scale test of the forage maturation hypothesis in a partially migratory ungulate population. *Ecological Monographs* 78:141–166.

Hector, A., Joshi, J., Lawler, E., Spehn, E., and Wilby, A. 2001. Conservation implications of the link between biodiversity and ecosystem functioning. *Oecologia* 129:624–628.

Henny, C.J., Byrd, M.A., Jacobs, J.A., McLain, P.D., Todd, M.R., and Halla, B.F. 1977. Mid-Atlantic coast osprey population: present numbers, productivity, pollutant contamination, and status. *Journal of Wildlife Management* 41:254–265.

Hersteinsson, P. and Macdonald, D.W. 1992. Interspecific competition and the geographical distribution of red and arctic foxes *Vulpes vulpes* and *Alopex lagopus*. *Oikos* 64:505–515.

Heske, E.J. and Campbell, M. 1991. Effects of an 11-year livestock exclosure on rodent and ant numbers in the Chihuahuan Desert, southeastern Arizona. *Southwestern Naturalist* 36:89–93.

Hickey, J.J. and Anderson, D.W. 1968. Chlorinated hydrocarbons and eggshell changes in raptorial and fish-eating birds. *Science* 162:271–273.

Hik, D.S. 1995. Does risk of predation influence population dynamics? Evidence from the cyclic decline of snowshoe hares. *Wildlife Research* 22:115–129.

Hik, D.S. and Jefferies, R.L. 1990. Increases in the net above-ground primary production of a salt-marsh forage grass: a test of the predictions of the herbivore-optimization model. *Journal of Ecology* 78:180–195.

Hilborn, R. and Mangel, M. 1997. *The Ecological Detective*. Princeton University Press, Princeton, NJ.

Hilborn, R., Branch T.A., Ernst, B., Magnusson, A. Minte-Vera, C.V., Scheurell, M.D., and Valero, J.L. 2003. State of the worlds' fisheries. *Annual Review of Environmental Resources* 28:15.1–15.40.

Hilborn, R., Stokes, K., Maguire, J.-J., Smith, T., Botsford, L.W., Mangel, M., Orensanz, J., Parma, A., Rice, J., Bell, J., Cochrane, K.L., Garcia, S., Hall, S.J., Kirkwood, G.P., Sainsbury, K., Stefansson, G., and Walters, C. 2004. When can marine reserves improve fisheries management? *Ocean and Coastal Management* 47:1970205.

Hill, C.J. 1995. Linear strips of rain forest vegetation as potential dispersal corridors for rain forest insects. *Conservation Biology* 9:1559–1566.

Hillebrand, H. and Blenckner, T. 2002. Regional and local impact on species diversity – from pattern to process. *Oecologia* 132:479–491.

Hilton-Taylor, C. 2001. *IUCN List of Threatened Species*. IUCN.

Hitch, A.T. and Leberg, P.L. 2007. Breeding distributions of North American bird species moving north as a result of climate change. *Conservation Biology* 21:534–539.

Hjeljord, O., Sæther B.-E., and Anderson R. 1994. Estimating energy intake of free-ranging moose cows and calves through collection of feces. *Canadian Journal of Zoology* 72:1409–1415.

Hobbs, N.T., Baker, D.L., Bear, G.D., and Bowden, D.C. 1996. Ungulate grazing in sagebrush grassland: mechanisms of resource competition. *Ecological Applications* 6:200–217.

Hobbs, N.T., Gross, J.E., Shipley, L.A., Spalinger, D.E., and Wunder, B.A. 2003. Herbivore functional response in heterogeneous environments: a contest among models. *Ecology* 84:666–681.

Hobbs, R.J. 2001. Synergisms among habitat fragmentation, livestock grazing, and biotic invasions in Southwestern Australia. *Conservation Biology* 15:1522–1528.

Hobbs, T.N. 1996 Modification of ecosystems by ungulates. *Journal Wildlife Management* 60:695–713.

Hochachka, W.M. and Dhondt, A.A. 2000. Density-dependent decline of host abundance resulting from a new infectious disease *Proceedings of the National Academy of Sciences USA* 97:5303–5306.

Hodges, K.E. and Sinclair, A.R.E. 2003. Does predation risk cause snowshoe hares to modify their diets? *Canadian Journal of Zoology* 81:1973–1985.

Hodges, K.E., Krebs, C.J., Hik, D.S., Stefan, C.I., Gillis, E.A., and Doyle, C.E. 2001. Snowshoe hare demography. In: Krebs, C.J., Boutin, S. and Boonstra, R. (eds) *Ecosystem Dynamics of the Boreal Forest*, pp. 142–178. Oxford University Press, Oxford.

Hofmann, R.R. 1973. *The Ruminant Stomach: Stomach Structure and Feeding Habits of East African Game Ruminants*. East African Monographs in Biology, Vol. 2. East African Literature Bureau, Nairobi.

Holdo, R.M., Holt, R.D., and Fryxell, J.M. 2009. Opposing rainfall and nutrient gradients best explain the wildebeest migration in the Serengeti. *American Naturalist* 173:431–445.

Holekamp, K.E. 1986. Proximal causes of natal dispersal in Belding's ground squirrels (*Spermophilus beldingi*). *Ecological Monographs* 56:365–391.

Holling, C.S. 1959. The components of predation as revealed by a study of small-mammal predation of the European pine sawfly. *Canadian Entomologist* 91:293–320.

Holling, C.S. 1973. Resilience and stability of ecological systems. *Annual Review of Ecology and Systematics* 4:1–23.

Holmgren, N. 1995. The ideal free distribution of unequal competitors: predictions from a behaviour-based functional response. *Journal of Animal Ecology* 64:197–212.

Holt, R.D. 1977. Predation, apparent competition and the structure of prey communities. *Theoretical Population Biology* 12:197–229.

Holt, R.D. 1984. Spatial heterogeneity, indirect interactions, and the coexistence of prey species. *American Naturalist* 124:377–406.

Holyoak, M. and Lawler, S.P. 1996. Persistence of an extinction-prone predator–prey interaction through metapopulation dynamics. *Ecology* 77:1867–1879.

Holyoak, M., Liebold, M.A., and Holt, R.D. 2005. *Metacommunities*. University of Chicago Press, Chicago, IL.

Hone J. 1994. *Analysis of Vertebrate Pest Control*. Cambridge University Press, Cambridge.

Hone, J. 2004. *Wildlife Damage Control. Principles for the Management of Damage by Vertebrate Pests*. CSIRO Publishing, Melbourne.

Hone, J., Pech, R., and Yip, P. 1992. Estimation of the dynamics and rate of transmission of classical swine fever (hog cholera) in wild pigs. *Epidemiological Infection* 108:377–386.

Hoogland, J.L. 1982. Prairie dogs avoid extreme inbreeding. *Science* 215:1639–1641.

Hooper, P.T., Lunt, R.A. Gould, A.R., Hyatt, A.D., Russell, G.M., Kattenbelt, J.A., Blacksell, S.D., Reddacliff, L.A., Kirkland, P.D., Davis, R.J., Durham, P.J.K., Bishop, A.S.L., and Waddington, J. 1999. Epidemic blindness in kangaroos – evidence of a viral aetiology. *Australian Veterinary Journal* 77:529–536.

Hope, J.H. 1972. Mammals of the Bass Strait islands. *Proceedings of the Royal Society of Victoria* 85:163–195.

Horne, J.S., Garton, E.O., Krone, S.M., and Lewis, J.S. 2007. Analyzing animal movements using Brownian bridges. *Ecology* 88:2354–2363.

Hornicke, H. and Björnhag, G. 1980. Coprophagy and related strategies for digesta utilization. In: Ruckebusch, Y. and Thivend, P. (eds) *Digestive Physiology and Metabolism in Ruminants. Proceedings of 5th International Symposium on Ruminant Physiology, Clermont Ferrand*, France, pp. 707–730.

Hoskinson, R.L. and Mech, L.D. 1976. White-tailed deer migration and its role in wolf predation. *Journal of Wildlife Management* 40:429–441.

Houston, D.B. 1982. *The Northern Yellowstone Elk: Ecology and Management*. Macmillan, New York, NY.

Howery, L.D. and Pfister, J.A. 1990. Dietary and fecal concentrations of nitrogen and phosphorus in penned white-tailed deer does. *Journal of Wildlife Management* 54:383–389.

Hubbell, S.P. 2001. *The Unified Neutral Theory of Biodiversity and Biogeography*. Princeton University Press, Princeton, NJ.

Hudson, P.J. 1986. The effect of a parsitic nematode on the breeding production of red grouse. *Journal of Animal Ecology* 55:85–92.

Hudson, P.J. and Dobson, A.P. 1991. Control of parasites in natural populations: nematode and virus infections of red grouse. In: Perrins, C.M., Lebreton, J.D., and Hirons, G.J.M. (eds) *Bird Population Studies*, pp.413–432. Oxford University Press, Oxford.

Hudson, P. and Greenman, J. 1998 Competition mediated by parasites: biological and theoretical progress. *Trends in Ecology and Evolution* 13:387–390.

Hudson, P.J., Dobson A.P., and Newborn, D. 1992a. Do parasites make prey vulnerable to predation? Red grouse and parasites. *Journal of Animal Ecology* 61:681–692.

Hudson, P.J., Newborn, D., and Dobson, A.P. 1992b. Regulation and stability of a free-living host-parasite system: *Trichostrongylus tenuis* in red grouse. I. Monitoring and parasite reduction experiments. *Journal of Animal Ecology* 61:477–486.

Hudson, P.J., Norman, R., Laurenson, M.K., Newborn, D., Gaunt, M. Jones, L., Reid, H., Gould, E., Bowers, R., and Dobson, A. 1995. Persistence and transmission of tick-borne viruses: *Ixodes ricinus* and louping-ill virus in red grouse populations. *Parasitology* 111:S49–S58.

Hudson, P.J., Dobson A.P., and Newborn D. 1998. Prevention of populationc cycles by parasite removal. *Science* 282:2256–2258.

Hudson, R.J., Drew, K.R., and Baskin, L.M. (eds). 1989. *Wildlife Production Systems*. Cambridge University Press, Cambridge.

Huffaker, C.B. 1958. Experimental studies on predation: dispersion factors and predator–prey oscillations. *Hilgardia* 27:343–383.

Hume, I.D. and Warner, C.I. 1980. Evolution of microbial digestion in mammals. In: Ruckebusch, Y. and Thivend, P. (eds) *Digestive Physiology and Metabolism in Ruminants. Proceedings of the 5th International Symposium on Ruminant Physiology, Clermont Ferrand*, France, pp. 669–684.

Huntley, B., Altwegg, R., Barnard, P., Collingham, Y.C., and Hole, D.G. 2012. Modelling relationships between species spatial abundance patterns and climate. *Global Ecology and Biogeography* 21:668–681.

Huntley, B. and Barnard, P. 2012. Potential impacts of climate change on southern African birds of fynbos and grassland biodiversity hotspots. *Diversity and Distributions* 18:769–781.

Huntly, N. and Inouye, R. 1988. Pocket gophers in ecosystems: patterns and mechanisms. *Bioscience* 38:786–793.

Hurlbert, S.H. 1984. Pseudoreplication and the design of ecological field experiments. *Ecological Monographs* 54:187–211.

Hutchinson, G.E. 1957. Concluding remarks. *Cold Spring Harbor Symposia on Quantitative Biology* 22:415–427.

Hutton, I. 1991. *Birds of Lord Howe Island Past and Present*. Ian Hutton, Coffs Harbour Plaza, Australia.

Hutton, I. 1998. *The Australian Geographic Book of Lord Howe Island*. Australian Geographic Pty, Terrey Hills, Australia.

Huyser, O., Ryan, P.G., and Cooper, J. 2000. Changes in population size, habitat use and breeding biology of lesser sheathbills (*Chionis minor*) at Marion island: impacts of cats, mice and climate change? *Biological Conservation* 92:299–310.

Ihlow, F., Dambach, J., Engler, J.O., Flecks, M., Hartmann, T., Nekum, S., Rajaei, H., and Rodder, E. 2012. On the brink of extinction? How climate change may affect global chelonian species richness and distribution. *Global Change Biology* 18:1520–1530.

Imbeau, L. and Desrochers, A. 2002. Area sensitivity and edge avoidance: the case of the three-toed woodpecker (*Picoides tridactylus*) in a managed forest. *Forest Ecology and Management* 164:249–256.

Ims, R.A. 1990. On the adaptive value of reproductive synchrony as a predator-swamping strategy. *American Naturalist* 136:485–498.

Ims, R.A., Bondrup-Nielsen, S., and Stenseth, N. C. 1988. Temporal patterns of breeding events in small rodent populations. *Oikos* 53:229–234.

Inchausti, P. and Ginzburg, L.R. 1998. Small mammals cycles in northern Europe: patterns and evidence for a maternal effects hypothesis. *Journal of Animal Ecology* 67:180–194.

Inman, A.J. 1990. Group foraging in starlings – distribution of unequal competitors. *Animal Behaviour* 40:801–810.

Innes J., Hay, R., Flux, I., Bradfield, P., Speed, H. and Jansen, P. 1999. Successful recovery of North Island kokako (*Callaeas cinerea wilsoni*) populations, by adaptive management. *Biological Conservation* 87:201–214.

IPCC. 2007. *Climate Change 2007: Synthesis Report. Contributions of working groups I, II and III to the Fourth Assessment Report of the Intergovernmental Panel on Climate Change*. Edited by Pachurin R.K. and Reisinger, A. IPCC, Geneva, Switzerland.

Isenberg, A.C. 2000. *The Destruction of the Bison*. Cambridge University Press, Cambridge.

Jachmann, H., Berry, P.S.M., and Imae, H. 1995. Tusklessness in African elephants – a future trend. *African Journal of Ecology* 33:230–235.

Jackson, J.B.C., Kirby, M.X., Berger, W.H., Bjorndal, K.A., Botsford, L.W., Bourque, B.J., Bradbury, R.H., Cooke, R., Erlandson, J., Estes, J.A., Hughes, T.P., Kidwell, S., Lange, C.B., Lenihan, H.S., Pandolfi, J.M., Peterson, C.H., Steneck, R.S., Tegner, M.J., and Warner, R.R. 2001. Historical overfishing and the recent collapse of coastal ecosystems. *Science* 293:629–638.

Janzen, D.H. 1970. Herbivores and the number of tree species in tropical forests. *American Naturalist* 104:501–528.

Jarman, P.J. 1973. The free water intake of impala in relation to the water content of their food. *East African Agricultural and Forestry Journal* 38:343–351.

Jarman, P.J. 1974. The social organisation of antelope in relation to their ecology. *Behaviour* 48:215–266.

Jarman, P.J. and Sinclair, A.R.E. 1979. Feeding strategy and the pattern of resource partitioning in ungulates. In: Sinclair, A.R.E. and Norton-Griffiths, M. (eds) *Serengeti: Dynamics of an Ecosystem*, pp. 130–163. University of Chicago Press, Chicago, IL.

Jefferies, R.L. and Gottlieb, L.D. 1983. Genetic variation within and between populations of the asexual plant *Puccinellia* X *phryganodes*. *Canadian Journal of Botany* 61:774–779.

Jeffries, R.L., Rockwell, R.F., and Abraham, K.F. 2004. Agricultural food subsidies , migratory connectivity, and large-scale disturbance in Arctic coastal systems: a case study. *Integrative and Comparative Biology* 44:130–139.

Jenkins, S.H. 1980. A size-distance relation in food selection by beavers. *Ecology* 61:740–746.

Jenks, J.A., Soper, R.B., Lochmiller, R.L. and Leslie, D.M. Jr., 1990. Effect of exposure on nitrogen and fiber characteristics of white-tailed deer feces. *Journal of Wildlife Management* 54:389–391.

Jensen, A.L. 2000. Sex and age structured matrix model applied to harvesting a white tailed deer population. *Ecological Modelling* 128:245–249.

Johnson, C.N. 1994. Nutritional ecology of a mycophagous marsupial in relation to production of hypogeous fungi. *Ecololgy.* 75:2015–2021.

Johnson, D.H. 1980. The comparison of usage and availability measurements for evaluating resource preference. *Ecology* 61:65–71.

Johnson, D.H. 1999. The insignificance of statistical significance testing. *Journal of Wildlife Management* 63:763–772.

Johnson, D.H. 2002. The importance of replication in wildlife research. *Journal of Wildlife Management* 66:919–932.

Johnson, D.H., Krapu, G.L., Reinecke, K.J., and Jorde, D.G. 1985. An evaluation of condition indices for birds. *Journal of Wildlife Management* 49:569–575.

Johnson, J.B. and Omland, K.S. 2004. Model selection in ecology and evolution. *Trends in Ecology and Evolution* 19:101–108.

Johnson, M.L. and Gaines, M.S. 1990. Evolution of dispersal: theoretical models and empirical tests using birds and mammals. *Annual Review of Ecology and Systematics* 21:449–480.

Jolly, G.M. 1969. Sampling methods for aerial censuses of wildlife populations. *East African Agricultural Forestry Journal* 33:46–49.

Joly, D.O. and Messier, F. 2004. Testing hypotheses of bison population decline (1970–1999) in Wood Buffalo National Park: synergism between exotic disease and predation. *Canadian Journal of Zoology* 82:1165–1176.

Jones, C. 2003. Safety in numbers for secondary prey populations: an experimental test using egg predation by small mammals in New Zealand. *Oikos* 102:57–66.

Jones, D.M. 1982. Conservation in relation to animal diseases in Africa and Asia. In: Edwards, M.A. and McDonnell, U. (eds) *Animal Disease in Relation to Animal Conservation, Symposium of the Zoological Society of London*, No. 50, pp. 271–285. Academic Press, London.

Jones, M.M. 1980. Nocturnal loss of muscle protein from house sparrows (*Passer domesticus*). *Journal of Zoology (London)* 192:33–39.

Jones, W.T. 1987. Dispersal patterns in kangaroo rats (*Dipodomys spectabilis*). In: Chepko-Sade, B.D. and Halpin, Z.T. (eds) *Mammalian Dispersal Patterns: The Effects of Social Structure on Population Genetics*, pp. 119–127. University of Chicago Press, Chicago, IL.

Jørgenson, T. 1990. Long-term changes in age at sexual maturity of Northeast Arctic cod (*Gadus morhua* L.). *Journal du Conseil International pour l'Exploration de la Mer* 46:235–248.

Jorgenson, J.T., Festa-Bianchet, M., Gaillard, J.-M., and Wishart, W.D. 1997. Effects of age, sex, disease, and density on survival of bighorn sheep. *Ecology* 78:1019–1032.

Joubert, S.C.J. 1983. A monitoring programme for an extensive national park. In: Owen-Smith, R.N. (ed.) *Management of Large mammals in African Conservation Areas*, pp. 201–212. HAUM Educational Publishers, Pretoria.

Kâllén, A., Arcuri, P., and Murray, J.D. 1985. A simple model for the spatial spread and control of rabies. *Journal of Theoretical Biology* 116:377–393.

Kat, P.W., Alexander, K.A., Smith J.S., and Munson, L. 1995. Rabies and African wild dogs in Kenya. *Proceedings of the Royal Society of London B* 262:229–233.

Keane, B., Creel, S.R., and Waser, P.M. 1996. No evidence of inbreeding avoidance or inbreeding depression in a social carnivore. *Behavioral Ecology* 7:480–489.

Keating, K.A. and Cherry, S. 2004. Use and interpretation of logistic regression in habitat selection studies. *Journal of Wildlife Management* 68:774–789.

Keen, C.L. and Graham, T.W. 1989. Trace elements. In: Kaneko, J.J. (ed.) *Clinical Biochemistry of Domestic Aniamls*, 4 edn. Academic Press, New York, NY.

Keith, L.B., Cary, J.R., Rongstad, O.J., and Brittingham, M.C. 1984. Demography and ecology of a declining snowshoe hare population. *Wildlife Monographs* 90:1–43.

Kelker, G.H. 1940. Estimating deer populations by a differential hunting loss in the sexes. *Proceedings of the Utah Academy of Sciences, Arts and Letters* 17:65–69.

Kelker, G.H. 1944. Sex ratio equations and formulas for determining wildlife populations. *Proceedings of the Utah Academy of Science, Arts and Letters* 20:189–198.

Keller, L.F., Arcese, P., Smith, J.N.M., Hochachka, W.M., and Stearns, S.C. 1994. Selection against inbred song sparrows during a natural population bottleneck. *Nature* 372:356–357.

Kelly, M. and Durant, S.M. 2000. Viability of the Serengeti cheetah population. *Conservation Biology* 14:786–797.

Kelsall, J.P. and Prescott, W. 1971. *Moose and Deer Behaviour in Snow in Fundy National Park, New Brunswick*. Canadian Wildlife Service Report Series no. 15. Information Canada, Ottawa, ON.

Kenagy, G.J. 1989. Daily and seasonal uses of energy stores in torpor and hibernation. In: Malan, A. and Canguilhem, B. (eds) *Second International Symposium on Living in the Cold*, pp. 17–24. John Libbey Eurotext, London.

Kenagy, G.J., Sharbaugh, S.M., and Nagy, K.A. 1989. Annual cycle of energy and time expenditure in a golden-mantled ground squirrel population. *Oecologia* 78:269–282.

Kendall, B.E., Bjørnstad, O.N., Bascompte, J., Keitt, T.H., and Fagan, W.F. 2000. Dispersal, environmental correlation, and spatial synchrony in population dynamics. *American Naturalist* 155:628–636.

Kennedy, S. 1990. A review of the 1988 European morbillivirus epizootic. *The Veterinary Record* 127:563–567.

Keymer, A.E. and Dobson, A.P. 1987. The ecology of helminthes in populations of small mammals. *Mammal Review* 17:105–116.

Kie, J.G., Matthiopolous, J., Fieberg, J., Powell, R.A., Cagnacci, F., Mitchell, M.S., Gaillard, J.-M., and Moorcroft., P.R. 2010. The home-range concept: are traditional estimators still relevant with modern telemetry technology? *Philosophical Transactions of the Royal Society (B)* 365:2221–2231.

Kiff, L.F. 1989. DDE and the California condor *Gymnogyps californianus*: the end of a story? In: Meyburg, B.U. and Chancellor, R.D. (eds) *Raptors in the Modern World* , pp. 477–480. World Working Group on Birds of Prey and Owls, Berlin.

King, A.A. and Schaffer, W.M. 2001. The geometry of a population cycle: a mechanistic model of snowshoe hare demography. *Ecology* 82:814–830.

King, C.M. 1983. The relationships between beech (*Nothofagus* sp.) seedfall and populations of mice (*Mus musculus*) and the demographic and dietary responses of stoats (*Mustela erminea*), in three New Zealand forests. *Journal of Animal Ecology* 52:141–166.

Kingsland, S.E. 1985. *Modeling nature*. University of Chicago Press, Chicago, IL.

Kinnear, J.E., Onus, M.L. and Sumner, N.R. 1998. Fox control and rock-wallaby population dynamics – II. An update. *Wildlife Research* 25:81–88.

Kinzig, A.P., Pacala, S.W., and Tilman D. (eds) 2001. *The Functional Consequences of Biodiversity*. Princeton University Press, Princeton, NJ.

Kirkpatrick, R.L. 1980. Physiological indices in wildlife management. In: S.D. Schemnitz. *Wildlife Management Techniques Manual*, 4 edn, pp. 99–112. The Wildlife Society, Washington, DC.

Kistner, T.P., Trainer, C.E., and Hartmann, N.A. 1980. A field technique for evaluating physical condition of deer. *Wildlife Society Bulletin* 8:11–17.

Kittle, A. 2014. The Impact of Prey, Habitat and Anthropogenic Disturbance on Space Use by Social Carnivores. Unpublished PhD Thesis, University of Guelph.

Kleiber, M. 1947. Body size and metabolic rate. *Physiological Reviews* 27:511–541.

Klein, D.R. and Olson, S.T. 1960. Natural mortality patterns of deer in southeast Alaska. *Journal of Wildlife Management* 24:80–88.

Knopf, F.L., Sedgwick, J.A., and Cannon, R.W. 1988. Guild structure of a riparian avifauna relative to seasonal cattle grazing. *Journal of Wildlife Management* 52:280–290.

Knowlton, N. 1992. Thresholds and multiple stable states in coral reef community dynamics. *American Zoologist* 32:674–682.

Kodric-Brown, A. and Brown, J.H. 1978. Influence of economics, interspecific competition, and sexual dimorphism on territoriality of migrant rufous hummingbirds. *Ecology* 59:285–296.

Koford, C.B. 1953. *The California Condor*. Research Report 4, National Audubon Society, New York, NY. 154pp.

Komdeur, J. 1992. Importance of habitat saturation and territory quality for evolution of cooperative breeding in the Seychelles warbler. *Nature* 358:493–495.

Komdeur, J. 1993. Fitness-related dispersal. *Nature* 366:23–24.

Koopman, M.E., Cypher, B.L.K., and Scrivner, J.H. 2000. Dispersal patterns of San Joaquin Kit foxes (*Vulpes macrotis mutica*). *Journal of Mammalogy* 81:213–222.

Korpimäki, E. and Norrdahl, K. 1991. Functional and numerical responses of kestrels, short-eared owls, and long-eared owls to vole densities. *Ecology* 72:814–826.

Koteen, L. 2002. Climate change, whitebark pine, and grizzly bears in the greater Yellowstone ecosystem. In: Schneider, S.H. and Root, T.L. *Wildlife Responses to Climate Change: North American Case Studies*, pp. 343–414. Island Press, Washington, DC.

Kraft, K.M., Johnson, D.H., Samuelson, J.M., and Allen, S.H. 1995. Using known populations of pronghorn to evaluate sampling plans and estimators. *Journal of Wildlife Management* 59:129–137.

Krause, R.M. 1992. The origin of plagues: old and new. *Science* 257:1073–1078.

Krebs, C.J. 1999. *Ecological Methodology*, 2 edn. Benjamin/Cummings, Menlo Park, CA.

Krebs, C.J. 2001. *Ecology: The Experimental Analysis of Distribution and Abundance*, 5 edn. Addison Wesley, San Francisco, CA.

Krebs, C.J. and Myers, J.H. 1974. Population cycles in small mammals. *Advances in Ecological Research* 8:267–399.

Krebs, C.J., Gilbert, B.S., Boutin, S., Sinclair, A.R.E., and Smith, J.N.M. 1986. Population biology of snowshoe hares: I. Demography of food-supplemented populations in the southern Yukon, 1976–84. *Journal of Animal Ecology* 55:963–982.

Krebs, C.J., Boutin, S., Boonstra, R., Sinclair, A.R.E., Smith, J.N.M., Dale, M.R.T., Martin, K., and Turkington, R. 1995. Impact of food and predation on the snowshoe hare cycle. *Science* 269:1112–1115.

Krebs, C.J., Dale, M.R.T., Vilis, O., Sinclair, A.R.E., and O'Donoghue, M. 2001a. Shrubs. In: Krebs, C.J., Boutin, S., and Boonstra R. (eds) *Ecosystem Dynamics of the Boreal Forest: The Kluane Project*, pp. 92–115. Oxford University Press, Oxford.

Krebs, C.J., Boutin, S., and Boonstra, R. (eds) 2001b. *Ecosystem Dynamics of the Boreal Forest The Kluane Project*. Oxford University Press, Oxford.

Krebs, J.R. 1971. Territory and breeding density in the great tit, *Parus major* L. *Ecology* 52:2–22.

Krebs, J.R., Erichsen, J.T., and Webber, M.I. 1977. Optimal prey selection in the great tit (*Parus major*). *Animal Behavior* 25:30–38.

Krivan, V. 1996. Optimal foraging and predator–prey dynamics. *Theoretical Population Biology* 49:265–290.

Krivan, V. 1997. Dynamic ideal-free distribution: effects of optimal patch choice on predator–prey dynamics. *American Naturalist* 149:164–178.

Kruuk, H., Conroy, J.W.H. and Moorhouse, A. 1987. Seasonal reporduction, mortality and food of otters (*Lutra lutra* L.) in Shetland. *Symposium Zoological Society London* 58:263–278.

Laake, J.L., Buckland, S.T. Anderson, D.R., and Burnham, K.P. 1993. *Distance User's Guide* Colorado Cooperative Fish and Wildlife Research Unit, Colorado State University, Fort Collins, CO.

Laake, J.L., Calambokdis, J. Osmek, S.D., and Rugh,. D.J. 1997. Probability of detecting harbor porpoise from aerial surveys: estimating g(0). *Journal of Wildlife Management* 61:63–75.

Laca, E.A., Distel, R.A., Griggs, T.C., Deo, G., and Demment, M.W. 1993. Field test of optimal foraging with cattle: the marginal value theorem successfully predicts patch selection and utilisation. *Proceedings of the 17th International Grassland Congress, New Zealand*, 709–710.

Lack, D. 1971. *Ecological Isolation in Birds*. Blackwell, Oxford.

Lafferty, K.D. and Morris, K. 1996. Altered behaviour of parasitized killifish increases susceptibility to predation by bird final hosts. *Ecology* 77:1390–1397.

Laikre, L. and Ryman, N. 1991. Inbreeding depression in a captive wolf (*Canis lupus*) population. *Conservation Biology* 5:33–40.

Lamberson, R.H., McKelvey, R., Noon, B.R., and Voss, C. 1992. A dynamic analysis of northern spotted owl viability in a fragmented forest landscape. *Conservation Biology* 6:505–512.

Lamberson, R.H., Noon, B.R., Voss, C., and McKelvey, K.S. 1994. Reserve design for territorial species: the effects of patch size and spacing on the viability of the northern spotted owl. *Conservation Biology* 8:185–195.

Lamprey, H.F. 1963. Ecological separation of the large mammal species in the Tarangire game reserve, Tanganyika. *East African Wildlife Journal* 1:63–92.

Lande, R. 1987. Extinction thresholds in demographic models of territorial populations. *American Naturalist* 130:624–635.

Lande, R. 1993. Risks of population extinction from demographic and environmental stochasticity and random catastrophes. *American Naturalist* 142:911–927.

Lande, R., Sæther, B., Filli, F., Matthysen, E., and Weimerskirch, H. 2002. Estiamting density dependence from population time series using demographic theory and life-history data. *American Naturalist* 159:321–337.

Lande, R., Engen, S., and Sæther, B.-E. 2003. *Stochastic Population Dynamics in Ecology and Conservation*. Oxford University Press, Oxford.

Landsberg, J. 1988. Dieback of rural eucalypts: Tree phenology and damage caused by leaf-feeding insects. *Australian Journal of Ecology* 13:251–267.

Lane, J.E., Kruuk, L.E.B., Charmantier, A., Murie, J.O., Coltman, D.W., Buoro, M., Raveh, S., and Dobson, F.S. 2011. A quantitative genetic analysis of hibernation emergence date in a wild population of Columbian ground squirrels. *Journal of Evolutionary Biology* 24:1949–1959.

Lane, J.E., Kruuk, L.E.B., Charmantier, A., Murie, J.O., and Dobson, F.S. 2012. Delayed phenology and reduced fitness associated with climate change in a wild hibernator. *Nature* 489:554–557.

Langvatn, R. and Hanley, T.A. 1993. Feeding-patch choice by red deer in relation to foraging efficiency. *Oecologia* 95:164–170.

Larter, N.C., Sinclair, A.R.E., Ellsworth, T., Nishi, J., and Gates, C.C. 2000. Dynamics of reintroduction in an indigenous large ungulate: the wood bison of northern Canada. *Animalonservation* 3:299–309.

La Polla, V.N. and Barrett, B.W. 1993. Effects of corridor width and presence on the population dynamics of the meadow vole (*Microtus pennsylvanicus*). *Landscape Ecology* 8:25–37.

Laurance, W.F. and Cochrane, M.A. 2001. Synergistic effects in fragmented landscapes. *Conservation Biology* 15:1488–1489.

Laurance, W.F., Lovejoy, T.E., Vasconcelos, H.L. Bruna, E.M., Didham R.K., Stouffer, P.C. Gascon, C., Bierregaard, R.O. Laurance, S.G., and Sampaio, E. 2002. Ecosystem decay of Amazonian forest fragments: a 22-year investigation. *Conservation Biology* 16:605–618.

Lavigne, D.M., Brooks, R.J., Rosen, D.A., and Galbraith, D.A. 1989. Cold, energetics and populations. In: Wange, L.C.H. (ed.) *Advances in Comparative and Environmental Physiology: Vol. 4, Animal Adaptation to Cold*, pp. 403–432. Springer-Verlag, Berlin.

Law, R. 2000. Fishing, selection, and phenotypic evolution. *ICES Journal of Marine Science* 57:659–668.

Law, R. 2007. Fisheries-induced evolution: present status and future directions. *Marine Ecology Progress Series* 335:271–277.

Laws, R.M. 1969. The Tsavo Research Project. *Journal of Reproduction and Fertility* 6(Suppl.):495–531.

Leader-Williams, N. 1988. *Reindeer on South Georgia*. Cambridge University Press, Cambridge.

Lebreton, J.-D., Burnham, K.P., Clobert, J., and Anderson, D.R. 1992. Modeling survival and testing biological hypotheses using marked animals: a unified approach with case studies. *Ecological Monographs* 62:67–118.

Lefkovitch, L.P. 1965. The study of population growth in organisms shaped by stages. *Biometrics* 21:1–18.

Lehtonen, A. 1998. Managing moose, *Alces alces*, populations in Finland: hunting virtual animals. *Annales Zoologici Fennici* 35:173–179.

Leibold, M.A. 1995. The niche concept revisited: mechanistic models and community context. *Ecology* 76:1371–1382.

Leibold, M.A., Holyoak, M., Mouquet, N., Amarasekare, P., Chase, J.M., Hoopes, M.F., Holt, R.D., Shurin, J.B., Law, R., Tilman, D., Loreau, M., and Gonzales, A. 2004. The metacommunity concept: a framework for multi-scale community ecology. *Ecology Letters* 7:601–613.

Lens, L. and Dhondt, A.A. 1994. Effects of habitat fragmentation on the timing of crested tit *Parus cristatus* natal dispersal. *Ibis* 136:147–152.

Lesica, P. and Alendorf, R.W. 1995. When are peripheral populations valuable for conservation? *Conservation Biology* 9:753–760.

Leslie, D.M. Jr, and Starkey, E.E. 1985. Fecal indices to dietary quality of cervids in old-growth forests. *Journal of Wildlife Management* 49:142–146.

Leslie, D.M. Jr,, Starkey, E.E., and Vavra, M. 1984. Elk and deer diets in old-growth forests in western Washington. *Journal of Wildlife Management* 48:762–775.

Leslie, P.H. 1945. On the use of matrices in certain population mathematics. *Biometrika* 33:183–212.

Leslie, P.H. 1948. Some further notes on the use of matrices in population mathematics. *Biometrika* 35:213–245.

Lester, S.E., Halpern, B.S., Grorud-Colvert, K., Lubchenco, J., Ruttenberg, B.I., Gaines, S.D., Airamé, S., and Warner, R.R. 2009. Biological effects within no-take marine reserves: a global synthesis. *Marine Ecology Progress Series* 384:33–46.

Letcher, A.J. and Harvey, P.H. 1994. Variation in geographical range size among mammals of the palearctic. *American Naturalist* 144:30–42.

Levey, D.J. 1990. Habitat-dependent fruiting behavior of an understory tree, *Miconia centrodesma*, and tropical treefall gaps as keystone habitats for frugivores in Costa Rica. *Journal of Tropical Ecology* 6:409–420.

Levins, R. 1969. Some demographic and genetic consequences of environmental heterogeneity for biological control. *Bulletin Entomological Society America* 15:237–240.

Lewis, M.A. and Murray, J.D. 1993. Modelling territoriality and wolf-deer interactions. *Nature* 366:738–740.

Lima, S.L. and Dill, L.M. 1990. Behavioral decisions made under the risk of predation: a review and prospectus. *Canadian Journal of Zoology* 68:619–640.

Lindenmayer, D.B. 1994. Wildlife corridors and the mitigation of logging impacts on fauna in wood-production in south-eastern Australia: a review. *Wildlife Research* 21:323–340.

Lindenmayer, D.B. and Possingham, H.P. 1996. Ranking conservation and timber management options for Leadbetter's possum in southeastern Australia using population viability analysis. *Conservation Biology* 10:235–251.

Lindenmayer, D.B., Possingham, H.P., Lacy, R.C., McCarthy, M.A., and Pope, M.L. 2003. How accurate are population models? Lessons from landscape-scale tests in a fragmented system. *Ecology Letters* 6:41–47.

Lindstrom, E.R., Andren, H., Angelstam, P. Cederlund, G., Hornfeldt, B., Jaderberg, L., Lemnell, P.-A., Martinsson, B., Skold, K., and Swenson J.E. 1994. Disease reveals the predator: sarcoptic mange, red fox predation, and prey populations. *Ecology* 75:1042–1049.

Lockyer, C. 1987. Evaluation of the role of fat reserves in relation to the ecology of North Atlantic fin and sei whales. In: Huntley, A.C., Costa, D.P., Worthy, G.A.G., and Castellini, M.A. (eds) *Approaches to Marine Mammal Energetics*, Special Publication No. 1, pp. 183–203. Society for Marine Mammalogy, Lawrence, KS.

Lomolino, M.V. and Channell, R. 1995. Splendid isolation: patterns of geographic range collapse in endangered mammals. *Journal of Mammalogy* 76:335–347.

Long, K. and Robley, A. 2004. *Effective Feral Animal Exclusion Fencing for Areas of High Conservation Value in Australia*. A report for the Commonwealth of Australia, Department of Environment and Heritage, Environment Australia, Canberra.

Loreau, M. 2000. Biodiversity and ecosystem functioning: recent theoretical advances. *Oikos* 91:3–17.

Loreau, M., Naeem, S., and Inchausti, P. (eds) 2002. *Biodiversity and Ecosystem Functioning*. Oxford University Press, Oxford.

Losos, J.B., Schoener, T.W., and Spiller, D.A. 2004. Predator-induced behaviour shifts and natural selection in field-experimental lizard populations. *Nature* 432:505–508.

Losos, J.B., Schoener, T.W., Langerhans, R.B., and Spiller, D.A. 2006. Rapid temporal reversal in predator-driven natural selection. *Science* 314:1111

Lotka, A.J. 1925. *Elements of Physical Biology*. Williams and Wilkins, Baltimore, MD.

Lovejoy, T.E. and Hannah, L. (eds) 2005. *Climate Change and Biodiversity*. Yale University Press, New Haven, CT.

Loye, J. and Carroll, S. 1995. Birds, bugs and blood: avian parasitism and conservation. *Trends in Ecological Evolution* 10:232–235.

Loyn, R.H., Runnalls, R.G., Forward, G.Y., and Tyers, J. 1983. Territorial bell miners and other birds affecting populations of insect prey. *Science* 221:1411–1412.

Lubina, J.A. and Levin, S.A. 1988. The spread of a reinvading species: range expansion in the California sea otter. *American Naturalist* 131:526–543.

Ludwig, D. 1999. Is it meaningful to estimate a probability of extinction? *Ecology* 80:298–310.

Ludwig, D., Jones, D.D., and Holling, C.S. 1978 Quantitative analysis of insect outbreak systems: the spruce budworm and forests. *Journal Animal Ecology* 47:315–332.

Lundberg, P. 1988. Functional response of a small mammalian herbivore: the disc equation revisited. *Journal of Animal Ecology* 57:999–1006.

Lyster, S. 1985. *International Wildlife Law*. Groteius Publications, Cambridge.

MacArthur, R.H. 1957. On the relative abundance of bird species. *Proceedings of the National Academy of Sciences USA* 43:293–295.

MacArthur, R.H. 1958. Population ecology of some warblers of northeastern coniferous forests. *Ecology* 39:599–619.

MacArthur, R.H. 1960. On the relative abundance of species. *American Naturalist* 94:25–36.

MacArthur, R.H. 1972. *Geographical Ecology Patterns in the Distribution of Species*. Harper and Row, New York, NY.

MacArthur, R.H. and Pianka, E.R. 1966. On optimal use of a patchy environment. *American Naturalist* 100:603–609.

MacArthur, R.H. and Levins, R. 1967. The limiting similarity, convergence and divergence of coexisting species. *American Naturalist* 101:377–385.

MacArthur, R.H. and Wilson, E.O. 1967. *The Theory of Island Biogeography*. Princeton University Press, Princeton, NJ.

MacLulich, D.A. 1937. *Fluctuations in the Numbers of the Varying Hare (Lepus americanus)*. The University of Toronto Press, Toronto, ON.

Mann, C.C. and Plummer, M.L. 1995. Are wildlife corridors the right path? *Science* 270:1428–1430.

Mann, M.E., Zhang, Z., Hughes, M.K., Bradley, R.S., Miller, S.K., Rutherford, S., and Ni, F. 2008. Proxy-based reconstructions of hemispheric and global surface temperature variations over the past two millennia. *Proceedings of the National Academy of Sciences USA*. 105:13 252–13 257.

Mantyka-Pringle, .S., Martin, T.G., and Rhodes, J.R. 2012. Interactions between climate and habitat loss effects on biodiversity: a systematic review and meta-analysis. *Global Change Biology* 18:1239–1252.

McCallum, H. and Dobson, A. 1995. Detecting disease and parasite threats to endangered species and ecosystems. *Trends in Ecology and Evolution* 10:190–194.

McCann, K. and Yodzis, P. 1994. Biological conditions for chaos in a three-species food chain. *Ecology* 75:561–564.

McCauley, E., Wilson, W.G., and de Roos, A.M. 1993. Dynamics of age-structured and spatially structured predator–prey interactions: individual-based models and population-level interactions. *American Naturalist* 142:412–442.

McCleery, R.H. and Perrins, C.M. 1985. Territory size, reproductive success and population dynamics in the great tit, *Parus major*. In: Sibly, R.M. and Smith, R.H. (eds) *Behavioural Ecology: Ecological Consequences of Adaptive Behaviour*, pp. 353–373. The 25th Symposium of the British Ecological Society. Blackwell Scientific Publications, Oxford.

McCullough, D.R. 1979. *The George Reserve Deer Herd: Population Ecology of a K-Selected Species*. University of Michigan Press, Ann Arbor, MI.

McCullough, D.R. 2001. Population manipulations of North American deer *Odocoileus* spp.: balancing high yield with sustainability. *Wildlife Biology* 7:161–170.

McCullough, D.R., Pel, K.C.J., and Wang, Y. 2000. Home range, activity patterns, and habitat relations of Reeves muntkacs in Taiwan. *Journal of Wildlife Management* 64:430–441.

McCutchen, A.A. 1938. Preliminary results of wildlife census based on actual counts compared to previous estimates on national forests. *Transactions of the North American Wildlife Conference* 3:407–414.

McFarlane, D.J., George, R.J., and Farrington, P. 1993. Changes in the hydrologic cycle. In: Hobbs R.J. and Saunders D.A. *Reintegrating Fragmented Landscapes*, pp. 147–186. Springer-Verlag, New York, NY.

McGinley, M.A. and T.G. Whitham. 1985. Central place foraging by beavers (*Castor canadensis*): a test of foraging predictions and the impact of selective feeding on the growth form of cottonwoods (*Populus fremontii*). *Oecologia* 66:558–562.

McKilligan, N.G. 1984. The food and feeding ecology of the cattle egret, *Ardeola ibis*, when nesting in South-East Queensland. *Australian Wildlife Research* 11:133–144.

McLaren, B.E. and Peterson, R.O. 1994. Wolves, moose, and tree rings on Isle Royale. *Science* 266:1555–1558.

McLellan, C.H., Dobson, A.P., Wilcove, D.S., and Lynch, J.F. 1986. Effects of forest fragmentation on New- and Old-World bird communities: Empirical observations and theoretical implications. In: Verner, J., Morrison, M.L., and Ralph, J.C. (eds) *Wildlife 2000: Modeling Habitat Relationships of Terrestiral Vertebrates*, pp. 305–313. University of Wisconsin Press, Madison, WI.

McLeod, M.N. 1974. Plant tannins – their role in forage quality. *Nutrition Abstracts and Reviews* 44:803–815.

McLoughlin, C.A. and Owen-Smith, N. 2003. Viability of a diminishing roan antelope population: predation is the threat. *Animal Conservation* 6:231–236.

McNab, B.K. 1988. Complications inherent in scaling the basal rate of metabolism in mammals. *The Quarterly Review of Biology* 63:25–54.

McNaughton, S.J. 1976. Serengeti migratory wildebeest: facilitation of energy flow by grazing. *Science (Washington)* 191:92–94.

McNaughton, S.J. 1986. On plants and herbivores. *American Naturalist* 128:765–770.

McNaughton, S.J., Banykwa, F.F., and McNaughton, M.M. 1997. Promotion of the cycling of diet-enhancing nutrients by African grazers. *Science* 278:1798–1800.

McShea, W., Underwood, H.B., and Rappole, J.H. (eds) 1997. *The Science of Overabundance*. Smithsonian Institute Press, Washington, DC.

McWilliams, S.R., Afik, D., and Secor S. 1997. Patterns and processes in the vertebrate digestive system. *Trends in Ecology and Evolution* 12:420–422.

Mace, G.M. 1979. *The Evolutionary Ecology of Small Mammals*. D.Phil. Thesis, University of Sussex.

Macnab, J. 1985. Carrying capacity and related slippery shibboleths. *Wildlife Society Bulletin* 13:403–410.

Magnusson, W.E., Caughley, G.J., and Grigg, G.C. 1978. A double-survey estimate of population size from incomplete counts. *Journal of Wildlife Management* 42:174–176.

Maloiy, G.M.O. 1973. The water metabolism of a small East African antelope: the dikdik. *Proceedings of the Royal Society of London B, Biological Sciences* 184:167–178.

Mangel, M. and Clark, C.W. 1986. Towards a unified foraging theory. *Ecology* 67:1127–1138.

Manly, B.F., McDonald, L.L., and Thomas, D.L. 1993. *Resource Selection by Animals*. Chapman and Hall, London.

Manseau, M. 1996. Relationships between the Rivière George Caribou Herd and its Summer Range. PhD Thesis, University of Laval.

Marcogliese, D.J. 2001. Implications of climate change for parasitism of animals in the aquatic environment. *Canadian Journal of Zoology* 79:1331–1352.

Margules, C., Higgs, A.J., and Rafe, R.W. 1982. Modern biogeographical theory: are there any lessons for nature reserve design? *Biological Conservation* 24:115–128.

Marsh, R.E. 1988. Chemosterilants for rodent control. In: Prakash, I. (ed.) *Rodent Pest Management*, pp. 353–367. CRC Press, Boca Raton, FL.

Martinez del Rio, C. 1990. Dietary, phylogenetic, and ecological correlates of intestinal sucrose and maltase activity in birds. *Physiological Zoology* 63:987–1011.

May, R.M. 1972. Limit cycles in predator–prey communities. *Science (Washington)* 177:900–902.

May, R.M. 1973. *Stability and Complexity in Model Systems*. Princeton University Press, Princeton, NJ.

May, R.M. 1976. Simple mathematical models with very complicated dynamics. *Nature (London)* 261:459–467.

May, R.M. 1977. Thresholds and breakpoints in ecosystems with a multiplicity of stable states. *Nature* 269:471–477.

May, R.M. and Oster, G. 1976. Bifurcations and dynamic complexity in simple ecological models. *American Naturalist* 110:573–599.

Maynard Smith, J. 1982. *Evolution and the Theory of Games*. Cambridge University Press, Cambridge.

Mduma, S.A.R., Sinclair, A.R.E. and Hilborn, R. 1999. Food regulates the Serengeti wildebeest population: a 40-year record. *Journal of Animal Ecology* 68:1101–1122.

Meagher, M. and Meyer, M.E. 1994. On the origin of brucellosis in bison of Yellowstone National Park: a review. *Conservation Biology* 8:645–653.

Mech, L.D. 1994. Buffer zones of territories of gray wolves as regions of intraspecific strife. *Journal of Mammalogy* 75:199–202.

Mech, L.D. and DelGiudice, G.D. 1985. Limitations of the marrow-fat technique as an indicator of body condition. *Wildlife Society Bulletin* 13:204–206.

Messier, F. 1994. Ungulate population models with predation: a case study with the North American moose. *Ecology* 75:478–488.

Messier, F., and Crete, M. 1985. Moose-wolf dynamics and the natural regulation of moose populations. *Oecologia* 65:503–512.

Messier, F. and Joly, D.O. 2000. Comment: regulation of moose populations by wolf predation. *Canadian Journal Zoology* 78:506–510.

Messier, F., Huot, J., LeHenaff, D., and Luttich, S. 1988. Demography of the George River caribou herd: evidence of population regulation by forage exploitation and range expansion. *Arctic* 41:279–287.

Meyer, M.E. and Meagher, M. 1995. Brucellosis in free-ranging bison (*Bison bison*) in Yellowstone, Grand Teton, and Wood Buffalo National Parks: a review. *Journal of Wildlife Diseases* 31:579–598.

Middleton, A.D., Kauffman, M.J., McWhirter, D.E., Cook, J.G., Cook, R.C., Nelson, A.A., Jimenez, M.D., and Klaver, R.W. 2013. Animal migration amid shifting patterns of phenology and predation: lessons from a Yellowstone elk herd. *Ecology* 94:1245–1256.

Milinski, M. 1979. An evolutionarily stable feeding strategy in sticklebacks. *Zeitschrift für Tierpsychologie* 51:36–40.

Miller, B., Ceballos, G., and Reading, R. 1994. The prairie dog and biotic diversity. *Conservation Biology* 8:677–681.

Miller, F.L., Edmonds, E.J., and Gunn, A. 1982. Foraging behaviour of Peary caribou in response to springtime snow and ice conditions. *Canadian Wildlife Service Occasional Paper* 48:1–39.

Mills, L.S., Soule, M.E., and Doak, D.F. 1993 The keystone-species concept in ecology and conservation. *BioScience* 43:219–224.

Milner, J.M., Nilsen, E.B., and Andreassen, H.P. 2007. Demographic side effects of selective hunting in ungulates and carnivores. *Conservation Biology* 21:36–47.

Milner-Gulland, E.J. 1993. An econometric analysis of consumer demand for ivory and rhino horn. *Envrionmental and Resource Economics* 3:73–95.

Milner-Gulland, E.J. 1997. A stochastic dynamic programming model for the management of the saiga antelope. *Ecological Applications* 7:130–142.

Milner-Gulland, E.J. and Mace, R. 1998. *Conservation of Biological Resources*. Blackwell Science, Oxford.

Milner-Gulland, E.J., Bukreeva, O.M., Coulson, T., Lushchekina, A.A. Kholodova, M.V., Bekenov, A.B., and Grachev, I.A. 2003. Reproductive collapse in saiga antelope harems. *Nature* 422:135.

Milton, K. 1979. Factors influencing leaf choice by howler monkeys: a test of some hypotheses of food selection by generalist herbivores. *American Naturalist* 114:362–378.

Minchella, D.J. and Scott, M.E. 1991. Parasitism: a criptic determinant of animal community structure. *Trends in Ecology and Evolution* 6:250–254.

Mitchell, M.S. and Powell, R.A. 2004. A mechanistic home range model for optimal use of spatially distributed resources. *Ecological Modelling* 177:209–232.

Mladenoff, D.J. and Sickley, T.A. 1998. Assessing potential gray wolf restoration in the northeastern United States: a spatial prediction of favorable habitat and potential population levels. *Journal of Wildlife Management* 62:1–10.

Mladenoff, D.J., Sickley, T.A., Haight, R.G., and Wydeven, A.P. 1995. A regional landscape analysis and prediction of favorable gray wolf habitat in the northern Great Lakes region. *Conservation Biology* 9:279–294.

Moberg, A., Sonechkin, D.M., Holmgren, K., Datsenko, N.M., and Karlen, W. 2005. Highly variable northern and southern hemisphere temperatures reconstructed from low- and high-resolution proxy data. *Nature* 433:613–617.

Montgomery, B.R., Kelly, D., and Ladley, J.J. 2001. Pollinator limitation of seed set in *Fuschia perscandens* (Onagraceae) on Banks Peninsula, South Island, New Zealand. *New Zealand Journal of Botany* 29:559–565.

Monello, R.J., Murray D.L., and Cassirer E.F. 2001. Ecological correlates of pneumonia epizootics in bighorn sheep herds. *Canadian Journal of Zoology* 79:1423–1432.

Monteith, K.L., Bleich, V.C., Stephenson, T.R., Pierce, B.M., Conner, M.M., Klaver, R.W., and Bowyer, R.T. 2011. Timing of seasonal migration in mule deer: effects of climate, plant morphology, and life-history characteristics. *Ecosphere* 2:1–34.

Monteith, K.L., Long, R.A., Bleich, V.C., Heffelfinger, J.R., Krausman, P.R., and Bowyer, R.T. 2013. Effects of harvest, culture, and climate on trends in size of horn-like structures in trophy ungulates. *Wildlife Monographs* 183:1–26.

Mooney, H.A. and Ehrlich, P.R. 1997. Ecosystem services: a fragmentary history. In: Daily, G.C. (ed.) *Nature's Services: Societal Dependence on Natural Ecosystems*, pp. 11–19. Island Press, Washington, DC.

Moorcroft, P.R. and Lewis, M. 2006. *Mechanistic Home Range Dynamics*. Princeton University Press, Princeton, NJ.

Moorcroft, P.R. and Barnett, A. 2008. Mechanistic home range models and resource selection functions: a reconciliation and unification. *Ecology* 89:1112–1119.

Moorcroft, P.R., Lewis, M.A., and Crabtree, R.L. 2006. Mechanistic home range models capture spatial patterns and dynamics of coyote territories in Yellowstone. *Proceedings of the Royal Society of London (B)* 273:1651–1659.

Morales, J.M., Haydon, D.T., Frair, J., Holsinger, K.E., and Fryxell, J.M. 2004. Extracting more out of relocation data: building movement models as mixtures of random walks. *Ecology* 85:2436–2445.

Moritz, C., Patton, J.L., Conroy, C., Parra, J.L., White, G.C., and Beissinger, S.R. 2008. Impact of a century of climate change on small-mammal communities in Yosemite National Park, USA. *Science* 322:261–264.

Morris, W.F. and Doak, D.F. 2002. *Quantitative Conservation Biology*. SinauerAssociates, Sunderland, MA.

Moss, R. 1974. Winter diets, gut lengths, and interspecific competition in Alaskan ptarmigan. *Auk* 91:737–746.

Mould, E.D. and Robbins, C.T. 1981. Nitrogen metabolism in elk. *Journal of Wildlife Management* 45:323–334.

Mould, E.D. and Robbins, C.T 1982. Digestive capabilities in elk compared to white-tailed deer. *Journal of Wildlife Management* 46:22–29.

Moulton, M.P. and Sanderson., J. 1999. *Wildlife Issues in a Changing World*, 2 edn. CRC Press, Boca Raton, FL.

Mountainspring, S. and Scott, J.M. 1985. Interspecific competition among Hawaiian forest birds. *Ecological Monographs* 55:219–239.

Mueller, T. and Fagan, W.F. 2008. Search and navigation in dynamic environments – from individual behaviours to population distributions. *Oikos* 117:654–664.

Mueller, T., Olson, K.A., Fuller, T.K., Schaller, G.B., Murray, M.G., and Leimgruber, P. 2007. In search of forage: predicting dynamic habitats of Mongolian gazelles using satellite-based estimates of vegetation productivity. *Journal of Applied Ecology* 45:649–658

Mueller, T., Olson, K.A., Dressler, G., Leimgruber, P., Fuller, T.K., Nicolson, C., Novaro, A.J., Bolgeri, M.J., Wattles, D., DeStefano, S., Calabrese, J.M., and Fagan, W.F. 2011. How landscape dynamics link individual to population-level movement pattern: a multispecies comparison of ungulate relocation data. *Global Ecology and Biogeography* 20:683–694.

Mundy, P.J. and Ledger, J.A. 1976. Griffon vultures, carnivores and bones. *South African Journal of Science* 72:106–110.

Müller, H.-G., Carey, J.R., Wu, D., Liedo, P., and Vaupel, J.W. 2001. Reproductive potential predicts longevity of female Mediterranean fruitflies. *Proceedings of the Royal Society of London (B)* 268:445–450.

Munger, J.C. and Brown, J.H. 1981. Competition in desert rodents: an experiment with semipermeable exclosures. *Science (Washington, DC)* 211:510–512.

Murphy, B.C. and Dowding, J.E. 1995. Ecology of the stoat in *Nothofagus* forest: home range, habitat use and diet at different stages of the beech mast cycle. *New Zealand Journal of Ecology* 19:97–109.

Murphy, D.D. and Noon, B.R. 1992. Integrating scientific methods with habitat conservation planning: reserve design for northern spotted owls. *Ecological Applications* 2:3–17.

Murphy, D.J. and Kelly D. 2001. Scarce or distracted? Bellbird (*Anthornis melanura*) foraging and diet in an area of inadequate mistletoe pollination. *New Zealand Journal of Ecology* 25:69–81.

Murray, D.L. 2002. Differential body condition and vulnerability to predation in snowshoe hares. *Journal of Animal Ecology* 71:614–625.

Murray, D.L., Cary J.R., and Keith L.B. 1997. Interactive effects of sublethal nematodes and nutritional status on snowshoe hare vulnerability to predation. *Journal of Animal Ecology* 66:250–264.

Murray, D.L., Kapke, C.A., Evermann, J.F., and Fuller, T.K. 1999. Infectious disease and the conservation of free-ranging carnivores. *Animal Conservation* 2:241–254.

Murton, R.K., Isaacson, A.J., and Westwood, N.J. 1966. The relationships between wood-pigeons and their clover food supply and the mechanism of population control. *Journal of Applied Ecology* 3:55–93.

Mutze, G., Cooke, B., and Alexander, P. 1998. The initial impact of rabbit hemorrhagic disease on European rabbit population in South Australia. *Journal of Wildlife Diseases* 34:221–227.

Myers, J.P., Connors, P.G., and Pitelka, F.A. 1979. Territory size in wintering sanderlings: the effects of prey abundance and intruder density. *Auk* 96:551–561.

Myers, K. and Bults, H.G 1977. Observations on changes in the quality of food eaten by the wild rabbit. *Australian Journal of Ecology* 2:215–229.

Mysterud, A. and Ims, R.A. 1998. Functional responses in habitat use: availability influences relative use in trade-off situations. *Ecology* 79:1435–1441.

Naeem, S. 1998. Species redundancy and ecosystem reliability. *Conservation Biology.* 12:39–45.

Naeem, S. 2002. Ecosystem consequences of biodiversity loss: the evolution of a paradigm. *Ecology* 83:1537–1552.

Naeem, S. and Li, S. 1997. Biodiversity enhances ecosystem reliability. *Nature* 390:507.

Naeem, S., Thompson, L.J., Lawler, S.P., Lawton, J.H., and Woodfin, R.M. 1995. Empirical evidence that declining species diversity may alter the performance of terrestrial ecosystems. *Proceedings Royal Society London (B).* 347:249–262.

Nagy, K.A. 1980. CO2 production in animals: analysis of potential errors in the doubly labeled water method. *American Journal of Physiology* 238:R466–R473.

Nagy, K.A. 1983. *The Doubly Labelled Water Method: A Guide to Its Use.* UCLA Publication No. 12–1417. University of California, Los Angeles, CA.

Nagy, K.A. 1989. Field bioenergetics: accuracy of models and methods. *Physiological Zoology* 62:237–252.

Nagy, K.A., and Peterson, C.C. 1988. *Scaling of Water Flux Rate in Animals.* University of California Publications in Zoology, Vol. 120.

Nathan, R., Getz, W.M., Revilla, E., Holyoak, M., Kadmon, R., Saltz, D., and Smouse, P. 2008. A movement ecology paradigm for unifying organismal movement research. *Proceedings of the National Academy of Sciences USA* 105:19 052–19 059.

Nee, S. and May, R.M. 1992. Dynamics of metapopulations: habitat destruction and competitive coexistence. *Journal of Animal Ecology* 61:37–40.

Nelson, M.E. and Mech, L.D. 1981. Deer social organization and wolf predation in northeastern Minnesota. *Wildlife Monographs* 77:1–53.

Neubert, M.G. 2003. Marine reserves and optimal harvesting. *Ecology Letters* 6:843–849.

Nevo, E., Beiles, A., and Ben-Shlomo, R. 1984. The evolutionary significance of genetic diversity: ecological, demographic and life history correlates. *Lecture Notes in Biomathematics* 53:13–213.

Newmark, W.D. 1987. A land-bridge island perspective on mammalian extinctions in western North American parks. *Nature* 325:430–432.

Newton, I. 1972. *Finches*. Collins, London.

Nichols, J.D. 1992. Capture-recapture models: using marked animals to study population dynamics. *Bioscience* 42:94–102.

Nichols, J.D., Johnson, F.A., and Willimas, B.K. 1995. Managing North American waterfowl in the face of uncertainty. *Annual Review of Ecology and Systematics* 26:177–199.

Nicholson, M.C., Bowyer, R.T., and Kie, J.G. 1999. Habitat selection and survival of mule deer: trade-offs associated with migration. *Journal of Mammology* 78:483–504.

Noon, B.R. and McKelvey, K.S. 1996. A common framework for conservation planning: linking individual and metapopulation models. In: McCullough, D.R. (ed.) *Metapopulations and Wildlife Conservation*, pp. 139–166. Island Press, Washington, DC.

Norbury, G. 2001. Conserving dryland lizards by reducing predator-mediated apparent competition and direct competition with introduced rabbits. *Journal of Applied Ecology* 38:1350–1361.

Norbury, G.L., Norbury, D.C., and Oliver, A.J. 1994. Facultative behaviour in unpredictable environments: mobility of red kangaroos in arid Western Australia. *Journal of Animal Ecology* 63:410–418.

Norbury, G.L., Norbury, D.C., and Heyward, R.P. 1998. Behavioral responses of two predator species to sudden declines in primary prey. *Journal of Wildlife Management* 62:45–58.

Norton, G.A. 1988. Philosophy, concepts and techniques. In: Norton, G.A. and Pech, R.P. (eds) *Vertebrate Pest Management in Australia: a Decision Analysis/Systems Analysis Approach*, pp. 1–17. Project Report No. 5, CSIRO, Australia.

Norton-Griffiths, M. 1973. Counting the Serengeti migratory wildebeest using two-stage sampling. *East African Wildlife Journal* 11:135–149.

Norton-Griffiths, M. 1978. Counting animals. In: Grimsdell, J.J.R. (ed.) *Counting Animals*, 2 edn. Handbook No 1., African Wildlife Leadership Foundation, Nairobi, Kenya.

Noss, R.F. 1987. Corridors in real landscapes: a reply to Simberloff and Cox. *Conservation Biology* 1:159–164.

Novoseltsev, V.N., Novoseltseva, J.A., Boyko, S.I., and Yashin, A.I. 2003. What fecundity patterns indicate about aging and longevity: insights from *Drosophila* studies. *Journal of Georntology* 58A:484–494.

Nudds, T.D. 1990. Retroductive logic in retrospect: the ecological effects of meningeal worms. *Journal of Wildlife Management* 54:396–402.

Nudds, T.D., Sjoberg, K., and Lundberg, P. 1994. Ecomorphological relationships among Palearctic dabbling ducks on Baltic coastal wetlands and a comparison with the Nearctic. *Oikos* 69:295–303.

Nulsen, R.A. 1993 Changes in soil properties. In: Hobbs, R.J. and Saunders, D.A. (eds) *Reintegrating Fragmented Landscapes*, pp. 107–145. Springer-Verlag, New York, NY.

O'Brien, S.J., Roelke, M.E., Marker, L., Newman, A., Winkler, C.A., Meltzer, D., Colly, L., Evermann, J.F., Bush, M., and Wildt, D.E. 1985. Genetic basis for species vulnerability in the cheetah. *Science (Washington)* 227:1428–1434.

O'Brien, S.J., Wildt, D.E., and Bush, M. 1986. The cheetah in genetic peril. *Scientific American* 254(5):68–76.

O'Brien, S.J.,Wildt, D.E., Bush, M., Caro, T.M., FitzGibbon, C., Ag-gundey, I., and Leakey, R.E. 1987. East African cheetahs: evidence two population bottlenecks? *Proceedings of the National Academy of Sciences USA* 84:508–511.

O'Donoghue, M. 1991. *Reproduction, Juvenile Survival and Movements of Snowshoe Hares at a Cyclic Population Peak.* MSc Thesis, University of British Columbia, Canada.

O'Kelly, J.C. 1973. Seasonal variations in the plasma lipids of genetically different types of cattle: steers on different diets. *Comparative Biochemistry and Physiology* 44:303–312.

Oksanen, L. 1990. Predation, herbivory, and plant strategies along gradients of primary productivity. In: Tilman, D. and Grace, J. (eds) *Perspectives on Plant Competition*, pp. 445–474. Academic Press, New York, NY.

Ojasti, J. 1983. Ungulates and large rodents of South America. In: Bourlière, F. (ed.) *Ecosystems of the World: 13. Tropical Savanna*, pp. 427–439. Elsevier Scientific Publishing Company, Amsterdam.

Olliff, S., Renkin, R.A., Reinhart, D.P., Legg, K.L., and Wellington, E.M. 2013. Exotic fungus acts with natural disturbance agents to alter whitebark pine communities. In: White, P.J., Garrott, R.A., and Plumb, G.E. (eds) *Yellowstone's Wildlife in Transition*, pp. 236–254. Harvard University Press, Cambridge, MA.

Orians, G.H. and Pearson, N.E. 1979. On the theory of central place foraging. In: Horn, D.J., Stairs, G.R., and Mitchell R.D (eds) *Analysis of Ecological Systems*, pp. 155–179. Ohio University Press, Columbus, OH.

Orians, G.H. and Willson, M.F. 1964. Interspecific territories of birds. *Ecology* 45:736–745.

Osborne, P. 1985. Some effects of Dutch elm disease on the birds of a dorset dairy farm. *Journal of Applied Ecology* 22:681–691.

Otis, D.L., Burnham, K.P., White, G.C., and Anderson, D.R. 1978. Statistical interference from capture data on closed animal populations. *Wildlife Monographs* 62:1–135.

Otterman, J. 1974. Baring high-albedo soils by overgrazing: a hypothesized desertification mechanism. *Science* 186:531–533.

Ovaskainen, O., Smith, A.D., Osbourne, J.L., Reynolds, D.R., Carreck, N.L., Martin, A.P., Niitepold, K., and Hanski, I. 2008. Tracking butterfly movements with harmonic radar reveals an effect of population age on movement distance. *Proceedings of the National Academy of Sciences USA* 105:19 090–19 095.

Owen, R.B. Jr, and Reinecke, K.J. 1979. Bioenergetics of breeding dabbling ducks. In: Bookhout, T.A. (ed.) *Waterfowl and Wetlands – An Integrated Review*, pp. 71–93. Proceedings of a symposium held at the 39th Midwestern Fish and Wildlife Conference, Madison, Wisconsin. The Wildlife Society, Washington DC.

Owen-Smith, N. 1988. *Megaherbivores. The Influence of Very Large Body Size on Ecology.* Cambridge, Cambridge University Press.

Owen-Smith, N. 1990. Demography of a large herbivore, the greater kudu *Tragelaphus strepsiceros*, in relation to rainfall. *Journal of Animal Ecology* 59:893–913.

Owen-Smith, N. and Cooper, S.M. 1987. Palatability of woody plants to browsing ruminants in a South African savanna. *Ecology* 68:319–331.

Owen-Smith, N. and Cooper, S.M. 1989. Nutritional ecology of a browsing ruminant, the kudu (*Tragelaphus strepsiceros*), through the seasonal cycle. *Journal of Zoology (London)* 219:29–43.

Packer, C., Hilborn, R., Mosser, A., Kissui, B., Borner, M., Hopcraft, G., Wilmshurst, J., Mduma, S., and Sinclair, A.R.E. 2005. Ecological change, group territoriality, and population dynamics in Serengeti lions. *Science* 307:390–393.

Paine, R.T. 1969 A note on trophic complexity and species diversity. *American Naturalist* 103:91–93.

Pagel, M., May, R.M., and Collie, A.R. 1991. Ecological aspects of the geographic distribution and diversity of mammal species. *American Naturalist* 137:791–815.

Palomares F., Gaona, P., Ferreras, P., and Delibes, M. 1995. Positive effects on game species of top predators by controlling smaller predator populations: an example with lynx, mongooses, and rabbits. *Conservation Biology* 9:295–305.

Parer, I. 1987. Factors influencing the distribution and abundance of rabbits (*Oryctolagus cuniculus*) in Queensland. *Proceedings of the Royal Society of Queensland* 98:73–82.

Parer, I., Conolly, D., and Sobey, W.R. 1985. Myxomatosis: the effects of annual introductions of an immunizing strain and a highly virulent strain of myxoma virus into rabbit populations at Urana, NSW. *Australian Wildlife Research* 12:407–423.

Parkes, J.P. 1984. Feral goats on Raoul Island. I. Effect of control methods on their density, distribution, and productivity. *New Zealand Journal of Ecology* 7:85–94.

Parkes, J.P. and Tustin, K.G. 1985. A reappraisal of the distribution and dispersal of femal Himilayan tahr in New Zealand. *New Zealand Journal of Ecology* 8:5–10.

Parkes, J.P., Macdonald, N., and Leaman, G. 2002. An attempt to eradicate feral goats from Lord Howe Island. In: Veitch, C.R. and Clout, M.N. (eds) *Turning the Tide: The Eradication of Invasive Species*, pp. 233–239. Occasional Paper of the IUCN Species Survival Commission No. 27 IUCN – The World Conservation Union 2002, Gland, Switzerland and Cambridge, UK.

Parmesan, C. 2006. Ecological and evolutionary responses to recent climate change. *Annual Review of Ecology, Evolution, and Systematics* 37:637–669.

Parmesan, C. and Yohe, G. 2003. A globally coherent fingerprint of climate change impacts across natural systems. *Nature* 421:37–42.

Pastor, J., Dewey, B., Naiman, R.J., McInnes, P.F., and Cohen Y. 1993. Moose browsing and soil fertility in the boreal forests of Isle Royale National Park. *Ecology* 74:467–480.

Patino-Martinez, J., Marco, A., Quinones, L., and Hawkes, L. 2012. A potential tool to mitigate the impacts of climate change to the Caribbean leatherback sea turtle. *Global Change Biology* 18:401–411.

Patterson, I.J. 1965. Timing and spacing of broods in the black-headed gull. *Ibis* 107:433–459.

Paulus, S.L. 1982. Gut morphology of gadwalls in Louisiana in winter. *Journal of Wildlife Management* 46:483–489.

Pearson, R. 2011. *Driven to Extinction. The impact of Climate Change on Biodiversity*. Natural History Museum, London.

Pease, J.L., Vowles, R.H., and Keith, L.B. 1979. Interaction of snowshoe hares and woody vegetation. *Journal of Wildlife Management* 43:43–60.

Pech, R.P. 2000. Biological control of vertebrate pests. *Proceedings of the 19th Vertebrate Pest Conference*, pp. 206–211. University of California, Davis, CA.

Pech, R.P. and McIlroy, J.C. 1990. A model of the velocity of advance of foot and mouth disease in feral pigs. *Journal of Applied Ecology* 27:635–650.

Pech, R.P., Sinclair, A.R.E., Newsome, A.E., and Catling, P.C. 1992. Limits to predator regulation of rabbits in Australia: evidence from predator-removal experiments. *Oecologia* 89:102–112.

Peery, M.Z., Gutierrez, R.J., Kirby, R., Ledee, O.E., and Lahaye, W. 2012. Climate change and spotted owls: potentially contrasting responses in the Southwestern United States. *Global Change Biology* 18:865–880.

Pehrsson, O. 1984. Relationship of food to spatial and temporal breeding strategies of mallards in Sweden. *Journal of Wildlife Management* 48:322–339.

Pelletier. F., Clutton-Brock, T., Pemberton, J., Tuljapurkar, S., and Coulson, T. 2007. The evolutionary demography of ecological change: linking trait variation and population growth. *Science* 315:1571–1574.

Pelletier, L. and Krebs, C.J. 1997. Line-transect sampling for estimating ptarmigan (*Lagopus* spp.) density. *Canadian Journal of Zoology* 75:1185–1192.

Penning, S.C., Nadeau, M.T., and Paul, V.J. 1993. Selectivity and growth of the gneralist herbivore *Dolabella auricularia* feeding upon complementary resources. *Ecology* 74:879–890.

Perrins, C.M. 1970. The timing of birds' breeding seasons. *Ibis* 112:242–255.

Petersen, C.W. and Levitan, D.R. 2001. The Allee effect: a barrier to recovery by exploited species. In: Reynolds, J.D., Mace, G.M. Redford, K.H., and Robinson, J.G. (eds) *Conservation of Exploited Species*, pp. 281–300.Cambridge University Press, Cambridge.

Peterson, A.T., Ortega-Huerta, M.A., Bartley, J., Sánchez-Cordero, V., Soberón, J., Buddemeier, R.H., and Stockwell, D.R.B. 2002. Future projections for Mexican faunas under global climate change scenarios. *Nature* 416:626–629.

Peterson, M.J., Grant, W.E., and Davis, D.S. 1991. Bison-brucellosis management: simulation of alternative strategies. *Journal of Wildlife Management* 55:205–213

Peterson, R.O. 1999. Wolf-moose interaction on Isle Royale: the end of natural regulation? *Ecological Applications* 9:10–16.

Peterson, R.O. and Vucetich, J.A. 2003 *Ecological Studies of Wolves on Isle Royale: Annual Report 2002–2003*. Michigan Technological University, Houghton, MI.

Peterson, R.O., Thomas, N.J., Thurber, J.M., Vucetich, J.A., and Waite, T.A. 1998. Population limitation and the wolves of Isle Royale. *Journal of Mammalogy* 79:828–841.

Philipp, D.P., Cooke, S.J., Claussen, J.E., Koppelman, J.B., Suski, C.D., and Burkett, D.P. 2009. Selection for vulnerability to angling in largemouth bass. *Transactions of the American Fisheries Society* 138:189–199.

Pianka, E.R., Huey, R.B. and Lawlor, L.P. 1979. Niche segregation in desert lizards. In: Horn, D.J., Stairs, G.R., and Mitchell, R.D. (eds) *Analysis of Ecological Systems*, pp. 67–116. Ohio State University Press, Columbus, OH.

Pickett, S.T.R. and White, P.S. (eds) 1985. *The Ecology of Natural Disturbance and Patch Dynamics*. Academic Press, Orlando, FL.

Pierce, G.J. and Ollason, J.G. 1987. Eight reasons why optimal foraging is a complete waste of time. *Oikos* 49:111–118.

Piersma, T and Lindstrom, A. 1997. Rapid reversible changes in organ size as a component of adaptive behaviour. *Trends in Ecology and Evolution* 12:134–138.

Pimm, S.L. and Pimm, J.W. 1982. Resource use, competition and resource availability in Hawaiian honeycreepers. *Ecology* 63:1468–1480.

Plowright, W. 1982. The effects of rinderpest and rinderpest control on wildlife in Africa. *Symposium Zoology Society London* 50:1–28.

Pojar, T.M., Bowden, D.C., and Gill, R.B. 1995. Aerial counting experiments to estimate pronghorn density and herd structure. *Journal of Wildlife Management* 59:117–128.

Polis, G. A. 1999. Why are parts of the world green: multiple factors control productivity and the distribution of biomass. *Oikos* 86:3–15.

Pollock, K.H. 1974. *The Assumption of Equal Catchability in Tag-Recapture Experiments*. PhD Dissertation, Cornell University, Ithaca, NY.

Pollock, K.H., Winterstein, S.R., Bunck, C.M., and Curtis, P.D. 1989. Survival analysis in telemetry studies: the staggered entry design. *Journal of Wildlife Management* 53:7–15.

Porter, R.N. 1977. *Wildlife Management Objectives and Practices for the Hluhluwe Game Reserve and the Northern Corridor*. Natal Parks Board (unpublished).

Post, E. and Stenseth, N.C. 1998. Large-scale fluctuation and population dynamics of moose and white-tailed deer. *Journal of Animal Ecology* 67:537–543.

Post, E., Peterson, R.O., Stenseth, N.C., and McLaren, B.E. 1999. Ecosystem consequences of wolf behavioural response to climate. *Nature* 401:905–907.

Post, J.R., Sullivan, M., Cox, S., Lester, N.P., Walters, C.J., Parkinson, E.A., Paul, A.J., Jackson, L., and Shuter, B.J. 2002. Canada's recreational fisheries: the invisible collapse? *Fisheries* 27:6–17.

Poulin, R. 1999 The functional importance of parasites in animal communities: many roles at many levels? *International Journal for Parasitology* 29:903–914.

Pounds, J.A. and Crump, M.L. 1994. Amphibian declines and climate disturbance: the case of the golden toad and the harlequin frog. *Conservation Biology* 8:72–85.

Pounds, J.A., Fogden, M.P.L., and Campbell, J.H. 1999. Biological response to climate change on a tropical mountain. *Nature* 398:611–615.

Pounds, J.A., Bustamante, M.R., Coloma, L.A., Consuegra, J.A., Fogden, M.P.L., Foster, P.N., La Marca, E., Masters, K.L., Merino-Viteri, A., Puschendorf, R., Ron, S.R., Sánchez-Azofeifa, G.A., Still, C.J., and Young, B.E. 2006. Widespread amphibian extinctions from epidemic disease driven by global warming. *Nature* 439:161–167.

Power, M.E., Tilman, D., Estes, J.A., Menge, B.A., Bond, W.J., Mills, S., Daily, G., Castilla, J.C., Lubchenco, J., and Paine, R.T. 1996. Challenges in the quest for keystones. *BioScience* 46:609–620.

Probst, J.R. 1986. A review of factors limiting the Kirtland's warbler on its breeding grounds. *American Midland Naturalist* 116:87–100.

Prenzlow, D.M. and Lovvorn, J.R. 1996. Evaluation of visibility correction factors for waterfowl surveys in Wyoming. *Journal of Wildlife Management* 60:286–297.

Provenza, F.D., Burritt, E.A., Clausen, T.P., Bryant, J.P., Reichardt, P.B., and Distel, R.A. 1990. Conditioned flavour aversion: a mechanism for goats to avoid condensed tannins in blackbrush. *American Naturalist* 136:810–828.

Pulliam, H.R. 1974. On the theory of optimal diets. *American Naturalist* 108:59–75.

Pulliam H.R. 1988. Sources, sinks, and population regulation. *American Naturalist* 132:652–661.

Pulliam, H.R. and Danielson, B.J. 1991. Sources, sinks and habitat selection: a landscape perspective. *American Naturalist* 137:50–66.

Pusey, A.E. 1987. Sex-biased dispersal and inbreeding avoidance in birds and mammals. *Trends in Ecology and Evolution* 2:295–299.

Pusey, A.E. 1992. The primate perspective on dispersal. In: Stenseth, N.C. and Lidicker W.Z. Jr., *Animal Dispersal*. Chapman and Hall, London.

Pusey, A.E. and Packer, C. 1994. Infanticide in lions: consequences and counterstrategies. In: Parmigiani, S. and vom Saal, F.S. (eds). *Infanticide and and Parental Care*, pp. 277–299. Harwood Academic, London.

Raj, D. and Khamis, S.H. 1958. Some remarks on sampling with replacement. *Annals of Mathematical Statistics* 29:550–557.

Ralls, K., Ballou, J.D., and Templeton, A. 1988. Estimates of lethal equivalents and the cost of inbreeding in mammals. *Conservation Biology* 2:185–193.

Ramakrishnan, U., Coss, R.G., and Pelkey, N.W. 1999. Tiger decline caused by the reduction of large ungulate prey: evidence from a study of leopard diets in southern India. *Biological Conservation* 89:113–120.

Ramsey, D.S.L. 2000. The effect of fertility control on the population dynamics and behavior of brushtail possums (*Trichosurus vulpecula*) in New Zealand. *Proceedings of the 19th Vertebrate Pest Conference*, pp. 206–211. University of California, Davis, CA.

Ransom, A.B. 1965. Kidney and marrow fat as indicators of white-tailed deer condition. *Journal of Wildlife Management* 29:397–399.

Rapoport, E.H. 1982. *Aereography*. Pergamon Press, Oxford.

Rasmussen, D.I. and Doman, E.R. 1943. Census methods and their application in the management of mule deer. In: *Transactions 8th North American Wildlife Conference*, pp. 369–380. American Wildlife Institute, Washington, DC.

Ratcliffe, D.A. 1970. Changes attributable to pesticides in egg breakage frequency and eggshell thickness in some British birds. *Journal of Applied Ecology* 7:67–115.

Raxworthy, C.J., Pearson, R.G., Rabibisoa, N., Rakotondrazafy, A.M., Ramanamanjato, J-B., Raselimanana, A.P., Wu, S., Nussbaum, R.A., and Stone, D.A. 2008. Extinction vulnerability of tropical montane endemism from warming and upslope displacement: a preliminary appraisal for the highest massif in Madagascar. *Global Change Biology* 14:1–18.

Reale, D., Reader, S.M., Sol, D., McDougall, P.T., and Dingemanse, N.J. 2007. Integrating animal temperment within ecology and evolution. *Biological Reviews* 82:291–318.

Redfield, J.A., Krebs, C.J., and Taitt, M.J. 1977. Competition between *Peromyscus maniculatus* and *Microtus townsendii* in grasslands of coastal British Columbia. *Journal of Animal Ecology* 46:607–616.

Redhead, T.D. 1982. *Reproduction, Growth and Population Dynamics of House Mice in Irrigated and Non-Irrigated Cereal Farms in New South Wales.* PhD thesis, Australian National University, Canberra.

Redpath, S.M. and Thirgood, S.J. 1999. Numerical and functional responses in generalist predators: hen harriers and peregrines on Scottish grouse moors. *Journal of Animal Ecology* 68:879–892.

Redpath, S.M., Thirgood, S.J., and Leckie, F.M. 2001. Does supplementary feeding reduce predation of red grouse by hen harriers? *Journal of Applied Ecology* 38:1157–1168.

Reed, T.E., Jenouvrier, S., and Visser, M.E. 2013. Phenological mismatch strongly affects individual fitness but not population demography in a woodland passerine. *Journal of Animal Ecology* 82:131–144.

Relyea, R.A, Lawrence, R.K., and Demarias S. 2000. Home range of desert mule deer: testing the body-size and habitat-productivity hypothesis. *Journal of Wildlife Management* 64:146–153.

Rettie, W.J. and Messier, F. 1998. Dynamics of woodland caribou populations at the southern limit of their range in Saskatchewan. *Canadian Journal of Zoology* 76:251–259.

Rettie, W.J. and Messier, F. 2000. Hierarchical habitat selection by woodland caribou: its relationship to limiting factors. *Ecography* 23:466–478.

Reynolds, P.E. 1998. Dynamics and range expansion of a reestablished muskox population. *Journal of Wildlife Management* 62:734–744.

Reznick, D.N., Shaw, F.H., Rodd, F.H., and Shaw, R.G. 1997. Evaluation of the rate of evolution in natural populations of guppies (*Poecilia reticulata*). *Science* 275:1934–1937.

Reznick, D.N., Bryant, M.J., Roff, D., Ghalambor, C.K., and Ghalambor, D.E. 2004. Effect of extrinisic mortality on the evolution of scenescence in guppies. *Nature* 431:1095–1099.

Richards, S.A., Nisbet, R.M., Wilson, W.G., and Possingham, H.P. 2000. Grazers and diggers: exploitation competition and coexistence among foragers with different feeding strategies on a single resource. *The American Naturalist* 155:266–279.

Richman, L.K., Montali, R.J., Garber, R.L., Kennedy, M.A., Lehnhardt, J., Hidebrandt, T., Schmitt, D., Hardy, D., Alcendor, D.J., and Hayward, G.S. 1999. Novel endotheliotropic herpes virus fatal for Asian and African elephants. *Science* 283:1171–1176.

Ricker, W.E. 1954. Stock and recruitment. *Journal of the Fisheries Research Board of Canada* 11:559–623.

Ricklefs, R.E. and Miller, G.L. 2000. *Ecology*, 4 edn. W.H. Freeman and Company, New York, NY.

Ringelman, J.K. and Szymczak, M.R. 1985. A physiological condition index for wintering mallards. *Journal of Wildlife Management* 49:564–568.

Rivest, L.-P. and Crepeau, H. 1990. A two-stage sampling plan for the estimation of the size of a moose population. *Biometrics* 46:163–176.

Roberts, C.M. Hawkins, J.P., and Gell, F. 2005. The role of marine reserves in achieving sustainable fisheries. *Philosophical Transactions of the Royal Society of London (B)* 360:123–132.

Robbins, C.T. 1983. *Wildlife Feeding and Nutrition.* Academic Press, New York, NY.

Robbins, C.T. Hanley, T.A., Hagerman, A.E., Hjeljord, O., Baker, D.L., Schwartz, C.C., and Mautz, W.W. 1987. Role of tannins in defending plants against ruminants: reduction in protein availability. *Ecology* 68:98–107.

Robinson, S.K., Thompson, F.R., Donovan, T.M., Whitehead, D.R., and Faaborg, J. 1995. Regional forest fragmentation and the nesting success of migratory birds. *Science* 267:1987–1990.

Robertshaw, D. and Taylor, C.R. 1969. A comparison of sweat gland activity in eight species of East African bovids. *Journal of Physiology* 203:135–143.

Robertson, G. 1987. Plant dynamics. In: Caughley, G., Shepherd, N., and Short, J. (eds) *Kangaroos: Their Ecology and Management in the Sheep Rangelands of Australia*, pp. 50–68. Cambridge University Press, Cambridge.

Robertson, G., Short, J., and Wellard, G. 1987. The environment of the Australian sheep range-lands. In: Caughley, G. Shepherd, N., and Short, J. (eds) *Kangaroos: Their Ecology and Management in the Sheep Rangelands of Australia*, pp. 14–34. Cambridge University Press, Cambridge.

Robinette, W.L., Jones, D.A., and Loveless, C.M. 1974. Field tests of strip census methods. *Journal of Wildlife Management* 38:81–96.

Robley, A., Short, J., and Bradley, S. 2001. Dietary overlap between the burrowing bettong (*Bettongia lesurur*) and the European rabbit (*Oryctolagus cuniculus*) in semi-arid coastal Western Australia. *Wildlife Research* 28:341–349.

Robson, D.S. and Whitlock, J.H. 1964. Estimation of a truncation point. *Biometrika* 51:33–39.

Rockwell, R.F. and Gormezano, L.J. 2009. The early bear gets the goose: climate change, polar bears and lesser snow geese in western Hudson Bay. *Polar Biology* 32:539–547.

Rodda, G.H., Fritts, T.H., Campbell III,, E.W., Dean-Bradley, K., Perry G., and Qualls, C.P. 2002. Practical concerns in the eradication of island snakes. In: Veitch, C.R. and Clout, M.N. (eds) *Turning the Tide: The Eradication of Invasive Species*, pp. 260–265. Occasional Paper of the IUCN Species Survival Commission No. 27 IUCN – The World Conservation Union, Gland, Switzerland and Cambridge, UK.

Rodgers, J.A., Linda, S.B., and Nesbitt, S.A. 1995. Comparing aerial estimates with ground counts of nests in Wood Stork colonies. *Journal of Wildlife Management* 59:656–666.

Roelke-Parker, M.E., Munson, L., Packer, C., Kock, R., Cleaveland, S., Carpenter, M., O'Brien S.J., Pospischill, A., Hofmann-Lehmann, R., Lutz, H., Mwamengele G.L.M., Mgasa, M.N., Machange, G.A., Summers, B.A., and Appel, M.J.G. 1996. A canine distemper virus epidemic in Serengeti lions (*Panthera leo*). *Nature* 379:441–445.

Roemer, G.W., Donlan, C.J., and Courchamp, F. 2002. Golden eagles, feral pigs, and insular carnivores: how exotic species turn native predators into prey. *Proceedings National Academy Science USA* 99:791–796

Rogers, L.L., Mech, L.D., Dawson, D.K., Peek, J.M., and Korb, M. 1980. Deer distribution in relation to wolf pack territory edges. *Journal of Wildlife Management* 44:253–258.

Rohner, C., Doyle, F.I., and Smith J.N.M. 2001. Great horned owls. In: Krebs, C.J., Boutin, S., and Boonstra, R. (eds) *Ecosystem Dynamics of the Boreal Forest: the Kluane Project*, pp. 339–376. Oxford University Press, Oxford.

Rolley, R.E. and Keith, L.B. 1980. Moose population dynamics and winter habitat use at Rochester, Alberta, 1965–1979. *Canadian Field-Naturalist* 94:9–18.

Rominger, E.M., Whitlaw, H.A., Weybright, D.L., Dunn, W.C., and Ballard, W.B. 2004. The influence of mountain lion predation on bighorn sheep translocations. *Journal of Wildlife Management* 68:993–999.

Root, A. 1972. Fringe-eared oryx digging for tubers in the Tsavo National Park (East). *East African Wildlife Journal* 10:155–157.

Root, A. 1988. *Atlas of Wintering North American Birds: An Analysis of Christmas Bird Count Data*. University of Chicago Press, Chicago, IL.

Root, T.L., Price, J.T., Hall, K.R., Schneider, S.H., Rosenzweig, C., and Pounds, J.A. 2003. Fingerprints of wild animals and plants. *Nature* 421:57–60.

Rosenberg, A.A., Fogarty, M.J., Sissenwine, M.P., Beddington, J.R., and Shepherd, J.G. 1993. Achieving sustainable use of renewable resources. *Science* 262:828–829.

Rosenzweig, M.L. 1971. Paradox of enrichment: destabilization of exploitation ecosystems in ecological time. *Science* 171:385–387.

Rosenzweig, M.L. 1981. A theory of habitat selection. *Ecology* 62:327–335.

Rosenzweig, M.L. 1991 Habitat selection and population interactions: the search for mechanism. *American Naturalist* 137:S5–S28.

Rosenzweig, M.L. and MacArthur, R.H. 1963. Graphical representation and stability conditions of predator–prey interaction. *American Naturalist* 97:209–223.

Ross, J. 1982. Myxamotosis: the natural evolution of the disease. In: Edwards M.A. and McDonnell, U. (eds) *Animal Disease in Relation to Animal Conservation*, pp. 77–95. Academic Press, London.

Rossiter, P. 2001. Rinderpest. In: Williams, E.S. and Barker, I.K. *Infectious Diseases of Wild Mammals*, pp. 39–45. Manson Publishing/The Veterinary Press.

Rothstein, S.I. 1994. The cowbird's invasion of the far west: history, causes and consequences experienced by host species. *Studies in Avian Biology* 15:201–315.

Routledge, R.D. 1981. The unreliability of population estimates from repeated, incomplete aerial surveys. *Journal of Wildlife Management* 45:997–1000.

Routledge, R.D. 1982. The method of bounded counts: when does it work? *Journal of Wildlife Management* 46:757–761.

Rubsamen, K., Heller, R., Lawrenz, H., and Engelhardt, W.V. 1979. Water and energy metabolism in the rock hyrax (*Procavia habessinica*). *Journal of Comparative Physiology B* 131:303–309.

Rudolph, B.A., Porter, W.F., and Underwood, H.B. 2000. Evaluating immunocontraception for managing suburban white-tailed deer in Irondequoit, New York. *Journal of Wildlife Management* 64:463–473.

Rupprecht, C.E., Smith J.S., Fekadu, M., and Childs, J.E. 1995. The ascension of wildlife rabies: a cause for public health concern or intervention? *Emerging Infectious Diseases* 4:107–113.

Ruxton, G.D., Gurney, W.S.C., and de Roos, A.M. 1992. Interference and generation cycles. *Theoretical Population Biology* 42:235–253.

Ryder, O.A. 1993. Prsewalski's horse: prospects for reintroduction into the wild. *Conservation Biology* 7:13–15.

Sæther, B.E., Ringsby, T.H., Bakke, Ø., and Solberg, E.J. 1999. Spatial and temporal variation in emography of a house sparrow metapopulation. *Journal of Animal Ecology* 68:628–637.

Sæther, B.-E., Engen, S., Lande, R., Arcese, P., and Smith, J.N.M. 2000. Estimating the time to extinction in an island population of song sparrows. *Proceedings of the Royal Society of London (B)* 267:621–626.

Sæther, B.E., Engen, S., and Solberg, J.2001. Optimal harvest of age-structured populations of moose *Alces alces* in a fluctuating environment. *Wildlife Biology* 7:171–179.

Samuel, W.M., Pybus, M.J., Welch, D.A., and Wilke, C.J. 1992. Elk as a potential host for meningeal worm: implications for translocation. *Journal of Wildlife Management* 56:629–639.

Sauer, J.R., Dolton, D.D., and Droege, S. 1994. Mourning dove population trend estimates from call-count and North American breeding bird surveys. *Journal of Wildlife Management* 58:506–515.

Saunders, D.A. and Hobbs, R.J. (eds) 1991. *Nature Conservation: 2. The Role of Corridors*. Surrey Beatty & Sons, New South Wales, Australia.

Saunders, D.A. Hobbs, R.J., and Arnold G.W. 1993. The Kellerberrin project on fragmented landscapes: a review of current information. *Biological Conservation* 64:185–192.

Saunders, G. and McIlroy, J. 2001. Fertility control of foxes using immunocontraception – ecologically feasible? *12th Australasian Vertebrate Pest Conference, Melbourne*, pp. 169–172.

Savidge, J. 1987. Extinction of an island forest avifauna by an introduced snake. *Ecology* 68:660–668.

Sawyer, H., Kauffman, M.J., Nielson, R.M., and Horne, J.S. 2009. Identifying and prioritizing ungulate migration routes for landscape-level conservation. *Ecological Applications* 19:2016–2025.

Schall, J.J. 1992. Parsite-mediated competition in *Anolis* lizards. *Oecologia* 92:58–64.

Schall, J.J. and Pianka, E.R. 1978. Geographical trends in numbers of species. *Science* 201:679–686.

Schaller, G.B. 1972. *The Serengeti Lion*. University of Chicago Press, Chicago, IL.

Schaller, G.B., Jinchu, H., Wenski, P., and Jing, Z. 1985. *The Giant Pandas of Wolong*. University of Chicago Press, Chicago, IL.

Scheffer, M, Carpenter, S.R., Foley, J.A., Folke, C., and Walker, B. 2001. Catastrophic shifts in ecosystems. *Nature* 413:591–596.

Schluter, D. 1981. Does the theory of optimal diets apply in complex environments? *American Naturalist* 118:139–147.

Schluter, D. 1988. The evolution of finch communities on islands and continents: Kenya vs. Galapagos. *Ecological Monographs* 58:229–249.

Schluter, D. 2000. *The Ecology of Adaptive Radiation*. Oxford University Press, Oxford.

Schluter, D. and Ricklefs, R.E. 1993. Species diversity. An introduction to the Problem. In: Ricklefs, R.E. and Schluter, D. (eds) *Species Diversity in Ecological Communities*, pp. 1–10. University of Chicago Press, Chicago, IL.

Schluter, D., Price, T.D., and Grant, P.R. 1985. Ecological character discplacement in Darwin's finches. *Science* 227:1056–1059.

Schoener, T.W. 1971. Theory of feeding strategies. *Annual Review of Ecology and Systematics* 2:369–404.

Schoener, T.W. 1983. Simple models of optimal feeding-territory size: a reconciliation. *American Naturalist* 121:608–629.

Schoener, T.W. 1989. The ecological niche In: Cherrett, J.M. (ed.) *Ecological Concepts: The Contribution of Ecology to an Understanding of the Natural World*, pp. 79–113. The 29th Symposium of the British Ecological Society. Blackwell Scientific Publication, Oxford.

Schoener, T.W. 2011. The newest synthesis: understanding the interplay of evolutionary and ecological dynamics. *Science* 331:426–429.

Schoener, T.W. and Spiller, D.A. 1987. High population persistence in a system with high turnover. *Nature (London)* 330:474–477.

Schoener, WT.W., Spiller, D.A, and Losos, J.B. 2001. Predators increase the risk of catastrophic extinction of prey populations. *Nature* 412:183–186.

Schmitz, O.J., Beckerman, A.P., and O'Brien, K.M. 1997. Behaviorally mediated trophic cascades: effects of predation risk on food web interactions. *Ecology* 78:1388–1399.

Schmitz, O.J. and Nudds T.D. 1994. Parasite-mediated competition in deer and moose: how strong is the effect of meningeal worm on moose? *Ecological Applications* 4:91–103.

Schneider, S.H. and Root, T.L. (eds) 2002. *Wildlife Responses to Climate Change*. Island Press, Washington, DC.

Schulze, E.D. and Mooney, H.A. (eds) 1993. *Biodiversity and Ecosystem Function*. Springer-Verlag, Berlin.

Schullery, P. 1984. *Mountain Time*. Nick Lyons Books, New York, NY.

Schwagmeyer, P.L. and Woontner, S.J. 1985. Mating competition in an a social ground squirrel, *Spermophilus tridecemlineatus*. *Behavioural Ecology and Sociobiology* 17:291–296.

Schwartz, C.C., Nagy, J.G., and Regelin, W.L. 1980. Juniper oil yield, terpenoid concentration, and antimicrobial effects on deer. *Journal of Wildlife Management* 44:107–113.

Schwartz, M.W., Brigham, C.A., Hoeksema, J.D., Lyons, K.G., Mills, M.H., and van Mantgem, P.J. 2000. Linking biodiversity to ecosystem function: implications for conservation ecology. *Oecologia* 122:297–305.

Scott, J.M., Mountainspring, S., Ramsey, F.L., and Kepler, C.B. 1986. *Forest Bird Communities of the Hawaiian Islands: Their Dynamics, Ecology, and Conservation*. Studies in Avian Biology No. 9, Cooper Ornithological Society, University of California, Los Angeles, CA.

Scott, M.E. 1987. Regulation of mouse colony abundance by *Heligmosomoides polygyrus* (Nematoda). *Parasitology* 95:111–124.

Scott, M.E. and Dobson, A. 1989. The role of parasites in regulating host abundance. *Parsitology Today* 5:176–183.

Scott, M.E. and Lewis, J.W. 1987. Population dynamics of helminth parasites in wild and laboratory rodents. *Mammal Review* 17:95–103.

Seal, U.S. Thorne, E.T., Bogan, M.A., and Anderson, S.H. 1989. *Conservation Biology and the Black-Footed Ferret*. Yale University Press, New Haven, CT.

Seber, G.A.F. 1982. *The Estimation of Animal Abundance and Related Parameters*, 2 edn. Macmillan, New York, NY.

Seip, D.R. 1991. Predation and caribou populations. Proceedings of the Fifth North American Caribou Workshop, Yellowknife, Northwest Territories, Canada. *Rangifer*, Special Issue Number 7:46–52.

Seip, D.R. 1992. Factors limiting woodland caribou populations and their inter-relationships with wolves and moose in southeastern British Columbia. *Canadian Journal of Zoology* 70:1494–1503.

Semel, B. and Andersen, D.C. 1988. Vulnerability of acorn weevils (Coleoptera : Curculionidae) and attractiveness of weevils and infested *Quercus alba* acorns to *Peromyscus leucopus* and *Blarina brevicauda. American Midland Naturalist* 119:385–393.

Semlitsch, R.D., Todd, B.D., Blomquist, S.M., Calhoun, A.J.K., Gibbons, J.W., Gibbs, J.P., Graeter, G.J., Harper, E.B., Hocking, D.J., Hunter, M.L., Patrick, D.A., Rittenhouse, T.A.G., and Rothermel, B.B. 2009. Effects of timber harvest on amphibian populations: understanding mechanisms from forest experiments. *Bioscience* 59:853–862.

Serena, M. (ed.) 1994. *Reintroduction Biology of Australian and New Zealand Fauna*. Surrey Beatty & Sons, New South Wales, Australia.

Shaw, P. 1985. Late quaternary landforms and environmental changes in northwest Botswana: the evidence of Lake Ngami and Mababe Depression. *Transactions of the Institute of British Geography, New Series* 10:333–346.

Shellam, G.R. 1994. The potential of murine cytomegalovirus as a viral vector for immunocontraception. *Reproduction, Fertility and Development* 6:401–409.

Shepherd, N. and Caughley, G. 1987. Options for management of kangaroos. In: Caughley, G., Shepherd, N., and Short, J. (eds) *Kangaroos: Their Ecology and Management in the Sheep Rangelands of Australia*, pp. 188–219. Cambridge University Press, Cambridge.

Short, J. 1987. Factors affecting food intake of rangelands herbivores. In: Caughley, G., Shepherd, N., and Short, J. (eds) *Kangaroos: Their Ecology and Management in the Sheep Rangelands of Australia*, pp. 84–99. Cambridge University Press, Cambridge.

Short, J. and Smith, A. 1994. Mammal decline and recovery in Australia. *Journal of Mammalogy* 75:288–297.

Short, J., Bradshaw, S.D., Giles, J., Prince, R.I.T., and Wilson, G.R. 1992. Reintroduction of macropods (Marsupialia : Macropodoidea) in Australia – a review. *Biological Conservation* 62:189–204.

Short, J., Kinnear J.E., and Robley A. 2000. Surplus killing by introduced predators in Australia – evidence for ineffective anti-predator adaptations in native prey species? *Biological Conservation* 103:283–301.

Shurin, J.B. and Srivastava, D.S. 2005. New perspectives on local and regional diversity: beyond saturation In: Holyoak, M., Leibold, M.A., and Hold, R.D. (eds) *Metacommunities* pp (in press). University of Chicago Press, Chicago, IL.

Sibly, R.M. 1981. Strategies of digestion and defecation In: Townsend, C.R. and Calow, P. (eds) *Physiological Ecology: An Evolutionary Approach to Resource Use*, pp. 109–139. Blackwell Scientific Publishers, Oxford.

Sih, A. 1980. Optimal behavior: can foragers balance two conflicting demands? *Science* 210:1041–1042.

Sih, A. and Christensen, B. 2001. Optimal diet theory: when does it work and when and why does it fail? *Animal Behavior* 61:379–390.

Sih, A., Bell, A., and Johnson, J.C. 2004. Behavioral syndromes: an ecological and evolutionary overview. *Trends in Ecology and Evolution* 19:372–378.

Simberloff, D. 1988. The contribution of population and community biology to conservation science. *Annual Review of Ecology and Systematics* 19:473–511.

Simberloff, D. and Cox, J. 1987. Consequences and costs of conservation corridors. *Conservation Biology* 1:63–71.

Simberloff, D.S. and Wilson, E.O. 1970. Experimental zoogeography of islands: a two-year record of colonization. *Ecology* 51:934–937

Simberloff, D., Farr, J.A., Cox, J., and Mehlman, D.W. 1992. Movement corridors: conservation bargains or poor investments? *Conservation Biology* 6:493–504.

Sinclair, A.R.E. 1973 Population increases of buffalo and wildebeest in the Serengeti. *East African Wildlife Journal* 11:93–107.

Sinclair, A.R.E. 1977. *The African Buffalo: A Study of the Resource Limitation of Populations.* University of Chicago Press, Chicago, IL.

Sinclair, A.R.E. 1978. Factors affecting the food supply and breeding season of resident birds and movements of palaearctic migrants in a tropical African savannah. *Ibis* 120:480–497.

Sinclair, A.R.E. 1979a. Dynamics of the Serengeti ecosystem. Process and pattern. In: Sinclair, A.R.E. and Norton-Griffiths, M. (eds) *Serengeti: Dynamics of an Ecosystem*, pp. 1–30. University of Chicago Press, Chicago, IL.

Sinclair, A.R.E. 1979b. The eruption of the ruminants. In: Sinclair, A.R.E. and Norton-Griffiths, M. (eds) *Serengeti: Dynamics of an Ecosystem*, pp. 82–103. University of Chicago Press, Chicago, IL.

Sinclair, A.R.E. 1983. The adaptations of African ungulates and their effects on community function. In: Bourlière, F. (ed.) *Ecosystems of the World: 13. Tropical Savannas*, pp. 401–426. Elsevier Scientific Publishing Company, Amsterdam.

Sinclair, A.R.E. 1985. Does interspecific competition or predation shape the African ungulate community? *Journal of Animal Ecology* 54:899–918.

Sinclair, A.R.E. 1989. Population regulation in animals. In: Cherrett, J.M. (ed.) *Ecological Concepts: The Contribution of Ecology to an Understanding of the Natural World*, pp. 197–241. The 29th Symposium of the British Ecological Society Blackwell Scientific Publication, Oxford.

Sinclair, A.R.E. 1992. Do large mammals disperse like small mammals? In: Stenseth, N.C. and Lidicker W.Z. Jr., *Animal Dispersal* Chapman and Hall, London.

Sinclair, A.R.E. 1995. Serengeti Past and Present. In: Sinclair, A.R.E. and Arcese, P. (eds) *Serengeti II Dynamics, Management and Conservation of an Ecosystem*, pp. 3–30. University of Chicago Press, Chicago, IL.

Sinclair, A.R.E. 1996.Mammal populations: fluctuation, regulation, life history theory and their implications for conservation. In: Floyd, R.B. and Sheppard, A.W (eds) *Frontiers and Applications of Population Ecology*, pp. 127–154. CSIRO, Melbourne.

Sinclair. A.R.E. 1998. Natural regulation of ecosystems in protected areas as ecological baselines. *Wildlife Society Bulletin* 26:399–409.

Sinclair, A.R.E. 2003. Mammal population regulation, keystone processes and ecosystem dynamics. *Philosphical Transactions Royal Society London B.* 358:1729–1740.

Sinclair, A.R.E. and Arcese, P. 1995. Population consequences of predation-sensitive foraging: the Serengeti wildebeest. *Ecology* 76:882–891.

Sinclair, A.R.E. and Fryxell, J.M. 1985. The Sahel of Africa: ecology of a disaster. *Canadian Journal of Zoology* 63:987–994.

Sinclair, A.R.E. and Krebs, C.J. 2002. Complex numerical responses to top-down and bottom-up processes in vertebrate populations. *Philosphical Transactions of the Royal Society London (B).* 357:1221–1231.

Sinclair, A.R.E. and Smith, J.N.M. 1984. Do plant secondary compounds determine feeding preferences of snowshoe hares? *Oecologia* 61:403–410.

Sinclair, A.R.E., Krebs, C.J., and Smith, J.N.M. 1982. Diet quality and food limitation in herbivores: the case of the snowshoe hare. *Canadian Journal of Zoology* 60:889–897.

Sinclair, A.R.E., Dublin, H., and Borner, M. 1985. Population regulation of Serengeti wildebeest: a test of the food hypothesis. *Oecologia* 65:266–268.

Sinclair, A.R.E., Krebs, C.J., Smith, J.N.M., and Boutin, S. 1988. Population biology of snowshoe hares: III. Nutrition, plant secondary compounds and food limitation. *Journal of Animal Ecology* 57:787–806.

Sinclair, A.R.E., Olsen, P.D., and Redhead, T.D. 1990. Can predators regulate small mammal populations? Evidence from house mouse outbreaks in Australia. *Oikos* 59:382–392.

Sinclair, A.R.E., Gosline, J.M., Holdsworth, G., Krebs, C.J., Boutin, S., Smith, J.N.M., and Dale, M. 1993. Can the solar cycle and climate synchronize the snowshoe hare cycle in Canada? Evidence from tree rings and ice cores. *American Naturalist* 141:173–198.

Sinclair, A.R.E., Pech, R.P., Dickman, C.R., Hik, D., Mahon, P. and Newsome, A.E. 1998. Predicting effects of predation on conservation of endangered prey. *Conservation Biology* 12:564–575.

Sinclair, A.R.E., Mduma, S.A.R., and Arcese, P. 2000. What determines the phenology and synchrony of births in Serengeti ungulates? *Ecology* 81:2100–2111.

Sinclair, A.R.E, Mduma, S.A.R., and Brashares, J.S. 2003. Patterns of predation in a diverse predator–prey system. *Nature* 425:288–290.

Singer, F.J., Swift, D.M., Coughenour, M.B., and Varley, J.D. 1998. Thunder on the Yellowstone revisited: an assessment of native ungulates by natural regulation, 1968–1993. *Wildlife Society Bulletin* 26:375–390.

Singleton, G.R. and Spratt, D.M. 1986. The effects of *Capillaria hepatica* (Nematoda) on natality and survival to weaning in BALB/c mice. *Australian Journal of Zoology* 34:677–681.

Sjögren Gulve, P. 1994. Distribution and extinction patterns within a northern metapopulation: case of the pool frog, *Rana lessonae*. *Ecology* 75:1357–1367.

Skellam, J.G. 1951. Random dispersal in theoretical populations. *Biometrika* 38:196–218.

Skelly, D.K., Joseph, L.N., Possingham, H.P., Freidenburg, L.K., Farrugia, T.J., Kinneson, M.T., and Hendry, A.P. 2007. Evolutionary responses to climate change. *Conservation Biology* 21:1353–1355.

Skogland, T. 1985. The effects of density-dependent resource limitations on the demography of wild reindeer. *Journal of Animal Ecology* 54:359–374.

Skole, D. and Tucker, C. 1993. Tropical deforestation and habitat fragmentation in the Amazon: satellite data from 1978 to 1988. *Science* 260:1905–1910.

Smith, A.T. and Gilpin, M.E. 1997. Spatially correlated dynamics in a pika metapopulation. In: Hanski, I. and Gilpin, M.E. (eds) *Metapopulation Biology*, pp. 407–428. Academic Press, San Diego, CA.

Smith, A.P. and Quin, D.G. 1996. Patterns and causes of extinction and decline in Australian conilurine rodents. *Biological Conservation* 77:243–267.

Smith, F.D.M., May, R.M., and Harvey, P.H. 1994. Geographical ranges of Australian mammals. *Journal of Animal Ecology* 63:441–450.

Smith, J.N.M. 1974. The food searching behaviour of European thrushes II: the adaptedness of search patterns. *Behaviour* 49:1–61.

Smith, J.N.M., Krebs, C.J., Sinclair, A.R.E., and Boonstra, R. 1988. Population biology of snowshoe hares: II. Interactions with winter food plants. *Journal of Animal Ecology* 57:269–286.

Smith, J.N.M., Cook, T.L., Rothstein, S.I., Robinson, S.K., and Sealy, S.G. (eds) 2000. *Ecology and Management of Cowbirds and their Hosts*. University of Texas Press, Austin, TX.

Smith, J.N.M., Marr, A.B., Arces, P., and Keller, L. 2006. Fluctuations in numbers: population regulation and catastrophic mortality. In: Smith, J.N.M., Keller, L.F., Marr, A.B., and Arcese, P. (eds) *Conservation and Biology of Small Populations*, pp. 43–64. Oxford University Press, Oxford.

Smith, N.S. 1970. Appraisal of condition estimation methods for East African ungulates. *East African Wildlife Journal* 8:123–129.

Smuts, G.L. 1978. Interrelations between predators, prey, and their environment. *BioScience* 28:316–320.

Snell, G.P. and Hlavachick, B.D. 1980. Control of prairie dogs – the easy way. *Rangelands* 2:239–240.

Snyder, N.F.R. and Johnson, E.V. 1985. Photographic censusing of the 1982–1983 California condor population. *Condor* 87:1–13.

Snyder, N.F.R., Wiley, J.W., and Kepler, C.B. 1987. *The Parrots of Luqillo: Natural History and Conservation of the Puerto Rican Parrot.* Western Foundation of Vertebrate Zoology, Los Angeles, CA.

Solomon, M.E. 1949. The natural control of animal populations. *Journal Animal Ecology* 18:1–35.

Spinage, C.A. 1973. The role of photoperiodism in the seasonal breeding of tropical African ungulates. *Mammal Reveiw* 3:71–84.

Spowart, R.A. and Thompson Hobbs, N. 1985. Effect of fire on diet overlap between mule deer and mountain sheep. *Journal of Wildlife Management* 49:942–946.

Spratt, D.M. 1990. The role of helminths in the biological control of mammals. *International Journal for Parasitology* 20:543–550.

Spratt, D.M. and Presidente, P.J.A. 1981. Prevalence of *Fasciola hepatica* infection in native mammals in southeastern Australia. *Australian Jounral of Experimental Biology and Medical Science* 59:713–721.

Spratt, D.M. and Singleton, G.R. 1986. Studies on the life cycle, infectivity and clinical effects of *Capillaria hepatica* (Bancroft) (Nematoda) in mice, *Mus musculus. Australian Journal of Zoology* 34:663–675.

Spray, C.J., Crick, H.Q.P., and Hart, A.D.M. 1987. Effects of aerial applications of fenitrothion on bird populations of a Scottish pine plantation. *Journal of Applied Ecology* 24:29–47.

Springer, A.M., Estes, J.A., van Vliet, G.B., Williams, T.M., Doak, D.F., Danner, E.M. Forney, K.A., and Pfister, B. 2003. Sequential megafaunal collapse in the North Pacific Ocean: an ongoing legacy of industrial whaling? *Proceedings of the National Academy of Science USA* 100:12 223–12 228.

Spurr, E.B. 1994. Review of the impacts on non-target species of sodium monofluoroacetate (1080) in baits used for brushtail possum control in New Zealand. In: Seawright, A.A and Eason, C.T. (eds) *Proceedings of the Science Workshop on 1080.* Royal Society of New Zealand Miscellaneous Series 28:124–133.

Spurr, E.B. 2000. Impacts of possum control on non-target species. In: Montague, T.L. (ed.) *The Brushtail Possum – Biology, Impact and Management of an Introduced Marsupial,* pp. 175–186. Manaaki Whenua Press, Lincoln, New Zealand.

Srivastava, D.S. 1999. Using local-regional richness plots to test for species saturation: pitfalls and potentials. *Journal of Animal Ecology* 68:1–16.

Srivastava, D.S. and Vellend, M. 2005. Biodiversity-ecosystem function research: is it relevant to conservation? *Annual Reviews of Ecology and Systematics* 36:267–294.

Stanley Price, M.R. 1989. *Animal Re-Introductions: The Arabian Oryx in Oman.* Cambridge University Press, Cambridge.

Stenseth N.C. and Lidicker, W.J. Jr., 1992. Presaturation and saturation dispersal 15 years later: some theoretical considerations. In: Stenseth, N.C. and Lidicker, W.Z. Jr., *Animal Dispersal.* Chapman and Hall, London.

Stenseth, N.C., Mysterud, A., Ottersen, G., Hurrell, J.W., Chan, K.-S., and Lima, M. 2002. Ecological effects of climate fluctuations. *Science* 297:1292–1296.

Stephens, D.W. and Dunbar, S.R. 1993. Dimensional analysis in behavioral ecology. *Behavioral Ecology* 4:172–183.

Stephens, D.W. and Krebs, J. 1986. *Foraging Theory.* Princeton University Press, Princeton, NJ.

Stirling, I. and Derocher, A.E. 2012. Effects of climate warming on polar bears: a review of the evidence. *Global Change Biology* 18:2694–2706.

Stith, B.M., Fitzpatrick, J.W., Woolfenden, G.E., and Pranty, B. 1996. Classification and conservation of metapopulations: a case study of the Florida scrub jay. In: McCullough, D.R. (ed.) *Metapopulations and Wildlife Conservation,* pp. 187–216. Island Press, Washington, DC.

Stonehouse, B. 1967. The general biology and thermal balances of penguins. In: Cragg, J.B. (ed.) *Advances in Ecological Research,* Vol. 4, pp. 131–196. Academic Press, London.

Stouffer, P.C. and Bierregaard, R.O. 1995. Use of Amazonian forest fragments by understoary insectivorous birds. *Ecology* 76:2429–2445.

Stroeve, J., Holland, M.M., Meir, W., Scambos, T., and Serreze, M., 2007. Arctic sea ice decline: faster than forecast. *Geophysical Research Letters* 34:L09501. Doi: 10.1029/2007GL029703.

Stuart, S.M., Chanson, J.S., Cox, N.A., Young, B.E., Rodrigues, A.S.L., Fishman, D.L., and Waller, R.W. 2004. Status and trends of amphibian declines and extinctions worldwide. *Science* 306:1783–1786.

Sullivan, T.P. and Klenner, W. 1993. Influence of diversionary food on red squirrel populations and crop tree damage in young lodgepole pine forest. *Ecological Applications* 3:708–718.

Sutherland, W.J. 1996. *From Individual Behaviour to Population Biology.* Oxford University Press, Oxford.

Swain, D.P., Sinclair, A.F., and Hanson, J.M. 2007. Evolutionary response to size-selective mortality in an exploited fish population. *Proceedings of the Royal Society of London (B)* 274:1873–1880.

Swenson, J.E. 2003. Implications of sexually selected intanticide for the hunting of large carnivores. In: Festa-Bianchet, M. and Appolonia, M. (eds) *Animal Behavior and Wildlife Conservation*, pp. 171–189. Island Press, Washington, DC.

Swenson, J.E., Sandegren, F., Söderberg, A., Bjärvall, A., Franzén, R., and Wabakken, P. 1997. Infanticide caused by hunting of male bears. *Nature* 386:450–451.

Taber, R.D. 1956. Deer nutrition and population dynamics in the north coast range of California. *Transactions of the North American Wildlife Conference* 21:159–172.

Taitt, M.J. and Krebs, C.J. 1981. The effect of extra food on small rodent populations: II. Voles (*Microtus townsendii*). *Journal of Animal Ecology* 50:125–137.

Talbot, L.M. and Stewart, D.R.M. 1964. First wildlife census of the entire Serengeti-Mara Region, East Africa. *Journal of Wildlife Management* 28:815–827.

Taper, M.L. and Gogan, P.J.P. 2002. The northern Yellowstone elk: density dependence and climatic conditions. *Journal of Wildlife Management* 66:106–122.

Taylor, C.R. 1968a. The minimum water requirements of some East African bovids. In: Crawford, M.A. (ed.) *Comparative Nutrition of Wild Animals*, pp. 195–206. Symposia of the Zoological Society of London. Academic Press, London.

Taylor, C.R. 1968b. Hygroscopic food: a source of water for desert antelopes? *Nature (London)* 219:181–182.

Taylor, C.R. 1969. Metabolism, respiratory changes, and water balance of an antelope, the eland. *American Journal of Physiology* 217:317–320.

Taylor, C.R. 1970a. Strategies of temperature regulation: effect on evaporation in East African ungulates. *American Journal of Physiology* 219:1131–1135.

Taylor, C.R. 1970b. Dehydration and heat: effects on temperature regulation of East African ungulates. *American Journal of Physiology* 219:1136–1139.

Taylor, C.R. 1972. The desert gazelle: a paradox resolved. In: Maloiy, G.M.O. (ed.) *Symposia of the Zoological Society of London*, No 31, pp. 215–227. Academic Press, London.

Taylor, C.R., Spinage, C.A. and Lyman, C.P. 1969a. Water relations of the waterbuck, an East African antelope. *American Journal of Physiology* 217:630–634.

Taylor, C.R., Robertshaw, D. and Hofmann, R. 1969b. Thermal panting: a comparison of wildebeest and zebu cattle. *American Journal of Physiology* 217:907–910.

Taylor, R.H. 1979. How the Macquarie Island parakeet became extinct. *New Zealand Journal of Ecology* 2:42–45.

Tedman, R. and Green, B. 1987. Water and sodium fluxes and lactational energetics in suckling pups of Weddell seals (*Leptonychotes weddelli*). *Journal of Zoology (London)* 212:29–42.

Telfer, E.S. 1970. Winter habitat selection by moose and white-tailed deer. *Journal of Wildlife Management* 34:553–559.

Temple, S.A. 1977. Plant-animal mutualism: coevolution with dodo leads to near extinction of plant. *Science* 197:885–886.

Temple S.A. 1986. Predicting impacts of habitat fragmentation on forest birds: a comparison of two models. In: Verner, J., Morrison, M.L., and Ralph, J.C. (eds) *Wildlife 2000: Modeling Habitat*

Relationships of Terrestiral Vertebrates, pp.301–304. University of Wisconsin Press, Madison, WI.

Temple, S.A. 1991. The role of dispersal in the maintenance of bird populations in a fragmented landscape. *Acta XX Congressus Internationalis Ornithologici*, pp. 2298–2305. New Zealand Ornithological Congress Trust Board Wellington, New Zealand.

Temple, S.A. and Cary, J.R. 1988. Modeling dynamics of habitat-bird populations in fragmented landscapes. *Conservation Biology* 2:340–427.

Terborgh, J. 1988. The big things that run the world – a sequel to E.O. Wilson. *Conservation Biology* 2:402–403.

Terborgh, J. 1989. *Where Have All the Birds Gone? Essays on the Biology and Conservation of Birds that Migrate to the American Tropics.* Princeton University Press, Princeton, NJ.

Terborgh, J. 1992. Why American songbirds are vanishing. *Scientific American* 266(5):56–62.

Terborgh, J. and Estes, J.A. 2010. *Trophic Cascades.* Island Press, Washington, DC.

Terborgh, J.W. and Janson, C.H. 1986. The socioecology of primate groups. *Annual Review of Ecology and Systematics* 17:111–135.

Terborgh, J., Lopez, L., Nuñez, P., Rao, M., Shahabuddin, G., Orihuela, G., Riveros, M., Ascanio, R., Adler, G., Lambert, T., and Balbas, L. 2001. Ecological meltdown in predator-free forest fragments. *Science* 294:1923–1926.

Tewksbury, J.J., Levey, D.J., Haddad, N.M., Sargent, S., Orrock, J.L., Weldon, A., Danielson, B.J., Brinkerhoff, J., Damschen, E.I., and Townsend, P. 2002. Corridors affect plants, animals, and their interactions in fragmented landscapes. *Proceedings of the National Academy of Sciences USA* 99:12 923–12 926.

Thill, R.E. 1984. Deer and cattle diets on Louisiana pine-hardwood sites. *Journal of Wildlife Management* 48:788–798.

Thirgood, S.J., Redpath, S.M., Rothery, P., and Aebischer, N.J. 2000. Raptor predation and population limitation in red grouse. *Journal of Animal Ecology* 69:504–516.

Thomas, C.D. and Lennon, J.J. 1999. Birds extend their ranges northwards. *Nature* 399:213.

Thomas, C.D., Cameron, A., Green, R.E., Bakkenes, M., Beaumont, L.J., Collingham, Y.C., Erasmus, B.F.N., de Siqueira, M.F., Grainger, A., Hannah, L., Hughes, L., Huntley, B., van Jaarsveld, A.S., Midgley, G.F., Miles, L., Ortega-Huerta, M.A., Peterson, A.T., Phillips, O.L., and Williams, S.E. 2004. Extinction risk from climate change. *Nature* 427:145–148.

Thomas, V.G. 1988. Body condition, ovarian hierarchies, and their relation to egg formation in Anseriform and Galliform species. *Proceedings of the International Ornithological Congress* 19:353–363.

Thorne, E.T. and Williams, E.S. 1988. Disease and endangered species: the black-footed ferret as a recent example. *Conservation Biology* 2:66–74.

Thorne, E.T., Morton, J.K., Blunt, F.M., and Dawson, H.A. 1978. Brucellosis in elk. II. Clinical effects and means of transmission as determined through artifical infections. *Journal of Wildlife Diseases* 14:280–291.

Thornhill, N.W. 1993. *The Natural history of inbreeding and outbreeding. Theoretical and Empirical Perspectives.* University of Chicago Press, Chicago, IL.

Tilman, D. 1994. Competition and biodiversity in spatially structured habitats. *Ecology* 75:2–16.

Tilman, D. 1999. The ecological consequences of changes in biodiversity: a search for general principles. *Ecology* 80:1455–1474.

Tompkins, D.M. and Wilson, K. 1998. Wildlife disease ecology: from theory to policy. *Trends in Ecology and Evolution* 13:476–478.

Torbit, S.C., Carpenter, L.H., Bartmann, R.M., Alldredge, A.W., and White, C.G. 1988. Calibration of carcass fat indices in wintering mule deer. *Journal of Wildlife Management* 52:582–588.

Traveset, A. 1998. Effect of seed passage through vertebrate frugivores' guts on germination: a review. *Perspectives in Plant Ecology, Evolution and Systematics* 1:151–190.

Trites, A.W. and Donnelly, C.P. 2003. The decline of Steller sea lions *Eumetopias jubatus* in Alaska: a review of the nutritional stress hypothesis. *Mammal Review* 33:3–28.

Trostel, K., Sinclair, A.R.E., Walters, C.J. and Krebs, C.J. 1987. Can predation cause the 10-year hare cycle? *Oecologia* 74:185–192.

Turchin, P. 1998. *Quantitative Analysis of Movement.* Sinauer Associates, Sunderland, MA.

Turchin, P. 2003. *Complex Population Dynamics.* Princeton University Press, Princeton, NJ.

Turchin, P. and Batzli, G. 2001. Availability of food and the population dynamics of arvicoline rodents. *Ecology* 82:1521–1534.

Turchin, P. and Ellner, S.P. 2000. Living on the edge of chaos: population dynamics of Fennoscandian voles. *Ecology* 81:3099–3116.

Turchin, P. and Hanski, I. 1997. An empirically-based model for the latitudinal gradient in vole population dynamics. *American Naturalist* 149:842–874.

Turchin, P. and Hanski, I. 2001. Contrasting alternative hypotheses about rodent cycles by translating them into parameterized models. *Ecology Letters* 4:267–276.

Turner, J.W. and Kirkpatrick, J.F. 1991. New developments in feral horse contraception and their potential application to wildlife. *Wildlife Society Bulletin* 19:350–359.

Twigg, L.E. and Williams, C.K. 1999. Fertility control of overabundant species: can it work for rabbits? *Ecology Letters* 2:281–285.

Twigg, L.E., Lowe, T.J., Martin, G.R., Wheeler, A.G., Gray, G.S. Griffing, S.L., O'Reilly, C.M., Robinson, D.J., and Hubach, P.H. 2000. Effects of surgically imposed sterility on free-ranjng rabbit populations. *Journal of Applied Ecology.* 37:16–39.

Tyndale-Biscoe, C.H. 1994. Virus-vectored immunocontraception of feral mammals. *Reproduction, Fertility and Development* 6:281–287.

Urquhart, D. and Farnell, R. 1986. *The Fortymile Herd.* Department of Renewable Resources, Whitehorse, Yukon.

Vander Lee, B.A., Lutz, F.S., Hansen, L.A., and Mathews, N.E. 1999. Effects of supplemental prey, vegetation, and time on success of artifical nests. *Journal of Wildlife Management* 63:1299–1305.

Van der Wal, R., Madan, N., Van Lieshout, S., Dormann, C., Langvatn, R., and Albon, S.D. 2000. Trading forage quality versus quantity? Plant phenology and patch choice by an Arctic ungulate. *Oecologia (Berlin)* 123:108–115.

Van Horne, B., Schooley, R.L., Knick S.T., Olson G.S., and Burnham, K.P. 1997. Use of burrow entrances to indicate densities of Townsend's ground squirrels. *Journal of Wildlife Management* 61:92–101.

Van Houtan, K.S., Pimm, S.L., Halley, J.M., Bierregaard, R.O., and Lovejoy, T.E. 2007. Dispersal of Amazonian birds in continuous and fragmented forest. *Ecology Letters* 10:219–229.

Van Riper, C. III, and van Riper, S.G. 1986. The epizootiology and ecological significance of malaria in Hawaiian land birds. *Ecological Monographs* 56:327–344.

Varnham, K.J., Roy, S.S., Seymour, A., Mauremootoo, J., Jones, C.G., and Harris, S. 2002. Eradicating Indian musk shrews (*Suncus murinus*, Sloricidae) from Mauritian offshore islands. In: Veitch, C.R. and Clout, M.N. (eds) *Turning the Tide: The Eradication of Invasive Species*, pp. 342–349. Occasional Paper of the IUCN Species Survival Commission No. 27 IUCN – The World Conservation Union 2002, Gland, Switzerland and Cambridge, UK.

Vásárhelyi, C. and Thomas, V.G. 2003. Analysis of Canadian and American legislation controlling exotic species in the Great Lakes. *Aquatic Conservation: Marine and Freshwater Ecosystems* 13:417–427.

Veitch, C.R. and Clout, V.N. 2002. *Turning the Tide: The Eradication of Invasive Species.* Occasional Paper of the IUCN Species Survival Commission No. 27 IUCN – The World Conservation Union 2002, Gland, Switzerland and Cambridge, UK.

Veloso, C. and Bozinovic, F. 1993. Dietary and digestive constraints on basal energy metabolism in a small herbivorous rodent. *Ecology* 74:2003–2010.

Verme, L.J. and Holland, J.C. 1973. Reagent-dry assay of marrow fat in white-tailed deer. *Journal of Wildlife Management* 37:103–105.

Vesey-Fitzgerald, D.F. 1960. Grazing succession among East African game animals. *Journal of Mammalogy* 41:161–172.

Vincent, J.-P., Hewison, A.J.M., Angibault, J.-M., and Cargnelutti, B. 1996. Testing density estimators on a fallow deer population of known size. *Journal of Wildlife Management* 60:18–28.

Visser, M. and Both, C. 2005. Shifts in phenology due to global climate change: the need for a yardstick. *Proceedings of the Royal Society of London B* 272:2561–2569.

Visser, M. and Holleman, J.M. 2001. Warmer springs disrupt the synchrony of oak and winter moth phenology. *Proceedings of the Royal Society of London B* 268:289–294.

Visser, M., Holleman, J.M., and Gienapp, P. 2006. Shifts in caterpillar biomass phenology due to climate change and its impact on the breeding biology of an insectivorous bird. *Oecologia* 147:164–172.

Vitousek, P.M. and Hooper, D.U. 1993. Biological diversity and terrestrial ecosystem biogeochemistry. In: Schulze, E.D. and Mooney H.A. (eds) *Biodiversity and Ecosystem Function*, pp. 3–14. Springer-Verlag, Berlin.

Vitousek, P.M., D'Antonio, C.M., Loope, L.L., and Westbrooks, R. 1996. Biological invasions as global environmental change. *American Scientist* 84:468–478.

Vivås, H.J. and Sæther, B.-E. 1987. Interactions between a generalist herbivore, the moose *Alces alces* and its food resources: an experimental study of winter foraging behaviour in relation to browse availability. *Journal of Animal Ecology* 56:509–520.

Volterra V. 1926a Variations and fluctuations of the numbers of individuals in animal species living together. In: Chapman, R.N. (ed.) *Animal Ecology*, McGraw-Hill, New York, NY.

Volterra, V. 1926b. Fluctuations in the abundance of a species considered mathematically. *Nature* 118:558–560.

Voyles, J., Young, S., Berger, L., Campbell, C., Voyles, W.F., Dinudom, A., Cook, D., Webb, R., Alford, R.A., Skerratt, R.F., and Speare, R. 2009. Pathogenesis of Chytridiomycosis, a cause of catastrophic amphibian declines. *Science* 326:582–585.

Wake, D.B. and Vredenburg, V.T. 2008. Are we in the midst of the sixth mass extinction? A view from the world of amphibians. *Proceedings of the National Academy of Sciences USA* 105:11 466–11 473.

Walker, B.H. 1992. Biological diversity and ecological redundancy. *Conservation Biology* 6:18–23.

Walker, B.H. 1995. Conserving biological diversity through ecosystem resilience. *Conservation Biology* 9:747–752.

Walker, B.H., Ludwig, D., Holling, C.S., and Peterman, R.M. 1981. Stability of semi-arid savanna grazing systems. *Journal of Ecology* 69:473–498.

Walsh, J. 1987. War on cattle disease divides the troops. *Science* 237:1289–1291.

Walters, C. 1986. *Adaptive Management of Renewable Resources*. Macmillan, New York, NY.

Walters, C.J. and Holling, C.S. 1990. Large-scale management experiments and learning by doing. *Ecology* 71:2060–2068.

Ward, D. and Salz, D. 1994. Foraging at different spatial scales: dorcas gazelles foraging for lilies in the Negev Desert. *Ecology* 75:48–58.

Ward, P. 1969. The annual cycle of the yellow-vented bulbul *Pycnonotus goiavier* in a humid equatorial environment. *Journal of Zoology (London)* 157:25–45.

Warnecke, L. , Turner, J.M., Bollinger, T.K., Lorch, J.M., Nisra, V., Cryan, P.M., Wibbelt, G., Blehert, D.S., and Willis, C.K.R. 2012. Inoculation of bats with European *Geomyces destructans* supports the novel pathogen hypothesis for the origin of white-nose syndrome. *Proceedings of the National Academy of Sciences USA* 109:6999–7003.

Waser, P. 1996. Patterns and consequences of dispersal in gregarious carnivores. In: Gittleman, J.L. (ed.) *Carnivore Behavior, Ecology, and Evolution*, pp. 267–295. Cornell University Press, Ithaca, NY and London.

Warner, R.E. 1968. The role of introduced diseases in the extinction of the endemic Hawaiian avifauna. *Condor* 70:101–120.

Watson, A. and Moss, R. 1971. Spacing as affected by territorial behavior, habitat, and nutrition in red grouse (*Lagopus l. scoticus*). In: Esser, A.H. (ed.) *Behavior and Environment: The Use of Space by Animals and Men*, pp. 92–111. Plenum Press, New York, NY.

Wauters, L.A. and Gurnell, J. 1999. The mechanism of replacement of red squirrels by grey squirrels: a test of the interference competition hypothesis. *Ethology* 105:1053–1071.

Wayne, R.K., Lehman, N., Girman, D., Gogan, P.J.P., Gilbert, D.A., Hansen, K., Peterson, R.O., Seal, U.S., Eisenhawer, A., Mech, L.D., and Krumenaker, R.J. 1991. Conservation genetics of the endangered Isle Royale gray wolf. *Conservation Biology* 5:41–51.

Webb, C.J. and Kelly, D.K. 1993. The reproductive biology of the New Zealand flora. *Trends in Ecology and Evolution* 8:442–447.

Wehausen, J.D. 1995. Fecal measures of diet quality in wild and domestic ruminants. *Journal of Wildlife Management* 59:816–823.

Weir, J.S. 1972. Spatial distribution of elephants in an African National Park in relation to environmental sodium. *Oikos* 23:1–13.

Werner, E.E., Mittelbach, G.G., Hall, D.J., and Gilliam, J.F. 1983. Experimental tests of optimal habitat use in fish: the role of relative habitat profitability. *Ecology* 64:1525–1539.

Western, D. 1975. Water availability and its influence on the structure and dynamics of a savannah large mammal community. *East African Wildlife Journal* 13:265–286.

White, C.A., Olmsted, C.E., and Kay, C.E. 1998. Aspen, elk, and fire in the Rocky Mountain National Parks of North America. *Wildlife Society Bulletin* 26:449–462.

White, G.C. and Burnham, K.P. 1999. Program MARK: survival rate estimation from both live and dead encounters. *Bird Study* 46(Suppl.):S120–S139.

White, G.C., Anderson, D.R., Burnham, K.P., and Otis, D.L. 1982. *Capture-Recapture and Removal Methods for Sampling Closed Populations*. Report LA-8787-NERP, Los Alamos National Laboratory, Los Alamos, NM.

White, R.G., Bunnell, F.L., Gaare, E., Skogland, T., and Hubert, B. 1981. Ungulates on arctic ranges. In: Bliss, L.C., Heal, O.W., and Moore, J.J. (eds) *Tundra Ecosystems: A Comparative Analysis*, pp. 397–483. Cambridge University Press, Cambridge.

Whitfield, S.M., Bell, K.E., Philippi, T., Sasa, M., Bolanos, F., Chaves, G., Savage, J.M., and Donnelly, M.A. 2007. Amphibian and reptile declines over 35 years at La Selva, Costa Rica. *Proceedings of the National Academy of Sciences USA* 104:8352–8356.

Whitman, K., Starfield, A.M. Quadling, H.S., and Packer, C. 2004. Sustainable trophy hunting of African lions. *Nature* 428:175–178.

Whittaker, R.H. 1967. Gradient analysis of vegetation. *Biological Reviews* 42:207–264.

Whyte, R.J. and Bolen, E.G. 1985. Variation in mallard digestive organs during winter. *Journal of Wildlife Management* 49:1037–1040.

Wiemeyer, S.N., Scott, J.M., Anderson, M.P., Bloom, P.H., and Stafford, C.J. 1988. Environmental contaminants in California condors. *Journal of Wildlife Management* 52:238–247.

Wiens, J.A. 1977. On competition and variable environments. *American Scientist* 65:590–597.

Wilbur, S.R., Carrier, W.D., and Borneman, J.C. 1974. Supplemental feeding program for California condors. *Journal of Wildlife Management* 38:343–346.

Wilcove, D.S. 1985. Nest predation in forest tracts and the decline of migratory songbirds. *Ecology* 66:1211–1214.

Wilcove, D.S., Rothstein, D., Dubow, J., Philips, A., and Losos, E. 1998. Quantifying threats to imperiled species in the United States. *Bioscience* 48:607–615.

Williams, B.K., Johnson, F.A., and Wilkins, K. 1996. Uncertainty and the adaptive management of waterfowl harvests. *Journal of Wildlife Management* 60:223–232.

Williams, C.K. and Twigg, L.E. 1996. Responses of wild rabbit populations to imposed sterility. In: Floyd, R.B., Sheppard, A.W., and De Barro, P.J. (eds) *Frontiers of Population Ecology*, pp. 547–560. CSIRO Publishing, Australia.

Williams, M. 1979. The status and management of Black Swans *Cygnus atratus*, Lathum at Lake Ellesmere since the "Wahine" Storm, April 1968. *New Zealand Journal of Ecology* 2:34–41.

Williams, S.E., Bolitho, E.E., and Fox, S. 2003. Climate change in Australian tropical rainforests: an impending environmental catastrophe. *Proceedings of the Royal Society of London B.* 270:1887–1892.

Wilmshurst, J.F., Fryxell, J.M., and Hudson, R.J. 1995. Forage quality and patch choice by wapiti (*Cervus elaphus*). *Behavioral Ecology* 6:209–217.

Wilson D.J. and Jeffries, R.L. 1996. Nitrogen mineralization, plant growth and goose herbivory in an arctic coastal system. *Journal of Ecology* 85:841–851.

Wilson, R.J. and Gutierrez, D. 2012. Effect of climate change on the elevational limits of species ranges. In: Beever, E.A. and Belant, J.L. (eds) *Ecological Consequences of Climate Change*, pp. 107–132. CRC Press, Boca Raton, FL.

Windsor, D.A. 1998. Most of the species on Earth are parasites. *International Journal for Parasitolgy* 28:1939–1941.

Wingfield, J.C. 2008. Comparative endocrinology, environment and global change. *General and Comparative Endocrinology* 157:207–216.

Wingfield, J.C., Kelley, J.P., Angelier, F., Chastel, O., Lei, F., Lynn, S.E., Miner, B., Davis, J.E., Li, D., and Wang, D. 2011. Organism-environment interactions in a changing world: A mechanistic approach. *Journal of Ornithology* 152(Suppl. 1):S279–S288.

Wittemyer, G., Polansky, L., Douglas-Hamilton, I., and Getz, W.M. 2008. Disentangling the effects of forage, social rank, and risk on movement autocorrelation of elephants using Fourier and wavelet analysis. *Proceedings of the National Academy of Sciences USA* 105:19 108–19 113.

Wittmer, H.U., Sinclair, A.R.E., and McLellan, B.N. 2005. The role of predation in the decline and extirpation of woodland caribou. *Oecologia* 144:257–267.

Woolfenden, G.E. and Fitzpatrick, J.W. 1984. The Florida scrub-jay: demogaphy of a cooperative-breeding bird. Princeton University Press, Princeton, NJ.

Woolnough, A.P., Foley, J.F., Johnson, C.N., and Evans, M. 1997. Evaluation of techniques for indirect measurement of body composition in a free-ranging large herbivore, the southern hairy-nosed wombat. *Wildlife Research* 24:649–660.

Wootton J. T. 1994a. Predicting direct and indirect effects: an integrated approach using experiments and path analysis. *Ecology* 75:151–165.

Wootton, J.T. 1994b. The nature and consequences of indirect effects in ecological communities. *Annual Review of Ecology and Systematics* 24:443–466.

Worthy, T.H. and Holdaway, R.N. 2002. *The Lost World of the Moa*. Indiana University Press, Bloomington, IN.

Wu, X.B., Thurow, T.L., and Whisenant, S.G. 2000. Fragmentation and changes in hydrologic function of tiger bush landscapes, south-west Niger. *Journal of Animal Ecology* 88:790–800.

Wydeven, A.P. and Dahlgren, R.B. 1985. Ungulate habitat relationships in Wind Cave National Park. *Journal of Wildlife Management* 49:805–813.

Yodzis, P. and Innes, S. 1988. Body size and consumer-resource dynamics. *American Naturalist* 139:1151–1175.

Yuill, T.M. 1987. Diseases as components of mammalian ecosystems: mayhem and subtlety. *Canadian Journal of Zoology* 65:1061–1066.

Zar, J.H. 1996. *Biostatistical Analysis*, 3 edn. Prentice-Hall, London.

Zeleny, L. 1976. *The Bluebird: How You Can Help its Fight for Survival*. Indiana University Press, Bloomington, IN.

Zhang, Z., Pech, R., Davis, S., Shi, D., Wan, X., and Zhong, W. 2003. Extrinsic and intrinsic factors determine the eruptive dynamics of Brandt's voles *Microtus brandti* in Inner Mongolia, China. *Oikos* 100:299–310.

Index

Wildlife Ecology, Conservation, and Management, Third Edition.
John M. Fryxell, Anthony R. E. Sinclair and Graeme Caughley.
© 2014 John Wiley & Sons, Ltd. Published 2014 by John Wiley & Sons, Ltd.
Companion Website: www.wiley.com/go/Fryxell/Wildlife